S0-BRZ-365

Elementary
Multivariable
Calculus

Elementary Multivariable Calculus

——————————BERNARD KOLMAN AND WILLIAM F. TRENCH

Drexel University

ACADEMIC PRESS
New York and London

Copyright © 1971, by Academic Press, Inc.

All rights reserved.
No part of this book may be reproduced in any form,
by photostat, microfilm, retrieval system, or any
other means, without written permission from
the publishers.

ACADEMIC PRESS, INC.
111 Fifth Avenue, New York, New York 10003

United Kingdom Edition published by
ACADEMIC PRESS, INC. (LONDON) LTD.
Berkeley Square House, London W1X 6BA

Library of Congress Catalog Card Number: 73-156267
AMS (MOS) 1970 Subject Classification: 26A60

First Printing May 1971
Second Printing August 1971

Printed in the United States of America

3 3001 00577 1638

Contents

176338

Preface

This book presents an elementary treatment of the multivariable calculus. It is designed to be taught at the sophomore level for engineering, science, and mathematics majors who have completed a year of calculus and a linear algebra course. More than enough material has been included here for a four-hour one-quarter course or a three-hour one-semester course, even if the review sections are omitted.

We present a modern, but not extreme, treatment of the subject. Our choice of topics has been influenced, but not dictated, by the recommendations of the Committee on Undergraduate Preparation in Mathematics. Linear algebra is used as an integral part of the subject. The main emphasis throughout the book is on maintaining a sophomore level of understanding. Thus, we have not hesitated to omit proofs of difficult theorems, preferring instead to illustrate their meanings with numerous examples. In particular, there are few "epsilon–delta" arguments in this book. Most theorems and definitions are followed by worked out illustrative examples. Each section contains, in addition to an ample number of routine exercises of varying degrees of difficulty, a set of theoretical exercises designed to fill in the gaps in proofs and extend results obtained in the text. These exercises can also be used to raise the level of the book. Answers to odd-numbered routine exercises are given in the back of the book.

Chapter 1 deals with vectors and analytic geometry in n-space and includes a review of linear algebra, lines and planes, motion in R^3, and parametrically defined curves. The well-prepared student can omit some of this material. Chapter 2 treats the calculus of real-valued functions of several variables; the concepts of continuity and differentiability are developed and related to their familiar counterparts from the single variable calculus. Chapter 3 treats vector-valued functions as ordered m-tuples of real-valued functions so that the results of Chapter 2 can be directly applied. Chapter 4 deals with integration (line, surface, and multiple integrals) and includes a discussion of Green's and Stokes's theorems, and a section on physical applications.

It should again be emphasized that linear algebra is used freely and frequently in this book; the material on linear algebra included is only by way of review and is not intended to provide a first exposure to the subject.

Acknowledgments

We wish to express our thanks to Miss Susan R. Gershuni, who typed most of the manuscript, and to Mrs. Marjorie M. Bawduniak, who also helped with the typing; to Albert J. Herr and Robert L. Higgins, who provided solutions and corrections to some of the problems; to Mrs. Anna Hernandez for the art work; and to the staff of Academic Press, for their interest, encouragement, and cooperation.

We also wish to thank three unknown reviewers for their helpful suggestions for improving the manuscript.

Elementary
Multivariable
Calculus

Chapter 1

Vectors and Analytic Geometry

1.1 The Vector Space R^n

n-Vectors

In the first course in calculus and analytic geometry the real numbers are represented on a line and the set of ordered pairs (x, y) of real numbers is represented on a plane. In the calculus of several variables we deal with ordered n-tuples of real numbers.

Definition 1.1 An *n-vector* is an ordered *n*-tuple

$$\mathbf{X} = \begin{bmatrix} x_1 \\ \vdots \\ x_n \end{bmatrix}$$

of real numbers $x_1, \ldots, x_n$ which are called the *components* of $\mathbf{X}$. If

$$\mathbf{Y} = \begin{bmatrix} y_1 \\ \vdots \\ y_n \end{bmatrix},$$

then $\mathbf{X} = \mathbf{Y}$ if and only if $x_i = y_i$ $(1 \leq i \leq n)$. For a given n the set of all n-vectors is denoted by R^n.

When it is unnecessary to be specific about n, we shall refer to n-vectors simply as *vectors*. We identify the 1-vectors $[x]$ with the real number x; thus R^1 is simply the set of real numbers. Real numbers are also called *scalars*.

Example 1.1 Examples of vectors in R^2:

$$\begin{bmatrix} 0 \\ 1 \end{bmatrix}, \quad \begin{bmatrix} 1 \\ 0 \end{bmatrix}, \quad \begin{bmatrix} 1 \\ 2 \end{bmatrix}, \quad \begin{bmatrix} e \\ \pi \end{bmatrix}, \quad \begin{bmatrix} -\frac{4}{3} \\ -\frac{5}{4} \end{bmatrix}.$$

Note that

$$\begin{bmatrix} 0 \\ 1 \end{bmatrix} \neq \begin{bmatrix} 1 \\ 0 \end{bmatrix}.$$

Example 1.2 Examples of vectors in R^3:

$$\begin{bmatrix} 1 \\ 0 \\ 0 \end{bmatrix}, \quad \begin{bmatrix} 0 \\ 1 \\ 0 \end{bmatrix}, \quad \begin{bmatrix} 0 \\ 0 \\ 1 \end{bmatrix}, \quad \begin{bmatrix} -\frac{1}{3} \\ 2 \\ \frac{1}{4} \end{bmatrix}, \quad \begin{bmatrix} \frac{2}{3} \\ -\frac{3}{4} \\ -\frac{5}{8} \end{bmatrix}.$$

Vectors can be added and multiplied by scalars according to the following definitions.

Definition 1.2 (Scalar Multiplication) If

$$\mathbf{X} = \begin{bmatrix} x_1 \\ \vdots \\ x_n \end{bmatrix}$$

is an n-vector and c a scalar, then $c\mathbf{X}$ is the n-vector

$$c\mathbf{X} = \begin{bmatrix} cx_1 \\ \vdots \\ cx_n \end{bmatrix};$$

it is called a *scalar multiple* of $\mathbf{X}$.

Example 1.3 In R^3 let

$$\mathbf{X} = \begin{bmatrix} 1 \\ -2 \\ 3 \end{bmatrix}$$

and $c = 2$; then

$$c\mathbf{X} = 2\begin{bmatrix} 1 \\ -2 \\ 3 \end{bmatrix} = \begin{bmatrix} 2 \\ -4 \\ 6 \end{bmatrix}.$$

Definition 1.3 (Vector Addition)   If

$$\mathbf{X} = \begin{bmatrix} x_1 \\ \vdots \\ x_n \end{bmatrix} \qquad \text{and} \qquad \mathbf{Y} = \begin{bmatrix} y_1 \\ \vdots \\ y_n \end{bmatrix},$$

then their *vector sum* (or simply *sum*) is defined by

$$\mathbf{X} + \mathbf{Y} = \begin{bmatrix} x_1 + y_1 \\ \vdots \\ x_n + y_n \end{bmatrix}.$$

Example 1.4   In R^3 let

$$\mathbf{X} = \begin{bmatrix} 1 \\ -2 \\ 3 \end{bmatrix} \qquad \text{and} \qquad \mathbf{Y} = \begin{bmatrix} 0 \\ 2 \\ -4 \end{bmatrix};$$

then

$$\mathbf{X} + \mathbf{Y} = \begin{bmatrix} 1 \\ 0 \\ -1 \end{bmatrix}.$$

Theorem 1.1   If $\mathbf{X}$, $\mathbf{Y}$, and $\mathbf{Z}$ are in R^n and a and b are scalars, then:

(a) $\mathbf{X} + \mathbf{Y} = \mathbf{Y} + \mathbf{X}$.
(b) $(\mathbf{X} + \mathbf{Y}) + \mathbf{Z} = \mathbf{X} + (\mathbf{Y} + \mathbf{Z})$.
(c) $(a + b)\mathbf{X} = a\mathbf{X} + b\mathbf{X}$.
(d) $a(\mathbf{X} + \mathbf{Y}) = a\mathbf{X} + a\mathbf{Y}$.
(e) $a(b\mathbf{X}) = (ab)\mathbf{X}$.
(f) $1\mathbf{X} = \mathbf{X}$.
(g) There is a unique vector $\mathbf{0}$ in R^n, namely

$$\mathbf{0} = \begin{bmatrix} 0 \\ \vdots \\ 0 \end{bmatrix},$$

such that $\mathbf{X} + \mathbf{0} = \mathbf{0} + \mathbf{X} = \mathbf{X}$; $\mathbf{0}$ is called the *zero vector*.
   (h) There is a unique vector $-\mathbf{X}$ such that $\mathbf{X} + (-\mathbf{X}) = \mathbf{0}$; in fact

$$-\mathbf{X} = \begin{bmatrix} -x_1 \\ \vdots \\ -x_n \end{bmatrix} = (-1)\mathbf{X}.$$

We shall write $\mathbf{X} + (-\mathbf{Y})$ simply as $\mathbf{X} - \mathbf{Y}$.

Theorem 1.1 follows from the definitions of vector addition and scalar multiplication, and familiar properties of the real numbers. We shall prove only (d), leaving the rest of the proof for the student (Exercise T-1).

If $\mathbf{X}$ and $\mathbf{Y}$ are in R^n, then

$$a(\mathbf{X} + \mathbf{Y}) = a \begin{bmatrix} x_1 + y_1 \\ \vdots \\ x_n + y_n \end{bmatrix} = \begin{bmatrix} a(x_1 + y_1) \\ \vdots \\ a(x_n + y_n) \end{bmatrix}$$

$$= \begin{bmatrix} ax_1 + ay_1 \\ \vdots \\ ax_n + ay_n \end{bmatrix} = \begin{bmatrix} ax_1 \\ \vdots \\ ax_n \end{bmatrix} + \begin{bmatrix} ay_1 \\ \vdots \\ ay_n \end{bmatrix}$$

$$= a \begin{bmatrix} x_1 \\ \vdots \\ x_n \end{bmatrix} + a \begin{bmatrix} y_1 \\ \vdots \\ y_n \end{bmatrix} = a\mathbf{X} + a\mathbf{Y}.$$

A *vector space over the real numbers* is a set S of elements, called *vectors*, together with two operations, called *addition* and *scalar multiplication*, such that the sum of any two vectors is in S, scalar multiples of vectors are in S, and (a) through (h) hold. Thus, R^n is a vector space over the reals.

Subspaces of R^n

Theorem 1.2 If $\mathbf{X}$ is in R^n and c is a scalar then $0\mathbf{X} = c\mathbf{0} = \mathbf{0}$. If $c\mathbf{X} = \mathbf{0}$ then $c = 0$ or $\mathbf{X} = \mathbf{0}$.

We leave the proof for the student (Exercise T-2).

Definition 1.4 If $c_1, \ldots, c_m$ are scalars and $\mathbf{X}_1, \ldots, \mathbf{X}_m$ are in R^n, then the vector $\mathbf{X} = c_1\mathbf{X}_1 + \cdots + c_m\mathbf{X}_m$ is called a *linear combination* of $\mathbf{X}_1, \ldots, \mathbf{X}_m$.

Example 1.5 Let

$$\mathbf{X}_1 = \begin{bmatrix} 1 \\ 0 \\ 0 \\ 2 \end{bmatrix}, \quad \mathbf{X}_2 = \begin{bmatrix} 0 \\ 0 \\ 1 \\ 3 \end{bmatrix}, \quad \text{and} \quad \mathbf{X}_3 = \begin{bmatrix} 1 \\ 1 \\ 1 \\ 0 \end{bmatrix};$$

then

$$\begin{bmatrix} 5 \\ 3 \\ 2 \\ 1 \end{bmatrix} = 2\mathbf{X}_1 - \mathbf{X}_2 + 3\mathbf{X}_3$$

is a linear combination of $\mathbf{X}_1$, $\mathbf{X}_2$, and $\mathbf{X}_3$.

Definition 1.5   A nonempty subset H of R^n is said to be a *subspace* of R^n if $\mathbf{X} + \mathbf{Y}$ is in H whenever $\mathbf{X}$ and $\mathbf{Y}$ are, and $c\mathbf{X}$ is in H whenever $\mathbf{X}$ is in H and c is a scalar.

Example 1.6   The set containing only $\mathbf{0}$ is a subspace of R^n called the *zero subspace*, and R^n is a subspace of itself. If $\mathbf{X}_0$ is a fixed vector, then the set of scalar multiples of $\mathbf{X}_0$ is a subspace of R^n.

Example 1.7   Let H be the set of 2-vectors of the form

$$\mathbf{X} = \begin{bmatrix} x \\ y_0 \end{bmatrix},$$

where y_0 is a constant. Since

$$\begin{bmatrix} x_1 \\ y_0 \end{bmatrix} + \begin{bmatrix} x_2 \\ y_0 \end{bmatrix} = \begin{bmatrix} x_1 + x_2 \\ 2y_0 \end{bmatrix},$$

H cannot be a subspace of R^2 unless $y_0 = 0$. If $y_0 = 0$, then H is a subspace, since

$$\begin{bmatrix} x_1 \\ 0 \end{bmatrix} + \begin{bmatrix} x_2 \\ 0 \end{bmatrix} = \begin{bmatrix} x_1 + x_2 \\ 0 \end{bmatrix}$$

and

$$c\begin{bmatrix} x_1 \\ 0 \end{bmatrix} = \begin{bmatrix} cx_1 \\ c0 \end{bmatrix} = \begin{bmatrix} cx_1 \\ 0 \end{bmatrix}.$$

A subspace H of R^n is also a vector space over the reals, for if $\mathbf{X}$, $\mathbf{Y}$, and $\mathbf{Z}$ are restricted to H then (a) through (h) still hold, and the vectors appearing in them are all in H (Exercise T-13).

The next definition gives another example of a subspace; we shall see below (Theorem 1.6) that all subspaces of R^n are of this kind.

Definition 1.6 Let $X_1, \ldots, X_m$ be fixed vectors in R^n. The *subspace spanned* by $X_1, \ldots, X_m$ is the set of all linear combinations of $X_1, \ldots, X_m$. It is denoted by $S[X_1, \ldots, X_m]$. Thus X is in $S[X_1, \ldots, X_m]$ if and only if it can be written as

$$X = c_1 X_1 + \cdots + c_m X_m$$

for some choice of scalars $c_1, \ldots, c_m$. The student should verify that $S[X_1, \ldots, X_m]$ is a subspace (Exercise T-3).

Example 1.8 The subspace spanned by

$$X_1 = \begin{bmatrix} 1 \\ 0 \\ 0 \end{bmatrix}, \qquad X_2 = \begin{bmatrix} 0 \\ 0 \\ 1 \end{bmatrix}, \qquad \text{and} \qquad X_3 = \begin{bmatrix} 1 \\ 0 \\ 1 \end{bmatrix}$$

is the set of vectors of the form

$$X = \begin{bmatrix} x_1 \\ 0 \\ x_3 \end{bmatrix}.$$

Each such X can be written in infinitely many ways as a linear combination of X_1, X_2, and X_3; thus

$$\begin{bmatrix} x_1 \\ 0 \\ x_3 \end{bmatrix} = c_1 X_1 + c_2 X_2 + c_3 X_3,$$

provided

$$c_1 + c_3 = x_1$$

and

$$c_2 + c_3 = x_3.$$

Here any one of c_1, c_2, or c_3 can be chosen arbitrarily and then the other two are determined uniquely. Note that since each vector

$$X = \begin{bmatrix} x_1 \\ 0 \\ x_3 \end{bmatrix}$$

is of the form $X = x_1 X_1 + x_3 X_2$ the pair $\{X_1, X_2\}$ already spans the subspace. In fact, any pair chosen from X_1, X_2, and X_3 already spans $S[X_1, X_2, X_3]$. (Prove.)

Example 1.9   The vectors

$$E_1 = \begin{bmatrix} 1 \\ 0 \\ 0 \end{bmatrix}, \qquad E_2 = \begin{bmatrix} 0 \\ 1 \\ 0 \end{bmatrix}, \qquad \text{and} \qquad E_3 = \begin{bmatrix} 0 \\ 0 \\ 1 \end{bmatrix}$$

span R^3 since

$$\begin{bmatrix} x_1 \\ x_2 \\ x_3 \end{bmatrix} = x_1 E_1 + x_2 E_2 + x_3 E_3;$$

thus $S[E_1, E_2, E_3] = R^3$. Similarly,

(1)
$$E_1 = \begin{bmatrix} 1 \\ 0 \\ \vdots \\ 0 \end{bmatrix}, \qquad E_2 = \begin{bmatrix} 0 \\ 1 \\ 0 \\ \vdots \\ 0 \end{bmatrix}, \ldots, \qquad E_n = \begin{bmatrix} 0 \\ \vdots \\ 0 \\ 1 \end{bmatrix}$$

span R^n, since

$$\begin{bmatrix} x_1 \\ \vdots \\ x_n \end{bmatrix} = x_1 E_1 + \cdots + x_n E_n.$$

Linear Dependence

Definition 1.7   The vectors $X_1, \ldots, X_m$ are said to be *linearly dependent* if there are scalars $c_1, \ldots, c_m$, not all zero, such that

(2)
$$c_1 X_1 + \cdots + c_m X_m = 0.$$

Otherwise $X_1, \ldots, X_m$ are said to be *linearly independent*; in this case (2) implies that $c_1 = c_2 = \cdots = c_m = 0$. We shall also say that the set $\{X_1, \ldots, X_m\}$ is linearly dependent or independent, as the case may be.

Example 1.10   The vectors X_1, X_2, and X_3 defined in Example 1.8 are linearly dependent, since

$$\mathbf{X}_1 + \mathbf{X}_2 - \mathbf{X}_3 = \mathbf{0}.$$

However, any pair chosen from these three vectors is linearly independent. (Verify.)

Example 1.11 The vectors $\mathbf{E}_1$, $\mathbf{E}_2$, and $\mathbf{E}_3$ of Example 1.9 are linearly independent, since

$$c_1 \begin{bmatrix} 1 \\ 0 \\ 0 \end{bmatrix} + c_2 \begin{bmatrix} 0 \\ 1 \\ 0 \end{bmatrix} + c_3 \begin{bmatrix} 0 \\ 0 \\ 1 \end{bmatrix} = \begin{bmatrix} c_1 \\ c_2 \\ c_3 \end{bmatrix} = \begin{bmatrix} 0 \\ 0 \\ 0 \end{bmatrix}$$

if and only if $c_1 = c_2 = c_3 = 0$. Similarly $\mathbf{E}_1, \ldots, \mathbf{E}_n$ as defined by (1) are linearly independent in R^n.

Any set of vectors which contains $\mathbf{0}$ must be linearly dependent; if $\mathbf{X}_i = \mathbf{0}$ in Definition 1.7, then (2) is satisfied with $c_i = 1$ and $c_j = 0$ if $j \neq i$.

Theorem 1.3 The vectors $\mathbf{X}_1, \ldots, \mathbf{X}_m$ are linearly dependent if and only if one of them is a linear combination of the others.

Proof. If $V = \{\mathbf{X}_1, \ldots, \mathbf{X}_m\}$ is a linearly dependent set then (2) holds with some $c_i \neq 0$; hence we can divide by c_i and solve for $\mathbf{X}_i$ as a linear combination of the remaining vectors in V. Conversely, if $\mathbf{X}_i$ is a linear combination of the other members of V then (2) holds for some $c_1, \ldots, c_m$ with $c_i = -1$, and V is linearly dependent.

Theorem 1.4 If $V = \{\mathbf{X}_1, \ldots, \mathbf{X}_m\}$ is a linearly dependent set of vectors, not all zero, then it is possible to choose linearly independent vectors from V which span the same subspace as V.

Proof. By Theorem 1.3 one of the $\mathbf{X}_i$ is a linear combination of the others. We can assume that $\mathbf{X}_m$ is a linear combination of $\mathbf{X}_1, \ldots, \mathbf{X}_{m-1}$. (If this is not the case, we simply renumber the elements of V to make it so.) Then

$$\mathbf{X}_m = a_1 \mathbf{X}_1 + \cdots + a_{m-1} \mathbf{X}_{m-1},$$

and if

$$\mathbf{X} = c_1 \mathbf{X}_1 + \cdots + c_m \mathbf{X}_m,$$

then $\mathbf{X}$ can be written as a linear combination of $\mathbf{X}_1, \ldots, \mathbf{X}_{m-1}$:

$$\mathbf{X} = (c_1 + a_1 c_m) \mathbf{X}_1 + \cdots + (c_{m-1} + a_{m-1} c_m) \mathbf{X}_{m-1}.$$

Therefore

$$S[\mathbf{X}_1, \ldots, \mathbf{X}_{m-1}] = S[\mathbf{X}_1, \ldots, \mathbf{X}_m].$$

If $\mathbf{X}_1, \ldots, \mathbf{X}_{m-1}$ are linearly independent, we are finished. If they are not, we repeat the argument, eliminating one $\mathbf{X}_i$ each time, until those remaining are linearly independent. This must occur after fewer than m steps, since some $\mathbf{X}_i \neq \mathbf{0}$.

Example 1.12 In Example 1.8, $V = \{\mathbf{X}_1, \mathbf{X}_2, \mathbf{X}_3\}$ is a linearly dependent set. Three linearly independent subsets of V span $S[\mathbf{X}_1, \mathbf{X}_2, \mathbf{X}_3]$: $\{\mathbf{X}_1, \mathbf{X}_2\}$, $\{\mathbf{X}_1, \mathbf{X}_3\}$, and $\{\mathbf{X}_2, \mathbf{X}_3\}$.

Bases

Definition 1.8 A set $V = \{\mathbf{X}_1, \ldots, \mathbf{X}_m\}$ of vectors is called a *basis* for a subspace H of R^n if V is linearly independent and spans H.

Theorem 1.5 If $V = \{\mathbf{X}_1, \ldots, \mathbf{X}_m\}$ is a basis for a subspace H, then no linearly independent set of vectors in H contains more than m vectors, and every basis contains exactly m vectors.

Proof. Let $U = \{\mathbf{Y}_1, \ldots, \mathbf{Y}_k\}$ be a linearly independent set of vectors in H and suppose $k > m$. Since V is a basis for H we can write

$$\mathbf{Y}_1 = a_1\mathbf{X}_1 + \cdots + a_m\mathbf{X}_m,$$

where $a_i \neq 0$ for some i. Suppose $a_m \neq 0$ (if this is not so, simply renumber the elements of V to make it so); then

$$\mathbf{X}_m = \frac{\mathbf{Y}_1}{a_m} - \frac{1}{a_m}(a_1\mathbf{X}_1 + \cdots + a_{m-1}\mathbf{X}_{m-1})$$

and $\{\mathbf{X}_1, \ldots, \mathbf{X}_{m-1}, \mathbf{Y}_1\}$ spans H. Therefore we can write

$$\mathbf{Y}_2 = b_1\mathbf{X}_1 + \cdots + b_{m-1}\mathbf{X}_{m-1} + c\mathbf{Y}_1,$$

where $b_i \neq 0$ for at least one i, since $b_1 = \cdots = b_{m-1} = 0$ would imply that $\mathbf{Y}_2 = c\mathbf{Y}_1$, a contradiction. Again we assume (possibly after renumbering $\mathbf{X}_1, \ldots, \mathbf{X}_{m-1}$) that $b_{m-1} \neq 0$; then

$$\mathbf{X}_{m-1} = \frac{\mathbf{Y}_2}{b_{m-1}} - \frac{1}{b_{m-1}}(b_1\mathbf{X}_1 + \cdots + b_{m-2}\mathbf{X}_{m-2} + c\mathbf{Y}_1).$$

This implies that $\{\mathbf{X}_1, \ldots, \mathbf{X}_{m-2}, \mathbf{Y}_1, \mathbf{Y}_2\}$ spans H. Carrying out this argument a total of m times we arrive at the conclusion that $\{\mathbf{Y}_1, \ldots, \mathbf{Y}_m\}$ spans H.

In particular, if $k > m$, then

$$\mathbf{Y}_{m+1} = d_1 \mathbf{Y}_1 + \cdots + d_m \mathbf{Y}_m,$$

which contradicts the assumption that U is a linearly independent set; hence $k \leq m$. If U is also a basis for H, the same argument with U and V interchanged implies that $k \geq m$. Hence two bases have the same number of elements, and the proof of the theorem is complete.

Definition 1.9 The number of vectors in a basis for a subspace is called the *dimension of H*. If this number equals m, then H is said to be *m-dimensional*.

Thus R^n is n-dimensional, since it is spanned by the linearly independent vectors $\mathbf{E}_1, \ldots, \mathbf{E}_n$ defined in (1). This basis is called the *natural basis for R^n*. The subspace consisting of $\mathbf{0}$ alone, which we call the *zero subspace*, has no linearly independent vectors. It is therefore zero-dimensional.

Theorem 1.6 Every nonzero subspace of R^n has a basis.

Proof. A linearly independent set of vectors in H is also linearly independent in R^n. Therefore Theorem 1.5 implies that no linearly independent set in H contains more than n elements. If m is the maximum number of vectors in any linearly independent set in H, then any linearly independent set in H with m elements is a basis for H. We leave the proof of this statement to the student (Exercise T-5).

If $V = \{\mathbf{X}_1, \ldots, \mathbf{X}_k\}$ is a basis for H, then every vector in H can be written uniquely as a linear combination of vectors in V (Exercise T-6).

Vectors and Points

When dealing with vectors in R^2 and R^3, we shall usually write

$$\mathbf{X} = \begin{bmatrix} x \\ y \end{bmatrix} \quad \text{and} \quad \mathbf{X} = \begin{bmatrix} x \\ y \\ z \end{bmatrix}$$

to avoid unnecessary subscripts. This notation also suggests the familiar

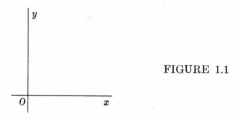

FIGURE 1.1

and natural correspondence between vectors in R^2 and R^3 and points in the plane and space, respectively. For example, let us recall the situation in the plane. We choose an arbitrary point O, the *origin*, and draw through it two perpendicular lines (Fig. 1.1), the *x-axis* and the *y-axis*, assigning a positive direction to each. Through an arbitrary point P in the plane we draw lines perpendicular to the x- and y-axes and denote the points where these lines intersect the axes by P' and P'', respectively (Fig. 1.2). Then the directed distance from P' to O (positive in Fig. 1.2a, negative in Fig. 1.2b) is the x-coordinate of P, and the directed distance from P'' to O (positive in Figs. 1.2a and 1.2b) is the y-coordinate of P. We say that P *has coordinates* (x, y) and write $P = (x, y)$.

Thus there is a one-to-one correspondence between vectors in R^2 and points in the plane:

(3)
$$\begin{bmatrix} x \\ y \end{bmatrix} \leftrightarrow (x, y).$$

In fact, we shall identify vectors and points and view R^2 both as a vector space, as defined at the beginning of this section, and as the collection of points in the plane. More generally we shall think of R^n as a vector space on the one hand, and as a collection of points on the other. We shall not

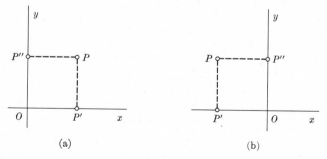

(a)   (b)

FIGURE 1.2

always distinguish between

$$\mathbf{X} = \begin{bmatrix} x_1 \\ \vdots \\ x_n \end{bmatrix},$$

and

$$\mathbf{X} = (x_1, \ldots, x_n);$$

both will stand for vectors and points in R^n. However we shall usually employ the vertical display when we wish to emphasize that $\mathbf{X}$ is a vector and the horizontal when this is not necessary.

When it is necessary to distinguish between the vector $\mathbf{X}$ and the point $\mathbf{X}$ we shall refer to the latter as the point P with coordinates $(x_1, \ldots, x_n)$.

If $\mathbf{X}$ and $\mathbf{Y}$ are points in R^n then the vector $\mathbf{X} - \mathbf{Y}$ is called the *position vector of $\mathbf{X}$ with respect to $\mathbf{Y}$*; in the special case $\mathbf{Y} = \mathbf{0}$ we say that $\mathbf{X}$ is the position vector of the point $(x_1, \ldots, x_n)$ (with respect to the origin).

We associate with each vector in R^2 the directed line segment from the origin to the point $\mathbf{X}$. We assume that the student is familiar with the parallelogram law for vector addition in R^2, which says that if $\mathbf{X}_1$ and $\mathbf{X}_2$ are two vectors such that their associated line segments are nonparallel (thus $\mathbf{X}_1$ and $\mathbf{X}_2$ are linearly independent), then $\mathbf{X}_2 + \mathbf{X}_1$ is associated with one diagonal of the parallelogram in Fig. 1.3, and $\mathbf{X}_2 - \mathbf{X}_1$ is associated with a line segment through the origin and parallel to the other diagonal.

We cannot visualize R^n if $n > 3$. Nevertheless important geometrical ideas such as distance, direction, line, and plane can be generalized to R^n.

In R^2 the distance from the origin to a point P with coordinates $(x) y$, is, according to the Pythagorean theorem,

$$| \overrightarrow{OP} | = \sqrt{x^2 + y^2}$$

(Fig. 1.4). This is also the *length of the vector* $\begin{bmatrix} x \\ y \end{bmatrix}$. We generalize this to define distance and length in R^n.

FIGURE 1.3

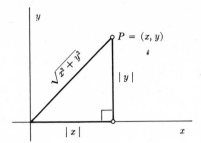

FIGURE 1.4

Inner Product

Definition 1.10   The *length*, or *magnitude*, of the vector

$$\mathbf{X} = \begin{bmatrix} x_1 \\ \vdots \\ x_n \end{bmatrix}$$

and the *distance from the origin to the point* $\mathbf{X} = (x_1, \ldots, x_n)$ are both defined by

$$|\mathbf{X}| = \sqrt{x_1^2 + \cdots + x_n^2}.$$

If $\mathbf{Y}$ is a second point in R^n then the *distance between the points* $\mathbf{X}$ *and* $\mathbf{Y}$ is defined to be the length of the vector $\mathbf{X} - \mathbf{Y}$:

$$|\mathbf{X} - \mathbf{Y}| = \sqrt{(x_1 - y_1)^2 + \cdots + (x_n - y_n)^2}.$$

Example 1.13   If

$$\mathbf{X} = \begin{bmatrix} 1 \\ 2 \\ -1 \\ 3 \end{bmatrix} \quad \text{and} \quad \mathbf{Y} = \begin{bmatrix} 3 \\ 0 \\ 2 \\ -2 \end{bmatrix},$$

then

$$|\mathbf{X}| = \sqrt{1^2 + 2^2 + (-1)^2 + 3^2} = \sqrt{15},$$
$$|\mathbf{Y}| = \sqrt{3^2 + 0^2 + 2^2 + (-2)^2} = \sqrt{17},$$

and

$$|\mathbf{X} - \mathbf{Y}| = \sqrt{(1 - 3)^2 + (2 - 0)^2 + (-1 - 2)^2 + (3 + 2)^2} = \sqrt{42}.$$

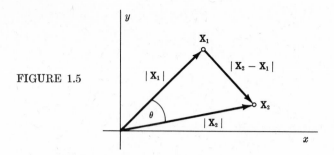

FIGURE 1.5

The angle between two nonzero vectors

$$\mathbf{X_1} = \begin{bmatrix} x_1 \\ y_1 \end{bmatrix} \quad \text{and} \quad \mathbf{X_2} = \begin{bmatrix} x_2 \\ y_2 \end{bmatrix}$$

in R^2 is defined to be the angle between the directed line segments from the origin to the points $\mathbf{X_1}$ and $\mathbf{X_2}$ (Fig. 1.5). Applying the law of cosines to the triangle in Fig. 1.5 yields

(4)
$$|\mathbf{X_2} - \mathbf{X_1}|^2 = |\mathbf{X_1}|^2 + |\mathbf{X_2}|^2 - 2|\mathbf{X_1}||\mathbf{X_2}|\cos\theta.$$

But

$$|\mathbf{X_2} - \mathbf{X_1}|^2 = (x_2 - x_1)^2 + (y_2 - y_1)^2$$

$$= x_2^2 + y_2^2 + x_1^2 + y_1^2 - 2(x_1x_2 + y_1y_2)$$

$$= |\mathbf{X_2}|^2 + |\mathbf{X_1}|^2 - 2(x_1x_2 + y_1y_2).$$

Substituting this in (4) yields

(5)
$$\cos\theta = \frac{x_1x_2 + y_1y_2}{|\mathbf{X_1}||\mathbf{X_2}|}.$$

To define the angle between two vectors in R^n, we need the notion of the inner product of two n-vectors.

Definition 1.11 The *inner product* of two n-vectors

$$\mathbf{X} = \begin{bmatrix} x_1 \\ \vdots \\ x_n \end{bmatrix} \quad \text{and} \quad \mathbf{Y} = \begin{bmatrix} y_1 \\ \vdots \\ y_n \end{bmatrix}$$

is defined by

$$\mathbf{X} \cdot \mathbf{Y} = x_1y_1 + \cdots + x_ny_n.$$

In R^2 and R^3 the inner product is also called the *scalar product* or *dot product*.

Example 1.14   For **X** and **Y** as defined in Example 1.13,

$$\mathbf{X \cdot Y} = (1)(3) + 2(0) + (-1)(2) + 3(-2)$$

$$= -5.$$

The next theorem follows directly from the definition of inner product.

Theorem 1.7   If **X**, **Y**, and **Z** are n-vectors and c is a scalar, then

(a) $\mathbf{X \cdot Y} = \mathbf{Y \cdot X}$;
(b) $\mathbf{X \cdot (Y + Z)} = \mathbf{X \cdot Y} + \mathbf{X \cdot Z}$;
(c) $(c\mathbf{X}) \cdot \mathbf{Y} = \mathbf{X \cdot} (c\mathbf{Y}) = c(\mathbf{X \cdot Y})$;
(d) $\mathbf{X \cdot X} = |\mathbf{X}|^2 \geq 0$, with equality if and only if $\mathbf{X = 0}$.

We leave the proof to the student (Exercise T-8).

Schwarz's inequality, which we prove next, occurs in many branches of mathematics. It is also known as Cauchy's inequality, or as the Cauchy–Schwarz inequality.

Theorem 1.8 (Schwarz's Inequality)   If **X** and **Y** are n-vectors then

(6) $$|\mathbf{X \cdot Y}| \leq |\mathbf{X}| \, |\mathbf{Y}|,$$

with equality if and only if one of the vectors is a scalar multiple of the other.

(Note that on the left $| \; |$ stands for the absolute value of a scalar, while on the right it stands for the length of a vector. We shall use the same symbol for these two quantities because it is convenient to do so, and it is always possible from the context to decide which meaning is intended.)

Proof.   If $\mathbf{X = 0}$, then $\mathbf{X \cdot Y} = |\mathbf{X}| \, |\mathbf{Y}| = 0$, and $\mathbf{X} = 0\mathbf{Y}$ and we are finished. Now assume **X** and **Y** are nonzero and let a be a real parameter. From the properties listed in Theorem 1.7,

(7)
$$0 \leq |\mathbf{X} - a\mathbf{Y}|^2 = (\mathbf{X} - a\mathbf{Y}) \cdot (\mathbf{X} - a\mathbf{Y})$$
$$= |\mathbf{X}|^2 - 2a(\mathbf{X \cdot Y}) + |\mathbf{Y}|^2 a^2.$$

For fixed **X** and **Y** the last expression in (7) is a quadratic polynomial $p(a)$ which is nonnegative for all real values of a, and therefore cannot have distinct real roots. (If it had distinct roots, it would have to assume negative values between them, as shown in Fig. 1.6.) Since the roots of $p(a)$

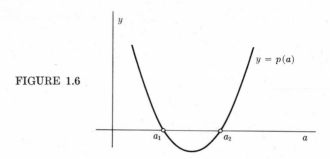

FIGURE 1.6

are, by the quadratic formula,

$$a = (\mathbf{X}\cdot\mathbf{Y}) \pm \sqrt{(\mathbf{X}\cdot\mathbf{Y})^2 - |\mathbf{X}|^2|\mathbf{Y}|^2},$$

this means that

$$(\mathbf{X}\cdot\mathbf{Y})^2 - |\mathbf{X}|^2|\mathbf{Y}|^2 \leq 0,$$

which implies (6). If equality occurs in (6) then $p(a)$ has the repeated real root $a_0 = (\mathbf{X}\cdot\mathbf{Y})$ and, from (7), $|\mathbf{X} - a_0\mathbf{Y}| = 0$ and $\mathbf{X} = a_0\mathbf{Y}$. We leave it to the student to show that if $\mathbf{X} = a\mathbf{Y}$ for some real a, then the equality occurs in (6).

Example 1.15 If

$$\mathbf{X} = \begin{bmatrix} 2 \\ -1 \\ 3 \\ 4 \end{bmatrix}, \quad \mathbf{Y} = \begin{bmatrix} 1 \\ 3 \\ -1 \\ 0 \end{bmatrix}, \quad \text{and} \quad \mathbf{Z} = \begin{bmatrix} 4 \\ -2 \\ 6 \\ 8 \end{bmatrix},$$

then

$$|\mathbf{X}| = \sqrt{30}, \quad |\mathbf{Y}| = \sqrt{11}, \quad |\mathbf{Z}| = \sqrt{120},$$

and

$$\mathbf{X}\cdot\mathbf{Y} = -4, \quad \mathbf{X}\cdot\mathbf{Z} = 60.$$

Thus

$$|\mathbf{X}\cdot\mathbf{Y}| = 4 < \sqrt{330} = |\mathbf{X}||\mathbf{Y}|$$

and

$$|\mathbf{X}\cdot\mathbf{Z}| = |60| = \sqrt{3600} = |\mathbf{X}||\mathbf{Z}|.$$

From the last equality we know that $\mathbf{Z}$ is a scalar multiple of $\mathbf{X}$; in fact $\mathbf{Z} = 2\mathbf{X}$.

In R^2 the points $\mathbf{X}$, $\mathbf{Y}$ and $\mathbf{X} + \mathbf{Y}$ are vertices of a parallelogram with sides OX and OY as shown in Fig. 1.7; thus there is a triangle with sides of length $|\mathbf{X}|$, $|\mathbf{Y}|$, and $|\mathbf{X} + \mathbf{Y}|$. Since the length of a side of a triangle

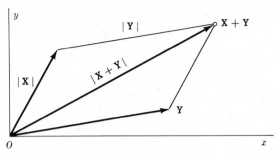

FIGURE 1.7

cannot exceed the sum of lengths of the other two, it follows that

$$|\mathbf{X} + \mathbf{Y}| \le |\mathbf{X}| + |\mathbf{Y}|.$$

That this inequality holds in R^n for any n is shown in the next theorem.

Theorem 1.9 (Triangle Inequality)   If $\mathbf{X}$ and $\mathbf{Y}$ are n-vectors then

(8) $$|\mathbf{X} + \mathbf{Y}| \le |\mathbf{X}| + |\mathbf{Y}|,$$

and equality holds if and only if $\mathbf{X} = \mathbf{0}$ or $\mathbf{Y} = a\mathbf{X}$ with $a \ge 0$.

Proof.   From (7) with $a = -1$,

$$|\mathbf{X} + \mathbf{Y}|^2 = |\mathbf{X}|^2 + 2(\mathbf{X} \cdot \mathbf{Y}) + |\mathbf{Y}|^2$$

and Schwarz's inequality implies that

$$|\mathbf{X} + \mathbf{Y}|^2 \le |\mathbf{X}|^2 + 2|\mathbf{X}||\mathbf{Y}| + |\mathbf{Y}|^2$$
$$= (|\mathbf{X}| + |\mathbf{Y}|)^2,$$

from which (8) can be obtained by taking square roots. We leave the proof of the assertion about equality to the student.

Example 1.16   Let $\mathbf{X}$, $\mathbf{Y}$, and $\mathbf{Z}$ be as defined in Example 1.15. Then

$$|\mathbf{X} + \mathbf{Y}| = \sqrt{33} < \sqrt{30} + \sqrt{11} = |\mathbf{X}| + |\mathbf{Y}|$$

and

$$|\mathbf{X} + \mathbf{Z}| = \sqrt{270} = \sqrt{30} + \sqrt{120} = |\mathbf{X}| + |\mathbf{Z}|.$$

(Note that $\mathbf{Z} = 2\mathbf{X}$.)

Definition 1.12   The *cosine of the angle* between two n-vectors $\mathbf{X}$ and $\mathbf{Y}$ is defined by

(9) $$\cos \theta = \frac{\mathbf{X} \cdot \mathbf{Y}}{|\mathbf{X}||\mathbf{Y}|}.$$

This definition generalizes (5), to which it reduces when $n=2$; it is also equivalent to the usual geometric definition for $n=3$. If we were not sure that the magnitude of the right side of (9) is bounded by unity for any n-vectors $\mathbf{X}$ and $\mathbf{Y}$, then we would not have the right to write it as $\cos \theta$; however, Schwarz's inequality resolves any doubts on that score.

In R^2 two vectors $\mathbf{X}$ and $\mathbf{Y}$ are parallel if and only if $|\cos \theta| = 1$; they are perpendicular, or orthogonal, if and only if $\cos \theta = 0$. This geometric language is useful in R^n.

Definition 1.13   The n-vectors $\mathbf{X}$ and $\mathbf{Y}$ said to be *parallel* if $|\mathbf{X} \cdot \mathbf{Y}| = |\mathbf{X}| |\mathbf{Y}|$, and *in the same direction* if $\mathbf{X} \cdot \mathbf{Y} = |\mathbf{X}| |\mathbf{Y}|$. They are said to be *orthogonal* if $\mathbf{X} \cdot \mathbf{Y} = 0$.

From Theorem 1.8, two n-vectors are parallel if and only if one is a scalar multiple of the other, and in the same direction if and only if the scalar is nonnegative. The zero vector is both parallel and perpendicular to every vector.

Example 1.17   If

$$\mathbf{X} = \begin{bmatrix} 1 \\ 0 \\ 1 \\ 0 \end{bmatrix} \quad \text{and} \quad \mathbf{Y} = \begin{bmatrix} 1 \\ 0 \\ 0 \\ 1 \end{bmatrix},$$

then

$$|\mathbf{X}| = \sqrt{2}, \qquad |\mathbf{Y}| = \sqrt{2}, \qquad \mathbf{X} \cdot \mathbf{Y} = 1,$$

and

$$\cos \theta = \tfrac{1}{2}.$$

The vector

$$\mathbf{Z} = \begin{bmatrix} 0 \\ 1 \\ 0 \\ 0 \end{bmatrix}$$

is orthogonal to $\mathbf{X}$ and $\mathbf{Y}$, since

$$\mathbf{X} \cdot \mathbf{Z} = \mathbf{Y} \cdot \mathbf{Z} = 0,$$

and

$$W = \begin{bmatrix} 2 \\ 0 \\ 2 \\ 0 \end{bmatrix}$$

is in the same direction as **X** since

$$\mathbf{W} \cdot \mathbf{X} = 4 = \sqrt{2} \sqrt{8} = |\mathbf{W}||\mathbf{X}|;$$

of course, $\mathbf{W} = 2\mathbf{X}$.

Example 1.18   A *unit vector* **U** is a vector of unit length:

$$|\mathbf{U}| = 1.$$

If **X** is an arbitrary nonzero vector then

$$\mathbf{U} = \left[\frac{1}{|\mathbf{X}|} \right] \mathbf{X} = \frac{\mathbf{X}}{|\mathbf{X}|}$$

is the unique unit vector in the direction of **X**. For example, if

$$\mathbf{X} = \begin{bmatrix} 1 \\ -1 \\ 2 \\ 3 \end{bmatrix}, \qquad \text{then} \qquad \mathbf{U} = \frac{1}{\sqrt{15}} \begin{bmatrix} 1 \\ -1 \\ 2 \\ 3 \end{bmatrix}$$

is the unit vector in the direction of **X**.

EXERCISES 1.1

1. If

$$\mathbf{X} = \begin{bmatrix} 1 \\ 2 \\ 3 \\ 4 \end{bmatrix}, \qquad \mathbf{Y} = \begin{bmatrix} 1 \\ 2 \\ z \\ w \end{bmatrix}, \qquad \text{and} \quad \mathbf{Z} = \begin{bmatrix} 2 \\ 4 \\ 4 \\ 6 \end{bmatrix},$$

find z and w such that

(a) $\mathbf{X} = \mathbf{Y}$;   (b) $\mathbf{X} + \mathbf{Y} = 2\mathbf{X}$;   (c) $2\mathbf{Y} - \mathbf{X} = 3\mathbf{X} - \mathbf{Z}$.

2. If

$$\mathbf{X} = \begin{bmatrix} 1 \\ 2 \\ 3 \end{bmatrix}$$

find $\mathbf{Y}$ such that

(a) $\mathbf{X} = \frac{2}{3}\mathbf{Y}$;   (b) $2\mathbf{X} + \mathbf{Y} = \mathbf{X}$;   (c) $\mathbf{X} + 2\mathbf{Y} = \mathbf{X} - \mathbf{Y}$.

3. Find $\mathbf{X} + \mathbf{Y}$, $\mathbf{X} - \mathbf{Y}$, $-3\mathbf{X}$, and $2\mathbf{X} - 3\mathbf{Y}$ if

(a) $\mathbf{X} = \begin{bmatrix} -1 \\ 2 \end{bmatrix}$, $\mathbf{Y} = \begin{bmatrix} 3 \\ -4 \end{bmatrix}$;   (b) $\mathbf{X} = \begin{bmatrix} 2 \\ 3 \\ 0 \end{bmatrix}$, $\mathbf{Y} = \begin{bmatrix} 0 \\ 1 \\ 2 \end{bmatrix}$;

(c) $\mathbf{X} = \begin{bmatrix} -1 \\ 0 \\ -2 \\ 4 \end{bmatrix}$, $\mathbf{Y} = \begin{bmatrix} -4 \\ -3 \\ 0 \\ 1 \end{bmatrix}$.

4. Repeat Exercise 3 for

(a) $\mathbf{X} = \begin{bmatrix} -1 \\ 2 \end{bmatrix}$, $\mathbf{Y} = \begin{bmatrix} 2 \\ 0 \end{bmatrix}$;   (b) $\mathbf{X} = \begin{bmatrix} 1 \\ -3 \\ 4 \end{bmatrix}$, $\mathbf{Y} = \begin{bmatrix} 0 \\ 1 \\ -3 \end{bmatrix}$;

(c) $\mathbf{X} = \begin{bmatrix} 1 \\ 2 \\ -1 \\ 3 \end{bmatrix}$, $\mathbf{Y} = \begin{bmatrix} 1 \\ -3 \\ -4 \\ 1 \end{bmatrix}$.

5. Verify (a) through (e) of Theorem 1.1 for

$$\mathbf{X} = \begin{bmatrix} 1 \\ 2 \\ 3 \\ 4 \end{bmatrix}, \quad \mathbf{Y} = \begin{bmatrix} 3 \\ -1 \\ 2 \\ 5 \end{bmatrix}, \quad \mathbf{Z} = \begin{bmatrix} 7 \\ -2 \\ -1 \\ 3 \end{bmatrix},$$

$a = -2$ and $b = 3$.

6. Repeat Exercise 5 for

$$\mathbf{X} = \begin{bmatrix} 2 \\ 5 \\ 3 \\ 1 \end{bmatrix}, \quad \mathbf{Y} = \begin{bmatrix} 1 \\ -2 \\ 7 \\ -1 \end{bmatrix}, \quad \mathbf{Z} = \begin{bmatrix} 3 \\ 2 \\ -4 \\ 1 \end{bmatrix},$$

$a = 2$ and $b = -1$.

7. Find which of the following vectors are linear combinations of

$$\mathbf{X}_1 = \begin{bmatrix} 1 \\ -1 \\ 3 \end{bmatrix}, \quad \mathbf{X}_2 = \begin{bmatrix} -1 \\ 0 \\ 2 \end{bmatrix}, \quad \text{and} \quad \mathbf{X}_3 = \begin{bmatrix} 0 \\ -1 \\ 5 \end{bmatrix}.$$

(a) $\begin{bmatrix} 9 \\ 2 \\ 14 \end{bmatrix}$;   (b) $\begin{bmatrix} 1 \\ -3 \\ 13 \end{bmatrix}$;   (c) $\begin{bmatrix} 0 \\ 1 \\ 2 \end{bmatrix}$;   (d) $\begin{bmatrix} -1 \\ -2 \\ 12 \end{bmatrix}$.

8. Write the following vectors as linear combinations of

$$\mathbf{X}_1 = \begin{bmatrix} 1 \\ 0 \\ 3 \end{bmatrix}, \quad \mathbf{X}_2 = \begin{bmatrix} -1 \\ 2 \\ -3 \end{bmatrix}, \quad \text{and} \quad \mathbf{X}_3 = \begin{bmatrix} 0 \\ 1 \\ 2 \end{bmatrix}.$$

(a) $\begin{bmatrix} 1 \\ 3 \\ 5 \end{bmatrix}$;   (b) $\begin{bmatrix} -2 \\ 6 \\ 1 \end{bmatrix}$;   (c) $\begin{bmatrix} -1 \\ 4 \\ 5 \end{bmatrix}$;   (d) $\begin{bmatrix} 0 \\ 3 \\ 2 \end{bmatrix}$.

9. State which of the following are subspaces of R^3. The set of all vectors of the form

(a) $\begin{bmatrix} x \\ y \\ 1 \end{bmatrix}$;   (b) $\begin{bmatrix} x \\ 0 \\ z \end{bmatrix}$;   (c) $\begin{bmatrix} x \\ y \\ x+y \end{bmatrix}$;   (d) $\begin{bmatrix} x \\ y \\ x+2 \end{bmatrix}$.

10. State which of the following are subspaces of R^3. The set of all vectors of the form

(a) $\begin{bmatrix} x \\ x \\ 1 \end{bmatrix}$;   (b) $\begin{bmatrix} x \\ x \\ 0 \end{bmatrix}$;   (c) $\begin{bmatrix} x \\ y \\ z \end{bmatrix}$ $(z > 0)$;   (d) $\begin{bmatrix} x^2 \\ y \\ z \end{bmatrix}$.

11. If

$$\mathbf{X}_1 = \begin{bmatrix} 1 \\ -2 \\ 3 \end{bmatrix}, \quad \mathbf{X}_2 = \begin{bmatrix} -2 \\ 3 \\ -4 \end{bmatrix}, \quad \text{and} \quad \mathbf{X}_3 = \begin{bmatrix} 4 \\ -5 \\ 6 \end{bmatrix},$$

show that $S[\mathbf{X}_1, \mathbf{X}_2, \mathbf{X}_3] = S[\mathbf{X}_1, \mathbf{X}_2]$. (It is sufficient to show that $\mathbf{X}_3$ is a linear combination of $\mathbf{X}_1$ and $\mathbf{X}_2$. Why?)

12. Let $\mathbf{X}_1$, $\mathbf{X}_2$, and $\mathbf{X}_3$ be as defined in Exercise 11. If

$$\mathbf{Y}_1 = \begin{bmatrix} -1 \\ 1 \\ -1 \end{bmatrix} \quad \text{and} \quad \mathbf{Y}_2 = \begin{bmatrix} -3 \\ 5 \\ -7 \end{bmatrix},$$

show that $S[\mathbf{X}_1, \mathbf{X}_2, \mathbf{X}_3] = S[\mathbf{Y}_1, \mathbf{Y}_2]$.

13. State which sets span R^2.

(a) $\begin{bmatrix} 1 \\ 1 \end{bmatrix}$, $\begin{bmatrix} 0 \\ 1 \end{bmatrix}$;   (b) $\begin{bmatrix} 1 \\ -1 \end{bmatrix}$, $\begin{bmatrix} 2 \\ -2 \end{bmatrix}$, $\begin{bmatrix} 0 \\ 0 \end{bmatrix}$;

(c) $\begin{bmatrix} 0 \\ 1 \end{bmatrix}$, $\begin{bmatrix} 1 \\ 0 \end{bmatrix}$, $\begin{bmatrix} 1 \\ 1 \end{bmatrix}$.

14. State which sets span R^3:

(a) $\begin{bmatrix} 1 \\ 0 \\ 1 \end{bmatrix}$, $\begin{bmatrix} 1 \\ 1 \\ 1 \end{bmatrix}$;   (b) $\begin{bmatrix} 0 \\ 0 \\ 0 \end{bmatrix}$, $\begin{bmatrix} 1 \\ 2 \\ 1 \end{bmatrix}$, $\begin{bmatrix} 1 \\ -1 \\ 2 \end{bmatrix}$, $\begin{bmatrix} 0 \\ 0 \\ 1 \end{bmatrix}$;

(c) $\begin{bmatrix} 1 \\ -1 \\ 2 \end{bmatrix}$, $\begin{bmatrix} 1 \\ 2 \\ 3 \end{bmatrix}$, $\begin{bmatrix} 1 \\ 4 \\ -2 \end{bmatrix}$;   (d) $\begin{bmatrix} 1 \\ 0 \\ 0 \end{bmatrix}$, $\begin{bmatrix} 1 \\ -1 \\ 2 \end{bmatrix}$, $\begin{bmatrix} 0 \\ 1 \\ 0 \end{bmatrix}$, $\begin{bmatrix} 1 \\ 1 \\ 1 \end{bmatrix}$.

15. Which of the sets in Exercise 13 are bases for R^2?

16. Which of the sets in Exercise 14 are bases for R^3?

17. Express

$$\mathbf{Y} = \begin{bmatrix} 2 \\ 3 \end{bmatrix}$$

in terms of the following bases for R^2:

(a) $\mathbf{X_1} = \begin{bmatrix} 1 \\ 2 \end{bmatrix}$, $\mathbf{X_2} = \begin{bmatrix} 1 \\ -3 \end{bmatrix}$;   (b) $\mathbf{X_1} = \begin{bmatrix} 0 \\ 1 \end{bmatrix}$, $\mathbf{X_2} = \begin{bmatrix} 1 \\ 1 \end{bmatrix}$;

(c) $\mathbf{X_1} = \begin{bmatrix} -1 \\ 2 \end{bmatrix}$, $\mathbf{X_2} = \begin{bmatrix} -1 \\ 1 \end{bmatrix}$;   (d) $\mathbf{X_1} = \begin{bmatrix} 0 \\ 1 \end{bmatrix}$, $\mathbf{X_2} = \begin{bmatrix} 1 \\ 0 \end{bmatrix}$.

18. Express

$$\mathbf{Y} = \begin{bmatrix} 2 \\ 1 \\ 3 \end{bmatrix}$$

in terms of the following bases for R^3:

(a) $\mathbf{X_1} = \begin{bmatrix} 1 \\ 1 \\ 1 \end{bmatrix}$, $\mathbf{X_2} = \begin{bmatrix} 1 \\ 2 \\ 3 \end{bmatrix}$, $\mathbf{X_3} = \begin{bmatrix} 0 \\ 1 \\ 0 \end{bmatrix}$;

(b) $\mathbf{X_1} = \begin{bmatrix} 1 \\ 2 \\ 1 \end{bmatrix}$, $\mathbf{X_2} = \begin{bmatrix} 2 \\ 1 \\ 3 \end{bmatrix}$, $\mathbf{X_3} = \begin{bmatrix} 1 \\ 0 \\ 1 \end{bmatrix}$;

(c) $\mathbf{X_1} = \begin{bmatrix} 1 \\ 2 \\ 1 \end{bmatrix}$, $\mathbf{X_2} = \begin{bmatrix} 1 \\ 1 \\ 4 \end{bmatrix}$, $\mathbf{X_3} = \begin{bmatrix} 1 \\ 5 \\ 1 \end{bmatrix}$.

19. Find the dimensions of the subspaces of R^2 spanned by the sets of vectors in Exercise 13.

20. Find a basis for the subspace of R^3 spanned by

(a) $\mathbf{X}_1 = \begin{bmatrix} 1 \\ 2 \\ 3 \end{bmatrix}$, $\mathbf{X}_2 = \begin{bmatrix} 2 \\ 0 \\ 1 \end{bmatrix}$, $\mathbf{X}_3 = \begin{bmatrix} 3 \\ 2 \\ 4 \end{bmatrix}$;

(b) $\mathbf{X}_1 = \begin{bmatrix} 1 \\ -1 \\ 2 \end{bmatrix}$, $\mathbf{X}_2 = \begin{bmatrix} 4 \\ 4 \\ 7 \end{bmatrix}$, $\mathbf{X}_3 = \begin{bmatrix} 1 \\ 2 \\ 1 \end{bmatrix}$, $\mathbf{X}_4 = \begin{bmatrix} 2 \\ 3 \\ 4 \end{bmatrix}$.

21. Find the lengths of the following vectors.

(a) $\begin{bmatrix} 1 \\ 2 \\ 3 \end{bmatrix}$; (b) $\begin{bmatrix} 0 \\ 0 \\ 0 \end{bmatrix}$; (c) $\begin{bmatrix} 1 \\ 0 \\ 1 \end{bmatrix}$; (d) $\begin{bmatrix} 1 \\ -2 \\ -3 \end{bmatrix}$.

22. Find the distance between the following pairs of points.

(a) $\mathbf{X} = (1, 2, 3)$, $\mathbf{Y} = (-2, -3, 4)$;
(b) $\mathbf{X} = (1, 2, 3)$, $\mathbf{Y} = (3, 2, -5)$;
(c) $\mathbf{X} = (-1, -2, -3)$, $\mathbf{Y} = (4, 5, 6)$.

23. Find $\mathbf{X} \cdot \mathbf{Y}$:

(a) $\mathbf{X} = \begin{bmatrix} 1 \\ 2 \\ -3 \end{bmatrix}$, $\mathbf{Y} = \begin{bmatrix} -2 \\ -3 \\ 4 \end{bmatrix}$; (b) $\mathbf{X} = \begin{bmatrix} 0 \\ 1 \\ 0 \end{bmatrix}$, $\mathbf{Y} = \begin{bmatrix} 1 \\ 0 \\ 0 \end{bmatrix}$;

(c) $\mathbf{X} = \begin{bmatrix} 2 \\ 2 \\ -3 \end{bmatrix}$, $\mathbf{Y} = \begin{bmatrix} 3 \\ -4 \\ 2 \end{bmatrix}$; (d) $\mathbf{X} = \begin{bmatrix} 1 \\ 1 \\ 1 \end{bmatrix}$, $\mathbf{Y} = \begin{bmatrix} 2 \\ -1 \\ 3 \end{bmatrix}$.

24. Find $|\mathbf{X} - \mathbf{Y}|$ for the pairs of vectors in Exercise 23.

25. Find the cosine of the angle between $\mathbf{X}$ and $\mathbf{Y}$ for each pair of vectors in Exercise 23.

26. Verify Schwarz's inequality for

(a) $\mathbf{X} = \begin{bmatrix} 1 \\ 2 \\ 3 \end{bmatrix}$, $\mathbf{Y} = \begin{bmatrix} -1 \\ 2 \\ -1 \end{bmatrix}$; (b) $\mathbf{X} = \begin{bmatrix} 1 \\ -1 \\ 2 \end{bmatrix}$, $\mathbf{Y} = \begin{bmatrix} 3 \\ -3 \\ 6 \end{bmatrix}$.

27. Verify the triangle inequality for

(a) $\mathbf{X} = \begin{bmatrix} 1 \\ 2 \\ 3 \end{bmatrix}$, $\mathbf{Y} = \begin{bmatrix} -1 \\ 2 \\ -1 \end{bmatrix}$, $\mathbf{Z} = \begin{bmatrix} 0 \\ 1 \\ 2 \end{bmatrix}$;

(b) $\mathbf{X} = \begin{bmatrix} 1 \\ 3 \\ 1 \end{bmatrix}$, $\mathbf{Y} = \begin{bmatrix} -1 \\ 0 \\ 2 \end{bmatrix}$, $\mathbf{Z} = \begin{bmatrix} -3 \\ -3 \\ 3 \end{bmatrix}$.

28. Find a unit vector in the direction of $\mathbf{X}$:

(a) $\mathbf{X} = \begin{bmatrix} 1 \\ 2 \end{bmatrix}$; (b) $\mathbf{X} = \begin{bmatrix} 2 \\ -3 \\ 4 \end{bmatrix}$; (c) $\mathbf{X} = \begin{bmatrix} 1 \\ 2 \\ 1 \\ -1 \end{bmatrix}$; (d) $\mathbf{X} = \begin{bmatrix} 0 \\ 0 \\ 1 \end{bmatrix}$.

THEORETICAL EXERCISES

T-1. Complete the proof of Theorem 1.1.

T-2. Prove Theorem 1.2.

T-3. Prove that $S[\mathbf{X}_1, \ldots, \mathbf{X}_m]$ is a subspace of R^n.

T-4. Let S_1 and S_2 be sets of vectors in R^n, and let S_1 be a subset of S_2. Prove:
   (a) If S_1 is linearly dependent, so is S_2.
   (b) If S_2 is linearly independent, so is S_1.

T-5. Prove: If m is the maximum number of vectors in any linearly independent set in a subspace H, then any linearly independent set in H with m elements is a basis for H.

T-6. Let $\mathbf{X}_1, \ldots, \mathbf{X}_k$ be a basis for a subspace H. Show that every vector in H can be written uniquely as a linear combination of $\mathbf{X}_1, \ldots, \mathbf{X}_k$.

T-7. Prove: If H is an m-dimensional subspace of R^n, then $m \leq n$ and no set of fewer than m vectors spans H.

T-8. Prove Theorem 1.7.

T-9. If $\mathbf{X}$ is orthogonal to $\mathbf{Y}$ and $\mathbf{Z}$, show that $\mathbf{X}$ is orthogonal to $S[\mathbf{Y}, \mathbf{Z}]$.

T-10. Let $\mathbf{X}_0$ be a fixed vector in R^n. Prove that the set of vectors orthogonal to $\mathbf{X}_0$ is a subspace of R^n.

T-11. (a) Show that if $\mathbf{X} \cdot \mathbf{Y} = 0$ for all $\mathbf{X}$ in R^n, then $\mathbf{Y} = \mathbf{0}$.
   (b) Show that if $\mathbf{X} \cdot \mathbf{Y} = \mathbf{X} \cdot \mathbf{Z}$ for all $\mathbf{X}$ in R^n, then $\mathbf{Y} = \mathbf{Z}$.

T-12. Show that

$$| \mathbf{X}_1 + \cdots + \mathbf{X}_n | \leq | \mathbf{X}_1 | + \cdots + | \mathbf{X}_n |,$$

and give conditions for equality.

T-13. Prove that a subset H of R^n is a subspace (Definition 1.5) of R^n if and only if H is a vector space over the real numbers.

1.2 Lines and Planes

The only subspaces of R^3, other than the zero subspace and R^3 itself, are those of dimensions one and two. By definition, a one-dimensional subspace consists of all multiples of a fixed nonzero vector, and a two-dimensional subspace consists of all linear combinations of two linearly independent vectors. These subspaces have familiar geometric interpretations.

Lines

Suppose H is a one-dimensional subspace of R^3. Then any nonzero vector

$$\mathbf{U} = \begin{bmatrix} u \\ v \\ w \end{bmatrix}$$

in H is a basis for H; thus $\mathbf{X}$ is in H if and only if

$$(1) \qquad\qquad \mathbf{X} = t\mathbf{U} \qquad (-\infty < t < \infty).$$

The points whose position vectors with respect to the origin are in H lie on a *line L through the origin in the direction of* $\mathbf{U}$; (1) is a parametric equation of this line. In general the *line L through* $\mathbf{X}_0 = (x_0, y_0, z_0)$ *parallel to* $\mathbf{U}$ consists of the points whose position vectors with respect to $\mathbf{X}_0$ are in H (Fig. 2.1); it is given parametrically by

$$\mathbf{X} = \mathbf{X}_0 + t\mathbf{U}$$

or in terms of components by

$$
\begin{aligned}
x &= x_0 + tu, \\
(2) \qquad\qquad y &= y_0 + tv, \\
z &= z_0 + tw.
\end{aligned}
$$

If u, v, and w are all nonzero, then we can solve each equation in (2) for t and equate the results to obtain the *symmetric* form of the equation of L:

$$(3) \qquad\qquad \frac{x - x_0}{u} = \frac{y - y_0}{v} = \frac{z - z_0}{w}.$$

If one or two of the components of $\mathbf{U}$ vanish, then (3) indicates division by zero, which is never legitimate. Nevertheless it is convenient to adopt

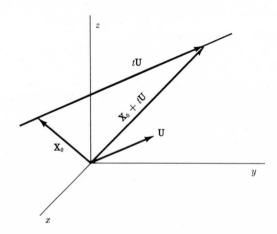

FIGURE 2.1

the convention that (3) stands for the line through $\mathbf{X}_0$ in the direction of $\mathbf{U}$ if at least one of the denominators is nonzero.

Thus

$$\frac{x - x_0}{0} = \frac{y - y_0}{v} = \frac{z - z_0}{w}$$

is the equation of a line along which x is constant ($x = x_0$) and, if u and v are both nonzero,

$$\frac{y - y_0}{v} = \frac{z - z_0}{w}.$$

Example 2.1 The line L through the origin in the direction of

$$\mathbf{U} = \begin{bmatrix} 1 \\ 2 \\ -1 \end{bmatrix}$$

is given parametrically by

$$\begin{bmatrix} x \\ y \\ z \end{bmatrix} = t \begin{bmatrix} 1 \\ 2 \\ -1 \end{bmatrix} \qquad (-\infty < t < \infty);$$

its symmetric equation is

$$x = \tfrac{1}{2}y = -z.$$

Example 2.2 The set of points

$$\mathbf{X} = \begin{bmatrix} 3 \\ 1 \\ -2 \end{bmatrix} + t \begin{bmatrix} 0 \\ -1 \\ 2 \end{bmatrix} \qquad (-\infty < t < \infty)$$

is the line L through $\mathbf{X}_0 = (3, 1, -2)$ in the direction of

$$\mathbf{U} = \begin{bmatrix} 0 \\ -1 \\ 2 \end{bmatrix};$$

its symmetric equation is

$$\frac{x-3}{0} = \frac{(y+1)}{-1} = \frac{z+2}{2}.$$

The vectors from $\mathbf{X}_0$ to points on L are the elements of the subspace spanned by $\mathbf{U}$; that is, a point $\mathbf{X}$ is on L if and only if the vector $\mathbf{X} - \mathbf{X}_0$ is parallel to $\mathbf{U}$.

Definition 2.1 Let $\mathbf{X}_0 = (\bar{x}_1, \ldots, \bar{x}_n)$ and

$$\mathbf{U} = \begin{bmatrix} u_1 \\ \vdots \\ u_n \end{bmatrix} \neq 0.$$

The line L through $\mathbf{X}_0$ in the direction of $\mathbf{U}$ is the set of points given parametrically by

$$(4) \qquad \mathbf{X} = \mathbf{X}_0 + t\mathbf{U} \qquad (-\infty < t < \infty),$$

or in symmetric form by

$$(5) \qquad \frac{x_1 - \bar{x}_1}{u_1} = \frac{x_2 - \bar{x}_2}{u_2} = \cdots = \frac{x_n - \bar{x}_n}{u_n}.$$

If the components of $\mathbf{U}$ are all nonzero then (5) can be obtained by rewriting (4) as

$$x_i = \bar{x}_i + tu_i \qquad (1 \leq i \leq n),$$

solving each equation for t, and equating the results. If one or more of the components of $\mathbf{U}$ vanishes then (5) must be interpreted as symbolically representing the line defined by (4).

Example 2.3 In R^5 the line through $\mathbf{X}_0 = (2, 3, -1, 4, 5)$ in the direction of

$$\mathbf{U} = \begin{bmatrix} 1 \\ 2 \\ -1 \\ 3 \\ 4 \end{bmatrix} \quad \text{is given parametrically by} \quad \mathbf{X} = \begin{bmatrix} 2 \\ 3 \\ -1 \\ 4 \\ 5 \end{bmatrix} + t \begin{bmatrix} 1 \\ 2 \\ -1 \\ 3 \\ 4 \end{bmatrix},$$

and in symmetric form by

$$\frac{x_1 - 2}{1} = \frac{x_2 - 3}{2} = \frac{(x_3 + 1)}{-1} = \frac{x_4 - 4}{3} = \frac{x_5 - 5}{4}.$$

Example 2.4 The symmetric equation

$$\frac{x_1 - 1}{0} = \frac{x_2 - 3}{4} = \frac{x_3 + 5}{0} = \frac{x_4 - 5}{2} = \frac{x_5}{1}$$

represents the line in R^5 through $\mathbf{X}_0 = (1, 3, -5, 5, 0)$ in the direction of

$$\mathbf{U} = \begin{bmatrix} 0 \\ 4 \\ 0 \\ 2 \\ 1 \end{bmatrix}.$$

A line is uniquely determined by requiring that it pass through two points. Thus, if $\mathbf{X}_0$ and $\mathbf{X}_1$ are distinct points on L, then L is the line through $\mathbf{X}_0$ in the direction of $\mathbf{X}_1 - \mathbf{X}_0$. It is given parametrically by

$$(6) \qquad\qquad \mathbf{X} = \mathbf{X}_0 + t(\mathbf{X}_1 - \mathbf{X}_0);$$

in Fig. 2.2 this line is depicted in R^3.

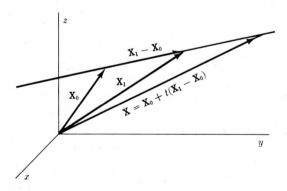

FIGURE 2.2

Example 2.5   In R^4 let $\mathbf{X}_0 = (1, 2, -1, 3)$, and $\mathbf{X}_1 = (4, 5, -2, 1)$. Then

$$\mathbf{X}_1 - \mathbf{X}_0 = \begin{bmatrix} 3 \\ 3 \\ -1 \\ -2 \end{bmatrix};$$

hence the line through $\mathbf{X}_0$ and $\mathbf{X}_1$ is given by

$$\mathbf{X} = \begin{bmatrix} 1 \\ 2 \\ -1 \\ 3 \end{bmatrix} + t \begin{bmatrix} 3 \\ 3 \\ -1 \\ -2 \end{bmatrix}.$$

Clearly any line L can be represented in infinitely many ways in parametric (or symmetric) form: its equation can be written in the form (4), where $\mathbf{X}_0$ in any point on L and $\mathbf{U}_0$ is any vector in the direction of L, or in the form (6), where $\mathbf{X}_1$ and $\mathbf{X}_0$ are any two distinct points on L. We shall give criteria for deciding when lines represented by distinct equations are identical. However, before doing this it is convenient to consider other relationships between lines.

Definition 2.2   If $\mathbf{X}_0$ is common to lines L_1 and L_2, then the lines are said to *intersect* at $\mathbf{X}_0$, which is called their *point of intersection*.

Definition 2.3   If nonzero vectors in the directions of lines L_1 and L_2 are parallel, then L_1 and L_2 are said to be *parallel*.

Definition 2.4   If nonzero vectors in the directions of lines L_1 and L_2 are orthogonal, then L_1 and L_2 are said to be *perpendicular*, or *orthogonal*.

Definition 2.5   Lines L_1 and L_2 are said to be *skew* if they are not parallel and do not intersect.

Theorem 2.1   Let L_1 and L_2 be given parametrically by

(7) $$\mathbf{X} = \mathbf{X}_0 + s\mathbf{U}$$

and

(8)
$$\mathbf{X} = \mathbf{X}_1 + t\mathbf{V}.$$

Then:

(a) L_1 and L_2 are parallel if and only if $\mathbf{U} = a\mathbf{V}$ for some scalar a.

(b) L_1 and L_2 are perpendicular if and only if $\mathbf{U}\cdot\mathbf{V} = 0$.

(c) L_1 and L_2 intersect if and only if $\mathbf{X}_1 - \mathbf{X}_0$ is a linear combination of $\mathbf{U}$ and $\mathbf{V}$.

Proof. Assertions (a) and (b) follow from Definitions 2.3 and 2.4. For (c), suppose L_1 and L_2 intersect at $\bar{\mathbf{X}}$; then there are scalars s_0 and t_0 such that

(9)
$$\bar{\mathbf{X}} = \mathbf{X}_0 + s_0\mathbf{U} = \mathbf{X}_1 + t_0\mathbf{V}.$$

Thus

(10)
$$\mathbf{X}_1 - \mathbf{X}_0 = s_0\mathbf{U} - t_0\mathbf{V}$$

is a linear combination of $\mathbf{U}$ and $\mathbf{V}$. Conversely, if $\mathbf{X}_1 - \mathbf{X}_0$ is of the form (10), then the last two members of (9) are equal, and the common point that they represent is on both lines. This completes the proof of Theorem 2.1.

Example 2.6 The equations

$$\mathbf{X} = \mathbf{X}_0 + s\mathbf{U} \qquad\qquad \mathbf{X} = \mathbf{X}_1 + t\mathbf{V}$$

$$= \begin{bmatrix} -1 \\ 0 \\ 2 \end{bmatrix} + s \begin{bmatrix} 1 \\ 2 \\ 0 \end{bmatrix} \quad\text{and}\quad = \begin{bmatrix} 3 \\ 1 \\ 2 \end{bmatrix} + t \begin{bmatrix} 2 \\ 4 \\ 0 \end{bmatrix}$$

define parallel lines, since $\mathbf{V} = 2\mathbf{U}$.

Example 2.7 The equations

$$\mathbf{X} = \mathbf{X}_0 + s\mathbf{U} \qquad\qquad \mathbf{X} = \mathbf{X}_1 + t\mathbf{V}$$

$$= \begin{bmatrix} -1 \\ 0 \\ 0 \end{bmatrix} + s \begin{bmatrix} 1 \\ 2 \\ 0 \end{bmatrix} \quad\text{and}\quad = \begin{bmatrix} 0 \\ 0 \\ z_0 \end{bmatrix} + t \begin{bmatrix} 1 \\ 0 \\ 2 \end{bmatrix}$$

are not parallel, since $\mathbf{U}$ and $\mathbf{V}$ are linearly independent. They intersect if and only if

$$\begin{bmatrix} -1 \\ 0 \\ 0 \end{bmatrix} + s \begin{bmatrix} 1 \\ 2 \\ 0 \end{bmatrix} = \begin{bmatrix} 0 \\ 0 \\ z_0 \end{bmatrix} + t \begin{bmatrix} 1 \\ 0 \\ 2 \end{bmatrix}$$

has a solution $s = s_0$ and $t = t_0$. The conditions on (s_0, t_0) are

$$s_0 - t_0 = 1,$$

$$2s_0 = 0,$$

$$2t_0 + z_0 = 0.$$

Since the first two equations imply that $s_0 = 0$ and $t_0 = -1$, this system has a solution if and only if $z_0 = 2$; thus the lines intersect if and only if $z_0 = 2$, in which case the point of intersection is $\mathbf{X}_0 = (-1, 0, 0)$.

Theorem 2.2 The lines L_1 and L_2 defined by (7) and (8) are identical if and only if $\mathbf{X}_1 - \mathbf{X}_0$ and $\mathbf{U}$ are both parallel to $\mathbf{V}$.

We leave the proof of this theorem to the student (Exercise T-1).

Planes

Now suppose H is the two-dimensional subspace of R^3 consisting of vectors of the form

(11) $$\mathbf{X} = s\mathbf{U} + t\mathbf{V} \qquad (-\infty < s, t < \infty),$$

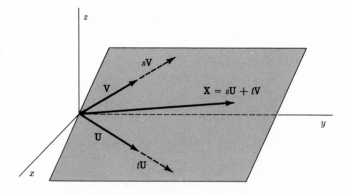

FIGURE 2.3

where **U** and **V** are linearly independent. The points whose position vectors with respect to the origin are in H form a plane through the origin (Fig. 2.3); (11) is a parametric equation of this plane, which we call the *plane of* **U** *and* **V**.

In this text we shall not use parametric equations for planes. It is intuitively evident that a point in R^3 is in the plane shown in Fig. 2.2 if and only if its position vector with respect to the origin is perpendicular to **N**. We shall use this fact to derive a nonparametric equation for the plane defined by (11).

The vector

$$\mathbf{N} = \begin{bmatrix} u_2v_3 - u_3v_2 \\ u_3v_1 - u_1v_3 \\ u_1v_2 - u_2v_1 \end{bmatrix}$$

is perpendicular to both **U** and **V**. This is easily checked by noting that if

$$\mathbf{W} = \begin{bmatrix} w_1 \\ w_2 \\ w_3 \end{bmatrix}$$

is an arbitrary vector then

$$\mathbf{W} \cdot \mathbf{N} = \begin{vmatrix} w_1 & w_2 & w_3 \\ u_1 & u_2 & u_3 \\ v_1 & v_2 & v_3 \end{vmatrix},$$

as can be seen by expanding the determinant with respect to its first row. Hence

$$\mathbf{U} \cdot \mathbf{N} = \mathbf{V} \cdot \mathbf{N} = 0,$$

because a determinant with two identical rows vanishes. Furthermore, $\mathbf{N} = \mathbf{0}$ would imply that **U** and **V** are parallel (Exercise T-2), contrary to assumption; hence $\mathbf{N} \neq \mathbf{0}$.

The vector **N** is called the *cross product* of **U** and **V** and is usually denoted by

$$\mathbf{N} = \mathbf{U} \times \mathbf{V}.$$

We shall discuss it further in Section 1.3, but for now we regard it merely as a particular vector perpendicular to **U** and **V**. Since it is also perpendicular to any linear combination of **U** and **V** (Exercise T-5), we say it is *perpendicular to the plane of* **U** *and* **V**.

Example 2.8   Let

$$\mathbf{U} = \begin{bmatrix} 2 \\ 3 \\ 1 \end{bmatrix} \quad \text{and} \quad \mathbf{V} = \begin{bmatrix} 3 \\ 1 \\ 0 \end{bmatrix}.$$

Then

$$\mathbf{N} = \mathbf{U} \times \mathbf{V}$$

$$= \begin{bmatrix} (3)(0) - (1)(1) \\ (1)(3) - (2)(0) \\ (2)(1) - (3)(3) \end{bmatrix} = \begin{bmatrix} -1 \\ 3 \\ -7 \end{bmatrix},$$

and by direct computation $\mathbf{N} \cdot \mathbf{U} = \mathbf{N} \cdot \mathbf{V} = 0$.

Theorem 2.3   A vector **X** is in the subspace of R^3 spanned by two linearly independent vectors **U** and **V** if and only if $\mathbf{X} \cdot (\mathbf{U} \times \mathbf{V}) = 0$.

Proof.   If **U** and **V** are linearly independent then **U**, **V**, and $\mathbf{N} = \mathbf{U} \times \mathbf{V}$ are also linearly independent (Exercise T-3); therefore they form a basis for R^3. An arbitrary **X** in R^3 can be written in the form

$$\mathbf{X} = s\mathbf{U} + t\mathbf{V} + a\mathbf{N},$$

where s, t, and a are scalars. Thus

$$\mathbf{X} \cdot \mathbf{N} = s\mathbf{U} \cdot \mathbf{N} + t\mathbf{V} \cdot \mathbf{N} + a\mathbf{N} \cdot \mathbf{N}$$

$$= a \mid \mathbf{N} \mid^2.$$

Thus $\mathbf{X} \cdot \mathbf{N} = 0$ if and only if $a = 0$; that is, if and only if **X** is in the subspace spanned by **U** and **V**.

We now define planes in R^n.

Definition 2.6   If **N** is a nonzero n-vector, then the set of points in $R^n (n \geq 3)$ which satisfy

$$\mathbf{X} \cdot \mathbf{N} = 0$$

is called *the plane through the origin normal to* **N**. More generally, the set of points $\mathbf{X}$ which satisfy

(12)
$$(\mathbf{X} - \mathbf{X}_0) \cdot \mathbf{N} = 0$$

for a fixed $\mathbf{X}_0$ is called the *plane through* $\mathbf{X}_0$ *normal to* $\mathbf{N}$.

Example 2.9 In R^4 the set of points whose coordinates satisfy

$$2(x_1 - 1) + (x_2 - 3) - 3x_3 + 3(x_4 + 2) = 0$$

is the plane through $\mathbf{X}_0 = (1, 3, 0, -2)$ normal to

$$\mathbf{N} = \begin{bmatrix} 2 \\ 1 \\ -3 \\ 3 \end{bmatrix}.$$

The equation of a plane can also be written as

$$\mathbf{X} \cdot \mathbf{N} = d,$$

where d is a constant. Comparing this equation with (12) shows that $d = \mathbf{X}_0 \cdot \mathbf{N}$, where $\mathbf{X}_0$ is any point on the plane.

Example 2.10 The equation of the plane in Example 2.10 can be written as
$$2x_1 + x_2 - 3x_3 + 3x_4 = -1.$$

In Section 1.3 we shall consider planes in R^3 in more detail. We conclude our general discussion with the following theorem:

Theorem 2.4 A plane H through the origin of R^n is an $(n - 1)$-dimensional subspace of R^n.

Proof. Let $\mathbf{N} \neq \mathbf{0}$ be normal to H and suppose $V = \{\mathbf{X}_1, \ldots, \mathbf{X}_n\}$ is a basis for R^n. Then $\mathbf{N}$ is a linear combination of $\mathbf{X}_1, \ldots, \mathbf{X}_n$:

$$\mathbf{N} = a_1\mathbf{X}_1 + \cdots + a_n\mathbf{X}_n.$$

We can assume, possibly at the expense of renumbering the basis vectors, that $a_n \neq 0$. Then

$$\mathbf{X}_n = \frac{1}{a_n}(\mathbf{N} - a_1\mathbf{X}_1 - \cdots - a_{n-1}\mathbf{X}_{n-1}),$$

and $V' = \{\mathbf{X}_1, \ldots, \mathbf{X}_{n-1}, \mathbf{N}\}$ spans R^n; hence V' is linearly independent and therefore a basis for R^n. Now consider the vectors

$$(13) \qquad \mathbf{Y}_i = \mathbf{X}_i - \frac{(\mathbf{X}_i \cdot \mathbf{N})}{|\mathbf{N}|^2} \mathbf{N} \qquad (1 \le i \le n - 1);$$

direct calculation shows that

$$(14) \qquad \mathbf{Y}_i \cdot \mathbf{N} = 0,$$

and $V'' = \{\mathbf{Y}_1, \ldots, \mathbf{Y}_{n-1}, \mathbf{N}\}$ is linearly independent, for if

$$b_1 \mathbf{Y}_1 + \cdots + b_{n-1} \mathbf{Y}_{n-1} + b_n \mathbf{N} = \mathbf{0},$$

then from (13),

$$b_1 \mathbf{X}_1 + \cdots + b_{n-1} \mathbf{X}_{n-1} + \left(b_n - b_1 \frac{(\mathbf{X}_1 \cdot \mathbf{N})}{|\mathbf{N}|^2} - \cdots - b_{n-1} \frac{(\mathbf{X}_{n-1} \cdot \mathbf{N})}{|\mathbf{N}|^2} \right) \mathbf{N} = \mathbf{0}$$

and the linear independence of V' implies that $b_1 = \cdots = b_n = 0$.

Now, $\mathbf{X}$ is in H if and only if

$$\mathbf{X} \cdot \mathbf{N} = 0;$$

on the other hand, an arbitrary $\mathbf{X}$ in R^n can be written as

$$(15) \qquad \mathbf{X} = a_1 \mathbf{Y}_1 + \cdots + a_{n-1} \mathbf{Y}_{n-1} + d\mathbf{N},$$

and therefore, from (14),

$$\mathbf{X} \cdot \mathbf{N} = d |\mathbf{N}|^2.$$

Thus $\mathbf{X}$ is in H if and only if $d = 0$ in (15). This implies that H is an $(n - 1)$-dimensional subspace of R^n.

EXERCISES 1.2

1. State which of the following points are on the line

$$\mathbf{X} = \begin{bmatrix} 2 \\ -1 \\ -2 \\ 3 \end{bmatrix} + t \begin{bmatrix} 0 \\ -1 \\ 4 \\ -3 \end{bmatrix}.$$

(a) $\begin{bmatrix} 2 \\ -3 \\ 6 \\ -3 \end{bmatrix}$; (b) $\begin{bmatrix} 2 \\ 4 \\ 4 \\ 2 \end{bmatrix}$; (c) $\begin{bmatrix} 2 \\ 1 \\ -10 \\ 9 \end{bmatrix}$; (d) $\begin{bmatrix} 2 \\ -2 \\ 2 \\ 0 \end{bmatrix}$.

2. State which of the following points are on the line
$$\frac{x-2}{3} = \frac{y-4}{-2} = \frac{z+5}{4}.$$

(a) $\begin{bmatrix} -1 \\ 6 \\ -9 \end{bmatrix}$;   (b) $\begin{bmatrix} 5 \\ 2 \\ -1 \end{bmatrix}$;   (c) $\begin{bmatrix} 1 \\ 2 \\ -1 \end{bmatrix}$;   (d) $\begin{bmatrix} 5 \\ 2 \\ 0 \end{bmatrix}$.

3. Write the equations of the following lines in symmetric form.

(a) $\mathbf{X} = \begin{bmatrix} 2 \\ -1 \\ 3 \\ 3 \end{bmatrix} + t \begin{bmatrix} 0 \\ -1 \\ 4 \\ -3 \end{bmatrix}$;   (b) $\mathbf{X} = \begin{bmatrix} 2 \\ 1 \\ 0 \end{bmatrix} + t \begin{bmatrix} 1 \\ -1 \\ 2 \end{bmatrix}$.

4. Write parametric equations for the following lines.

(a) $\dfrac{x-2}{4} = \dfrac{y+2}{0} = \dfrac{z-3}{-2}$;

(b) $\dfrac{x_1 - 4}{0} = \dfrac{x_2 + 3}{3} = \dfrac{x_3 - 2}{2} = \dfrac{x_4 - 3}{-3}$.

5. Find the equation of the line through $\mathbf{X}_0 = (1, 2, -1)$ in the direction of
$$\mathbf{U} = \begin{bmatrix} 4 \\ -2 \\ 5 \end{bmatrix}.$$

6. Repeat Exercise 5 for $\mathbf{X}_0 = (3, 2, 1, 0)$ and
$$\mathbf{U} = \begin{bmatrix} 0 \\ 4 \\ 0 \\ -3 \end{bmatrix}.$$

7. Find the equation of the line through $\mathbf{X}_1 = (2, 1, 0, 3)$, and $\mathbf{X}_2 = (-1, -1, 2, 0)$.

8. Repeat Exercise 7 for $\mathbf{X}_1 = (-3, 1, 0)$ and $\mathbf{X}_2 = (3, 4, -1)$.

9. Find the equation of the line through $(4, -2, 5, 2)$ parallel to the line

$$\mathbf{X} = \begin{bmatrix} 1 \\ -2 \\ 3 \\ 2 \end{bmatrix} + t \begin{bmatrix} 0 \\ 1 \\ -2 \\ 4 \end{bmatrix}.$$

10. Find the equation of the line through $(3, -1, 2)$ and parallel to the line

$$\frac{x-2}{-2} = \frac{y+3}{1} = \frac{z-3}{0}.$$

11. Are the points $(1, -2, 3)$, $(2, -3, 5)$, and $(0, -1, 1)$ on the same line?

12. Are the points $(0, 1, 1, -2)$, $(3, 2, -3, 1)$, and $(6, 3, -8, 4)$ on the same line?

13. Find the point of intersection of the lines

$$\mathbf{X} = \begin{bmatrix} 2 \\ 2 \\ 2 \end{bmatrix} + s \begin{bmatrix} 1 \\ -1 \\ 2 \end{bmatrix} \quad \text{and} \quad \mathbf{X} = \begin{bmatrix} 3 \\ 11 \\ -1 \end{bmatrix} + t \begin{bmatrix} 3 \\ -1 \\ 5 \end{bmatrix}.$$

14. Find the point of intersection of the lines

$$\mathbf{X} = \begin{bmatrix} 1 \\ 3 \\ -1 \\ 2 \end{bmatrix} + s \begin{bmatrix} 0 \\ 1 \\ 2 \\ -3 \end{bmatrix} \quad \text{and} \quad \mathbf{X} = \begin{bmatrix} 2 \\ 1 \\ 4 \\ 5 \end{bmatrix} + t \begin{bmatrix} 1 \\ -2 \\ 5 \\ 3 \end{bmatrix}.$$

15. Show that the lines

$$\mathbf{X} = \begin{bmatrix} 2 \\ 3 \\ 1 \\ 4 \end{bmatrix} + s \begin{bmatrix} -2 \\ 1 \\ 0 \\ 1 \end{bmatrix} \quad \text{and} \quad \mathbf{X} = \begin{bmatrix} 0 \\ 1 \\ 2 \\ 3 \end{bmatrix} + t \begin{bmatrix} -1 \\ 3 \\ 2 \\ 1 \end{bmatrix}$$

are skew.

16. Are the lines

$$\mathbf{X} = \begin{bmatrix} 1 \\ 2 \\ -1 \end{bmatrix} + s \begin{bmatrix} 1 \\ -2 \\ 3 \end{bmatrix} \quad \text{and} \quad \mathbf{X} = \begin{bmatrix} 0 \\ 2 \\ 3 \end{bmatrix} + t \begin{bmatrix} 0 \\ 3 \\ 2 \end{bmatrix}$$

perpendicular?

17. Are the lines

$$\mathbf{X} = \begin{bmatrix} 0 \\ 2 \\ 1 \\ 4 \end{bmatrix} + s \begin{bmatrix} 1 \\ -1 \\ -2 \\ 2 \end{bmatrix} \quad \text{and} \quad \mathbf{X} = \begin{bmatrix} 3 \\ 1 \\ -2 \\ -4 \end{bmatrix} + t \begin{bmatrix} 4 \\ 2 \\ 2 \\ -1 \end{bmatrix}$$

perpendicular?

18. Show that

$$\mathbf{X} = \begin{bmatrix} 0 \\ 1 \\ -2 \end{bmatrix} + s \begin{bmatrix} 1 \\ -3 \\ 4 \end{bmatrix} \quad \text{and} \quad \mathbf{X} = \begin{bmatrix} 3 \\ -8 \\ 10 \end{bmatrix} + t \begin{bmatrix} 3 \\ -9 \\ 12 \end{bmatrix}$$

define the same line.

19. Show that

$$\mathbf{X} = \begin{bmatrix} 1 \\ -2 \\ 3 \\ -1 \end{bmatrix} + s \begin{bmatrix} 0 \\ 1 \\ 2 \\ -3 \end{bmatrix} \quad \text{and} \quad \mathbf{X} = \begin{bmatrix} 1 \\ 0 \\ 7 \\ -7 \end{bmatrix} + t \begin{bmatrix} 0 \\ 2 \\ 4 \\ -6 \end{bmatrix}$$

define the same line.

20. Are the lines

$$\mathbf{X} = \begin{bmatrix} 2 \\ -1 \\ 3 \end{bmatrix} + s \begin{bmatrix} 5 \\ 3 \\ -2 \end{bmatrix} \quad \text{and} \quad \mathbf{X} = \begin{bmatrix} 4 \\ -2 \\ 6 \end{bmatrix} + t \begin{bmatrix} -10 \\ -6 \\ 4 \end{bmatrix}$$

parallel?

21. Are the lines

$$\mathbf{X} = \begin{bmatrix} 1 \\ -2 \\ 3 \\ 4 \end{bmatrix} + s \begin{bmatrix} 1 \\ -2 \\ -3 \\ 0 \end{bmatrix} \quad \text{and} \quad \mathbf{X} = \begin{bmatrix} 3 \\ 2 \\ -5 \\ 1 \end{bmatrix} + t \begin{bmatrix} -2 \\ 4 \\ 6 \\ 0 \end{bmatrix}$$

parallel?

22. Find the equation of the line through $(1, 0, 3)$, parallel to the line determined by $(3, 1, -2)$ and $(2, 1, -3)$.

23. Let L be defined by

$$\mathbf{X} = \begin{bmatrix} 1 \\ 0 \\ 2 \end{bmatrix} + t \begin{bmatrix} 3 \\ -1 \\ 2 \end{bmatrix}.$$

Find the equation of the line through $(3, 5, 3)$ that intersects L at right angles.

24. Find the cosine of the angle of intersection between the lines defined in Exercise 13.

25. Find the cosine of the angle of intersection between the lines defined in Exercise 14.

26. If

$$\mathbf{U} = \begin{bmatrix} 1 \\ 3 \\ -4 \end{bmatrix} \quad \text{and} \quad \mathbf{V} = \begin{bmatrix} -2 \\ 0 \\ 3 \end{bmatrix},$$

find $\mathbf{N} = \mathbf{U} \times \mathbf{V}$ and verify that $\mathbf{N}$ is perpendicular to $\mathbf{U}$ and $\mathbf{V}$.

27. Repeat Exercise 26 for

$$\mathbf{U} = \begin{bmatrix} -1 \\ 0 \\ 2 \end{bmatrix} \quad \text{and} \quad \mathbf{V} = \begin{bmatrix} 0 \\ -3 \\ -4 \end{bmatrix}.$$

28. Find the equation of the plane through $(4, -2, 1)$, perpendicular to

$$\mathbf{N} = \begin{bmatrix} 3 \\ -2 \\ 4 \end{bmatrix}.$$

29. Find the equation of the plane through $(1, -2, 3, 0)$, orthogonal to

$$\mathbf{N} = \begin{bmatrix} 2 \\ -1 \\ 4 \\ -5 \end{bmatrix}.$$

30. Find the equation of the plane that contains the points $(1, 3, 2)$, $(-1, 2, 3)$, and $(0, 2, 1)$.

31. Find the equation of the plane that contains $(1, -1, 2)$ and the line

$$\mathbf{X} = \begin{bmatrix} 4 \\ -1 \\ 2 \end{bmatrix} + t \begin{bmatrix} 0 \\ 2 \\ 1 \end{bmatrix}.$$

32. Find the equation of the plane that contains the lines defined in Exercise 13.

33. Find the line through $(1, 3, -2, 4)$, perpendicular to the plane defined by

$$2x_1 + 3x_2 + 3x_3 - 2x_4 = 5.$$

34. Find the equation of the plane through $(3, -2, 4, 2)$ and perpendicular to the line determined by $(1, 0, 2, -3)$ and $(3, -2, 4, 2)$.

THEORETICAL EXERCISES

T-1. Prove Theorem 2.2.

T-2. Prove: If $\mathbf{U} \times \mathbf{V} = \mathbf{0}$ then $\mathbf{U}$ and $\mathbf{V}$ are parallel.

T-3. Prove: If $\mathbf{U}$ and $\mathbf{V}$ are linearly independent, then $\mathbf{U}$, $\mathbf{V}$ and $\mathbf{U} \times \mathbf{V}$ form a basis for R^3.

T-4. (Cross product in R^n). Let $n \geq 2$ and

$$\mathbf{U}_1 = \begin{bmatrix} u_{11} \\ u_{12} \\ \vdots \\ u_{1n} \end{bmatrix}, \quad \mathbf{U}_2 = \begin{bmatrix} u_{21} \\ u_{22} \\ \vdots \\ u_{2n} \end{bmatrix}, \ldots, \quad \mathbf{U}_{n-1} = \begin{bmatrix} u_{n-1,1} \\ u_{n-1,2} \\ \vdots \\ u_{n-1,n} \end{bmatrix},$$

and define

$$\mathbf{N} = \begin{bmatrix} v_1 \\ v_2 \\ \vdots \\ v_n \end{bmatrix},$$

where v_i is the cofactor of x_i in the determinant

$$D = \begin{bmatrix} x_1 & x_2 \cdots x_n \\ u_{11} & u_{12} \cdots u_{1n} \\ \vdots & \vdots \\ u_{n-1,1} & u_{n-2,2} \cdots u_{n-1,n} \end{bmatrix}.$$

(a) Show that $\mathbf{N} \cdot \mathbf{U}_i = 0$ $(1 \leq i \leq n - 1)$.

(b) Show that $\mathbf{N} = \mathbf{0}$ if and only if $\mathbf{U}_1, \ldots, \mathbf{U}_{n-1}$ are linearly dependent.

T-5. Show that if **X** is perpendicular to each of $\mathbf{X}_1, \mathbf{X}_2, \ldots, \mathbf{X}_k$, then **X** is perpendicular to every vector in $S[\mathbf{X}_1, \mathbf{X}_2, \ldots, \mathbf{X}_k]$.

1.3 Vectors in R^3

Because of its special importance as a mathematical model of the three-dimensional world in which we live, we pay particular attention to R^3. Throughout this section we shall assume that all vectors, lines, and planes are in R^3. It is important to remember this, because some of our statements will be incorrect or meaningless in R^n for $n \neq 3$.

Coplanar and Collinear Vectors

Vectors **U**, **V**, and **W** are linearly dependent if there are constants a, b, and c, not all zero, such that

$$a\mathbf{U} + b\mathbf{V} + c\mathbf{W} = 0.$$

If $c \neq 0$, this equation can be rewritten as

$$\mathbf{W} = s\mathbf{U} + t\mathbf{V},$$

from which it follows that directed line segments drawn from the origin to represent **U**, **V**, and **W** lie in a plane (Fig. 3.1). Therefore we say that three vectors are *coplanar* if they are linearly dependent, and *noncoplanar* if they are not.

Two vectors **U** and **V** are linearly dependent if and only if line segments drawn from the origin in the directions of **U** and **V** are on the same line; hence two vectors are said to be *collinear* if they are parallel, and *noncollinear* if they are not.

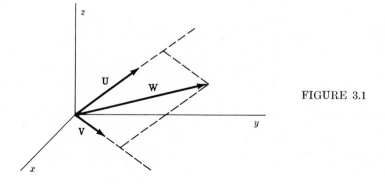

FIGURE 3.1

The Natural Basis for R^3

The natural basis for R^3 is

$$
E_1 = \begin{bmatrix} 1 \\ 0 \\ 0 \end{bmatrix}, \qquad E_2 = \begin{bmatrix} 0 \\ 1 \\ 0 \end{bmatrix}, \qquad \text{and} \qquad E_3 = \begin{bmatrix} 0 \\ 0 \\ 1 \end{bmatrix}.
$$

It is traditional to denote these vectors as $E_1 = i$, $E_2 = j$, and $E_3 = k$; then an arbitrary vector

$$
(1) \qquad \qquad X = \begin{bmatrix} x \\ y \\ z \end{bmatrix}
$$

can be written as

$$
X = xi + yj + zk.
$$

We shall use this representation for vectors in R^3 when it is convenient; however, we shall also continue to use (1).

Example 3.1

$$
\begin{bmatrix} 1 \\ 2 \\ 3 \end{bmatrix} = i + 2j + 3k, \qquad \begin{bmatrix} -3 \\ 0 \\ 1 \end{bmatrix} = -3i + k,
$$

and

$$
\begin{bmatrix} \frac{2}{3} \\ 5 \\ -1 \end{bmatrix} = \tfrac{2}{3}i + 5j - k.
$$

The vectors i, j, and k can be represented as directed line segments of unit length along the positive directions of the x-, y-, and z-axes, respectively (Fig. 3.2). They form an orthonormal basis for R^3; that is,

$$
i \cdot i = j \cdot j = k \cdot k = 1
$$

and

$$
i \cdot j = j \cdot k = k \cdot i = 0.
$$

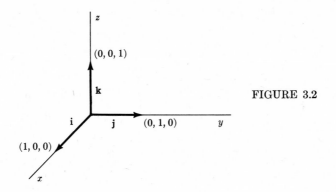

FIGURE 3.2

Cross Product of Two Vectors

Definition 3.1   The cross product $\mathbf{U} \times \mathbf{V}$ of two vectors $\mathbf{U} = u_1\mathbf{i} + u_2\mathbf{j} + u_3\mathbf{k}$ and $\mathbf{V} = v_1\mathbf{i} + v_2\mathbf{j} + v_3\mathbf{k}$ is defined by

$$(2) \qquad \mathbf{U} \times \mathbf{V} = (u_2v_3 - u_3v_2)\mathbf{i} + (u_3v_1 - u_1v_3)\mathbf{j} + (u_1v_2 - u_2v_1)\mathbf{k}.$$

It can be represented symbolically as a determinant,

$$(3) \quad \mathbf{U} \times \mathbf{V} = \begin{vmatrix} \mathbf{i} & \mathbf{j} & \mathbf{k} \\ u_1 & u_2 & u_3 \\ v_1 & v_2 & v_3 \end{vmatrix} = \begin{vmatrix} u_2 & u_3 \\ v_2 & v_3 \end{vmatrix}\mathbf{i} - \begin{vmatrix} u_1 & u_3 \\ v_1 & v_3 \end{vmatrix}\mathbf{j} + \begin{vmatrix} u_1 & u_2 \\ v_1 & v_2 \end{vmatrix}\mathbf{k}.$$

thus (2) is the formal expansion of (3) according to the elements of its first row.

Example 3.2   Let $\mathbf{U} = \mathbf{i} + \mathbf{j} - 2\mathbf{k}$, $\mathbf{V} = \mathbf{i} + 2\mathbf{j} + 3\mathbf{k}$, and $\mathbf{W} = 2\mathbf{i} + 4\mathbf{j} + 6\mathbf{k}$. Then

$$\mathbf{U} \times \mathbf{V} = \begin{vmatrix} \mathbf{i} & \mathbf{j} & \mathbf{k} \\ 1 & 1 & -2 \\ 1 & 2 & 3 \end{vmatrix} = 7\mathbf{i} - 5\mathbf{j} + \mathbf{k}$$

and

$$\mathbf{V} \times \mathbf{W} = \begin{vmatrix} \mathbf{i} & \mathbf{j} & \mathbf{k} \\ 1 & 2 & 3 \\ 2 & 4 & 6 \end{vmatrix} = \mathbf{0}.$$

Example 3.3 It is easily verified that

$$\mathbf{i} \times \mathbf{i} = \mathbf{j} \times \mathbf{j} = \mathbf{k} \times \mathbf{k} = 0;$$

(4) $$\mathbf{i} \times \mathbf{j} = \mathbf{k}, \qquad \mathbf{j} \times \mathbf{k} = \mathbf{i}, \qquad \mathbf{k} \times \mathbf{i} = \mathbf{j}$$

and

(5) $$\mathbf{j} \times \mathbf{i} = -\mathbf{k}, \qquad \mathbf{k} \times \mathbf{j} = -\mathbf{i}, \qquad \mathbf{i} \times \mathbf{k} = -\mathbf{j}.$$

To remember (4) and (5), it is helpful to refer to Fig. 3.3: if, when reading the two basis vectors in the order in which they appear in the cross product, one reads clockwise in Fig. 3.3, then the cross product is the third basis vector; if one reads counterclockwise, then the cross product is the negative of the third vector.

FIGURE 3.3

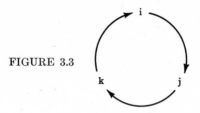

Parts (a) and (b) of the next theorem were established in Section 1.2. The remaining parts follow from the definition of $\mathbf{U} \times \mathbf{V}$; we leave it to the student to verify them.

Theorem 3.1 If $\mathbf{U}$, $\mathbf{V}$, and $\mathbf{W}$ are vectors then:

(a) $\mathbf{U} \times \mathbf{V} = 0$ if and only if $\mathbf{U}$ and $\mathbf{V}$ are parallel.

(b) If $\mathbf{U}$ and $\mathbf{V}$ are linearly independent then $(\mathbf{U} \times \mathbf{V}) \cdot \mathbf{W} = 0$ if and only if $\mathbf{W}$ is in the plane of $\mathbf{U}$ and $\mathbf{V}$; in particular $\mathbf{U} \times \mathbf{V}$ is perpendicular to $\mathbf{U}$ and $\mathbf{V}$.

(c) $\mathbf{V} \times \mathbf{U} = -\mathbf{U} \times \mathbf{V}$.

(d) $\mathbf{U} \times (\mathbf{V} + \mathbf{W}) = \mathbf{U} \times \mathbf{V} + \mathbf{U} \times \mathbf{W}$.

(e) $c(\mathbf{U} \times \mathbf{V}) = (c\mathbf{U}) \times \mathbf{V} = \mathbf{U} \times (c\mathbf{V})$.

The Right-Hand Rule

From (3) and the identity

$$(u_2v_3 - u_3v_2)^2 + (u_3v_1 - u_1v_3)^2 + (u_1v_2 - u_2v_1)^2$$

$$= (u_1{}^2 + u_2{}^2 + u_3{}^2)(v_1{}^2 + v_2{}^2 + v_3{}^2) - (u_1v_1 + u_2v_2 + u_3v_3)^2$$

it follows that

$$|\mathbf{U} \times \mathbf{V}|^2 = |\mathbf{U}|^2 |\mathbf{V}|^2 - (\mathbf{U}\cdot\mathbf{V})^2$$

(6)
$$= |\mathbf{U}|^2 |\mathbf{V}|^2 \left(1 - \frac{(\mathbf{U}\cdot\mathbf{V})^2}{|\mathbf{U}|^2 |\mathbf{V}|^2}\right)$$

$$= |\mathbf{U}|^2 |\mathbf{V}|^2 (1 - \cos^2\theta),$$

where θ is the angle between $\mathbf{U}$ and $\mathbf{V}$. Of course there are two such angles, as shown in Fig. 3.4. (We ignore the infinitely many additional ones that could be obtained by adding and subtracting multiples of 2π to those shown in Fig. 3.4). We remove this ambiguity by stipulating that θ is always chosen to be the smaller angle between $\mathbf{U}$ and $\mathbf{V}$; then $0 \le \theta \le \pi$, $\sin\theta \ge 0$ and (6) implies that

$$|\mathbf{U} \times \mathbf{V}| = |\mathbf{U}||\mathbf{V}|\sin\theta.$$

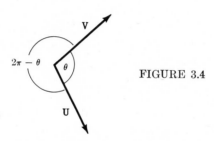

FIGURE 3.4

From Theorem 3.1, if $\mathbf{U} \times \mathbf{V} \ne 0$ then $\mathbf{U} \times \mathbf{V}$ and $-\mathbf{U} \times \mathbf{V}$ are normal to the plane of $\mathbf{U}$ and $\mathbf{V}$ and in opposite directions. There is a convenient geometric rule for distinguishing between them. In Fig. 3.1 we have labeled the coordinate axes so that if a right-hand screw is turned in the direction required to rotate the x-axis through a right angle into coincidence with the y-axis, then the screw will move in the positive z-direction (Fig. 3.5). A coordinate system with this property is said to be *right handed*; by contrast, the system shown in Fig. 3.6 is *left handed*.

Henceforth we shall require that our rectangular coordinate system in R^3 be right handed. With this convention the direction of $\mathbf{U} \times \mathbf{V}$ is that in which a right hand screw perpendicular to the plane of $\mathbf{U}$ and $\mathbf{V}$ would move if it were rotated through an acute angle from $\mathbf{U}$ to $\mathbf{V}$ (Fig. 3.7). From Example 3.3 it is easy to verify this statement if $\mathbf{U}$ and $\mathbf{V}$ are distinct elements of the natural basis for R^3, and a proof is indicated in Exercise T-3

FIGURE 3.5   FIGURE 3.6

for the case in which **U** and **V** lie in one of the coordinate planes. A general proof, which we only sketch here, can be obtained as follows:

(a) If **U** and **V** are linearly independent vectors and θ is the acute angle between them, define **N**(**U**, **V**) to be the vector of length $|\,\mathbf{U}\,|\,|\,\mathbf{V}\,|\sin\theta$, perpendicular to the plane of **U** and **V** and in the direction toward which a right-hand screw would move if rotated through the angle θ from **U** to **V**. (Fig. 3.7). If **U** and **V** are parallel, define **N**(**U**, **V**) = 0. From this definition and Example 3.3 it can be shown that if **U** and **V** are vectors in the natural basis $\{\mathbf{i},\,\mathbf{j},\,\mathbf{k}\}$, then

$$\mathbf{N}(\mathbf{U},\,\mathbf{V}) = \mathbf{U} \times \mathbf{V}.$$

(b) Show geometrically that

$$\mathbf{N}(\mathbf{U},\,\mathbf{V} + \mathbf{W}) = \mathbf{N}(\mathbf{U},\,\mathbf{V}) + \mathbf{N}(\mathbf{U},\,\mathbf{W})$$

and

$$\mathbf{N}(\mathbf{U},\,\mathbf{V}) = -N(\mathbf{V},\,\mathbf{U}).$$

(c) Use the results of (a) and (b) to show that

$$\mathbf{N}(\mathbf{U},\,\mathbf{V}) = \mathbf{N}(u_1\mathbf{i} + u_2\mathbf{j} + u_3\mathbf{k},\; v_1\mathbf{i} + v_2\mathbf{j} + v_3\mathbf{k})$$

$$= (u_2v_3 - u_3v_2)\mathbf{i} + (u_3v_1 - u_1v_3)\mathbf{j} + (u_1v_2 - u_2v_1)\mathbf{k}$$

$$= \mathbf{U} \times \mathbf{V}$$

for all **U** and **V**.

The cross product and scalar product have applications in analytic geometry. We shall present a few of these.

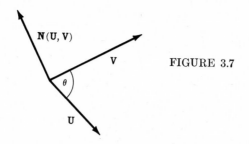

N(U, V)

V

θ

U

FIGURE 3.7

Area of a Triangle and a Parallelogram

The area A_T of a triangle with base b and height h is

$$A_T = \tfrac{1}{2}bh.$$

Consider the triangle with vertices $\mathbf{X}_1$, $\mathbf{X}_2$, and $\mathbf{X}_3$ (Fig. 3.8); if we take the line segment connecting $\mathbf{X}_1$ and $\mathbf{X}_2$ to be its base then

$$b = |\,\mathbf{X}_2 - \mathbf{X}_1\,|$$

and

$$h = |\,\mathbf{X}_3 - \mathbf{X}_1\,|\sin\theta,$$

where θ is the angle between $\mathbf{X}_1 - \mathbf{X}_2$ and $\mathbf{X}_1 - \mathbf{X}_3$; thus

$$A_T = \tfrac{1}{2}\,|\,\mathbf{X}_2 - \mathbf{X}_1\,|\,|\,\mathbf{X}_3 - \mathbf{X}_1\,|\sin\theta$$

$$= \tfrac{1}{2}\,|\,(\mathbf{X}_2 - \mathbf{X}_1)\,\times\,(\mathbf{X}_3 - \mathbf{X}_1)\,|.$$

The area A_P of the parallelogram with adjacent sides $\mathbf{X}_2 - \mathbf{X}_1$ and $\mathbf{X}_3 - \mathbf{X}_1$ is $2A_T$; hence

$$A_P = |\,(\mathbf{X}_2 - \mathbf{X}_1)\,\times\,(\mathbf{X}_3 - \mathbf{X}_1)\,|.$$

Example 3.4 If $\mathbf{X}_1 = (1, 2, 3)$, $\mathbf{X}_2 = (2, -1, -3)$, and $\mathbf{X}_3 = (0, 1, 2)$,

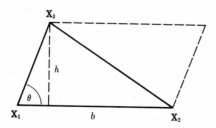

$\mathbf{X}_3$

h

θ

$\mathbf{X}_1$

b

$\mathbf{X}_2$

FIGURE 3.8

then

$$\mathbf{X}_2 - \mathbf{X}_1 = \mathbf{i} - 3\mathbf{j} - 6\mathbf{k},$$

$$\mathbf{X}_3 - \mathbf{X}_1 = -(\mathbf{i} + \mathbf{j} + \mathbf{k}),$$

and

$$(\mathbf{X}_2 - \mathbf{X}_1) \times (\mathbf{X}_3 - \mathbf{X}_1) = \begin{vmatrix} \mathbf{i} & \mathbf{j} & \mathbf{k} \\ 1 & -3 & -6 \\ -1 & -1 & -1 \end{vmatrix}$$

$$= -3\mathbf{i} + 7\mathbf{j} - 4\mathbf{k};$$

thus the area of the triangle with vertices $\mathbf{X}_1$, $\mathbf{X}_2$, and $\mathbf{X}_3$ is

$$A_{\mathrm{T}} = \tfrac{1}{2}\sqrt{(-3)^2 + 7^2 + (-4)^2} = \tfrac{1}{2}\sqrt{74}.$$

The Equation of a Plane through Three Points

The plane in R^3 which contains three noncollinear points $\mathbf{X}_1$, $\mathbf{X}_2$, and $\mathbf{X}_3$ (Fig. 3.9) has the equation

(7) $$\mathbf{N} \cdot (\mathbf{X} - \mathbf{X}_1) = 0,$$

where $\mathbf{N}$ is perpendicular to $\mathbf{X}_2 - \mathbf{X}_1$ and $\mathbf{X}_3 - \mathbf{X}_1$ (since $\mathbf{X}_2$ and $\mathbf{X}_3$ satisfy (7)). Therefore we can take

$$\mathbf{N} = (\mathbf{X}_2 - \mathbf{X}_1) \times (\mathbf{X}_3 - \mathbf{X}_1).$$

Example 3.5 Let $\mathbf{X}_1 = (1, 2, 1)$, $\mathbf{X}_2 = (2, -1, 3)$, and $\mathbf{X}_3 = (0, 4, 3)$. Then

$$\mathbf{X}_2 - \mathbf{X}_1 = \mathbf{i} - 3\mathbf{j} + 2\mathbf{k},$$

$$\mathbf{X}_3 - \mathbf{X}_1 = -\mathbf{i} + 2\mathbf{j} + 2\mathbf{k},$$

and

$$(\mathbf{X}_2 - \mathbf{X}_1) \times (\mathbf{X}_3 - \mathbf{X}_1) = -10\mathbf{i} - 4\mathbf{j} - \mathbf{k};$$

FIGURE 3.9

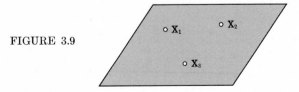

thus the equation of the plane is

$$10(x - 1) + 4(y - 2) + (z - 1) = 0$$

or

$$10x + 4y + z = 19.$$

The Equation of a Line through a Given Point and Intersecting a Given Line at Right Angles

Let a line L_0 be given by

$$\mathbf{X} = \mathbf{X}_0 + t\mathbf{U}_0$$

and suppose we wish to find the equation of the line L_1 that passes through a point $\mathbf{X}_1$ and intersects L_0 at right angles (Fig. 3.10). Let $\mathbf{X}_2$ be the unknown point of intersection of L_1 and L_0 and define $\mathbf{U}_1 = \mathbf{X}_2 - \mathbf{X}_1$. From Fig. 3.10,

(8) $$\mathbf{U}_1 = \mathbf{X}_0 - \mathbf{X}_1 + t_0\mathbf{U}_0$$

and therefore

(9) $$\mathbf{U}_1 \cdot \mathbf{U}_0 = (\mathbf{X}_0 - \mathbf{X}_1) \cdot \mathbf{U}_0 + t_0 \,|\, \mathbf{U}_0 \,|^2.$$

Since $\mathbf{U}_1$ and $\mathbf{U}_0$ are perpendicular, $\mathbf{U}_1 \cdot \mathbf{U}_0 = 0$, and (9) yields

$$t_0 = \frac{(\mathbf{X}_1 - \mathbf{X}_0) \cdot \mathbf{U}_0}{|\, \mathbf{U}_0 \,|^2}.$$

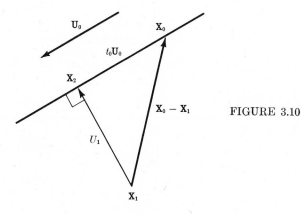

FIGURE 3.10

To obtain $\mathbf{U}_1$, substitute this in (8); then the equation of L_1 is

$$\mathbf{X} = \mathbf{X}_0 + s(a\mathbf{U}_1)$$

where a is any convenient nonzero scalar.

Example 3.6 Let it be required to find the equation of the line L_1 that passes through $\mathbf{X}_1 = (3, 5, 3)$ and intersects the line given by

$$\mathbf{X} = \begin{bmatrix} 1 \\ 0 \\ 2 \end{bmatrix} + t \begin{bmatrix} 3 \\ -1 \\ 2 \end{bmatrix}$$

at right angles. Thus $\mathbf{X}_0 = (1, 0, 2)$,

$$\mathbf{U}_0 = 3\mathbf{i} - \mathbf{j} + 2\mathbf{k},$$

$$\mathbf{X}_1 - \mathbf{X}_0 = 2\mathbf{i} + 5\mathbf{j} + \mathbf{k},$$

$$t_0 = \frac{(2\mathbf{i} + 5\mathbf{j} + \mathbf{k}) \cdot (3\mathbf{i} - \mathbf{j} + 2\mathbf{k})}{|\, 3\mathbf{i} - \mathbf{j} + 2\mathbf{k} \,|} = \frac{3}{14},$$

$$\mathbf{U}_1 = -(2\mathbf{i} + 5\mathbf{j} + \mathbf{k}) + \tfrac{3}{14}(3\mathbf{i} - \mathbf{j} + 2\mathbf{k})$$

$$= -\frac{(19\mathbf{i} + 73\mathbf{j} + 8\mathbf{k})}{14},$$

and L_1 is given by

$$\mathbf{X} = \begin{bmatrix} 1 \\ 0 \\ 2 \end{bmatrix} + s \begin{bmatrix} 19 \\ 73 \\ 8 \end{bmatrix}.$$

Line of Intersection of Two Planes

Let Γ_1 and Γ_2 be planes defined by

(10)
$$\mathbf{X} \cdot \mathbf{N}_1 = d_1$$

and

(11)
$$\mathbf{X} \cdot \mathbf{N}_2 = d_2.$$

If $\mathbf{N}_1$ and $\mathbf{N}_2$ are parallel, we say that Γ_1 *and* Γ_2 *are parallel* (Fig. 3.11). Parallel planes do not intersect unless $d_2\mathbf{N}_1 = d_1\mathbf{N}_2$, in which case they are identical (Exercise T-4).

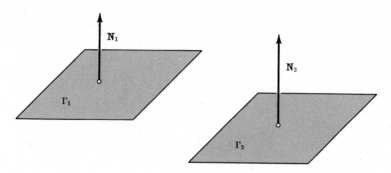

FIGURE 3.11

If Γ_1 and Γ_2 are not parallel, they intersect in a line. This is intuitively evident from Fig. 3.12. To prove it we observe that if $\mathbf{N}_1$ and $\mathbf{N}_2$ are not parallel then $\mathbf{N}_1$, $\mathbf{N}_2$ and $\mathbf{N}_1 \times \mathbf{N}_2$ form a basis for R^3, and any $\mathbf{X}$ in R^3 can be written as

$$\mathbf{X} = a\mathbf{N}_1 + b\mathbf{N}_2 + t(\mathbf{N}_1 \times \mathbf{N}_2)$$

for suitable scalars a, b, and t. Now $\mathbf{X}$ satisfies (10) and (11) if and only if

(12)
$$a \,|\, \mathbf{N}_1 \,|^2 + b(\mathbf{N}_1 \cdot \mathbf{N}_2) = d_1,$$
$$a(\mathbf{N}_1 \cdot \mathbf{N}_2) + b \,|\, \mathbf{N}_2 \,|^2 = d_2.$$

The determinant of this pair of equations in a and b is

$$|\, \mathbf{N}_1 \,|^2 \,|\, \mathbf{N}_2 \,|^2 - (\mathbf{N}_1 \cdot \mathbf{N}_2)^2 = |\, \mathbf{N}_1 \times \mathbf{N}_2 \,|^2 \neq 0 \qquad \text{(see (6))},$$

and therefore the system (12) has a unique solution. If

$$\bar{\mathbf{X}} = a\mathbf{N}_1 + b\mathbf{N}_2,$$

where a and b satisfy (12), then

$$\mathbf{X} = \bar{\mathbf{X}} + t(\mathbf{N}_1 \times \mathbf{N}_2)$$

is the equation of the required line of intersection. Of course, the equation of this line could just as well be written as

$$\mathbf{X} = \mathbf{X}_0 + at(\mathbf{N}_1 \times \mathbf{N}_2)$$

where $\mathbf{X}_0$ is any point common to Γ_1 and Γ_2, and a is any nonzero scalar. (Why?)

Example 3.7   The planes given by

(13)
$$\mathbf{N}_1 \cdot \mathbf{X} = 2x - 3y + 3z = 15$$

and

(14)
$$\mathbf{N}_2 \cdot \mathbf{X} = x + 2y - z = 4$$

are not parallel; hence they intersect in a line L which is parallel to

$$\mathbf{N_1} \times \mathbf{N_2} = -3\mathbf{i} + 5\mathbf{j} + 7\mathbf{k}.$$

To find the equation of the line we need only find a point common to the two planes, which can be done in this case by setting $z = 0$ in (13) and (14) and solving the resulting equations for x and y; this yields the common point $\mathbf{X_0} = (6, -1, 0)$. Thus the line of intersection is given by

$$\mathbf{X} = \begin{bmatrix} 6 \\ -1 \\ 0 \end{bmatrix} + t \begin{bmatrix} -3 \\ 5 \\ 7 \end{bmatrix}.$$

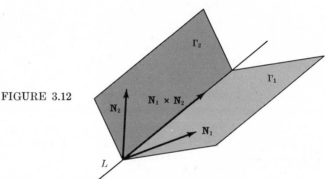

FIGURE 3.12

Distance from a Point to a Line

Suppose we wish to find the distance d from the point $\mathbf{X_1}$ to the line L given by

$$\mathbf{X} = \mathbf{X_0} + t\mathbf{U}.$$

Let θ be the angle between $\mathbf{U}$ and $\mathbf{X_1} - \mathbf{X_0}$ (Fig. 3.13); then

$$d = |\,\mathbf{X_1} - \mathbf{X_0}\,|\sin\theta$$
$$= \frac{|\,\mathbf{U} \times (\mathbf{X_1} - \mathbf{X_0})\,|}{|\,\mathbf{U}\,|}.$$

If L passes through the origin we may take $\mathbf{X_0} = \mathbf{0}$ to obtain

$$d = \frac{|\,\mathbf{U} \times \mathbf{X_1}\,|}{|\,\mathbf{U}\,|};$$

we shall use this result in Section 4.4.

Example 3.8   Let $\mathbf{X}_1 = (-2, 1, 0)$ and let L be given by

$$\mathbf{X} = \mathbf{X}_0 + t\mathbf{U} = \begin{bmatrix} -1 \\ 0 \\ 2 \end{bmatrix} + t \begin{bmatrix} 3 \\ 4 \\ 0 \end{bmatrix};$$

then

$$\mathbf{U} \times (\mathbf{X}_1 - \mathbf{X}_0) = -8\mathbf{i} + 6\mathbf{j} + 7\mathbf{k},$$
$$|\, \mathbf{U} \times (\mathbf{X}_1 - \mathbf{X}_0)\,| = \sqrt{149},$$
$$|\,\mathbf{U}\,| = 5,$$

and the distance from $\mathbf{X}_1$ to L is

$$d = \frac{\sqrt{149}}{5}.$$

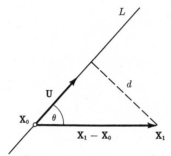

FIGURE 3.13

Distance between Two Lines

The distance d between two lines L_1 and L_2 given by

(15) $$\mathbf{X} = \mathbf{X}_1 + t\mathbf{U}_1$$

and

(16) $$\mathbf{X} = \mathbf{X}_2 + s\mathbf{U}_2$$

is defined to be the minimum value attained by $|\, \mathbf{Y}_1 - \mathbf{Y}_2\,|$ as $\mathbf{Y}_1$ and $\mathbf{Y}_2$ are allowed to vary over L_1 and L_2. If this minimum is attained with $\mathbf{Y}_1 = \bar{\mathbf{Y}}_1$ and $\mathbf{Y}_2 = \bar{\mathbf{Y}}_2$, then $\bar{\mathbf{Y}}_1 - \bar{\mathbf{Y}}_2$ is perpendicular to L_1 and L_2. (See Fig. 3.14 and Exercise T-5). Thus, if L_1 and L_2 are not parallel,

(17) $$\bar{\mathbf{Y}}_2 - \bar{\mathbf{Y}}_1 = c(\mathbf{U}_1 \times \mathbf{U}_2),$$

since any vector perpendicular to L_1 and L_2 is parallel to $\mathbf{U}_1 \times \mathbf{U}_2$. By taking inner products of both sides of (17) with $\mathbf{U}_1 \times \mathbf{U}_2$, we find that

$$(18) \qquad c = \frac{(\bar{\mathbf{Y}}_2 - \bar{\mathbf{Y}}_1) \cdot (\mathbf{U}_1 \times \mathbf{U}_2)}{|\, \mathbf{U}_1 \times \mathbf{U}_2 \,|^2}.$$

Since $\bar{\mathbf{Y}}_1$ and $\bar{\mathbf{Y}}_2$ are on L_1 and L_2, (15) and (16) imply that $\bar{\mathbf{Y}}_2 - \bar{\mathbf{Y}}_1$ is of the form

$$\bar{\mathbf{Y}}_2 - \bar{\mathbf{Y}}_1 = \mathbf{X}_2 - \mathbf{X}_1 + \bar{s}\mathbf{U}_2 - \bar{t}\mathbf{U}_1;$$

hence

$$(\bar{\mathbf{Y}}_2 - \bar{\mathbf{Y}}_1) \cdot (\mathbf{U}_1 \times \mathbf{U}_2) = (\mathbf{X}_2 - \mathbf{X}_1) \cdot (\mathbf{U}_1 \times \mathbf{U}_2)$$

and from (18),

$$c = \frac{(\mathbf{X}_2 - \mathbf{X}_1) \cdot (\mathbf{U}_1 \times \mathbf{U}_2)}{|\, \mathbf{U}_1 \times \mathbf{U}_2 \,|^2}.$$

It now follows from (17) that

$$(19) \qquad d = |\, c \,|\, |\, \mathbf{U}_1 \times \mathbf{U}_2 \,| = \frac{|\, (\mathbf{X}_2 - \mathbf{X}_1) \cdot (\mathbf{U}_1 \times \mathbf{U}_2) \,|}{|\, \mathbf{U}_1 \times \mathbf{U}_2 \,|}.$$

If L_1 and L_2 are parallel, then the distance between them is given by

$$d = \frac{|\, \mathbf{U}_1 \times (\mathbf{X}_2 - \mathbf{X}_1) \,|}{|\, \mathbf{U}_1 \,|}$$

(Exercise T-7).

Example 3.9 Let L_1 and L_2 be given by

$$\mathbf{X} = \mathbf{X}_1 + t\mathbf{U}_1 = \begin{bmatrix} -1 \\ 0 \\ 2 \end{bmatrix} + t \begin{bmatrix} 1 \\ -1 \\ 0 \end{bmatrix}$$

and

$$\mathbf{X} = \mathbf{X}_2 + s\mathbf{U}_2 = \begin{bmatrix} 0 \\ -1 \\ 0 \end{bmatrix} + s \begin{bmatrix} 1 \\ 0 \\ -1 \end{bmatrix};$$

then

$$\mathbf{X}_2 - \mathbf{X}_1 = \mathbf{i} - \mathbf{j} - 2\mathbf{k},$$

$$\mathbf{U}_1 \times \mathbf{U}_2 = \mathbf{i} + \mathbf{j} + \mathbf{k},$$

and

$$d = \frac{2}{\sqrt{3}}.$$

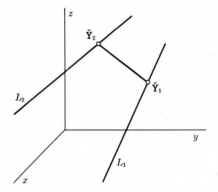

FIGURE 3.14

Distance from a Point to a Plane

Suppose we wish to find the distance d from a point $\mathbf{X}_1$ to a plane Γ defined by

$$\mathbf{N} \cdot (\mathbf{X} - \mathbf{X}_0) = 0.$$

Then

$$d = |\mathbf{X}_1 - \mathbf{X}_0| \, |\cos \theta|$$

where θ is the angle between $\mathbf{X}_1 - \mathbf{X}_0$ and $\mathbf{N}$ (Fig. 3.15); thus

$$\cos \theta = \frac{(\mathbf{X}_1 - \mathbf{X}_0) \cdot \mathbf{N}}{|\mathbf{X}_1 - \mathbf{X}_0| \, |\mathbf{N}|}$$

and

$$d = \frac{|(\mathbf{X}_1 - \mathbf{X}_0) \cdot \mathbf{N}|}{|\mathbf{N}|}.$$

Example 3.10   Let Γ be given by $2x - 3y + 4z = 6$, and $X_1 = (1, -2, 3)$. Then

$$\mathbf{N} = 2\mathbf{i} - 3\mathbf{j} + 4\mathbf{k};$$

therefore $\mathbf{X}_1 \cdot \mathbf{N} = 20$ and $\mathbf{X}_0 \cdot \mathbf{N} = 6$ for any $\mathbf{X}_0$ on Γ. Thus

$$d = \frac{20 - 6}{\sqrt{29}} = \frac{14}{\sqrt{29}}.$$

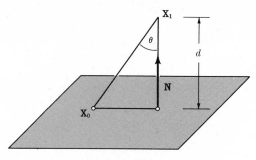

FIGURE 3.15

Scalar Triple Product

If $\mathbf{U}$, $\mathbf{V}$, and $\mathbf{W}$ are vectors in R^3, then $\mathbf{U} \cdot (\mathbf{V} \times \mathbf{W})$ is called a *scalar triple product*. Let $\mathbf{U} = u_1\mathbf{i} + u_2\mathbf{j} + u_3\mathbf{k}$, $\mathbf{V} = v_1\mathbf{i} + v_2\mathbf{j} + v_3\mathbf{k}$, and $\mathbf{W} = w_1\mathbf{i} + w_2\mathbf{j} + w_3\mathbf{k}$; since

$$(20) \qquad \mathbf{V} \times \mathbf{W} = (v_2w_3 - v_3w_2)\mathbf{i} + (v_3w_1 - v_1w_3)\mathbf{j} + (v_1w_2 - v_2w_1)\mathbf{k},$$

it follows that

$$\mathbf{U} \cdot (\mathbf{V} \times \mathbf{W}) = u_1(v_2w_3 - v_3w_2) + u_2(v_3w_1 - v_1w_3) + u_3(v_1w_2 - v_2w_1),$$

which can also be written as

$$(21) \qquad \mathbf{U} \cdot (\mathbf{V} \times \mathbf{W}) = \begin{vmatrix} u_1 & u_2 & u_3 \\ v_1 & v_2 & v_3 \\ w_1 & w_2 & w_3 \end{vmatrix}.$$

Transposition (interchange) of two rows of a determinant changes its sign. Since such a transposition is equivalent to interchanging the corresponding vectors on the left side of (21), it follows that

$$\mathbf{U} \cdot (\mathbf{V} \times \mathbf{W}) = \mathbf{V} \cdot (\mathbf{W} \times \mathbf{U}) = \mathbf{W} \cdot (\mathbf{U} \times \mathbf{V})$$

$$= -\mathbf{V} \cdot (\mathbf{U} \times \mathbf{W}) = -\mathbf{W} \cdot (\mathbf{V} \times \mathbf{U}) = -\mathbf{U} \cdot (\mathbf{W} \times \mathbf{V})$$

(Exercise T-10).

Example 3.11   Let

$$\mathbf{U} = \mathbf{i} + \mathbf{j} - \mathbf{k},$$
$$\mathbf{V} = 2\mathbf{i} + \mathbf{j} - 2\mathbf{k},$$
$$\mathbf{W} = \mathbf{i} - \mathbf{j};$$

then

(22) $$\mathbf{V} \times \mathbf{W} = -2\mathbf{i} - 2\mathbf{j} - 3\mathbf{k}$$

and

$$\mathbf{U} \cdot (\mathbf{V} \times \mathbf{W}) = (\mathbf{i} + \mathbf{j} + \mathbf{k}) \cdot (-2\mathbf{i} - 2\mathbf{j} - 3\mathbf{k}) = -1.$$

Also,

$$\mathbf{W} \times \mathbf{U} = \mathbf{i} + \mathbf{j} + 2\mathbf{k}$$

and

$$\mathbf{V} \cdot (\mathbf{W} \times \mathbf{U}) = (2\mathbf{i} + \mathbf{j} - 2\mathbf{k}) \cdot (\mathbf{i} + \mathbf{j} + 2\mathbf{k}) = -1$$
$$= \mathbf{U} \cdot (\mathbf{V} \times \mathbf{W}).$$

However,

$$\mathbf{V} \cdot (\mathbf{U} \times \mathbf{W}) = (2\mathbf{i} + \mathbf{j} - 2\mathbf{k}) \cdot (-\mathbf{i} - \mathbf{j} - 2\mathbf{k}) = 1$$
$$= -\mathbf{U} \cdot (\mathbf{V} \times \mathbf{W}).$$

Example 3.12 Consider a parallelepiped P with a vertex at the origin and let the edges of P which meet at the origin be associated with vectors **U**, **V**, and **W** as shown in Fig. 3.16. The volume v of P can be obtained by multiplying the area of the face containing **V** and **W**, which we have shown to be $|\mathbf{V} \times \mathbf{W}|$, by the distance d from this face to the face parallel to it. From Fig. 3.16,

$$d = |\mathbf{U}| |\cos \theta|,$$

where θ is the angle between **U** and $\mathbf{V} \times \mathbf{W}$; hence

$$v = |\mathbf{V} \times \mathbf{W}| |\mathbf{U}| |\cos \theta| = |\mathbf{U} \cdot (\mathbf{V} \times \mathbf{W})|.$$

For example, to find the volume of the parallelepiped P with vertices at $(0, 0, 0)$, $(1, 2, 1)$, $(2, -1, 1)$, and $(1, -2, 1)$, we take $\mathbf{U} = \mathbf{i} + 2\mathbf{j} + \mathbf{k}$,

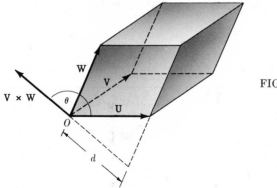

FIGURE 3.16

$V = 2\mathbf{i} - \mathbf{j} + \mathbf{k}$, and $W = \mathbf{i} - 2\mathbf{j} + \mathbf{k}$. Then

$$V \times W = \mathbf{i} - \mathbf{j} - 3\mathbf{k}$$

and

$$\mathbf{U} \cdot (\mathbf{V} \times \mathbf{W}) = (\mathbf{i} + 2\mathbf{j} + \mathbf{k}) \cdot (\mathbf{i} - \mathbf{j} - 3\mathbf{k}) = -4;$$

hence the volume of P is

$$v = |\, \mathbf{U} \cdot (\mathbf{V} \times \mathbf{W}) \,| = 4.$$

Vector Triple Product

Let $\mathbf{U}$, $\mathbf{V}$, and $\mathbf{W}$ be as defined in our discussion of the scalar triple product. Then $\mathbf{U} \times (\mathbf{V} \times \mathbf{W})$ is called a *vector triple product*. It can be expanded as follows:

$$(23) \quad \mathbf{U} \times (\mathbf{V} \times \mathbf{W}) = u_1(\mathbf{i} \times (\mathbf{V} \times \mathbf{W})) + u_2(\mathbf{j} \times (\mathbf{V} \times \mathbf{W}))$$

$$+ u_3(\mathbf{k} \times (\mathbf{V} \times \mathbf{W}));$$

from (20),

$$\mathbf{i} \times (\mathbf{V} \times \mathbf{W}) = (v_2 w_1 - v_1 w_2)\mathbf{j} + (v_3 w_1 - v_1 w_3)\mathbf{k}$$

$$= w_1(v_1\mathbf{i} + v_2\mathbf{j} + v_3\mathbf{k}) - v_1(w_1\mathbf{i} + w_2\mathbf{j} + w_3\mathbf{k})$$

$$= w_1\mathbf{V} - v_1\mathbf{W}.$$

Similarly,

$$\mathbf{j} \times (\mathbf{V} \times \mathbf{W}) = w_2\mathbf{V} - v_2\mathbf{W}$$

and

$$\mathbf{k} \times (\mathbf{V} \times \mathbf{W}) = w_3\mathbf{V} - v_3\mathbf{W}.$$

Substituting these three results into (23) yields

$$\mathbf{U} \times (\mathbf{V} \times \mathbf{W}) = (u_1 w_1 + u_2 w_2 + u_3 w_3)\mathbf{V} - (u_1 v_1 + u_2 v_2 + u_3 v_3)\mathbf{W};$$

thus

$$(24) \qquad \mathbf{U} \times (\mathbf{V} \times \mathbf{W}) = (\mathbf{U} \cdot \mathbf{W})\mathbf{V} - (\mathbf{U} \cdot \mathbf{V})\mathbf{W}.$$

Remembering this equation is made easier by observing that $\mathbf{U} \times (\mathbf{V} \times \mathbf{W})$ must be a linear combination of $\mathbf{V}$ and $\mathbf{W}$ because it is orthogonal to $\mathbf{V} \times \mathbf{W}$.

The triple vector product $(\mathbf{U} \times \mathbf{V}) \times \mathbf{W}$ is in general different from (24); in fact,

$$(\mathbf{U} \times \mathbf{V}) \times \mathbf{W} = -\mathbf{W} \times (\mathbf{U} \times \mathbf{V}) = \mathbf{W} \times (\mathbf{V} \times \mathbf{U}),$$

which can be expanded by interchanging **U** and **W** in (24). The result is

(25) $\qquad\qquad (\mathbf{U} \times \mathbf{V}) \times \mathbf{W} = (\mathbf{W}\cdot\mathbf{U})\mathbf{V} - (\mathbf{W}\cdot\mathbf{V})\mathbf{U}$

Example 3.13 Let **U**, **V**, and **W** be as defined in Example 3.10. From (22),

$$\mathbf{U} \times (\mathbf{V} \times \mathbf{W}) = \begin{vmatrix} \mathbf{i} & \mathbf{j} & \mathbf{k} \\ 1 & 1 & -1 \\ -2 & -2 & -3 \end{vmatrix} = -5\mathbf{i} + 5\mathbf{j}.$$

We can obtain the same result from (24); since $\mathbf{U}\cdot\mathbf{W} = 0$, and $\mathbf{U}\cdot\mathbf{V} = 5$,

$$\mathbf{U} \times (\mathbf{V} \times \mathbf{W}) = 0\mathbf{V} - 5\mathbf{W} = -5\mathbf{i} + 5\mathbf{j}.$$

From (25) we find that

$$(\mathbf{U} \times \mathbf{V}) \times \mathbf{W} = 0\mathbf{V} - 1\mathbf{U} = -\mathbf{i} - \mathbf{j} + \mathbf{k}.$$

EXERCISES 1.3

1. State which of the following sets of vectors are coplanar.

(a) $\begin{bmatrix} 0 \\ 7 \\ 5 \end{bmatrix}$, $\begin{bmatrix} -3 \\ 5 \\ -2 \end{bmatrix}$, $\begin{bmatrix} 2 \\ -1 \\ 3 \end{bmatrix}$; (b) $\begin{bmatrix} 1 \\ 2 \\ 3 \end{bmatrix}$, $\begin{bmatrix} 0 \\ 1 \\ 2 \end{bmatrix}$, $\begin{bmatrix} 1 \\ 0 \\ 2 \end{bmatrix}$;

(c) $\begin{bmatrix} -3 \\ -2 \\ 1 \end{bmatrix}$, $\begin{bmatrix} 1 \\ 0 \\ -1 \end{bmatrix}$, $\begin{bmatrix} 5 \\ 3 \\ -2 \end{bmatrix}$.

2. Write in terms of **i**, **j**, and **k**:

(a) $\begin{bmatrix} 0 \\ -2 \\ \frac{2}{5} \end{bmatrix}$; (b) $\begin{bmatrix} \frac{4}{5} \\ -7 \\ 5 \end{bmatrix}$; (c) $\begin{bmatrix} 2 \\ 0 \\ -4 \end{bmatrix}$; (d) $\begin{bmatrix} 0 \\ 0 \\ -5 \end{bmatrix}$.

3. Write in "column" form:
   (a) $2i - 3j;$ (b) $i - \frac{2}{3}j + \frac{4}{7}k;$
   (c) $2j;$ (d) $3i - j + k.$

4. Compute $W = U \times V$ and verify that W is perpendicular to U and V:

   (a) $U = \begin{bmatrix} 1 \\ 0 \\ 3 \end{bmatrix}$, $V = \begin{bmatrix} 1 \\ 2 \\ 3 \end{bmatrix}$;   (b) $U = V = \begin{bmatrix} 1 \\ 2 \\ 3 \end{bmatrix}$;

   (c) $U = \begin{bmatrix} 1 \\ 2 \\ 3 \end{bmatrix}$, $V = -U$;   (d) $U = \begin{bmatrix} -1 \\ 5 \\ 2 \end{bmatrix}$, $V = \begin{bmatrix} 2 \\ -1 \\ 3 \end{bmatrix}$.

5. Repeat Exercise 4 for the following pairs of vectors.
   (a) $U = 2i - 3j + 4k,$ (b) $U = 4i,$
      $V = 2i - j;$ $V = 3i + 2j - 5k;$
   (c) $U = 4i + 3j - k,$ (d) $U = 2i - j,$
      $V = i + j - 3k;$ $V = -3j + 5k.$

6. Verify the cross products in Example 3.3.

7. Verify (c), (d), and (e) of Theorem 3.1 for

   $$U = \begin{bmatrix} 1 \\ 2 \\ 3 \end{bmatrix}, \quad V = \begin{bmatrix} -1 \\ 3 \\ 4 \end{bmatrix}, \quad W = \begin{bmatrix} 0 \\ 2 \\ -1 \end{bmatrix}, \quad \text{and} \quad c = -3.$$

8. Verify that $|U \times V| = |U||V| \sin \theta$ for the pairs of vectors in Exercise 4.

9. Find the area of the triangle with vertices $X_1 = (1, 0, 2)$, $X_2 = (2, -1, 3)$, and $X_3 = (4, -3, 2)$.

10. Find the area of the triangle with vertices X_1, X_2, and X_3 such that $X_2 - X_1 = 3i + j - 2k$, and $X_3 - X_1 = -i + 3k$.

11. Find the area of the triangle with vertices X_1, X_2, and X_3, where $X_1 = (-1, 0, 2)$, $X_2 = (4, 1, 3)$, and $X_3 - X_1 = i + 2j - 3k$.

12. Find the area of the parallelogram with vertices at X_1, X_2, X_3, and X_4, where $X_2 - X_1 = 3i - 2j + k$, and $X_3 - X_1 = i + j - 3k$.

13. Find the plane determined by the points $(1, 2, 3)$, $(-2, 1, 4)$, and $(0, 1, 5)$.

14. Find the plane determined by the points $(1, -1, 2)$, $(3, -2, 5)$, and $(0, 0, 2)$.

15. Find the equation of the line through $(1, -1, 3)$ which intersects the line

$$\mathbf{X} = \begin{bmatrix} 1 \\ 3 \\ 5 \end{bmatrix} + t \begin{bmatrix} 1 \\ -1 \\ 3 \end{bmatrix}$$

at right angles.

16. Find the equation of the line through $(-2, 3, 4)$ which intersects the line

$$\frac{x-3}{0} = \frac{y-5}{2} = \frac{z-4}{-3}$$

at right angles.

17. Verify that the planes $x + 2y + 3z = -5$ and $2x + 4y + 6z = 8$ are parallel.

18. Find the equation of the line of intersection of the planes $2x + 3y - z = 8$ and $-x + y + 2z = 6$.

19. Repeat Exercise 18 for the planes $2x - y + 3z = 15$ and $x + y - z = 5$.

20. Find the distance d from $\mathbf{X}_1 = (2, 3, -1)$ to the line L defined by

$$\mathbf{X} = \begin{bmatrix} 0 \\ 1 \\ 2 \end{bmatrix} + t \begin{bmatrix} 1 \\ 2 \\ -3 \end{bmatrix}$$

and find $\bar{\mathbf{X}}$ on L such that $|\bar{\mathbf{X}} - \mathbf{X}_1| = d$.

21. Find the distance d from the point $\mathbf{X}_1 = (1, 1, 3)$ to the line L defined by

$$\frac{x-3}{2} = \frac{y+5}{0} = \frac{z-2}{-3},$$

and find $\bar{\mathbf{X}}$ on L such that $|\bar{\mathbf{X}} - \mathbf{X}_1| = d$.

22. Find the distance between the lines

$$\mathbf{X} = \begin{bmatrix} 1 \\ -2 \\ 3 \end{bmatrix} + s \begin{bmatrix} 2 \\ 2 \\ -1 \end{bmatrix} \quad \text{and} \quad \mathbf{X} = \begin{bmatrix} -2 \\ 3 \\ 4 \end{bmatrix} + t \begin{bmatrix} 2 \\ 3 \\ -2 \end{bmatrix}.$$

23. Find the distance between the lines

$$\mathbf{X} = \begin{bmatrix} 1 \\ 2 \\ -1 \end{bmatrix} + s \begin{bmatrix} 2 \\ -1 \\ 1 \end{bmatrix} \quad \text{and} \quad \mathbf{X} = \begin{bmatrix} 2 \\ 3 \\ 4 \end{bmatrix} + t \begin{bmatrix} -4 \\ 2 \\ -6 \end{bmatrix}.$$

24. Find the distance between the lines

$$\frac{x-3}{2} = \frac{y-4}{0} = \frac{z+5}{-2}$$

and

$$\frac{x+1}{3} = \frac{y-2}{-3} = \frac{z+1}{2}.$$

25. Find the distance from $(3, -1, 2)$ to the plane $2x + 3y - z = 5$.

26. Find the distance from $(2, 1, -3)$ to the plane $x - 2y + z = 2$.

27. Find the distance between the parallel planes $2x + y - z = 8$ and $2x + y - z = 12$.

28. Repeat Exercise 27 for the planes $x + 3y - 5z = 14$ and $2x + 6y - 10z = 12$.

29. Find the plane containing the line

$$\mathbf{X} = \begin{bmatrix} 4 \\ -2 \\ 5 \end{bmatrix} + t \begin{bmatrix} 1 \\ 3 \\ 5 \end{bmatrix}$$

and parallel to the line

$$\mathbf{X} = \begin{bmatrix} 2 \\ 3 \\ 4 \end{bmatrix} + s \begin{bmatrix} 2 \\ 4 \\ -3 \end{bmatrix}.$$

30. Find the plane containing the line

$$X = \begin{bmatrix} 1 \\ 2 \\ -3 \end{bmatrix} + t \begin{bmatrix} 0 \\ 2 \\ -3 \end{bmatrix}$$

and parallel to the line

$$X = \begin{bmatrix} 0 \\ 2 \\ 1 \end{bmatrix} + s \begin{bmatrix} -1 \\ 3 \\ 2 \end{bmatrix}.$$

31. Find the plane determined by the lines

$$X = \begin{bmatrix} 1 \\ 3 \\ -1 \end{bmatrix} + t \begin{bmatrix} 0 \\ 1 \\ 2 \end{bmatrix} \quad \text{and} \quad X = \begin{bmatrix} 2 \\ 1 \\ -5 \end{bmatrix} + s \begin{bmatrix} 1 \\ 2 \\ 4 \end{bmatrix}.$$

32. Find the plane passing through $(1, 2, -1)$ and $(2, 3, 4)$ and parallel to the line

$$X = \begin{bmatrix} 1 \\ 3 \\ 2 \end{bmatrix} + t \begin{bmatrix} 5 \\ 2 \\ -1 \end{bmatrix}.$$

33. Find the plane through $(1, 3, 4)$ and $(-5, 2, 1)$ and parallel to the line

$$\frac{x - 2}{3} = \frac{y - 5}{0} = \frac{z + 5}{-2}.$$

34. Find the plane through the line

$$X = \begin{bmatrix} 2 \\ 3 \\ -1 \end{bmatrix} + t \begin{bmatrix} 1 \\ 2 \\ -1 \end{bmatrix}$$

and perpendicular to the plane $3x + 2y - 5z = 8$.

35. Find all planes through the line

$$\frac{x-2}{3} = \frac{y+2}{-2} = \frac{z-5}{0}.$$

36. Find the plane through the line

$$\mathbf{X} = \begin{bmatrix} 1 \\ 2 \\ 3 \end{bmatrix} + t \begin{bmatrix} 1 \\ 2 \\ -1 \end{bmatrix}$$

and through the point $(2, 3, -1)$.

37. Find the equation of the plane through $(2, 3, -4)$ perpendicular to the planes $x + 2y - 3z = 5$, and $x - y + z = 7$.

38. Verify (21) for

$$\mathbf{U} = \begin{bmatrix} 1 \\ 2 \\ 3 \end{bmatrix}, \quad \mathbf{V} = \begin{bmatrix} -1 \\ 2 \\ 5 \end{bmatrix}, \quad \text{and} \quad \mathbf{W} = \begin{bmatrix} 2 \\ 3 \\ 5 \end{bmatrix}.$$

39. Find the volume of the parallelepiped with edges

$$\mathbf{U} = \begin{bmatrix} 1 \\ 2 \\ 3 \end{bmatrix}, \quad \mathbf{V} = \begin{bmatrix} 3 \\ -2 \\ 4 \end{bmatrix}, \quad \text{and} \quad \mathbf{W} = \begin{bmatrix} 4 \\ 2 \\ 5 \end{bmatrix}.$$

40. Compute $\mathbf{U} \times (\mathbf{V} \times \mathbf{W})$:

(a) $\mathbf{U} = \begin{bmatrix} 1 \\ 2 \\ -1 \end{bmatrix}, \quad \mathbf{V} = \begin{bmatrix} 2 \\ 3 \\ 4 \end{bmatrix}, \quad \mathbf{W} = \begin{bmatrix} 1 \\ 2 \\ 3 \end{bmatrix};$

(b) $\mathbf{U} = \begin{bmatrix} 1 \\ 3 \\ 2 \end{bmatrix}, \quad \mathbf{V} = \begin{bmatrix} 0 \\ 1 \\ 0 \end{bmatrix}, \quad \mathbf{W} = \begin{bmatrix} 0 \\ 1 \\ 1 \end{bmatrix}.$

41. Compute $\mathbf{U} \times (\mathbf{V} \times \mathbf{W})$:

(a) $\mathbf{U} = \begin{bmatrix} 1 \\ -3 \\ -2 \end{bmatrix}$, $\quad \mathbf{V} = \begin{bmatrix} 1 \\ 0 \\ 1 \end{bmatrix}$, $\quad \mathbf{W} = \begin{bmatrix} 2 \\ 5 \\ 7 \end{bmatrix}$;

(b) $\mathbf{U} = \begin{bmatrix} 0 \\ -1 \\ 2 \end{bmatrix}$, $\quad \mathbf{V} = \begin{bmatrix} 3 \\ 2 \\ 0 \end{bmatrix}$, $\quad \mathbf{W} = \begin{bmatrix} 5 \\ 2 \\ -1 \end{bmatrix}$.

THEORETICAL EXERCISES

T-1. Prove Theorem 3.1.

T-2. Verify the "right-hand rule" for the direction of $\mathbf{U} \times \mathbf{V}$ if $\mathbf{U}$ and $\mathbf{V}$ are distinct elements of the natural basis for R^3.

T-3. Verify the "right-hand rule" for the direction of $\mathbf{U} \times \mathbf{V}$ if $\mathbf{U}$ and $\mathbf{V}$ lie in one of the coordinate planes. Use the results of Exercise T-2.

T-4. Prove : if the planes defined by $\mathbf{X} \cdot \mathbf{N}_1 = d_1$ and $\mathbf{X} \cdot \mathbf{N}_2 = d_2$ are identical, then $d_1 \mathbf{N}_2 = d_2 \mathbf{N}_1$.

T-5. Let L be a line, let $\mathbf{X}_1$ be a point not on L, and let $\mathbf{X}_0$ be the point on L closest to $\mathbf{X}_1$. Show that the line connecting $\mathbf{X}_0$ and $\mathbf{X}_1$ is perpendicular to L.

T-6. Let L_1 and L_2 be nonintersecting parallel lines, and let the distance between them be $d = |\, \bar{\mathbf{Y}}_1 - \bar{\mathbf{Y}}_2 \,|$, where $\bar{\mathbf{Y}}_1$ is on L_1 and $\bar{\mathbf{Y}}_2$ is on L_2. Show that $\bar{\mathbf{Y}}_1 - \bar{\mathbf{Y}}_2$ is perpendicular to both L_1 and L_2.

T-7. Show that the distance between two parallel lines L_1 and L_2 is given by

$$d = \frac{|\, \mathbf{U} \times (\mathbf{X}_2 - \mathbf{X}_1) \,|}{|\, \mathbf{U} \,|},$$

where $\mathbf{U}$ is any nonzero vector parallel to L_1 and L_2, $\mathbf{X}_1$ is on L_1, and $\mathbf{X}_2$ is on L_2.

T-8. Show that (19) yields $d = 0$ if L_1 and L_2 intersect.

T-9.   Suppose $\mathbf{U} \neq \mathbf{0}$ and $\mathbf{X}_0$ is a point in R^3. What is the set of points $\mathbf{X}$ which satisfy $\mathbf{U} \times (\mathbf{X} - \mathbf{X}_0) = \mathbf{0}$?

T-10. Verify that

$$\mathbf{U} \cdot (\mathbf{V} \times \mathbf{W}) = \mathbf{V} \cdot (\mathbf{W} \times \mathbf{U}) = \mathbf{W} \cdot (\mathbf{U} \times \mathbf{V})$$
$$= -\mathbf{V} \cdot (\mathbf{U} \times \mathbf{W}) = -\mathbf{W} \cdot (\mathbf{V} \times \mathbf{U}) = -\mathbf{U} \cdot (\mathbf{W} \times \mathbf{V}).$$

T-11. Prove that

$$|\mathbf{U} \times \mathbf{V}|^2 + (\mathbf{U} \cdot \mathbf{V})^2 = |\mathbf{U}|^2 |\mathbf{V}|^2.$$

T-12. Prove that

(a) $\mathbf{U} \times (\mathbf{V} + \mathbf{W}) = \mathbf{U} \times \mathbf{V} + \mathbf{U} \times \mathbf{W}$;
(b) $(\mathbf{U} + \mathbf{V}) \times \mathbf{W} = \mathbf{U} \times \mathbf{W} + \mathbf{V} \times \mathbf{W}$;
(c) $\mathbf{U} \times \mathbf{V} = -(\mathbf{V} \times \mathbf{U})$.

T-13. If $\mathbf{U} \cdot \mathbf{V} = 0$ and $\mathbf{U} \times \mathbf{V} = \mathbf{0}$, show that $\mathbf{U} = \mathbf{0}$ or $\mathbf{V} = \mathbf{0}$.

T-14. Prove the *Jacobi identity*:

$$(\mathbf{U} \times \mathbf{V}) \times \mathbf{W} + (\mathbf{V} \times \mathbf{W}) \times \mathbf{U} + (\mathbf{W} \times \mathbf{U}) \times \mathbf{V} = \mathbf{0}.$$

1.4   Motion in R^3

In this section we discuss the use of vectors to study the motion of a particle in R^3. Since the emphasis is on applications rather than mathematical rigor, we shall be informal here; for example, we use terms like "curve" and "function" even though they are not defined until later, relying on the student's understanding of them on the basis of his first calculus course.

Position, Velocity, and Acceleration Vectors

The position of a particle moving in R^3 can be represented as a vector-valued function of time t:

$$\mathbf{X}(t) = \begin{bmatrix} x(t) \\ y(t) \\ z(t) \end{bmatrix} = x(t)\mathbf{i} + y(t)\mathbf{j} + z(t)\mathbf{k}.$$

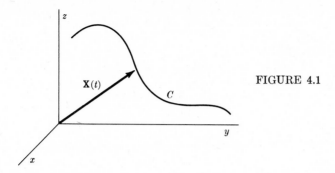

FIGURE 4.1

As t varies the particle moves along a curve C in R^3 (Fig. 4.1); we say that C is defined parametrically by $\mathbf{X} = \mathbf{X}(t)$.

If x, y, and z are continuous functions of t, we say that C is a *continuous curve*; physically this means that the particle does not undergo any sudden jumps.

The *velocity* of the particle is defined to be the vector valued function

$$\mathbf{V}(t) = \mathbf{X}'(t) = x'(t)\mathbf{i} + y'(t)\mathbf{j} + z'(t)\mathbf{k},$$

provided $x'(t)$, $y'(t)$, and $z'(t)$ exist. It has an important geometric interpretation. Suppose $h \neq 0$ and consider the vector

$$\frac{\mathbf{X}(t+h) - \mathbf{X}(t)}{h} = \frac{x(t+h) - x(t)}{h}\mathbf{i} + \frac{y(t+h) - y(t)}{h}\mathbf{j}$$

$$+ \frac{z(t+h) - z(t)}{h}\mathbf{k},$$

which is parallel to the secant vector $\mathbf{X}(t+h) - \mathbf{X}(t)$ (Fig. 4.2). Its components approach the components of $\mathbf{V}(t)$ as h approaches zero; therefore, if $\mathbf{V}(t) \neq 0$, its direction is that approached by the secant vector as h approaches zero; that is, $\mathbf{V}(t)$ is tangent to C at $\mathbf{X} = \mathbf{X}(t)$. The direction of $\mathbf{V}(t)$ is the *instantaneous direction of motion* and $|\mathbf{V}(t)|$ is the *instantaneous speed*.

The *acceleration vector* is defined by

$$\mathbf{A}(t) = \mathbf{V}'(t) = \mathbf{X}''(t) = x''(t)\mathbf{i} + y''(t)\mathbf{j} + z''(t)\mathbf{k},$$

provided $x''(t)$, $y''(t)$, and $z''(t)$ exist.

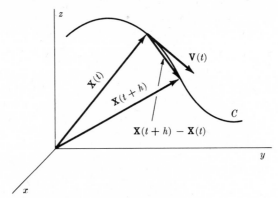

FIGURE 4.2

Example 4.1   Let the motion of a particle be described by

$$\mathbf{X}(t) = a \cos t \, \mathbf{i} + a \sin t \, \mathbf{j} + bt\mathbf{k} \qquad (-\infty < t < \infty);$$

then the curve traversed by the particle is a helix (Fig. 4.3). The velocity vector is

$$\mathbf{V}(t) = -a \sin t \, \mathbf{i} + a \cos t \, \mathbf{j} + b\mathbf{k}$$

and the speed is constant, since

$$|\mathbf{V}(t)| = \sqrt{(-a \sin t)^2 + (a \cos t)^2 + b^2} = \sqrt{a^2 + b^2}.$$

The acceleration vector is

$$\mathbf{A}(t) = -a \cos t \, \mathbf{i} - a \sin t \, \mathbf{j}.$$

The position, velocity and acceleration vectors are examples of *vector functions* of t. We can also define a general vector function $\mathbf{F}$ by

$$\mathbf{F}(t) = f(t)\mathbf{i} + g(t)\mathbf{j} + h(t)\mathbf{k},$$

where f, g, and h are scalar functions of t, and its derivatives by

$$\mathbf{F}'(t) = f'(t)\mathbf{i} + g'(t)\mathbf{j} + h'(t)\mathbf{k},$$

$$\mathbf{F}''(t) = f''(t)\mathbf{i} + g''(t)\mathbf{j} + h''(t)\mathbf{k},$$

etc., so long as f, g, and h possess the derivatives shown on the right.

If ρ is a scalar function of time then the vector function $\rho\mathbf{F}$ is defined by

(1) $$(\rho\mathbf{F})(t) = \rho(t)\mathbf{F}(t),$$

FIGURE 4.3

and if

$$\mathbf{F}_1(t) = f_1(t)\mathbf{i} + g_1(t)\mathbf{j} + h_1(t)\mathbf{k},$$

$$\mathbf{F}_2(t) = f_2(t)\mathbf{i} + g_2(t)\mathbf{j} + h_2(t)\mathbf{k},$$

then we define the sum $\mathbf{F}_1 + \mathbf{F}_2$ by

$$(\mathbf{F}_1 + \mathbf{F}_2)(t) = \mathbf{F}_1(t) + \mathbf{F}_2(t).$$

Similarly, we define $\mathbf{F}_1 \cdot \mathbf{F}_2$ and $\mathbf{F}_1 \times \mathbf{F}_2$ by

$$(2) \qquad\qquad (\mathbf{F}_1 \cdot \mathbf{F}_2)(t) = \mathbf{F}_1(t) \cdot \mathbf{F}_2(t)$$

and

$$(\mathbf{F}_1 \times \mathbf{F}_2)(t) = \mathbf{F}_1(t) \times \mathbf{F}_2(t).$$

Theorem 4.1 Let $\mathbf{F}_1$ and $\mathbf{F}_2$ be vector functions and ρ be a scalar function of t. Then

(a) $(\mathbf{F}_1 + \mathbf{F}_2)'(t) = \mathbf{F}_1'(t) + \mathbf{F}_2'(t);$

(b) $(\rho\mathbf{F}_1)'(t) = (\rho'\mathbf{F}_1 + \rho\mathbf{F}_1')(t);$

(c) $(\mathbf{F}_1 \cdot \mathbf{F}_2)'(t) = (\mathbf{F}_1' \cdot \mathbf{F}_2 + \mathbf{F}_1 \cdot \mathbf{F}_2')(t),$

and

(d) $(\mathbf{F}_1 \times \mathbf{F}_2)'(t) = (\mathbf{F}_1' \times \mathbf{F}_2 + \mathbf{F}_1 \times \mathbf{F}_2')(t),$

provided the derivatives on the right hand sides exist.

Proof. We prove (b) and (c), and leave the rest for the student. From (1),

$$(\rho \mathbf{F})'(t) = \frac{d}{dt} [\rho(t) \mathbf{F}(t)]$$

$$= \left[\frac{d}{dt} \rho(t) f(t)\right] \mathbf{i} + \left[\frac{d}{dt} \rho(t) g(t)\right] \mathbf{j} + \left[\frac{d}{dt} \rho(t) h(t)\right] \mathbf{k}$$

$$= [\rho'(t) f(t) + \rho(t) f'(t)] \mathbf{i} + [\rho'(t) g(t) + \rho(t) g'(t)] \mathbf{j}$$
$$+ [\rho'(t) h(t) + \rho(t) h'(t)] \mathbf{k}$$

$$= \rho'(t) [f(t) \mathbf{i} + g(t) \mathbf{j} + h(t) \mathbf{k}]$$
$$+ \rho(t) [f'(t) \mathbf{i} + g'(t) \mathbf{j} + h'(t) \mathbf{k}]$$

$$= \rho'(t) \mathbf{F}(t) + \rho(t) \mathbf{F}'(t)$$

$$= (\rho' \mathbf{F} + \rho \mathbf{F}')(t),$$

which proves (b). From (2)

$$(\mathbf{F}_1 \cdot \mathbf{F}_2)'(t) = \frac{d}{dt} (f_1(t) f_2(t) + g_1(t) g_2(t) + h_1(t) h_2(t))$$

$$= f_1'(t) f_2(t) + g_1'(t) g_2(t) + h_1'(t) h_2(t)$$
$$+ f_1(t) f_2'(t) + g_1(t) g_2'(t) + h_1(t) h_2'(t)$$

$$= \mathbf{F}_1'(t) \cdot \mathbf{F}_2(t) + \mathbf{F}_1(t) \cdot \mathbf{F}_2'(t)$$

$$= (\mathbf{F}_1' \cdot \mathbf{F}_2 + \mathbf{F}_1 \cdot \mathbf{F}_2')(t),$$

which proves (c).

Example 4.2 Let

$$\mathbf{F}_1(t) = \cos t \, \mathbf{i} + \sin t \, \mathbf{j} + t \mathbf{k},$$
$$\mathbf{F}_2(t) = t \mathbf{i} + t^2 \mathbf{j} + t^3 \mathbf{k},$$

and

$$\rho(t) = e^{t^2};$$

then

$$\mathbf{F}_1'(t) = -\sin t \, \mathbf{i} + \cos t \, \mathbf{j} + \mathbf{k},$$
$$\mathbf{F}_2'(t) = \mathbf{i} + 2t \mathbf{j} + 3t^2 \mathbf{k},$$

and

$$\rho'(t) = 2t e^{t^2}.$$

We illustrate Theorem 4.1 by verifying (a) through (d):

(a) $(\mathbf{F_1 + F_2})(t) = (\cos t + t)\mathbf{i} + (\sin t + t^2)\mathbf{j} + (t + t^3)\mathbf{k}$

and

$$(\mathbf{F_1 + F_2'})(t) = (-\sin t + 1)\mathbf{i} + (\cos t + 2t)\mathbf{j} + (1 + 3t^2)\mathbf{k}$$
$$= (-\sin t\,\mathbf{i} + \cos t\,\mathbf{j} + \mathbf{k}) + (\mathbf{i} + 2t\mathbf{j} + 3t^2\mathbf{k})$$
$$= \mathbf{F_1'}(t) + \mathbf{F_2'}(t).$$

(b) $(\rho \mathbf{F_1})(t) = (e^{t^2} \cos t)\mathbf{i} + (e^{t^2} \sin t)\mathbf{j} + (e^{t^2}t)\mathbf{k}$

and

$$(\rho \mathbf{F_1})'(t) = (2te^{t^2} \cos t - e^{t^2} \sin t)\mathbf{i} + (2te^{t^2} \sin t + e^{t^2} \cos t)\mathbf{j}$$
$$+ (2t^2 e^{t^2} + e^{t^2})\mathbf{k}$$
$$= 2te^{t^2}(\cos t\,\mathbf{i} + \sin t\,\mathbf{j} + t\mathbf{k})$$
$$+ e^{t^2}(-\sin t\,\mathbf{i} + \cos t\,\mathbf{j} + \mathbf{k})$$
$$= \rho'(t)\,\mathbf{F_1}(t) + \rho(t)\,\mathbf{F_1'}(t)$$
$$= (\rho'\mathbf{F_1} + \rho\mathbf{F_1'})(t).$$

(c) $(\mathbf{F_1 \cdot F_2})(t) = t \cos t + t^2 \sin t + t^4$

and

(3) $(\mathbf{F_1 \cdot F_2})'(t) = (\cos t - t \sin t) + (2t \sin t + t^2 \cos t) + 4t^3$
$$= \cos t + t \sin t + t^2 \cos t + 4t^3.$$

On the other hand,

$$(\mathbf{F_1' \cdot F_2} + \mathbf{F_1 \cdot F_2'})(t) = \mathbf{F_1'}(t) \cdot \mathbf{F_2}(t) + \mathbf{F_1}(t) \cdot \mathbf{F_2'}(t)$$
$$= (-t \sin t + t^2 \cos t + t^3) + (\cos t + 2t \sin t + 3t^3)$$
$$= \cos t + t \sin t + t^2 \cos t + 4t^3,$$

which agrees with (3).

(d) $(\mathbf{F_1 \times F_2})(t) = \begin{vmatrix} \mathbf{i} & \mathbf{j} & \mathbf{k} \\ \cos t & \sin t & t \\ t & t^2 & t^3 \end{vmatrix}$

$$= (t^3 \sin t - t^3)\mathbf{i} + (t^2 - t^3 \cos t)\mathbf{j}$$
$$+ (t^2 \cos t - t \sin t)\mathbf{k}$$

and

(4) $(\mathbf{F}_1 \times \mathbf{F}_2)'(t) = (3t^2 \sin t + t^3 \cos t - 3t^2)\mathbf{i}$
$$+ (2t - 3t^2 \cos t + t^3 \sin t)\mathbf{j}$$
$$+ (2t \cos t - t^2 \sin t - \sin t - t \cos t)\mathbf{k}.$$

On the other hand,

(5) $(\mathbf{F}_1' \times \mathbf{F}_2)(t) = \begin{vmatrix} \mathbf{i} & \mathbf{j} & \mathbf{k} \\ -\sin t & \cos t & 1 \\ t & t^2 & t^3 \end{vmatrix}$

$$= (t^3 \cos t - t^2)\mathbf{i} + (t + t^3 \sin t)\mathbf{j}$$
$$+ (-t^2 \sin t - t \cos t)\mathbf{k}$$

and

(6) $(\mathbf{F}_1 \times \mathbf{F}_2')(t) = \begin{vmatrix} \mathbf{i} & \mathbf{j} & \mathbf{k} \\ \cos t & \sin t & t \\ 1 & 2t & 3t^2 \end{vmatrix}$

$$= (3t^2 \sin t - 2t^2)\mathbf{i} + (t - 3t^2 \cos t)\mathbf{j}$$
$$+ (2t \cos t - \sin t)\mathbf{k}.$$

Adding the right hand sides of (4) and (5) yields the right hand side of (6).

Example 4.3   To calculate the derivative of $|\mathbf{F}(t)|$ we write

$$|\mathbf{F}(t)|^2 = \mathbf{F}(t) \cdot \mathbf{F}(t)$$

and differentiate both sides:

$$2|\mathbf{F}(t)|\frac{d}{dt}|\mathbf{F}(t)| = \mathbf{F}'(t) \cdot \mathbf{F}(t) + \mathbf{F}(t) \cdot \mathbf{F}'(t)$$

$$= 2\mathbf{F}(t) \cdot \mathbf{F}'(t).$$

Thus, if $\mathbf{F}(t) \neq \mathbf{0}$,

(7) $$\frac{d}{dt}|\mathbf{F}(t)| = \frac{\mathbf{F}'(t) \cdot \mathbf{F}(t)}{|\mathbf{F}(t)|}.$$

If $|\mathbf{F}(t)|$ is constant then $\mathbf{F}(t)$ is perpendicular to its derivative, since (7) becomes

$$0 = \mathbf{F}'(t) \cdot \mathbf{F}(t).$$

For example, if

$$\mathbf{F}(t) = \cos t\,\mathbf{i} + \sin t\,\mathbf{j} + \mathbf{k},$$

then

$$\mathbf{F}'(t) = -\sin t\,\mathbf{i} + \cos t\,\mathbf{j}$$

and

$$\mathbf{F}(t)\cdot\mathbf{F}'(t) = -\cos t \sin t + \sin t \cos t = 0,$$

which is to be expected, since $|\mathbf{F}(t)| \equiv \sqrt{2}$.

Example 4.4 If $\mathbf{F}(t) \neq \mathbf{0}$, define the unit vector

$$\mathbf{U}(t) = \frac{\mathbf{F}(t)}{|\mathbf{F}(t)|};$$

then, from (b) of Theorem 4.1,

$$\mathbf{U}'(t) = \frac{\mathbf{F}'(t)}{|\mathbf{F}(t)|} + \left[\frac{d}{dt}\left(\frac{1}{|\mathbf{F}(t)|}\right)\right]\mathbf{F}(t)$$

$$= \frac{\mathbf{F}'(t)}{|\mathbf{F}(t)|} - \left[\frac{d}{dt}|\mathbf{F}(t)|\right]\frac{\mathbf{F}(t)}{|\mathbf{F}(t)|^2}.$$

From (7) this can be rewritten as

$$\mathbf{U}'(t) = \frac{\mathbf{F}'(t)}{|\mathbf{F}(t)|} - \frac{(\mathbf{F}(t)\cdot\mathbf{F}'(t))}{|\mathbf{F}(t)|^2}\mathbf{F}(t)$$

$$= \frac{|\mathbf{F}(t)|^2\,\mathbf{F}'(t) - (\mathbf{F}(t)\cdot\mathbf{F}'(t))\mathbf{F}(t)}{|\mathbf{F}(t)|^3},$$

which can be expressed as a triple product:

(8) $$\frac{d}{dt}\frac{\mathbf{F}(t)}{|\mathbf{F}(t)|} = \frac{\mathbf{F}(t)\times(\mathbf{F}'(t)\times\mathbf{F}(t))}{|\mathbf{F}(t)|^3}.$$

Unit Tangent, Normal, and Binormal

Let us again consider a particle moving along a curve C. We have seen that its velocity vector $\mathbf{V}(t)$ is tangent to C at $\mathbf{X} = \mathbf{X}(t)$, provided $\mathbf{V}(t) \neq \mathbf{0}$ (which we assume throughout this section). The *unit tangent vector to C at $\mathbf{X} = \mathbf{X}(t)$* is defined by

(9) $$\mathbf{T}(t) = \frac{\mathbf{V}(t)}{|\mathbf{V}(t)|}.$$

From (8)

(10)
$$\mathbf{T}'(t) = \frac{\mathbf{V}(t) \times (\mathbf{A}(t) \times \mathbf{V}(t)}{|\mathbf{V}(t)|^3}.$$

If $\mathbf{A}(t)$ and $\mathbf{V}(t)$ are noncollinear then we can define $\mathbf{N}(t)$, the *unit principal normal vector* $\mathbf{N}$, to be the unit vector in the direction of $\mathbf{T}'(t)$; thus

(11)
$$\mathbf{N}(t) = \frac{\mathbf{T}'(t)}{|\mathbf{T}'(t)|}$$

$$= \frac{\mathbf{V}(t) \times (\mathbf{A}(t) \times \mathbf{V}(t))}{|\mathbf{V}(t)||\mathbf{A}(t) \times \mathbf{V}(t)|}.$$

The unit binormal vector to C at $\mathbf{X} = \mathbf{X}(t)$ is defined by

$$\mathbf{B}(t) = \mathbf{T}(t) \times \mathbf{N}(t);$$

from (9) and (11), it is given explicitly by

(12)
$$\mathbf{B}(t) = \frac{\mathbf{V}(t) \times \mathbf{A}(t)}{|\mathbf{V}(t) \times \mathbf{A}(t)|}$$

(Exercise T-2).

The plane Γ_t through $\mathbf{X} = \mathbf{X}(t)$ that contains $\mathbf{A}(t)$ and $\mathbf{V}(t)$ is the *osculating plane at* $\mathbf{X} = \mathbf{X}(t)$. From (9) and (11), Γ_t contains $\mathbf{T}(t)$ and $\mathbf{N}(t)$. The situation is shown in Fig. 4.4, where the solid curve is C and the dotted curve is the projection of C on Γ_t. If we consider Γ_t to be the plane of the page, then $\mathbf{B}(t)$ is perpendicular to the page and pointed toward the reader.

The equation of Γ_t is, according to Eq. (12), Section 1.2, and Eq. (12) above,

$$(\mathbf{X} - \mathbf{X}(t)) \cdot \mathbf{B}(t) = 0.$$

Example 4.5 Let us calculate $\mathbf{T}$, $\mathbf{N}$, and $\mathbf{B}$ and find the equation of the osculating plane to the helix defined by

$$\mathbf{X}(t) = \cos t\,\mathbf{i} + \sin t\,\mathbf{j} + e^t\mathbf{k}$$

at $\mathbf{X}(0) = (1, 0, 1)$. The velocity and acceleration vectors are

$$\mathbf{V}(t) = -\sin t\,\mathbf{i} + \cos t\,\mathbf{j} + e^t\mathbf{k}$$

and

$$\mathbf{A}(t) = -\cos t\,\mathbf{i} - \sin t\,\mathbf{j} + e^t\mathbf{k};$$

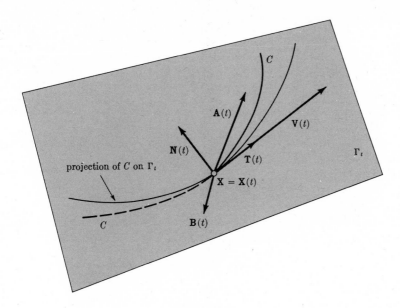

FIGURE 4.4

thus $\mathbf{V}(0) = \mathbf{j} + \mathbf{k}$ and $\mathbf{A}(0) = -\mathbf{i} + \mathbf{k}$. From (9),

$$\mathbf{T}(0) = \frac{\mathbf{V}(0)}{|\mathbf{V}(0)|} = \frac{1}{\sqrt{2}}(\mathbf{j} + \mathbf{k}).$$

Since

$$\mathbf{V}(0) \times (\mathbf{A}(0) \times \mathbf{V}(0)) = (\mathbf{j} + \mathbf{k}) \times [(-\mathbf{i} + \mathbf{k}) \times (\mathbf{j} + \mathbf{k})]$$

$$= (\mathbf{j} + \mathbf{k}) \times (-\mathbf{i} + \mathbf{j} - \mathbf{k})$$

$$= -2\mathbf{i} - \mathbf{j} + \mathbf{k},$$

it follows from (11) that

$$\mathbf{N}(0) = \frac{1}{\sqrt{6}}(-2\mathbf{i} - \mathbf{j} + \mathbf{k}).$$

Now, from (12),

$$\mathbf{B}(0) = \frac{1}{\sqrt{2}}(\mathbf{j} + \mathbf{k}) \times \frac{1}{\sqrt{6}}(-2\mathbf{i} - \mathbf{j} + \mathbf{k})$$

$$= \frac{1}{\sqrt{3}}(\mathbf{i} - \mathbf{j} + \mathbf{k}),$$

and the equation of the osculating plane is

$$[(x - 1)\mathbf{i} + y\mathbf{j} + (z - 1)\mathbf{k}] \cdot [\mathbf{i} - \mathbf{j} + \mathbf{k}] = 0,$$

or

$$x - y + z = 2.$$

Differentiation with Respect to Arc Length

We saw earlier that $|\mathbf{V}(t)|$ is the instantaneous speed of the particle at $\mathbf{X} = \mathbf{X}(t)$; thus the integral

$$\int_{t_0}^{t} |\mathbf{V}(\tau)| \, d\tau \qquad (t \geq t_0)$$

is the distance, or *arc length*, traversed by the particle during the time interval $t_0 \leq t \leq t$. We denote this distance by $s(t) - s(t_0)$; thus

$$s(t) = s(t_0) + \int_{t_0}^{t} |\mathbf{V}(\tau)| \, d\tau.$$

If $|\mathbf{V}(t)|$ is continuous (which we assume henceforth), then

(13) $$s'(t) = |\mathbf{V}(t)|.$$

The derivative of T with respect to arc length is defined by

$$\frac{d\mathbf{T}}{ds}(t) = \lim_{h \to 0} \frac{\mathbf{T}(t + h) - \mathbf{T}(t)}{s(t + h) - s(t)},$$

which can be rewritten as

$$\frac{d\mathbf{T}}{ds}(t) = \lim_{h \to 0} \frac{\dfrac{\mathbf{T}(t + h) - \mathbf{T}(t)}{h}}{\dfrac{s(t + h) - s(t)}{h}}$$

$$= \frac{\mathbf{T}'(t)}{|\mathbf{V}(t)|};$$

thus, from (10) and (11),

$$\frac{d\mathbf{T}}{ds}(t) = \frac{\mathbf{V}(t) \times (\mathbf{A}(t) \times \mathbf{V}(t))}{|\mathbf{V}(t)|^4} = \frac{|\mathbf{V}(t) \times \mathbf{A}(t)|}{|\mathbf{V}(t)|^3} \mathbf{N}(t).$$

Hence $(d\mathbf{T}/ds)(t)$ is a vector in the direction of $\mathbf{N}(t)$. We call its magnitude the *curvature* of C at $\mathbf{X} = \mathbf{X}(t)$ and denote it by $\kappa(t)$:

$$\kappa(t) = \frac{|\mathbf{V}(t) \times \mathbf{A}(t)|}{|\mathbf{V}(t)|^3}.$$

The *radius of curvature* of C at $\mathbf{X} = \mathbf{X}(t)$ is defined by

$$\rho(t) = \frac{1}{\kappa(t)}.$$

Example 4.6 For the helix considered in Example 4.1 the curvature is constant, since

$$(14) \qquad \mathbf{V}(t) \times \mathbf{A}(t) = \begin{vmatrix} \mathbf{i} & \mathbf{j} & \mathbf{k} \\ -a \sin t & a \cos t & b \\ -a \cos t & -a \sin t & 0 \end{vmatrix}$$

$$= ab \sin t\,\mathbf{i} - ab \cos t\,\mathbf{j} + a^2\mathbf{k}$$

and

$$\kappa(t) = \frac{a\sqrt{a^2 + b^2}}{(\sqrt{a^2 + b^2})^3}.$$

$$= \frac{a}{a^2 + b^2}.$$

Thus the radius of curvatures is also constant:

$$(15) \qquad \rho(t) = \frac{a^2 + b^2}{a}.$$

If $b = 0$, the helix reduces to a circle of radius a in the xy-plane; thus it follows from (15) that the radius of curvature of a circle is equal to its radius.

Example 4.7 From (9) and (13),

$$\mathbf{V}(t) = s'(t)\mathbf{T}(t);$$

hence utilizing (11),

$$\mathbf{A}(t) = s''(t)\mathbf{T}(t) + s'(t)\mathbf{T}'(t)$$

$$= s''(t)\mathbf{T}(t) + s'(t)\,|\,\mathbf{T}'(t)\,|\,\mathbf{N}(t).$$

According to (10), and because $\mathbf{V}(t)$ and $\mathbf{A}(t) \times \mathbf{V}(t)$ are perpendicular,

$$|\,\mathbf{T}'(t)\,| = \frac{|\,\mathbf{A}(t) \times \mathbf{V}(t)\,|}{|\,\mathbf{V}(t)\,|^2} = \frac{|\,\mathbf{A}(t) \times \mathbf{V}(t)\,|}{|\,\mathbf{V}(t)\,|^3}\,|\,\mathbf{V}(t)\,|$$

$$= \kappa(t)\,|\,\mathbf{V}(t)\,|$$

$$= \frac{s'(t)}{\rho(t)};$$

thus $A(t)$ can be written as

$$A(t) = A_N(t) + A_T(t),$$

where

$$A_T(t) = s''(t)T(t)$$

and

$$A_N(t) = \frac{(s'(t))^2}{\rho(t)} N(t)$$

are the *tangential* and *normal* components of acceleration, respectively.

The derivative of the unit binormal with respect to arc length is defined by

(16)
$$\frac{dB}{ds}(t) = \lim_{h \to 0} \frac{B(t+h) - B(t)}{s(t+h) - s(t)}$$

$$= \lim_{h \to 0} \frac{\dfrac{B(t+h) - B(t)}{h}}{\dfrac{s(t+h) - s(t)}{h}}$$

$$= \frac{B'(t)}{|V(t)|}.$$

Differentiating (12) with respect to t and using (8) with $F = V \times A$, noting that

$$(V \times A)' = (V \times A') + (V' \times A)$$

$$= (V \times A') + (A \times A) = V \times A',$$

we find that

$$B'(t) = \frac{(V(t) \times A(t)) \times [(V(t) \times A'(t)) \times (V(t) \times A(t))]}{|V(t) \times A(t)|^3}$$

$$= \frac{(V(t) \times A(t)) \times [(A(t) \times V(t)) \cdot A'(t)]V(t)}{|V(t) \times A(t)|^3}$$

$$= \frac{(A(t) \times V(t)) \cdot A'(t)}{|V(t) \times A(t)|^3} [V(t) \times (A(t) \times V(t))].$$

From this and (11) and (16),

$$\frac{dB}{ds}(t) = -\frac{(V(t) \times A(t)) \cdot A'(t)}{|V(t) \times A(t)|^2} N(t).$$

Thus $(d\mathbf{B}/ds)\,(t)$ is parallel to $\mathbf{N}(t)$. The quantity

$$\tau(t) = -\frac{|\,\mathbf{V}(t)\,|\,(\mathbf{V}(t)\,\times\,\mathbf{A}(t))\cdot\mathbf{A}'(t)}{|\,\mathbf{V}(t)\,\times\,\mathbf{A}(t)\,|^2}$$

is called the *torsion of C at* $\mathbf{X} = \mathbf{X}(t)$; it is identically zero if and only if C lies in a plane.

Example 4.8   For the helix of Example 4.1,

$$\mathbf{A}'(t) = a\sin t\,\mathbf{i} - a\cos t\,\mathbf{j}$$

and

$$\mathbf{V}(t)\,\times\,\mathbf{A}(t) = ab\sin t\,\mathbf{i} - ab\cos t\,\mathbf{j} + a^2\mathbf{k}$$

(see (14)). Therefore

$$\tau(t) = \frac{a^2 b}{a^4 + a^2 b^2}$$

$$= \frac{b}{a^2 + b^2}\,.$$

If $b = 0$, then C lies in the xy-plane and $\tau(t) \equiv 0$.

EXERCISES 1.4

In Exercises 1 through 4 find the velocity and acceleration vectors and the cosine of the angle between them. Also, find the instantaneous speed and direction of motion.

1. $\mathbf{X}(t) = t^3\mathbf{i} + t^2\mathbf{j} + \mathbf{k}.$

2. $\mathbf{X}(t) = \cos t\,\mathbf{i} + \sin t\,\mathbf{j} + \mathbf{k}.$

3. $\mathbf{X}(t) = e^t\mathbf{i} + \cos t\,\mathbf{j} + t^3\mathbf{k}.$

4. $\mathbf{X}(t) = (t^2 - t)\mathbf{i} + (t^3 + t^2)\mathbf{j}.$

5. Verify Theorem 4.1 for $\mathbf{F}_1(t) = e^t\mathbf{i} + e^{-t}\mathbf{j}$, $\mathbf{F}_2(t) = \cos t\,\mathbf{j} + \sin t\,\mathbf{k}$, and $\rho(t) = e^t.$

6. Verify Theorem 4.1 for $\mathbf{F}_1(t) = t\mathbf{i} + t^2\mathbf{j} + t^3\mathbf{k}$, $\mathbf{F}_2(t) = t^3\mathbf{i} + t^2\mathbf{j} + t\mathbf{k}$, and $\rho(t) = t.$

In Exercises 7 through 11 compute

$$\frac{d}{dt} \, | \, \mathbf{F}(t) \, | \qquad \text{and} \qquad \frac{d}{dt} \left(\frac{\mathbf{F}(t)}{| \, \mathbf{F}(t) \, |} \right)$$

directly and also by means of Eqs. (7) and (8), respectively.

7. $\mathbf{F}(t) = (t + 1)\mathbf{i} + (t - 1)\mathbf{j} + t^2\mathbf{k}$.

8. $\mathbf{F}(t) = e^t\mathbf{i} + e^{-t}\mathbf{j}$.

9. $\mathbf{F}(t) = e^t \sin t \, \mathbf{i} + e^t \cos t \, \mathbf{k}$.

10. $\mathbf{F}(t) = \mathbf{i} - \mathbf{j} + t\mathbf{k}$.

11. $\mathbf{F}(t) = \sqrt{2}t\mathbf{i} - t\mathbf{j} + \mathbf{k}$.

In Exercises 12 through 15 compute $\mathbf{T}(t_0)$, $\mathbf{N}(t_0)$, and $\mathbf{B}(t_0)$ and find the equation of the osculating plane to the curve at $\mathbf{X}(t_0)$.

12. $\mathbf{X}(t) = 2t\mathbf{i} + t^2\mathbf{j} + t^3\mathbf{k}; \qquad t_0 = 1$.

13. $\mathbf{X}(t) = \sin t \, \mathbf{i} + \cos t \, \mathbf{j} + 2\mathbf{k}; \qquad t_0 = \dfrac{\pi}{3}$.

14. $\mathbf{X}(t) = e^t \sin t \, \mathbf{i} + e^t \cos t \, \mathbf{j} + e^t\mathbf{k}; \qquad t_0 = 0$.

15. $\mathbf{X}(t) = t^2\mathbf{i} + (t + 1)\mathbf{j} + t\mathbf{k}; \qquad t_0 = 0$.

In Exercises 16 through 18 find the arc length of the curve defined by $\mathbf{X} = \mathbf{X}(t)$ $(t_0 \le t \le t_1)$.

16. $\mathbf{X}(t) = \cos 3t \, \mathbf{i} + \sin 3t \, \mathbf{j} + 4t\mathbf{k}; \qquad 0 \le t \le 2\pi$.

17. $\mathbf{X}(t) = t\mathbf{i} + \dfrac{t^2}{2}\mathbf{j} + \dfrac{2\sqrt{2}}{3} t^{3/2}\mathbf{k}; \qquad 1 \le t \le 2$.

18. $\mathbf{X}(t) = e^t\mathbf{i} + \sqrt{2}t\mathbf{j} - e^{-t}\mathbf{k}; \qquad 0 \le t \le 1$.

In Exercises 19 through 22 compute the curvature, radius of curvature, and torsion of the curve defined by $\mathbf{X} = \mathbf{X}(t)$ at the point $\mathbf{X}(t_0)$. Also find the tangential and normal components of acceleration at $t = t_0$.

19. $\mathbf{X}(t) = \cos t \, \mathbf{i} + \sin t \, \mathbf{j} + t\mathbf{k}; \qquad t_0 = \dfrac{\pi}{3}$.

20. $\mathbf{X}(t) = e^t \sin t \, \mathbf{i} + e^{-t} \cos t \, \mathbf{j} + e^{-t}\mathbf{k}; \qquad t_0 = 0$.

21. $\mathbf{X}(t) = t\mathbf{i} + t^2\mathbf{j} + (t^2 + 1)\mathbf{k}; \qquad t_0 = 0$.

22. $\mathbf{X}(t) = \sqrt{t}\mathbf{i} + t^{3/2}\mathbf{j} + t^2\mathbf{k}; \qquad t_0 = 1.$

23. A particle moves around a circle of radius r with constant speed v_0. Find its tangential and normal components of acceleration.

THEORETICAL EXERCISES

T-1. Prove (a) and (d) of Theorem 4.1.

T-2. Derive Eq. (12).

T-3. Show that the acceleration of a particle moving along a curve with constant speed is in the direction of the normal.

T-4. Suppose a particle moves so that its acceleration is always perpendicular to its velocity. Show that its speed is constant.

T-5. Suppose a particle moves along a curve C given by $\mathbf{X} = \mathbf{X}(t)$. Show that

$$\kappa(t) = \frac{\mid x'(t)y''(t) - y'(t)x''(t) \mid}{[(x'(t))^2 + (y'(t))^2]^{3/2}}.$$

T-6. Use the result of Exercise T-5 to show that the curvature of a curve defined by $y = f(x)$ is

$$\kappa(x) = \frac{\mid f''(x) \mid}{(1 + (f'(x))^2)^{3/2}}$$

at the point $(x, f(x))$.

T-7. *Prove:* If a particle moves so that $\mathbf{A}(t) \times \mathbf{V}(t)$ is constant, then the motion is in a plane.

1.5 Parametrically Defined Curves

In this section we generalize some of the ideas of Section 1.4. We consider *parametric functions*; that is, vector valued functions of a single real variable. A parametric function $\boldsymbol{\Phi}$ is a vector valued function of the form

$$\boldsymbol{\Phi}(t) = \begin{bmatrix} \Phi_1(t) \\ \vdots \\ \Phi_n(t) \end{bmatrix}$$

where $\Phi_1, \ldots, \Phi_n$ are continuous real valued functions defined on a closed interval $[a, b]$. As t varies over $[a, b]$, $\Phi(t)$ traverses a curve C in R^n. When we wish to specify $[a, b]$ and the dimension of the space containing C, we shall write

$$\Phi : [a, b] \to R^n.$$

Example 5.1   Let $\Phi : [0, \pi] \to R^2$ be defined by

$$\Phi(t) = \begin{bmatrix} \cos t \\ \sin t \end{bmatrix} \qquad (0 \leq t \leq \pi).$$

Since $\cos^2 t + \sin^2 t = 1$, $-1 \leq \cos t \leq 1$, and $0 \leq \sin t \leq 1$ for $0 \leq t \leq \pi$, it follows that C is the upper half of the unit circle (Fig. 5.1).

FIGURE 5.1

The parametric functions

$$\Psi(s) = \begin{bmatrix} -\sin 2s \\ \cos 2s \end{bmatrix} \qquad \left(-\frac{\pi}{4} \leq s \leq \frac{\pi}{4} \right)$$

and

$$\eta(r) = \begin{bmatrix} \sin r \\ \cos r \end{bmatrix} \qquad \left(-\frac{\pi}{2} \leq r \leq \frac{\pi}{2} \right)$$

define the same curve as Φ. As s and t increase over their respective intervals, $\Phi(t)$ and $\Psi(s)$ traverse the semicircle from $(1, 0)$ to $(-1, 0)$; however, as r varies from $-\frac{\pi}{2}$ to $\frac{\pi}{2}$, $\eta(r)$ moves in the opposite direction.

In connection with the line integral (Section 4.5), the manner in which $X = \Phi(t)$ traverses the curve C is important. Two parametric functions are *equivalent* if they not only define the same curve C, but prescribe the same order for traversing its points. Thus in Example 5.1, Φ and Ψ are equivalent, but Φ and η are not.

Definition 5.1 Let C be defined by $\mathbf{X} = \mathbf{\Phi}(t)$ $(a \le t \le b)$. Then $\mathbf{\Phi}(a)$ and $\mathbf{\Phi}(b)$ are respectively the *initial* and *final endpoints* of C; if $\mathbf{\Phi}(a) = \mathbf{\Phi}(b)$ then C is a *closed curve*. If $\mathbf{\Phi}(t_1)$ and $\mathbf{\Phi}(t_2)$ are distinct whenever $a \le t_1 < t_2 < b$ or $a < t_1 < t_2 \le b$, then C is *simple*. A *simple closed curve* is both closed and simple.

Example 5.2 The curve defined in Example 5.1 is simple. The curve defined by

$$\mathbf{X}(t) = \begin{bmatrix} \cos t \\ \sin t \end{bmatrix} \qquad (0 \le t \le 2\pi)$$

is the unit circle, traversed in the counterclockwise direction; it is a simple closed curve (Fig. 5.2).

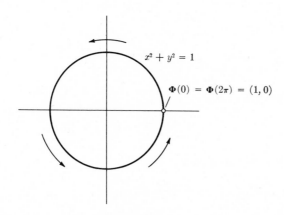

FIGURE 5.2

Example 5.3 The curve defined by

$$\mathbf{X}(t) = \begin{bmatrix} \cos t \\ \sin t \end{bmatrix} \qquad (0 \le t \le 4\pi)$$

is closed but not simple, and the curve defined by

$$\mathbf{X}(t) = \begin{bmatrix} \cos t \\ \sin t \end{bmatrix} \qquad (0 \le t \le 3\pi)$$

is neither closed nor simple. The curve shown in Fig. 5.3 is closed but not simple.

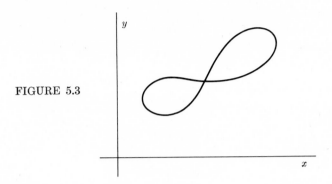

FIGURE 5.3

Definition 5.2 The *derivative* of a parametric function $\boldsymbol{\Phi}:[a,\ b] \rightarrow R^n$ is defined by

$$\boldsymbol{\Phi}' = \begin{bmatrix} \Phi_1' \\ \Phi_2' \\ \vdots \\ \Phi_n' \end{bmatrix},$$

whenever the vector on the right exists.

By $\Phi_i'(a)$ we mean a right hand derivative; thus

$$\Phi_i'(a) = \lim_{t \to a^+} \frac{\Phi_i(t) - \Phi_i(a)}{t - a}$$

where $\lim\limits_{t \to a^+}$ means that t approaches a from above. Similarly $\Phi_i'(b)$ is a left hand derivative:

$$\Phi_1'(b) = \lim_{t \to b^-} \frac{\Phi_i(t) - \Phi_i(b)}{t - b}.$$

Definition 5.3 The curve C defined by $\mathbf{X} = \boldsymbol{\Phi}(t)$ $(a \leq t \leq b)$ is *continuously differentiable* if Φ_i' exists and is continuous in $[a,\ b]$ for $i = 1$, $2, \ldots, n$; if in addition $\boldsymbol{\Phi}'$ does not vanish in $[a,\ b]$, then C is *smooth*. The *unit tangent vector* to C at a point $\mathbf{X}_0 = \boldsymbol{\Phi}(t_0)$ where $\boldsymbol{\Phi}'(t_0) \neq \mathbf{0}$ is defined by

$$\mathbf{T}(t_0) = \frac{\boldsymbol{\Phi}'(t_0)}{|\boldsymbol{\Phi}'(t_0)|};$$

thus a smooth curve has a continuously turning unit tangent vector.

This definition of the unit tangent is analogous to that given for $n = 3$ in Section 1.4.

Example 5.4 Let C be defined by

$$\mathbf{X} = \boldsymbol{\Phi}(t) = \begin{bmatrix} \cos t \\ \sin t \\ e^t \\ e^{-t} \end{bmatrix} \qquad (-1 < t < 1).$$

Then

$$\boldsymbol{\Phi}'(t) = \begin{bmatrix} -\sin t \\ \cos t \\ e^t \\ -e^{-t} \end{bmatrix}$$

and

$$| \boldsymbol{\Phi}'(t) | = \sqrt{(-\sin t)^2 + (\cos t)^2 + (e^t)^2 + (-e^{-t})^2}$$

$$= \sqrt{1 + 2 \cosh 2t} \neq 0;$$

thus $\boldsymbol{\Phi}'$ is continuous and nonvanishing in $[-1, 1]$ and therefore C is smooth. The unit tangent vector to C at $\mathbf{X} = \boldsymbol{\Phi}(t)$ is

$$\mathbf{T}(t) = \frac{1}{\sqrt{1 + 2 \cosh 2t}} \begin{bmatrix} -\sin t \\ \cos t \\ e^t \\ -e^{-t} \end{bmatrix}.$$

The requirement that a curve be smooth is quite stringent; many common and useful curves do not satisfy it, but are made up of smooth sections joined together.

Definition 5.4 A curve C defined by $\mathbf{X} = \boldsymbol{\Phi}(t)$ $(a \leq t \leq b)$ is *piecewise smooth* if there exist points $a = t_0 < t_1 < \cdots < t_n = b$ such that the curves C_i defined by $\mathbf{X} = \boldsymbol{\Phi}(t)$ $(t_{i-1} \leq t \leq t_i)$ are smooth $(i = 1, \ldots, n)$.

Example 5.5 Let C consist of the line segment from $(0, 0)$ to $(1, 1)$, followed by the line segment from $(1, 1)$ to $(1, 0)$ (Fig. 5.4). C is defined

by

(1) $$\mathbf{X}(t) = \begin{bmatrix} t \\ t \end{bmatrix} \qquad (0 \le t \le 1),$$

and

(2) $$\mathbf{X}(t) = \begin{bmatrix} 1 \\ 2 - t \end{bmatrix} \qquad (1 \le t \le 2).$$

Since C has no tangent at $(1, 1)$ it is not smooth. However C_1 and C_2, defined by (1) and (2), respectively, are smooth; hence C is piecewise smooth.

FIGURE 5.4

The following theorem, which we state without proof, gives a method for constructing parametric functions equivalent to a given one.

Theorem 5.1   If $\mathbf{\Psi}:[c, d] \to R^n$ defines a curve C and α is a strictly increasing continuous function which maps $[a, b]$ onto $[c, d]$, then

$$\mathbf{X} = \mathbf{\Phi}(t) = \mathbf{\Psi}(\alpha(t)) \qquad (a \le t \le b)$$

is an equivalent parametric representation of C. We say that $\mathbf{\Phi}$ is obtained from $\mathbf{\Psi}$ by the *change of parameter* $s = \alpha(t)$.

Example 5.6   Consider the change of parameter

$$s = \alpha(t) = \frac{t - a}{b - a} d + \frac{t - b}{a - b} c,$$

which maps $[a, b]$ onto $[c, d]$. If $\mathbf{\Psi}:[c, d] \to R^n$ defines a curve C then

$$\mathbf{X} = \mathbf{\Psi}(\alpha(t)) = \mathbf{\Psi}\left(\frac{t - a}{b - a} d + \frac{t - b}{a - b} c\right) \qquad (a \le t \le b)$$

is an equivalent representation of C. Setting $a = 0$ and $b = 1$ we find that any curve has a representation with parameter interval $[0, 1]$.

Example 5.7 In Example 5.1, Φ can be obtained from Ψ by the change of parameter $s = \alpha(t) = (t/2) - (\pi/4)$, since

$$
\Psi\left(\frac{t}{2} - \frac{\pi}{4}\right) =
\begin{bmatrix}
-\sin\left(t - \dfrac{\pi}{2}\right) \\
\\
\cos\left(t - \dfrac{\pi}{2}\right)
\end{bmatrix}
$$

$$
=
\begin{bmatrix}
\cos t \\
\sin t
\end{bmatrix}
= \Phi(t)
$$

and α is strictly increasing and maps $[0, \pi]$ onto $\left[-\dfrac{\pi}{4}, \dfrac{\pi}{4}\right]$.

Definition 5.5 The change of parameter $s = \alpha(t)$ $(a \leq t \leq b)$ is *smooth* if α is continuously differentiable and $\alpha'(t) > 0$ in $[a, b]$; it is *piecewise smooth* if there exist points $a = t_0 < t_1 < \cdots < t_n = b$ such that $s = \alpha(t)$ is smooth in $[t_{j-1}, t_j]$ $(j = 1, \ldots, n)$.

Example 5.8 The linear change of parameter discussed in Example 5.7 is smooth; so are

$$
s = \alpha(t) = \sin t \qquad \left(-\frac{\pi}{4} < t < \frac{\pi}{4}\right)
$$

and

$$
s = \alpha(t) = e^t \qquad (a \leq t \leq b; \ a, b \text{ arbitrary}).
$$

However,

$$
s = \alpha(t) = t^3 \qquad (-1 \leq t \leq 1)
$$

is not smooth, because $\alpha'(0) = 0$;

$$
s = \alpha(t) = \sin t \qquad \left(0 \leq t \leq \frac{\pi}{2}\right)
$$

is not smooth, because $\alpha'(\pi/2) = 0$. Finally

$$
s = \alpha(t) =
\begin{cases}
t & (0 \leq t \leq \frac{1}{2}), \\
2t - \frac{1}{2} & (\frac{1}{2} \leq t \leq 1)
\end{cases}
$$

is not smooth, because $\alpha'(\frac{1}{2})$ does not exist (Fig. 5.5); however, it is piecewise smooth.

FIGURE 5.5

If a parametric function Φ defines a smooth curve C, then we shall also say that Φ is smooth; similarly if C is piecewise smooth, we shall say that Φ is piecewise smooth. We then have the following theorem.

Theorem 5.2 If $\Psi:[c, d] \to R^n$ is a smooth parametric function and $s = \alpha(t)$ is a smooth change of parameter which maps $[a, b]$ onto $[c, d]$ then

$$(3) \qquad\qquad \Phi(t) = \Psi(\alpha(t)) \qquad (a \le t \le b)$$

is also smooth, and equivalent to Ψ. Conversely, if Φ and Ψ are equivalent smooth representations of a curve C then there is a smooth change of parameter $s = \alpha(t)$ which satisfies (3). These statements remain true if "smooth" is replaced by "piecewise smooth."

The proof of the first statement leans heavily on the relationship

$$(4) \qquad\qquad \Phi'(t) = \Psi'(\alpha(t))\alpha'(t),$$

which follows from the chain rule (Exercise T-1), and holds wherever the indicated derivatives exist.

The next definition generalizes the definition of arc length given in Section 1.4.

Definition 5.6 Let C be defined by a smooth parametric function $\mathbf{X} = \Phi(t)$ $(a \le t \le b)$. Then the *arc length* of C is the integral

$$L(C) = \int_a^b |\, \Phi'(t) \,|\, dt$$

$$= \int_a^b \sqrt{(\Phi_1'(t))^2 + \cdots + (\Phi_n'(t))^2} \, dt.$$

Example 5.9   If C is a curve in R^2 defined by

$$\mathbf{X}(x) = \begin{bmatrix} x \\ f(x) \end{bmatrix} \quad (a \le x \le b),$$

then

$$L(C) = \int_a^b \sqrt{1 + (f'(x))^2}\, dx,$$

which should be familiar to the student.

Example 5.10   If C is

$$\mathbf{X}(t) = \begin{bmatrix} \cos t \\ \sin t \end{bmatrix} \quad (0 \le t \le 2\pi),$$

the unit circle traversed counterclockwise, then

$$L(C) = \int_0^{2\pi} \sqrt{(-\sin t)^2 + (\cos t)^2}\, dt$$

$$= \int_0^{2\pi} dt = 2\pi,$$

again a familiar result.

Example 5.11   Let C be the line segment in R^n from $\mathbf{X}_1 = (a_1, \ldots, a_n)$ to $\mathbf{X}_2 = (b_1, \ldots, b_n)$. It can be represented by

$$\mathbf{X} = \mathbf{\Phi}(t) = \begin{bmatrix} a_1 + t(b_1 - a_1) \\ \vdots \\ a_n + t(b_n - a_n) \end{bmatrix} \quad (0 \le t \le 1);$$

hence

$$\mathbf{\Phi}'(t) = \begin{bmatrix} b_1 - a_1 \\ \vdots \\ b_n - a_n \end{bmatrix}$$

and

$$L(C) = \int_0^1 \sqrt{(b_1 - a_1)^2 + \cdots + (b_n - a_n)^2}\, dt$$

$$= \sqrt{(b_1 - a_1)^2 + \cdots + (b_n - a_n)^2} \int_0^1 dt$$

$$= \sqrt{(b_1 - a_1)^2 + \cdots + (b_n - a_n)^2} = |\,\mathbf{X}_2 - \mathbf{X}_1\,|,$$

which agrees with our earlier definition of length of a line segment (Section 1.1).

The following theorem shows that the arc length of a curve C does not depend upon the particular parametric representation chosen to represent C.

Theorem 5.3 If $\mathbf{X} = \mathbf{\Phi}(t)$ $(a \leq t \leq b)$ and $\mathbf{X} = \mathbf{\Psi}(s)$ $(c \leq s \leq d)$ are equivalent piecewise smooth representations of the same curve C, then

$$(5) \qquad \int_a^b | \, \mathbf{\Phi}'(t) \, | \, dt = \int_c^d | \, \mathbf{\Psi}'(s) \, | \, ds;$$

that is, $L(C)$ may be computed by means of any piecewise smooth representation of C.

Proof. Let $s = \alpha(t)$ be the change of variable relating $\mathbf{\Phi}$ and $\mathbf{\Psi}$ according to (3). The rule for change of variable of integration implies that

$$\int_c^d | \, \mathbf{\Psi}'(s) \, | \, ds = \int_a^b | \, \mathbf{\Psi}'(\alpha(t)) \, | \, \alpha'(t) \, dt$$

$$= \int_a^b | \, \mathbf{\Psi}'(\alpha(t)) \alpha'(t) \, | \, dt$$

since $| \, \alpha'(t) \, | = \alpha'(t) > 0$. Now (4) yields (5).

EXERCISES 1.5

In Exercises 1, 2, and 3 show that $\mathbf{X} = \mathbf{\Phi}(t)$ and $\mathbf{X} = \mathbf{\Psi}(s)$ define the same curve.

1. $\mathbf{\Phi}(t) = \begin{bmatrix} t \\ t^2 \\ t^3 \end{bmatrix}$ $(1 \leq t \leq 2)$; $\mathbf{\Psi}(s) = \begin{bmatrix} s^2 \\ s^4 \\ s^6 \end{bmatrix}$ $(1 \leq s \leq \sqrt{2})$.

2. $\mathbf{\Phi}(t) = \begin{bmatrix} e^t \\ e^{-t} \\ 2 \cosh t \end{bmatrix}$ $(0 \leq t \leq 1)$; $\mathbf{\Psi}(s) = \begin{bmatrix} s \\ \dfrac{1}{s} \\ s + \dfrac{1}{s} \end{bmatrix}$ $(1 \leq s \leq e)$.

Describe the curves in Exercises 3 through 7 as closed, simple, both, or neither.

3.

4. $\mathbf{X} = \mathbf{\Phi}(t) = \begin{bmatrix} \cos t \\ \sin t \\ -\sin t \\ -\cos t \end{bmatrix}$ $(0 \le t \le \pi)$.

5. $\mathbf{X} = \mathbf{\Phi}(t) = \begin{bmatrix} \cos t \\ \sin t \\ -\sin t \\ -\cos t \end{bmatrix}$ $(0 \le t \le 2\pi)$.

6.

7. $\mathbf{X} = \mathbf{\Phi}(t) = \begin{bmatrix} \cos t \\ \sin t \\ -\sin t \\ -\cos t \end{bmatrix}$ $(0 \le t \le 3\pi)$.

8. Show that the curve defined by

$$\mathbf{X} = \mathbf{\Phi}(t) = \begin{bmatrix} e^t \cos t \\ e^t \sin t \\ \sqrt{2}\, e^t \end{bmatrix} \quad (0 \le t \le \pi)$$

is continuously differentiable.

9. Show that the curve defined by

$$\mathbf{X} = \mathbf{\Phi}(t) = \begin{bmatrix} \dfrac{t^2}{2} \\[2mm] \dfrac{\sqrt{2}}{3} t^3 \\[2mm] \dfrac{t^4}{4} \end{bmatrix} \qquad (0 \le t \le 1)$$

is continuously differentiable.

10. Show that the curve C defined by

$$\mathbf{X} = \mathbf{\Phi}(t) = \begin{bmatrix} e^t \cos t \\[1mm] e^{-t} \sin t \\[1mm] e^t \\[1mm] e^{-t} \end{bmatrix} \qquad (-\pi \le t \le \pi)$$

is smooth and find the tangent vector to C at $\mathbf{X}_0 = \mathbf{\Phi}(0)$.

11. Show that the curve C defined by

$$\mathbf{X} = \mathbf{\Phi}(t) = \begin{bmatrix} t^2 + 1 \\[1mm] t^3 + t \\[1mm] t^4 \end{bmatrix} \qquad (0 \le t \le 2)$$

is smooth and find the tangent vector to C at $\mathbf{X}_0 = \mathbf{\Phi}(1)$.

12. Show that the curve defined by

$$\mathbf{X} = \mathbf{\Phi}(t) = \begin{bmatrix} |t| \\[1mm] |t - 1| \\[1mm] t^2 \end{bmatrix} \qquad (-2 \le t \le 2)$$

is piecewise smooth, but not smooth.

13. Show that the curve defined by

$$\mathbf{X} = \mathbf{\Phi}(t) = \begin{bmatrix} e^t \\ \sin t \\ |\cos t | \end{bmatrix} \qquad (-\pi \le t \le 3\pi)$$

is piecewise smooth, but not smooth.

In Exercises 14, 15, and 16 find two parametric functions equivalent to $\mathbf{\Phi}$.

14. $\mathbf{\Phi}(t) = \begin{bmatrix} t \\ t^2 \\ t^3 \\ t^4 \end{bmatrix} \qquad (0 \le t \le 1).$

15. $\mathbf{\Phi}(t) = \begin{bmatrix} -\sin t \\ \cos t \\ \sin t \end{bmatrix} \qquad \left(0 \le t \le \dfrac{\pi}{2} \right).$

16. $\mathbf{\Phi}(t) = \begin{bmatrix} e^t \sin t \\ e^t \cos t \\ e^{-t} \cos t \\ e^{-t} \sin t \end{bmatrix} \qquad (0 \le t \le \pi).$

17. Find a smooth change of parameter mapping $[-2, -1]$ onto $[1, 2]$. (See Example 5.6).

18. Find a parametric representation for the circle $x^2 + y^2 = 1$ traversed once in the counterclockwise direction, starting from $\left(\dfrac{1}{\sqrt{2}}, \dfrac{1}{\sqrt{2}} \right)$.

19. Find a parametric equation for the ellipse $(x^2/4) + (y^2/9) = 1$ traversed in the clockwise direction, starting from $(-2, 0)$.

20. Let C consist of the portion of the parabola $y = x^2$ from $(0, 0)$ to $(1, 1)$, followed by the line segment from $(1, 1)$ to $(1, 0)$, followed

by the line segment from $(1, 0)$ to $(0, 0)$. Show that C is piecewise smooth and compute its arc length.

21. Compute the arc length of the curve defined in Exercise 8.

22. Compute the arc length of the curve defined in Exercise 9.

23. Let C be the curve defined in Example 5.1. Verify Theorem 5.3 by computing the arc length of C using the equivalent parametric functions Φ and Ψ defined in Examples 5.1 and 5.7.

24. (a) Show that the function

$$\Psi(s) = \begin{bmatrix} \dfrac{s}{2} \\[2mm] \dfrac{\sqrt{2}}{3} s^{3/2} \\[2mm] \dfrac{s^2}{4} \end{bmatrix} \qquad (0 \le s \le 1)$$

is equivalent to the function Φ defined in Exercise 9.

(b) Verify Theorem 5.3 by computing the arc length of C using the representation $X = \Psi(s)$, and comparing the result with that of Exercise 9.

THEORETICAL EXERCISES

T-1. *Prove:* If Ψ and α (as defined in Theorem 5.2) are smooth, then $\Phi = \Psi \circ \alpha$ is smooth. (*Hint:* Use the chain rule.)

T-2. *Prove:* If $\Phi : [a, b] \to R^n$ and $\Psi : [c, d] \to R^n$, represent same smooth curve, then they are related by a smooth change of parameter. (*Hint:* Since Φ and Ψ are equivalent,

$$\Phi(t) = \Psi(\alpha(t)) \qquad (a \le t \le b)$$

where α is strictly increasing and continuous in $[a, b]$. By considering difference quotients, show that

$$\alpha'(t_0) = \frac{|\Phi'(t_0)|}{|\Psi'(\alpha(t_0))|}$$

for each t_0 in $[a, b]$.

1.6   Coordinate Systems in R^3

We have considered R^3 simply as a set of ordered triples on which certain operations such as addition and scalar multiplication are defined. However, we also have a physical conception of a three-dimensional universe which exists on its own, independent of our mathematical constructions. We can combine these two ideas by regarding the triples (x, y, z) as being assigned as addresses to points which already exist on their own. Thus, if P_0 is an arbitrary point in space we choose three mutually perpendicular directions through P_0 and associate the triple (x, y, z) with the point P that we reach by walking x units along the first direction, then y units along the second, and then z units along the third (Fig. 6.1).

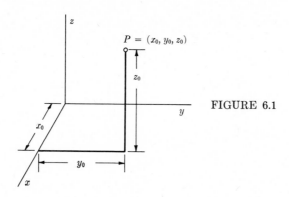

FIGURE 6.1

Thus we do not have to regard the point as being the same as the triple (x, y, z), although we have previously found it convenient to do so, and will again. Rather, we can think of the triple simply as the address of the point, in accordance with a particular kind of addressing method, called a *cartesian*, or *rectangular coordinate* system.

There are other coordinate systems for three-dimensional space. We shall consider two of the most important ones.

Cylindrical Coordinates

Let a cartesian coordinate system have its origin at a point P_0, and suppose P is a point with coordinates (x, y, z) with respect to this system. If ρ and θ satisfy

$$x = \rho \cos \theta$$

(1)

$$y = \rho \sin \theta,$$

then ρ, θ, and z are said to be *cylindrical coordinates* of P. There are infinitely many values of θ which satisfy (1), but we can specify θ uniquely, if $(x, y) \neq (0, 0)$, by stipulating that $\rho > 0$ and $0 \leq \theta < 2\pi$. Then (1) can be solved to obtain θ and ρ:

$$\rho = \sqrt{x^2 + y^2},$$

$$\theta = \cos^{-1}\frac{x}{\rho} = \sin^{-1}\frac{y}{\rho}$$

$$= \tan^{-1}\frac{y}{x} \qquad (0 \leq \theta < 2\pi).$$

If $x^2 + y^2 = 0$, then θ is arbitrary.

The student should recognize that ρ and θ are simply the polar coordinates of the projection P' of P on the xy-plane (Fig. 6.2).

FIGURE 6.2

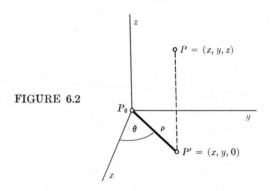

Example 6.1 If P has cylindrical coordinates $(\rho, \theta, z) = (3, \pi/6, -2)$ then

$$x = 3 \cos \frac{\pi}{6} = 3 \frac{\sqrt{3}}{2},$$

$$y = 3 \sin \frac{\pi}{6} = \frac{3}{2},$$

$$z = -2;$$

hence the position vector of P with respect to P_0 is

$$\overrightarrow{P_0P} = \frac{3\sqrt{3}}{2}\mathbf{i} + \frac{3}{2}\mathbf{j} - 2\mathbf{k}.$$

Example 6.2 If P has rectangular coordinates $(2, 2\sqrt{3}, -3)$ then

$$\rho = \sqrt{2^2 + (2\sqrt{3})^2 + (-3)^2} = 5,$$

(2) $$\qquad \theta = \tan^{-1}\sqrt{3},$$

$$z = -3;$$

hence its cylindrical coordinates are

$$(\rho, \theta, z) = \left(5, \frac{\pi}{3}, -3\right).$$

Notice that we have used more than (2) to determine θ: x and y are both positive, which implies that P is in the first quadrant. Without this information we could not decide whether $\theta = \pi/3$ or $\theta = 4\pi/3$.

Spherical Coordinates

Again let a rectangular coordinate system have its origin at P_0, and let P be an arbitrary point. Then the *spherical coordinates* of P are r, θ and ϕ, where r is the distance from P_0 to P, θ is the angle from the positive x-axis to the vector from P_0 to P', the projection of P on the xy-plane, and ϕ is the angle between the positive z-axis and the vector from P_0 to P. From Fig. 6.3,

$$x = r \sin \phi \cos \theta,$$

(3) $$\qquad y = r \sin \phi \sin \theta,$$

$$z = r \cos \phi.$$

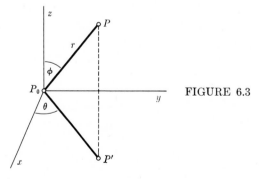

FIGURE 6.3

If $x^2 + y^2 \neq 0$, there is a unique triple (r, θ, ϕ) which satisfies (3) and the inequalities $r > 0$, $0 \le \phi \le \pi$, and $0 \le \theta < 2\pi$:

$$r = \sqrt{x^2 + y^2 + z^2},$$

$$\theta = \cos^{-1} \frac{x}{\sqrt{x^2 + y^2}} = \sin^{-1} \frac{y}{\sqrt{x^2 + y^2}}$$

$$= \tan^{-1} \frac{y}{x} \qquad (0 \le \theta < 2\pi),$$

$$\phi = \cos^{-1} \frac{z}{\sqrt{x^2 + y^2 + z^2}} \qquad (0 \le \phi \le \pi).$$

If $x^2 + y^2 = 0$, then θ can be chosen arbitrarily; if $x^2 + y^2 + z^2 = 0$, then both θ and ϕ can be chosen arbitrarily.

Example 6.3   If P has spherical coordinates $(r, \theta, \phi) = (3, \pi/3, \pi/4)$ then

$$x = 3 \sin \frac{\pi}{4} \cos \frac{\pi}{3}$$

$$= 3 \left(\frac{\sqrt{2}}{2}\right)\left(\frac{1}{2}\right) = \frac{3\sqrt{2}}{4},$$

$$y = 3 \sin \frac{\pi}{4} \sin \frac{\pi}{3}$$

$$= 3 \left(\frac{\sqrt{2}}{2}\right)\left(\frac{\sqrt{3}}{2}\right) = \frac{3\sqrt{6}}{4},$$

$$z = 3 \cos \frac{\pi}{4} = \frac{3\sqrt{2}}{2};$$

hence the position vector of P with respect to P_0 is

$$\overrightarrow{P_0 P} = \frac{3}{\sqrt{2}} \left(\frac{1}{2} \mathbf{i} + \frac{\sqrt{3}}{2} \mathbf{j} + \mathbf{k}\right).$$

Example 6.4 If P has rectangular coordinates $(-9/2,\ -3\sqrt{3}/2,\ 3)$ then

$$r = \sqrt{\left(-\frac{9}{2}\right)^2 + \left(-\frac{3\sqrt{3}}{2}\right)^2 + 3^2} = 6,$$

$$\theta = \tan^{-1}\left(\frac{-\dfrac{3\sqrt{3}}{2}}{-\dfrac{9}{2}}\right) = \tan^{-1}\frac{1}{\sqrt{3}} = \frac{7\pi}{6},$$

$$\phi = \cos^{-1}\frac{3}{6} = \frac{\pi}{3}.$$

In determining θ we have used the fact that x and y are both negative, which implies that θ is in the third quadrant.

Translation and Rotation of Rectangular Coordinate Systems

Suppose two rectangular coordinate systems have origins at P_0 and P_0' and let $(\mathbf{i},\ \mathbf{j},\ \mathbf{k})$ and $(\mathbf{i}',\ \mathbf{j}',\ \mathbf{k}')$ be unit vectors along their coordinate axes. Let P be a point with coordinates $(x,\ y,\ z)$ and $(x',\ y',\ z')$ with respect to the two systems; thus

(4) $$\overrightarrow{P_0P} = x\mathbf{i} + y\mathbf{j} + z\mathbf{k}$$

and

(5) $$\overrightarrow{P_0'P} = x'\mathbf{i}' + y'\mathbf{j}' + z'\mathbf{k}'.$$

We wish to find the relationship between $(x,\ y,\ z)$ and $(x',\ y',\ z')$. From Fig. 6.4,

(6) $$\overrightarrow{P_0'P} = -\overrightarrow{P_0P_0'} + \overrightarrow{P_0P}.$$

Suppose P_0' has coordinates $(x_0,\ y_0,\ z_0)$ with respect to the $(x,\ y,\ z)$ system; thus

$$\overrightarrow{P_0P_0'} = x_0\mathbf{i} + y_0\mathbf{j} + z_0\mathbf{k}.$$

Then from (4), (5) and (6)

(7) $$x'\mathbf{i}' + y'\mathbf{j}' + z'\mathbf{k}' = (x - x_0)\mathbf{i} + (y - y_0)\mathbf{j} + (z - z_0)\mathbf{k}.$$

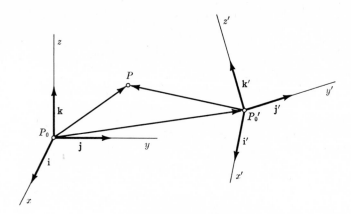

FIGURE 6.4

Since $\mathbf{i}'$, $\mathbf{j}'$, and $\mathbf{k}'$ form a basis for R^3, we can write

$$\mathbf{i} = l_1\mathbf{i}' + m_1\mathbf{j}' + n_1\mathbf{k}',$$

(8)
$$\mathbf{j} = l_2\mathbf{i}' + m_2\mathbf{j}' + n_2\mathbf{k}',$$

$$\mathbf{k} = l_3\mathbf{i}' + m_3\mathbf{j}' + n_3\mathbf{k}'.$$

Substituting this in (7) and collecting coefficients of $\mathbf{i}'$, $\mathbf{j}'$, and $\mathbf{k}'$ yields

$$x'\mathbf{i}' + y'\mathbf{j}' + z'\mathbf{k}' = [l_1(x - x_0) + l_2(y - y_0) + l_3(z - z_0)]\mathbf{i}'$$
$$+ [m_1(x - x_0) + m_2(y - y_0) + m_3(z - z_0)]\mathbf{j}'$$
$$+ [n_1(x - x_0) + n_2(y - y_0) + n_3(z - z_0)]\mathbf{k}'.$$

Thus, from the linear independence of $\mathbf{i}'$, $\mathbf{j}$, and $\mathbf{k}'$,

$$x' = l_1(x - x_0) + l_2(y - y_0) + l_3(z - z_0),$$
$$y' = m_1(x - x_0) + m_2(y - y_0) + m_3(z - z_0),$$
$$z' = n_1(x - x_0) + n_2(y - y_0) + n_3(z - z_0).$$

This can be rewritten in vector–matrix form as

(9)
$$\begin{bmatrix} x' \\ y' \\ z' \end{bmatrix} = \mathbf{M} \begin{bmatrix} x - x_0 \\ y - y_0 \\ z - z_0 \end{bmatrix},$$

where

$$\mathbf{M} = \begin{bmatrix} l_1 & l_2 & l_3 \\ m_1 & m_2 & m_3 \\ n_1 & n_2 & n_3 \end{bmatrix}.$$

Since $\mathbf{i}'$, $\mathbf{j}'$, and $\mathbf{k}'$ are orthonormal it follows from (8) that

$$l_r l_s + m_r m_s + n_r n_s = \begin{cases} 0 & \text{if } r \neq s; \\ 1 & \text{if } r = s. \end{cases}$$

Using this and the definition of matrix multiplication it is easy to verify that

(10) $$\mathbf{M}^{\mathrm{T}}\mathbf{M} = \mathbf{I},$$

where

$$\mathbf{I} = \begin{bmatrix} 1 & 0 & 0 \\ 0 & 1 & 0 \\ 0 & 0 & 1 \end{bmatrix}$$

and $\mathbf{M}^{\mathrm{T}}$ is the transpose of $\mathbf{M}$:

$$\mathbf{M}^{\mathrm{T}} = \begin{bmatrix} l_1 & m_1 & n_1 \\ l_2 & m_2 & n_2 \\ l_3 & m_3 & n_3 \end{bmatrix}.$$

A matrix having property (10) is said to be an *orthogonal* matrix; thus a matrix is orthogonal if and only if its transpose and inverse are identical (Exercise T-2).

We can visualize the change of coordinates (9) as occurring in two steps. Let

(11) $$\begin{bmatrix} \bar{x} \\ \bar{y} \\ \bar{z} \end{bmatrix} = \begin{bmatrix} x - x_0 \\ y - y_0 \\ z - z_0 \end{bmatrix};$$

then

$$(12) \qquad \begin{bmatrix} x' \\ y' \\ z' \end{bmatrix} = \mathbf{M} \begin{bmatrix} \bar{x} \\ \bar{y} \\ \bar{z} \end{bmatrix}.$$

Clearly $(\bar{x}, \bar{y}, \bar{z})$ are the coordinates of P with respect to a coordinate system with origin at P_0' and axes parallel to those of the unprimed system (Fig. 6.4). Thus the axes of the $(\bar{x}, \bar{y}, \bar{z})$ system can be obtained by translating those of the (x, y, z) system parallel to themselves from P_0 to P_0'. For this reason, (1) is called a *translation* of coordinate systems. The axes of the $(\bar{x}, \bar{y}, \bar{z})$ system can be brought into coincidence with those of the (x', y', z') system by a rotation; thus (12) is called a *rotation* of coordinates. (More precisely (12) is called a rotation if $\mathbf{M}$ is an orthogonal matrix.)

We have shown that the change of coordinates from one cartesian coordinate system to another can be considered as a translation followed by a rotation.

Example 6.5 Consider the change of coordinates

$$(13) \qquad \begin{aligned} x' &= \tfrac{2}{3}x + \tfrac{2}{3}y + \tfrac{1}{3}z - \tfrac{5}{3}, \\ y' &= -\tfrac{2}{3}x + \tfrac{1}{3}y + \tfrac{2}{3}z + \tfrac{5}{3}, \\ z' &= \tfrac{1}{3}x - \tfrac{2}{3}y + \tfrac{2}{3}z + \tfrac{2}{3}. \end{aligned}$$

We wish to verify that this change of coordinates can be viewed as a translation followed by a rotation. First, we determine the coordinates (x_0, y_0, z_0) of P_0', the origin of the (x', y', z') system with respect to P_0, the origin of the (x, y, z) system. The coordinates of P_0' with respect to the (x', y', z') system are $(0, 0, 0)$. Thus, from (13),

$$0 = \tfrac{2}{3}x_0 + \tfrac{2}{3}y_0 + \tfrac{1}{3}z_0 - \tfrac{5}{3}$$

$$0 = -\tfrac{2}{3}x_0 + \tfrac{1}{3}y_0 + \tfrac{2}{3}z_0 + \tfrac{5}{3}$$

$$0 = \tfrac{1}{3}x_0 - \tfrac{2}{3}y_0 + \tfrac{2}{3}x_0 + \tfrac{2}{3}z_0 + \tfrac{2}{3},$$

which must be solved for x_0, y_0, z_0. This system of equations can be written as

$$\mathbf{M} \begin{bmatrix} x_0 \\ y_0 \\ z_0 \end{bmatrix} = \begin{bmatrix} \tfrac{5}{3} \\ -\tfrac{5}{3} \\ -\tfrac{2}{3} \end{bmatrix},$$

where

$$\mathbf{M} = \begin{bmatrix} \frac{2}{3} & \frac{2}{3} & \frac{1}{3} \\[6pt] -\frac{2}{3} & \frac{1}{3} & \frac{2}{3} \\[6pt] \frac{1}{3} & -\frac{2}{3} & \frac{2}{3} \end{bmatrix}.$$

Hence

$$\begin{bmatrix} x_0 \\[6pt] y_0 \\[6pt] z_0 \end{bmatrix} = \mathbf{M}^{-1} \begin{bmatrix} \frac{5}{3} \\[6pt] -\frac{5}{3} \\[6pt] \frac{2}{3} \end{bmatrix}$$

$$= \mathbf{M}^{\mathrm{T}} \begin{bmatrix} \frac{5}{3} \\[6pt] -\frac{5}{3} \\[6pt] -\frac{2}{3} \end{bmatrix} = \begin{bmatrix} 2 \\[6pt] 1 \\[6pt] -1 \end{bmatrix}.$$

Thus, the change of coordinates consists of the translation

$$\begin{bmatrix} \bar{x} \\[6pt] \bar{y} \\[6pt] \bar{z} \end{bmatrix} = \begin{bmatrix} x - 2 \\[6pt] y - 1 \\[6pt] z + 1 \end{bmatrix}$$

followed by the rotation

$$\begin{bmatrix} x' \\[6pt] y' \\[6pt] z' \end{bmatrix} = \mathbf{M} \begin{bmatrix} \bar{x} \\[6pt] \bar{y} \\[6pt] \bar{z} \end{bmatrix}.$$

EXERCISES 1.6

1. Find cylindrical coordinates of the points with the following rectangular coordinates.

    (a) $(5\sqrt{3}, -5, 2)$;   (b) $(-2, -2, 5)$;

    (c) $(0, 1, 0)$;   (d) $(3, 3\sqrt{3}, -3)$.

2. Find rectangular coordinates of the points with the following cylindrical coordinates.

(a) $\left(2, \dfrac{3\pi}{4}, 3\right)$;   (b) $\left(4, \dfrac{5\pi}{3}, 4\right)$;

(c) $(2, 0, -3)$;   (d) $\left(5, \dfrac{\pi}{6}, 2\right)$.

3. Find spherical coordinates for the points with the following rectangular coordinates.

(a) $(\sqrt{2}, \sqrt{2}, 2)$;   (b) $(-3, 3\sqrt{3}, 0)$;

(c) $\left(\dfrac{1}{2}, \dfrac{-\sqrt{3}}{2}, -\sqrt{3}\right)$;   (d) $\left(\dfrac{-\sqrt{3}}{2}, \dfrac{1}{2}, -\sqrt{3}\right)$.

4. Find rectangular coordinates for the points with the following spherical coordinates.

(a) $\left(2, \dfrac{5\pi}{6}, \dfrac{\pi}{3}\right)$;   (b) $\left(4, \dfrac{3\pi}{4}, \dfrac{\pi}{2}\right)$;

(c) $\left(2, \dfrac{5\pi}{3}, \dfrac{\pi}{4}\right)$;   (d) $\left(3, \dfrac{\pi}{6}, \dfrac{3\pi}{4}\right)$.

5. Find spherical coordinates for the points with the following cylindrical coordinates.

(a) $\left(3, \dfrac{\pi}{3}, 2\right)$;   (b) $\left(4, \dfrac{2\pi}{3}, -3\right)$;

(c) $\left(2, \dfrac{5\pi}{6}, 4\right)$;   (d) $\left(2, \dfrac{3\pi}{4}, 3\right)$.

6. Find cylindrical coordinates for the points with the following spherical coordinates.

(a) $\left(3, \dfrac{2\pi}{3}, \dfrac{\pi}{4}\right)$;   (b) $\left(5, \dfrac{3\pi}{4}, \dfrac{\pi}{2}\right)$;

(c) $\left(4, \dfrac{5\pi}{4}, 0\right)$;   (d) $\left(2, \dfrac{7\pi}{6}, \dfrac{\pi}{3}\right)$.

7. In Eq. (9) let $(x_0, y_0, z_0) = (1, -2, 3)$ and

$$\mathbf{M} = \begin{bmatrix} 1 & 0 & 0 \\ 0 & \dfrac{1}{\sqrt{2}} & \dfrac{-1}{\sqrt{2}} \\ 0 & \dfrac{-1}{\sqrt{2}} & \dfrac{-1}{\sqrt{2}} \end{bmatrix}.$$

Find (x', y', z') for the points with the following (x, y, z) coordinates.

(a) $(1, 0, 2)$;   (b) $(-2, -3, 4)$;   (c) $(0, 0, 1)$;
(d) $(2, 3, 5)$.

8. Verify that

$$\mathbf{M} = \begin{bmatrix} \frac{2}{3} & -\frac{2}{3} & \frac{1}{3} \\ \frac{2}{3} & \frac{1}{3} & -\frac{2}{3} \\ \frac{1}{3} & \frac{2}{3} & \frac{2}{3} \end{bmatrix}$$

is an orthogonal matrix.

THEORETICAL EXERCISES

T-1. Verify Eq. (10).

T-2. Show that a matrix $\mathbf{M}$ is orthogonal if and only if $\mathbf{M}^{-1} = \mathbf{M}^T$.

T-3. Let P_0, P_1, and P_2 have coordinates (x_i, y_i, z_i), $i = 1, 2, 3$, with respect to a rectangular coordinate system, and let them have coordinates (x_i', y_i', z_i') with respect to a second system, related to the first by a translation and a rotation. Then there are two ways to compute the length of the vector $\overrightarrow{P_0 P_1}$:

$$|\overrightarrow{P_0 P_1}| = \sqrt{(x_1 - x_0)^2 + (y_1 - y_0)^2 + (z_1 - z_0)^2}$$

or

$$|\overrightarrow{P_0 P_1}| = \sqrt{(x_1' - x_0')^2 + (y_1' - y_0')^2 + (z_1' - z_0')^2}.$$

Show that the two results are the same. Show also that the corresponding ways for computing the cosine of the angle between $P_0 P_1$ and $P_0 P_2$ yield the same result.

(As a consequence of this problem, we say that distance and angle are *invariant under translation and rotation*.)

1.7 Surfaces in R^3

A *surface* S in R^3 is the set of points that satisfy

$$f(x, y, z) = 0,$$

where f is a function such as

$$f(x, y, z) = x + y + z$$

or

$$f(x, y, z) = x^2 + y^2 + z^2 - 1.$$

(In Section 2.1 we shall say more precisely what we mean by a function.) Perhaps the simplest surfaces in R^3 are the planes, which we have already studied; thus

$$Ax + By + Cz + D = 0$$

defines a plane if $A^2 + B^2 + C^2 \neq 0$.

In this section we present examples of surfaces which will be used for illustrative purposes throughout the rest of the book.

When sketching a surface S it is often useful to find the points, if any, where S intersects the coordinate axes and the coordinate planes. The x-coordinates of the points of intersection of S with the x-axis are called the *x-intercepts of S*; *y-intercepts* and *z-intercepts* are defined similarly. The curve in which S intersects a plane Γ is called *the trace of S in Γ*.

Cylinders

Let C be a curve in a plane Γ and L be a line, not in Γ, through a point of C. The set of points on lines through C parallel to L comprise a surface S, which is called a *cylinder* (Fig. 7.1). The curve C is the *directrix* of S and L is a *generator*; we can think of S as being obtained by moving L parallel to itself along C.

Example 7.1 Let Γ be the xy-plane, C the circle

(1) $$x^2 + y^2 = 1,$$

and L the line through $(0, 1, 0)$ parallel to the z-axis. As L is moved around C parallel to itself, it sweeps out the cylinder shown in Fig. 7.2. This is called a *right circular cylinder*, because its directrix is a circle and its generator is perpendicular to the plane of the directrix. In this case, (1) is also

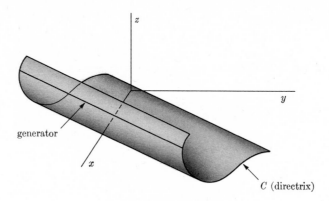

FIGURE 7.1

the equation of S, since a point (x, y, z) is on S if and only if x and y satisfy (1).

Example 7.2 The cylinders defined by

$$\frac{x^2}{a^2} + \frac{y^2}{b^2} = 1$$
$$(a, b > 0)$$
$$\frac{x^2}{a^2} - \frac{y^2}{b^2} = 1$$

and

$$y = ax^2 \qquad (a > 0)$$

FIGURE 7.2

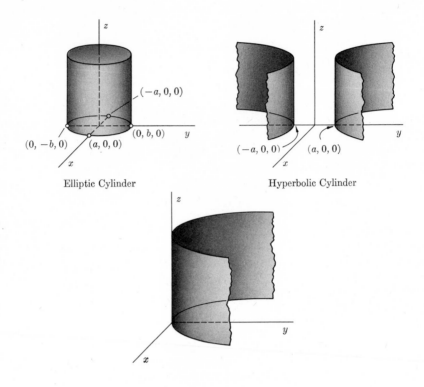

Elliptic Cylinder

Hyperbolic Cylinder

FIGURE 7.3

are named after their traces in the xy-plane; thus they are *elliptic, hyperbolic,* and *parabolic* cylinders, respectively. Portions of these surfaces above the xy-plane and below a plane parallel to it are shown in Fig. 7.3.

Quadric Surfaces

We devote the rest of this section to *quadric surfaces;* that is, surfaces defined by equations of the form

$$(2) \quad Ax^2 + By^2 + Cz^2 + 2Dxy + 2Exz + 2Fyz + Gx + Hy + Iz + J = 0,$$

where at least one of the coefficients A, B, C, D, E, and F is nonzero. The cylinders of Example 7.2 are surfaces of this kind. There are basically six other kinds of quadric surfaces, examples of which are given below. There are also degenerate cases, as in the following example.

Example 7.3 (Degenerate Quadric Surfaces)   The surface defined by (2) may be a plane; thus

$$x + y + z = 0,$$

which defines a plane, is of the form (2). It can also be written as

$$(x + y + z)^2 = x^2 + y^2 + z^2 + 2xy + 2xz + 2yz = 0,$$

which is also of the form (2). It is clearly preferable to study planes on their own, rather than as special cases of quadric surfaces; hence they are considered to be trivial, or *degenerate*, quadric surfaces. Other examples of degenerate quadric surfaces are as follows.

(a)   Pairs of planes: The locus of points satisfying

$$x^2 - y^2 = 0$$

comprises the two planes defined by

$$x - y = 0$$

and

$$x + y = 0.$$

(b)   Lines: The points satisfying

$$(x - x_0)^2 + (y - y_0)^2 = 0$$

comprise the line through $(x_0, y_0, 0)$ parallel to the z-axis.

(c)   Single points: Only the point (x_0, y_0, z_0) satisfies

$$(x - x_0)^2 + (y - y_0)^2 + (z - z_0)^2 = 0.$$

(d)   No locus: There are no points in R^3 that satisfy

$$x^2 + y^2 + z^2 + 1 = 0.$$

We now give simple examples of the six kinds of nondegenerate, noncylindrical quadric surfaces. In all cases we assume that a, b, and c are positive.

Example 7.4 (The Ellipsoid)   The surface S defined by

$$\frac{x^2}{a^2} + \frac{y^2}{b^2} + \frac{z^2}{c^2} = 1$$

is an *ellipsoid*; if $a = b = c$, then S is a sphere of radius a with center at the origin. The intercepts of S are $(\pm a, 0, 0)$, $(0, \pm b, 0)$, and $(0, 0, \pm c)$. The trace of S in any plane parallel to (and sufficiently close to) a coordinate plane is an ellipse. In particular the traces in the xy-, yz- and xz-planes are defined by

$$\frac{x^2}{a^2} + \frac{y^2}{b^2} = 1,$$

$$\frac{y^2}{b^2} + \frac{z^2}{c^2} = 1,$$

and

$$\frac{x^2}{a^2} + \frac{z^2}{c^2} = 1.$$

The ellipsoid and its traces in the coordinate planes are shown in Fig. 7.4.

FIGURE 7.4

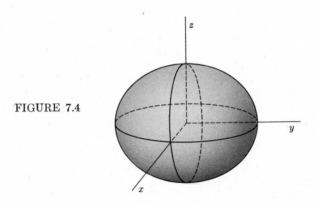

Example 7.5 (The Hyperboloid of One Sheet)   The x- and y-intercepts of the surface defined by

$$\frac{x^2}{a^2} + \frac{y^2}{b^2} - \frac{z^2}{c^2} = 1$$

are $(\pm a, 0, 0)$ and $(0, \pm b, 0)$; it has no z-intercepts. Its trace in the plane $z = k$ is the ellipse

$$\frac{x^2}{a^2} + \frac{y^2}{b^2} = 1 + \frac{k^2}{c^2},$$

while its trace in the plane $y = k$ is the hyperbola

$$\frac{x^2}{a^2} - \frac{z^2}{c^2} = 1 - \frac{k^2}{b^2},$$

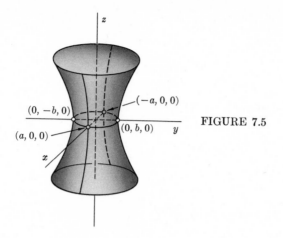

FIGURE 7.5

if $|k| < b$. Similarly, its trace in the plane $x = k$ is the hyperbola

$$\frac{y^2}{b^2} - \frac{z^2}{c^2} = 1 - \frac{k^2}{a^2}$$

if $|k| < a$. This surface, which is known as a *hyperboloid of one sheet*, is shown in Fig. 7.5.

Example 7.6 (The Hyperboloid of Two Sheets)   The surface defined by

$$\frac{x^2}{a^2} - \frac{y^2}{b^2} - \frac{z^2}{c^2} = 1$$

has no y- or z-intercepts; its x-intercepts are $(\pm a, 0, 0)$. Its traces in planes parallel to the xy- and xz-planes are hyperbolas. It has no trace in the plane $x = k$ unless $|k| > a$, in which case the trace is an ellipse. This surface has two halves which are mirror images of each other (Fig. 7.6). It is called a *hyperboloid of two sheets*.

FIGURE 7.6

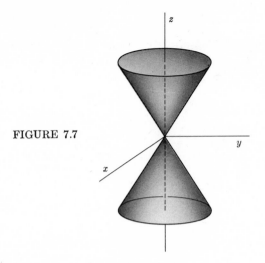

FIGURE 7.7

Example 7.7 (The Elliptic Cone)   The surface defined by

$$\frac{x^2}{a^2} + \frac{y^2}{b^2} = \frac{z^2}{c^2}$$

intersects the coordinate axes only at $(0, 0, 0)$, which is also its trace in the xy-plane. The trace in the plane $z = k \neq 0$ is an ellipse, while its traces in the planes $x = k$ and $y = k$ are hyperbolas which reduce to pairs of intersecting lines if $k = 0$. This surface is an *elliptic cone*; it is shown in Fig. 7.7.

Example 7.8 (The Elliptic Paraboloid)   The *elliptic paraboloid* defined by

$$\frac{x^2}{a^2} + \frac{y^2}{b^2} = \frac{z}{c},$$

passes through $(0, 0, 0)$, but does not intersect the coordinate axes elsewhere. Its traces in planes parallel to the yz- and xz-planes are parabolas, and its trace in the plane $z = k > 0$ is an ellipse; it does not intersect any plane $z = k < 0$. It is shown in Fig. 7.8.

Example 7.9 (The Hyperbolic Paraboloid)   The surface defined by

$$\frac{x^2}{a^2} - \frac{y^2}{b^2} = \frac{z}{c}$$

intersects the coordinate axes only at the origin. It intersects the plane $z = k$ in a hyperbola which reduces to the pair of lines $y = \pm(b/a)x$ if

FIGURE 7.8

$k = 0$. Its trace in the plane $y = k$ is a parabola opening upward, and its trace in plane $x = k$ is a parabola opening downward. Near the origin the surface looks like a saddle (Fig. 7.9); it is called a *hyperbolic paraboloid*.

We have said that Examples 7.1 and 7.3 through 7.9 contain all of the nondegenerate quadric surfaces. This does not mean that the equations given in those examples are the only ones of the form (2); this is obviously false, since none of the former contain xy, xz, or yz terms, while the latter may. Rather, it means that, if (2) defines a nondegenerate quadric surface S with respect to some rectangular coordinate system, then there is a rectangular coordinate system (x', y', z'), which can be obtained from the (x, y, z) system by a translation and a rotation, relative to which the equation of S takes one of the *standard forms* discussed in the examples.

Example 7.10 Consider the second-degree equation

$$36x^2 - 72x + 16y^2 + 64y + 9z^2 - 18z = 42,$$

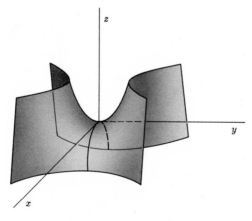

FIGURE 7.9

which can be rewritten as

$$36(x^2 - 2x) + 16(y^2 + 4y) + 9(z^2 - 2z) = 42.$$

Completing the squares yields

$$36(x^2 - 2x + 1) + 16(y^2 + 4y + 4)$$
$$+ 9(z^2 - 2z + 1) = 42 + 36 + 64 + 9 = 151$$

or

$$\frac{(x-1)^2}{6} + \frac{(y+2)^2}{9} + \frac{(z-1)^2}{16} = 1.$$

By setting $x' = x - 1$, $y' = y + 2$, and $z' = z - 1$, we obtain

$$\frac{(x')^2}{6} + \frac{(y')^2}{9} + \frac{(z')^2}{16} = 1,$$

which is the equation of an ellipsoid (Example 7.4).

Reduction to standard form in this example was very easy, requiring only a translation which was found by completing the squares. If (2) has mixed terms with nonzero coefficients, one must also find an appropriate rotation. We shall not go into this problem here.

EXERCISES 1.7

In Exercises 1 through 20 identify and sketch each surface.

1. $9x^2 - 4y^2 - 36z = 0$.

2. $y^2 - 16z^2 = 0$.

3. $x^2 - 25y^2 + 9z^2 = 225$.

4. $x^2 + 9y^2 + 16z^2 + 4 = 0$.

5. $25y^2 - 100z^2 = 400$.

6. $4x^2 + 9y^2 - 36z^2 = 0$.

7. $x^2 + 4z^2 = 0$.

8. $-4x^2 - 9y^2 + z^2 = 36$.

9. $y - 3z^2 = 0$.

10. $9x^2 + 4y^2 - 36z = 0$.

11. $x^2 + y^2 + z^2 = 36$.

12. $4x^2 + y^2 = 4$.

13. $4x^2 + 9y^2 - 16z^2 = 144$.

14. $x - 3z^2 = 0$.

15. $4x^2 + 9y^2 + 16z^2 = 144$.

16. $4x^2 - 10y^2 - 25z^2 = 100$.

17. $-4x^2 - 100y + 25z^2 = 0$.

18. $4x^2 - 100y + 25z^2 = 0$.

19. $16x^2 - 9z^2 = 144$.

20. $4x^2 - 16y^2 + z^2 = 0$.

In Exercises 21 through 30 translate axes and identify and sketch the surface.

21. $4x^2 + 9y^2 + z^2 + 8x - 18y - 4z - 19 = 0.$

22. $-5y^2 + 20y + z - 23 = 0.$

23. $-x^2 + 16y^2 - 4z^2 - 4x - 96y - 16z + 62 = 0.$

24. $y^2 - z^2 - 9x - 4y + 8z - 39 = 0.$

25. $x^2 - 4z^2 - 4x + 8z = 0.$

26. $x^2 + 3y^2 + 2z^2 - 6x - 6y + 4z + 14 = 0.$

27. $x^2 + 4y^2 + 4x + 16y - 16z + 4 = 0.$

28. $4x^2 - y^2 + z^2 - 16x + 8y - 6z + 9 = 0.$

29. $x^2 + y^2 + z^2 - 2x - 4y + 6z + 5 = 0.$

30. $4x^2 - 9y^2 + z^2 + 8x + 18y - 6z + 4 = 0.$

Chapter 2

Differential Calculus of Real-Valued Functions

2.1 Functions, Limits, and Continuity

Functions

Let D be a subset of R^n. A rule f which assigns a unique real number to each point of D will be called a *real valued function on* D. The value that f assigns to

$$\mathbf{X} = (x_1, \ldots, x_n)$$

will be denoted by

$$f(\mathbf{X}) = f(x_1, \ldots, x_n),$$

except that for $n = 1$, 2, and 3 we shall often write $f(x), f(x, y)$, and $f(x, y, z)$; $f(\mathbf{X})$ is a real number which we call the *value of* f *at* $\mathbf{X}$.

We distinguish between the function, which is f, and its *value* at a point $\mathbf{X}$, which is the real number $f(\mathbf{X})$, except where it is too clumsy to do so. For instance, if c is a fixed real number and f is the function whose value for every $\mathbf{X}$ in R^n is c, we shall simply write $f = c$; in particular, $f = 0$ is the function whose value is zero for all $\mathbf{X}$. Notice that $f(\mathbf{X}) = c$ and $f = c$ do not mean the same thing; $f(\mathbf{X}) = c$ means that c is the value of f at a particular $\mathbf{X}$, while $f = c$ means that c is the value of f at every $\mathbf{X}$ in R^n.

The set D on which f is defined is called the *domain* of f; the set of values assumed by $f(\mathbf{X})$ as $\mathbf{X}$ varies over D is called the *range*, or *image* of f. The specification of the domain is an essential part of the definition of the function.

Example 1.1 Let $n = 2$ and

$$f(\mathbf{X}) = \begin{cases} 1 & (|\mathbf{X}| > 1), \\ -1 & (|\mathbf{X}| < 1). \end{cases}$$

Here D comprises all points of R^2 except those for which $|\mathbf{X}| = 1$, and the range consists of the two points ± 1.

Example 1.2 Let $D = R^2$ and

$$(1) \qquad\qquad f(x, y) = xy.$$

The range of f is R^1.

Example 1.3 Let D be all $\mathbf{X}$ in R^2 except those for which $x = y$ and

$$(2) \qquad\qquad f(x, y) = \frac{1}{x - y}.$$

The range of f is the set of nonzero real numbers.

Usually a function will be defined by a single formula valid over its entire domain D. For brevity we adopt the convention that if f is given by a formula, and D is unspecified, then D should be understood to be the largest subset of R^n for which the formula makes sense. Thus in Example 1.2 we would simply write (1); since the right side is defined for all $\mathbf{X}$ in R^2, it would be understood that $D = R^2$. The right side of (2) is undefined for $x = y$; hence the domain of f in Example 1.3 would be understood to be as given there, unless specifically stated to the contrary. Also, if f is defined simply by

$$f(x, y) = \sqrt{1 - x^2 - y^2}$$

then D is the set of all $\mathbf{X}$ such that $|\mathbf{X}| \leq 1$, since the right side defines a real number if and only if $\mathbf{X}$ is in this set.

Definition 1.1 Suppose f_1 and f_2 have domains D_1 and D_2 which have points in common, and let D be the set of common points. We define $f_1 + f_2$, $f_1 - f_2$, and $f_1 f_2$ for $\mathbf{X}$ in D by

$$(f_1 + f_2)(\mathbf{X}) = f_1(\mathbf{X}) + f_2(\mathbf{X}),$$

$$(f_1 - f_2)(\mathbf{X}) = f_1(\mathbf{X}) - f_2(\mathbf{X}),$$

and

$$(f_1 f_2)(\mathbf{X}) = f_1(\mathbf{X})f_2(\mathbf{X}).$$

We also define f_1/f_2 by

$$(f_1/f_2)(\mathbf{X}) = \frac{f_1(\mathbf{X})}{f_2(\mathbf{X})},$$

but the domain of f_1/f_2 is the set D' of points in D for which $f_2(\mathbf{X}) \neq 0$ (f_1/f_2 is not defined if $f_2(\mathbf{X}) = 0$ for every $\mathbf{X}$ in D).

Example 1.4 Let

$$f_1(x, y) = \sqrt{1 - x^2 - y^2}$$

and

$$f_2(x, y) = x\sqrt{1 - 2x^2 - 2y^2}.$$

Then D_1 is the set of (x, y) such that $x^2 + y^2 \leq 1$, and D_2 is the set of (x, y) such that $x^2 + y^2 \leq \frac{1}{2}$; hence $D = D_2$,

$$(f_1 + f_2)(x, y) = \sqrt{1 - x^2 - y^2} + x\sqrt{1 - 2x^2 - 2y^2},$$
$$(f_1 - f_2)(x, y) = \sqrt{1 - x^2 - y^2} - x\sqrt{1 - 2x^2 - 2y^2}$$

and

$$(f_1 f_2)(x, y) = x\sqrt{(1 - 2x^2 - 2y^2)(1 - x^2 - y^2)}$$

for all $\mathbf{X}$ in D_2. Since $f_2(0, y) = 0$, D' is the set of (x, y) such that $x^2 + y^2 \leq \frac{1}{2}$ and $x \neq 0$. If (x, y) is in D', then

$$(f_1/f_2)(x, y) = \frac{1}{x}\sqrt{\frac{1 - x^2 - y^2}{1 - 2x^2 - 2y^2}}.$$

Definition 1.2 Let g be a function of a single variable t with domain S on the real axis, and let f be a real valued function with domain D in R^n. If D_0 is the set of all $\mathbf{X}$ in D such that $f(\mathbf{X})$ is in S, then the *composite function* $g \circ f$ is defined by

$$(g \circ f)(\mathbf{X}) = g(f(\mathbf{X}))$$

for $\mathbf{X}$ in D_0. (Of course, we must assume that D_0 actually contains some points—that is, it is *nonempty*.) For $n = 2$ the situation is depicted in Fig. 1.1.

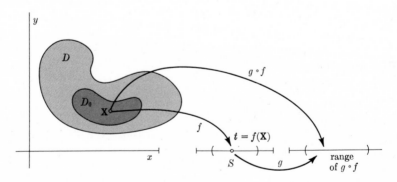

FIGURE 1.1

Example 1.5 Let

$$g(t) = \frac{1}{\sqrt{1 - t^2}}$$

and

$$f(x, y) = \sin (x^2 + y^2);$$

then S is the unit interval $-1 < t < 1$, and D is R^2. Since $|\sin \theta| < 1$ unless $\theta = (2k + 1)\pi/2$ $(k = 0, \pm 1, \ldots)$,

$$(g \circ f)(x, y) = \frac{1}{\sqrt{1 - \sin^2(x^2 + y^2)}},$$

provided $x^2 + y^2 \neq (2k + 1)\pi/2$ $(k = 0, 1, \ldots)$.

Limits

The student should recall the concepts of limit and continuity, which are fundamental to the calculus of functions of one variable. We now generalize these ideas to real valued functions of several variables; in Section 3.1 we shall extend them further to vector valued functions.

Definition 1.3 If $\mathbf{X}_0$ is a point in R^n and $\rho > 0$, the set of points $\mathbf{X}$ that satisfy $|\mathbf{X} - \mathbf{X}_0| < \rho$ will be called the *n-ball of radius ρ about $\mathbf{X}_0$*. The set obtained by excluding $\mathbf{X}_0$ from this n-ball will be called the *deleted n-ball of radius ρ about $\mathbf{X}_0$*; $\mathbf{X}$ is in this deleted n-ball if and only if $0 < |\mathbf{X} - \mathbf{X}_0| < \rho$.

When $n = 1$, 2, or 3, we shall also refer to the corresponding n-balls by their more familiar names: *interval, disk,* and *ball*.

Example 1.6 If $n = 3$, the set of points whose coordinates satisfy

$$\sqrt{x^2 + (y - 1)^2 + (z - 2)^2} < 5$$

is the 3-ball (or simply ball) of radius 5 about $\mathbf{X}_0 = (0, 1, 2)$. The deleted ball of radius 5 about $\mathbf{X}_0$ is obtained from this set by excluding $\mathbf{X}_0$; it consists of all (x, y) such that

$$0 < \sqrt{x^2 + (y - 1)^2 + (z - 2)^2} < 5.$$

A function f is said to *approach the limit L as $\mathbf{X}$ approaches $\mathbf{X}_0$*,

$$\lim_{\mathbf{X} \to \mathbf{X}_0} f(\mathbf{X}) = L,$$

if for any $\epsilon > 0$ all values of $f(\mathbf{X})$ for $\mathbf{X}$ in some deleted n-ball about $\mathbf{X}_0$ are in the interval $(L - \epsilon, L + \epsilon)$. More formally, for each $\epsilon > 0$ there exists a positive number δ such that

$$|f(\mathbf{X}) - L| < \epsilon$$

for all $\mathbf{X}$ satisfying

$$0 < |\mathbf{X} - \mathbf{X}_0| < \delta.$$

For $n = 1$ this definition reduces to the familiar definition studied in the first calculus course.

The essential idea here is that f has a limit L as $\mathbf{X}$ approaches $\mathbf{X}_0$ if it is possible, by restricting $\mathbf{X}$ to a sufficiently small deleted n-ball about $\mathbf{X}_0$, to restrict the values $f(\mathbf{X})$ to within any previously given open interval about the real number L. Notice that the value $f(\mathbf{X}_0)$ does not enter into the definition; in fact, f need not be defined at $\mathbf{X}_0$.

Example 1.7 Let $n = 1$ and

$$f(x) = x^2;$$

then $\lim\limits_{x \to 2} f(x) = 4$, because

$$
\begin{aligned}
|f(x) - 4| &= |x^2 - 4| \\
&= |x - 2||x + 2| \\
&= |x - 2||(x - 2) + 4| \\
&\leq |x - 2|(|x - 2| + 4).
\end{aligned}
$$

Given $\epsilon > 0$, take $\delta = \min(1, \epsilon/5)$. Then if $|x - 2| < \delta$,

$$|f(x) - 4| < \frac{\epsilon}{5}(1 + 4) = \epsilon.$$

Example 1.8 Let $n = 3$, $\mathbf{X}_0 = (2, 1, 1)$, and $f(x, y, z) = x + 3y + 2z$; then $\lim\limits_{\mathbf{X} \to \mathbf{X}_0} f(\mathbf{X}) = 7$. To verify this, we calculate as follows:

$$
\begin{aligned}
|f(\mathbf{X}) - 7| &= |x + 3y + 2z - 7| \\
&= |(x - 2) + 3(y - 1) + 2(z - 1)| \\
&\leq |x - 2| + 3|y - 1| + 2|z - 1| \\
&\leq |\mathbf{X} - \mathbf{X}_0| + 3|\mathbf{X} - \mathbf{X}_0| + 2|\mathbf{X} - \mathbf{X}_0| \\
&= 6|\mathbf{X} - \mathbf{X}_0|;
\end{aligned}
$$

given $\epsilon > 0$, take $\delta = \epsilon/6$.

The definition of limit implicitly assumes that f is defined for all points of a sufficiently small deleted n-ball about $\mathbf{X}_0$. We now reformulate the definition to avoid this restriction, which is too stringent for our purposes.

Definition 1.4 A point $\mathbf{X}_0$ is a *limit point* of a subset D of R^n if every deleted n-ball about $\mathbf{X}_0$, no matter how small, contains points of D.

A limit point of D need not be in D, nor is a point of D necessarily a limit point (see Example 1.9).

Definition 1.5 A point $\mathbf{X}_0$ in a subset D of R^n is an *interior* point of D if there is an n-ball about $\mathbf{X}_0$ which is contained entirely in D.

Notice that an interior point is necessarily a limit point, but a limit point need not be an interior point.

Example 1.9 Let $n = 2$ and D be those points (x, y) such that $0 \leq x < y$ (Fig. 1.2), together with $(x, y) = (1, 0)$.

Every point in D except $(1, 0)$ and those for which $x = 0$ is an interior point. Points on the positive y-axis $(x = 0, y > 0)$ are limit points of D which are contained in D, but are not interior points; points on the half line $x = y \geq 0$ are limit points which are not in D; $(1, 0)$ is not a limit point of D.

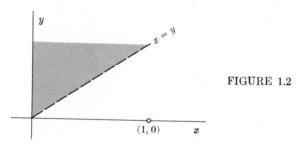

FIGURE 1.2

Definition 1.6 A set S is *open* if all of its points are interior points.

Example 1.10 The n-ball B of radius ρ about any point $\mathbf{X}_0$ is open. Let $\mathbf{X}_1$ be any point in B and $|\mathbf{X}_1 - \mathbf{X}_0| = \rho_1 < \rho$. If $\mathbf{X}$ is any point such that $|\mathbf{X} - \mathbf{X}_1| < \rho - \rho_1$, then the triangle inequality implies that

$$|\mathbf{X} - \mathbf{X}_0| = |(\mathbf{X} - \mathbf{X}_1) + (\mathbf{X}_1 - \mathbf{X}_0)|$$
$$\leq |\mathbf{X} - \mathbf{X}_1| + |\mathbf{X}_1 - \mathbf{X}_0|$$
$$< \rho - \rho_1 + \rho_1 = \rho;$$

hence the n-ball about $\mathbf{X}_1$ of radius $\rho - \rho_1$ is in B (Fig. 1.3). We have shown

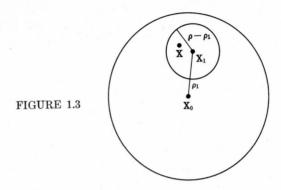

FIGURE 1.3

that X_1 is an interior point (Definition 1.5) of the n-ball B. Since X_1 was arbitrarily chosen, every point of B is an interior point. Hence, B is open, by Definition 1.6.

Example 1.11   The set D defined in Example 1.9 is not open, since $(1, 0)$ and points on the positive y-axis are in D, but are not interior points. However, the set D° consisting of all (x, y) such that $0 < x < y$ is open; it is called the *interior* of D. In general, the *interior* of a set S is the set of interior points of S; it is an open set. (Exercise T-13).

Definition 1.7   Let X_0 be a limit point of the domain D of a real valued function f. Then we say that $\lim_{X \to X_0} f(X) = L$ if for every $\epsilon > 0$ there exists a $\delta > 0$, depending upon ϵ, such that

$$|f(X) - L| < \epsilon$$

for all X in D that satisfy $0 < |X - X_0| < \delta$. We also say that f has *limit* L at X_0.

Example 1.12   Let $n = 2$ and

$$f(x, y) = \frac{x^2 - y^2}{x - y} ;$$

then f is not defined on the line $x = y$. However, these points are limit points of the domain of f. Let $X_0 = (c, c)$. If $x \neq y$, we can write

$$f(X) = x + y$$

and the student can verify that

$$\lim_{X \to X_0} f(X) = 2c.$$

It is essential to recognize that Definition 1.7 allows no restriction on the manner in which X approaches X_0, except that X be in the domain of f.

Example 1.13 Let $n = 1$ and

$$f(x) = \begin{cases} -1 + x & (x < 0), \\ 1 + x & (x > 0); \end{cases}$$

then f does not have a limit at $x = 0$, since in any open interval containing the origin there are points x_1 and x_2 such that $f(x_1) < -\frac{1}{2}$ and $f(x_2) > \frac{1}{2}$ (Fig. 1.4). If we were careless and examined only positive values of x, we might incorrectly conclude that $\lim_{x \to 0} f(x) = 1$; similarly, examining only negative values of x might lead to the conclusion that $\lim_{x \to 0} f(x) = -1$, also invalid. In fact, f has no limit at $x = 0$.

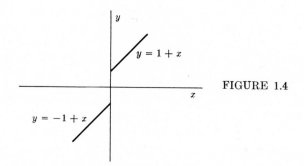

FIGURE 1.4

Example 1.14 Let $n = 2$, $X_0 = (0, 0)$, and

$$f(x, y) = 2xy \left(1 + \frac{1}{x^2 + y^2} \right), \qquad (x, y) \neq (0, 0).$$

Suppose X is constrained to approach the origin along the ray $x = \rho \cos \psi$, $y = \rho \sin \psi$, where $\rho > 0$ and ψ is fixed but arbitrary (Fig. 1.5). Then

$$f(\rho \cos \psi, \rho \sin \psi) = 2(1 + \rho^2) \sin \psi \cos \psi$$

$$= (1 + \rho^2) \sin 2\psi,$$

which approaches $\sin 2\psi$ as ρ approaches zero. Can we conclude that $\lim_{X \to 0} f(X) = \sin 2\psi$? Clearly not, since the value of the right side depends upon the direction of approach.

FIGURE 1.5

These two examples should convince the student that it is not safe to conclude that $\lim_{X \to X_0} f(X) = L$ after restricting the way in which X approaches X_0. One can, however, conclude that a limit does not exist if different limiting values are obtained for different paths of approach, as in the last two examples.

Example 1.15 Let

$$f(x, y) = \frac{1}{x^2 + y^2} \, ;$$

f has no limit at $(0, 0)$, because it grows beyond all bounds as (x, y) approaches $(0, 0)$. That is, given any M, no matter how large, $|f(x, y)| > M$ if $0 < x^2 + y^2 < 1/M$. We say in this case that

$$\lim_{(x,y) \to (0,0)} \frac{1}{x^2 + y^2} = \infty .$$

In general we say that

$$\lim_{X \to X_0} f(X) = \infty$$

if for every $M > 0$, no matter how large, there is a $\delta > 0$ such that $|f(X)| > M$ if $0 < |X - X_0| < \delta$. Thus

$$\lim_{(x,y) \to (0,0)} \frac{1}{\sin xy} = \infty$$

and

$$\lim_{(x,y,z) \to (0,0,1)} \frac{1}{x + y + z - 1} = \infty .$$

Theorem 1.1 If $\lim_{X \to X_0} f(X) = L_1$, and $\lim_{X \to X_0} f(X) = L_2$, then $L_1 = L_2$; that is, if a limit exists it is unique.

Proof. Suppose $L_1 \neq L_2$. From the triangle inequality,

$$| L_1 - L_2 | = | (L_1 - f(\mathbf{X})) + (f(\mathbf{X}) - L_2) |$$

$$\leq | L_1 - f(\mathbf{X}) | + | f(\mathbf{X}) - L_2 |.$$

From the definition of limit, each of the last two terms is less than

$$\frac{| L_1 - L_2 |}{4}$$

for $\mathbf{X}$ sufficiently close to $\mathbf{X}_0$ (take $\epsilon = | L_1 - L_2 |/4$ in Definition 1.7). This implies that

$$| L_1 - L_2 | < \frac{| L_1 - L_2 |}{2} ,$$

a contradiction; hence $L_1 = L_2$.

For more complicated functions the "epsilon–delta" argument required in the definition of limit becomes cumbersome, and it is often difficult to guess what the limit is in the first place. The following two theorems are useful for finding $\lim_{\mathbf{X} \to \mathbf{X}_0} f(\mathbf{X})$ if f can be obtained from simpler functions by a finite sequence of additions, subtractions, multiplications, divisions, and compositions.

Theorem 1.2 Let $\mathbf{X}_0$ be a limit point of the set of points common to the domains of f_1 and f_2. If $\lim_{\mathbf{X} \to \mathbf{X}_0} f_1(\mathbf{X}) = L_1$, and $\lim_{\mathbf{X} \to \mathbf{X}_0} f_2(\mathbf{X}) = L_2$, then

(a) $\lim_{\mathbf{X} \to \mathbf{X}_0} (f_1 + f_2)(\mathbf{X}) = L_1 + L_2, \lim_{\mathbf{X} \to \mathbf{X}_0} (f_1 - f_2)(\mathbf{X}) = L_1 - L_2;$

(b) $\lim_{\mathbf{X} \to \mathbf{X}_0} (f_1 f_2)(\mathbf{X}) = L_1 L_2;$ and

(c) if $L_2 \neq 0, \lim_{\mathbf{X} \to \mathbf{X}_0} (f_1/f_2)(\mathbf{X}) = L_1/L_2.$

Proof. The proof is analogous to that given in the first year calculus course; hence we shall prove only (b), leaving the rest for the exercises. Write

$f_1(\mathbf{X})f_2(\mathbf{X}) - L_1 L_2$

$$= (f_1(\mathbf{X}) - L_1)(f_2(\mathbf{X}) - L_2) + L_2(f_1(\mathbf{X}) - L_1) + L_1(f_2(\mathbf{X}) - L_2);$$

then

$$|f_1(\mathbf{X})f_2(\mathbf{X}) - L_1 L_2|$$

$$\leq |f_1(\mathbf{X}) - L_1||f_2(\mathbf{X}) - L_2| + |L_2||f_1(\mathbf{X}) - L_1| + |L_1||f_2(\mathbf{X}) - L_2|.$$

The terms on the right can be made as small as we please by taking $\mathbf{X}$ sufficiently close to $\mathbf{X}_0$, which proves (b).

Example 1.16   Let $\mathbf{X}_0 = (1, 2)$ and find $\lim\limits_{\mathbf{X}\to\mathbf{X}_0} f(x, y)$,   where $f(x, y) = xy/(x + y)$.

Solution.   Let $g_1(x, y) = x$, and $g_2(x, y) = y$; then $\lim\limits_{\mathbf{X}\to\mathbf{X}_0} g_1(x, y) = \lim\limits_{x\to 1} x = 1$ and $\lim\limits_{\mathbf{X}\to\mathbf{X}_0} g_2(x, y) = 2$. (We assume that the student can evaluate one dimensional limits, using techniques learned in his first calculus course). Now let $f_1 = g_1 g_2$, and $f_2 = g_1 + g_2$; then $f = f_1/f_2$. From Theorem 1.2

$$\lim_{\mathbf{X}\to\mathbf{X}_0} f_2(x, y) = \lim_{\mathbf{X}\to\mathbf{X}_0} g_1(x, y) + \lim_{\mathbf{X}\to\mathbf{X}_0} g_2(x, y) = 1 + 2 = 3,$$

$$\lim_{\mathbf{X}\to\mathbf{X}_0} f_1(x, y) = \left(\lim_{\mathbf{X}\to\mathbf{X}_0} g_1(x, y)\right)\left(\lim_{\mathbf{X}\to\mathbf{X}_0} g_2(x, y)\right) = 1\cdot 2 = 2,$$

and

$$\lim_{\mathbf{X}\to\mathbf{X}_0} f(x, y) = \frac{\lim\limits_{\mathbf{X}\to\mathbf{X}_0} f_1(x, y)}{\lim\limits_{\mathbf{X}\to\mathbf{X}_0} f_2(x, y)} = \frac{2}{3}.$$

Continuity

Definition 1.8   A real valued function f with domain D is said to be *continuous at a point* $\mathbf{X}_0$ if

(a) $\mathbf{X}_0$ is in D (thus $f(\mathbf{X}_0)$ is defined);
(b) $\mathbf{X}_0$ is a limit point of D; and
(c) $\lim\limits_{\mathbf{X}\to\mathbf{X}_0} f(\mathbf{X}) = f(\mathbf{X}_0)$.

We shall also agree, as a matter of convenience, that f is continuous at any point of D which is not a limit point of D. (Such points are *isolated points*; for example, $(1, 0)$ in Example 1.9.)

If f is continuous at every point of a set S, we say that f *is continuous on* S.

The student should convince himself that the conditions on $\mathbf{X}_0$ in (a) and (b) are merely those required so that (c) should make sense.

Often our understanding of a definition is aided by examining ways in which it can fail to be satisfied.

Example 1.17 Let $n = 2$ and

$$f(x, y) = \frac{1}{1 - x^2 - y^2} \qquad (0 < x^2 + y^2 < 1),$$

$$f(0, 0) = 2, \qquad f(0, 1) = 1, \qquad f(0, 2) = 5.$$

Then $\lim_{\mathbf{X} \to (0,0)} f(\mathbf{X}) = 1 \neq 2$; hence f is not continuous at $(0, 0)$ even though f is defined and has a limit there. At $(0, 1)$, f is defined but has no limit, and again is not continuous. The function f is continuous at $(0, 2)$, which is an isolated point of its domain. If $0 < |\mathbf{X}_0| < 1$, then f is continuous at $\mathbf{X}_0$ (Exercise 26).

Example 1.18 Let g be a function of the single variable x with domain D, and let

$$f(x, y) = g(x)$$

on D^*, the set of points $\mathbf{X} = (x, y)$ such that x is in D. If g is continuous at x_0, and y_0 is arbitrary, then f is continuous at $\mathbf{X}_0 = (x_0, y_0)$.

Proof. Given $\epsilon > 0$, there is a $\delta > 0$ such that

$$|g(x) - g(x_0)| < \epsilon$$

if $|x - x_0| < \delta$ and x is in D. Since $|x - x_0| \leq |\mathbf{X} - \mathbf{X}_0|$, it follows that

$$|f(x, y) - f(x_0, y_0)| = |g(x) - g(x_0)| < \epsilon$$

if $|\mathbf{X} - \mathbf{X}_0| < \delta$ and $\mathbf{X}$ is in D^*.

This result has a natural generalization. Suppose f is defined on a subset of R^n $(n > 1)$, but its values are independent of some of the variables, say the last $n - k$; that is,

$$f(x_1, \ldots, x_n) = g(x_1, x_2, \ldots, x_k).$$

Then if g is continuous at $(c_1, \ldots, c_k)$, under the definition of continuity for functions on R^k, it follows that f is continuous at $(c_1, \ldots, c_k, x_{k+1}, \ldots, x_n)$ for any $x_{k+1}, \ldots, x_n$, under the definition of continuity for functions on R^n.

Theorem 1.3 Let D be the set of points common to the domains of f_1 and f_2. Let $\mathbf{X}_0$ be in D and be a limit point of D. Then if f_1 and f_2 are continuous at

$\mathbf{X}_0$, so are $f_1 + f_2, f_1 - f_2$, and $f_1 f_2$. If $f_2(\mathbf{X}_0) \neq 0$, then f_1/f_2 is also continuous at $\mathbf{X}_0$.

Proof. Again we consider only the product, the other cases being left as an exercise (Exercise T-2). By assumption, the limits

$$\lim_{\mathbf{X} \to \mathbf{X}_0} f_1(\mathbf{X}) = L_1$$

and

$$\lim_{\mathbf{X} \to \mathbf{X}_0} f_2(\mathbf{X}) = L_2$$

exist. By Theorem 1.2,

(3) $$\lim_{\mathbf{X} \to \mathbf{X}_0} (f_1 f_2)(\mathbf{X}) = L_1 L_2.$$

Since f_1 and f_2 are continuous at $\mathbf{X}_0$, $L_1 = f(\mathbf{X}_0)$ and $L_2 = f(\mathbf{X}_0)$. Therefore (3) can be rewritten

$$\lim_{\mathbf{X} \to \mathbf{X}_0} (f_1 f_2)(\mathbf{X}) = f_1(\mathbf{X}_0)f_2(\mathbf{X}_0)$$

$$= (f_1 f_2)(\mathbf{X}_0);$$

hence $f_1 f_2$ is continuous at $\mathbf{X}_0$.

Example 1.19 By the same sequence of applications of the assertions of Theorem 1.3 that we performed with those of Theorem 1.2 in Example 1.16

$$f(x, y) = \frac{xy}{x + y}$$

is continuous at $\mathbf{X}_0 = (1, 2)$.

The following theorem states that "a continuous function of a continuous function is continuous."

Theorem 1.4 Let f and g be as in Definition 1.2. Assume that f is continuous at $\mathbf{X}_0$, that $t_0 = f(\mathbf{X}_0)$ is an interior point of the domain S of g, and that g is continuous at t_0. Then the composite function $g \circ f$ is continuous at $\mathbf{X}_0$.

Proof. Let $\epsilon > 0$ be given. Since t_0 is an interior point of S and g is continuous at t_0, there is a $\delta_1 > 0$ such that $g(t)$ is defined and $|g(t) - g(t_0)| < \epsilon$ if $|t - t_0| < \delta_1$. As $\lim_{\mathbf{X} \to \mathbf{X}_0} f(\mathbf{X}) = f(\mathbf{X}_0) = t_0$, there is a $\delta > 0$ such that $|f(\mathbf{X}) - t_0| < \delta_1$ if $|\mathbf{X} - \mathbf{X}_0| < \delta$. Hence, if $|\mathbf{X} - \mathbf{X}_0| < \delta$, then $|g(f(\mathbf{X})) - g(t_0)| < \epsilon$, which completes the proof. The situation is depicted in Fig. 1.6: if $\mathbf{X}$ is in the n-ball of radius δ about $\mathbf{X}_0$, then $t = f(\mathbf{X})$ is in the interval $(t_0 - \delta_1, t_0 + \delta_1)$ and therefore $g(f(\mathbf{X}))$ is in the interval $(g(t_0) - \epsilon, g(t_0) + \epsilon)$.

FIGURE 1.6

Example 1.20 Let $\mathbf{X}_0 = (1, 2)$,

$$f(x, y) = \frac{xy}{x + y}$$

and

$$g(t) = \sqrt{t}.$$

From Example 1.19, f is continuous at $\mathbf{X}_0$. Furthermore $t_0 = f(\mathbf{X}_0) = \frac{2}{3}$ is an interior point of the domain of g, and g is continuous at t_0. Consequently $g \circ f$ is continuous at $\mathbf{X}_0$. It follows that

$$\lim_{\mathbf{X} \to \mathbf{X}_0} \sqrt{\frac{xy}{x + y}} = g(t_0) = \sqrt{\frac{2}{3}}.$$

Example 1.18 and Theorems 1.3 and 1.4 allow us to build up an extensive collection of continuous functions by addition, subtraction, multiplication, division, and composition of continuous functions.

Example 1.21 Let

$$f(x, y, z) = z + \log [2 + \sin (x^2 + y^2)].$$

This is continuous on all of R^3 by the following argument:

(a) Let $f_1(x, y, z) = x$, $f_2(x, y, z) = y$, and $f_3(x, y, z) = z$; then f_1, f_2, and f_3 are continuous on R^3 (Example 1.18).

(b) $f_1^2 \; (=f_1 f_1)$, f_2^2, and $f_4 = f_1^2 + f_2^2$ are continuous on R^3 (Theorem 1.3):

$$f_4(x, y, z) = x^2 + y^2.$$

(c) Let $g_1(t) = \sin t$; then $f_5 = g_1 \circ f_4$ is continuous on R^3 (Theorem 1.4):

$$f_5(x, y, z) = \sin (x^2 + y^2).$$

(d) $f_6 = 2$ is continuous on R^3, hence $f_7 = f_5 + f_6$ is also:

$$f_7(x, y, z) = 2 + \sin (x^2 + y^2).$$

(e) $g_2(t) = \log t$ is continuous for $t \geq 0$; since the values of f_7 are positive, $f_8 = g_1 \cdot f_7$ is continuous on R^3 (Theorem 1.4).

(f) $f = f_3 + f_8$ is continuous on R^3 (Theorem 1.3).

EXERCISES 2.1

1. Find the domains of the following functions.

(a) $f(x, y, z) = \sqrt{x^2 + y^2 - 1}.$ (b) $f(x, y) = (xy)^{3/2}.$

(c) $f(x, y) = \dfrac{x^2 + y^2}{xy}.$ (d) $f(x, y) = \dfrac{1}{\sin (x^2 - y^2)}.$

(e) $f(x, y) = \dfrac{1}{\cos (x^2 - y^2)}.$ (f) $f(x, y) = \cos xy.$

2. Find the domains of the following functions.

(a) $f(x, y, z) = x^2 + y^2 + z.$ (b) $f(x, y) = \log(x - y).$

(c) $f(x, y, z) = x^2 - \dfrac{y}{z - 1}.$ (d) $f(x, y) = \sqrt{x^2 - 1}.$

(e) $f(x, y) = \dfrac{x}{y^2 - 1}.$ (f) $f(x, y) = e^{x^2 + y^2}.$

3. Let $f(x, y) = \sqrt{1 - x^2 - y^2}$ and $g(x, y) = \log(x - y)$. What are the domains of $f + g, f - g, fg,$ and f/g?

4. Let $g(t) = e^{t^2}$ and $f(x, y) = \sqrt{x^2 - y^2}.$

   (a) Find $(g \circ f)(x, y)$. What is the domain of $g \circ f$?

   (b) Find $(g \circ f)(3, 2).$

5. Let $g(t) = 1/(1 - t)$ and $f(x, y) = 1 + \cos(x + y).$

   (a) Find $(g \circ f)(x, y)$. What is the domain of $g \circ f$?

   (b) Find $(g \circ f)(\pi/2, 3\pi/2).$

6. (a) Sketch the disk of radius $\rho = 2$ about $\mathbf{X}_0 = (1, 2)$.

   (b) Repeat (a) for $\rho = 1$ and $\mathbf{X}_0 = (3, 4)$.

7. Let S be an n-ball of radius ρ about $\mathbf{X}_0$ and let $\mathbf{X}_1$ be in S. Find the radius of the largest n-ball about $\mathbf{X}_1$ that is contained entirely within S.

8. Find the limit points of the subsets of R^1 defined by

   (a) $0 \le x \le 1$;   (b) $0 < x \le 1$;

   (c) $|x| = 1$;   (d) $x = 1, \dfrac{1}{2}, \ldots, \dfrac{1}{n}, \ldots$

9. Find the limit points of the subsets of R^2 defined by

   (a) $x^2 + y^2 \le 1$;

   (b) $(x, y) = (0, 0)$ or $(x, y) = (1, 1)$;

   (c) $x^2 + y^2 \le 1, \quad x \ne 0, \quad y \ne 0$;

   (d) $0 < x \le y$;

   (e) $(x, y) \ne (0, 0)$;

   (f) $x^2 + y^2 = 1$ or $(x, y) = (1, 2)$.

10. For the sets in Exercise 8, state which points are (a) interior points, (b) limit points but not interior points, (c) *isolated points* (not interior or limit points).

11. Repeat Exercise 10 for the sets in Exercise 9.

12. For each of the following functions find $\lim\limits_{\mathbf{X} \to \mathbf{X}_0} f(x, y)$ if it exists. When there is a limit, prove it by an "epsilon–delta" argument.

    (a) $f(\mathbf{X}) = x + 2y, \qquad \mathbf{X}_0 = (1, 2)$.

    (b) $f(\mathbf{X}) = \dfrac{x^2 - y^2}{x + y}, \qquad \mathbf{X}_0 = (1, -1)$.

    (c) $f(\mathbf{X}) = \dfrac{x^2 y}{x^2 + y^2}, \qquad \mathbf{X}_0 = (0, 0)$.

    (d) $f(\mathbf{X}) = \sin \dfrac{1}{x} - y, \qquad \mathbf{X}_0 = (0, 3)$.

13. Repeat Exercise 12 for the following functions.

   (a) $f(\mathbf{X}) = 3x + 2y, \qquad \mathbf{X}_0 = (2, 1)$.

   (b) $f(\mathbf{X}) = \dfrac{x^2 + y^2}{x - y}, \qquad \mathbf{X}_0 = (1, 1)$.

   (c) $f(\mathbf{X}) = \dfrac{xy}{\sqrt{x^2 + y^2}}, \qquad \mathbf{X}_0 = (0, 0)$.

   (d) $f(x, y) = x^2 + y^2 + z^2, \qquad \mathbf{X}_0 = (-1, 0, 1)$.

14. Using Theorems 1.2 and 1.4, find $\lim\limits_{\mathbf{X} \to \mathbf{X}_0} f(\mathbf{X})$, when it exists, for each of the following functions.

   (a) $f(\mathbf{X}) = \log \sqrt{1 - x^2 - y^2 - z^2}, \qquad \mathbf{X}_0 = (0, 0, 0)$.

   (b) $f(\mathbf{X}) = \exp\left(\dfrac{1}{\sqrt{1 - \sin(1 + x - y)(\pi/2)}}\right), \qquad \mathbf{X}_0 = (0, 0)$.

   (c) $f(\mathbf{X}) = \cos^2(1 + 2x^2 + y^2) + \sin^2\dfrac{1}{1 - x^2}, \qquad \mathbf{X}_0 = (0, 0)$.

15. Repeat Exercise 14 for the following functions.

   (a) $f(\mathbf{X}) = \cos\dfrac{\pi}{1 - 3xyz}, \qquad \mathbf{X}_0 = (1, 1, 1)$.

   (b) $f(\mathbf{X}) = \log \exp(\sqrt{x^2 + 2xy + y^2}), \qquad \mathbf{X}_0 = (1, 1)$.

   (c) $f(\mathbf{X}) = \sin \log(|x| + |y|), \qquad \mathbf{X}_0 = (0, 0)$.

16. Let $g(t) = t^3$ and $f(x, y) = (x^2 - y^2)/(x + y)$. Find

$$\lim_{(x, y) \to (2, -2)} g(f(x, y)).$$

17. Let $g(t) = \sqrt{t}$ and $f(x, y) = \cos(x + y)$. Find

$$\lim_{(x, y) \to (2\pi, 4\pi)} g(f(x, y)).$$

18. Prove that f is continuous at $\mathbf{X}_0$:

   (a) $f(x, y, z) = x^2 + 2y^2 - z, \qquad \mathbf{X}_0 = (a, b, c)$.

   (b) $f(x, y) = \cos(x^2 + y^2), \qquad \mathbf{X}_0 = \left(\sqrt{\dfrac{\pi}{2}}, \sqrt{\pi}\right)$.

(c) $f(x, y, z) = \exp(x^2 + y + z^2)$, $\mathbf{X}_0 = (1, 0, 1)$.

(d) $f(x, y) = \log(x^2 + y^2)$, $\mathbf{X}_0 = (1, -1)$.

19. Where are the functions of Exercise 1 continuous?

20. Where are the functions of Exercise 2 continuous?

21. Where is $g \circ f$ of Exercise 4 continuous?

22. Where are the functions of Exercise 14 continuous?

23. Where are the functions of Exercise 15 continuous?

24. Where are the following functions continuous?

(a) $f(x, y) = \begin{cases} \dfrac{x + y}{x - y} & \text{if } x \neq y, \\[2ex] 0 & \text{if } x = y. \end{cases}$

(b) $f(x, y) = \begin{cases} x \sin \dfrac{1}{x} - y & \text{if } x \neq 0, \\[2ex] -y & \text{if } x = 0. \end{cases}$

(c) $f(x, y) = \begin{cases} \dfrac{x^2 - y^2}{x - y} & \text{if } x \neq y, \\[2ex] x + y & \text{if } x = y. \end{cases}$

25. How would you define f at $(0, 0)$ to make f, with your extended definition, continuous at $\mathbf{X}_0$?

(a) $f(x, y) = \dfrac{xy}{\sqrt{x^2 + y^2}}$.   (b) $f(x, y) = \dfrac{\sin xy}{xy}$.

(c) $f(x, y) = \dfrac{x^2 + y^2 - 2xy}{x - y}$.

THEORETICAL EXERCISES

T-1. Complete the proof of Theorem 1.2.

T-2. Complete the proof of Theorem 1.3.

T-3. A function f is said to be *bounded* near a limit point $\mathbf{X}_0$ of its domain D if there are positive numbers M and ϵ such that $|f(\mathbf{X})| < M$ whenever $\mathbf{X}$ is in D and $0 < |\mathbf{X} - \mathbf{X}_0| < \epsilon$. (If this is not the case, f is said to be *unbounded* near $\mathbf{X}_0$.) *Prove*: If $\lim_{\mathbf{X} \to \mathbf{X}_0} f(\mathbf{X}) = L$, then f is bounded near $\mathbf{X}_0$, and if $L \neq 0$, $1/f$ is also.

T-4. *Prove*: If f is continuous at $\mathbf{X}_0$, then it is bounded near $\mathbf{X}_0$.

T-5. If f is unbounded near $\mathbf{X}_0$, does it follow that $\lim_{\mathbf{X} \to \mathbf{X}_0} f(\mathbf{X}) = \infty$? (*Hint*: Consider $f(x) = (1/x)\sin(1/x)$.)

T-6. Prove that every point of an n-ball is an interior point.

T-7. A point $\mathbf{X}_0$ is a *boundary point* of a set S in R^n if every n-ball about $\mathbf{X}_0$ contains points in S and points not in S. Find all boundary points of the sets in Exercises 8 and 9.

T-8. Show that a boundary point of a set S is either a limit point or an isolated point of S.

T-9. Let f and g be continuous at $\mathbf{X}_0$; let c and d be real numbers. Show that $cf + dg$ is continuous at $\mathbf{X}_0$.

T-10. A function f is said to have a *removable discontinuity* at $\mathbf{X}_0$ if it is not continuous at $\mathbf{X}_0$, but can be defined (or possibly redefined) there by $f(\mathbf{X}_0) = y_0$ so that, with the new definition, f is continuous at $\mathbf{X}_0$. *Prove*: If f has a removable discontinuity, then y_0 is unique.

T-11. If $f_1, \ldots, f_n$ are functions, define the *sum* function $(f_1 + \cdots + f_n)$ and the *product* function $f_1 \cdots f_n$.

T-12. The *line segment connecting two points* $\mathbf{X}_1$ *and* $\mathbf{X}_2$ in R^n is defined to be the set of points of the form

$$\mathbf{X} = \mathbf{X}_1 + t(\mathbf{X}_2 - \mathbf{X}_1) \qquad (0 \leq t \leq 1).$$

A set D in R^n is *convex* if whenever it contains a pair of points it also contains the line segment connecting them. Show that an n-ball is convex.

T-13. Prove that the interior of a set is open.

2.2 Directional and Partial Derivatives

To prepare the way for the new ideas in this section, let us first recall the definition of the derivative of a function of one variable. If g is such a function and t_0 a point of its domain, consider the function G defined by

$$G(t) = \frac{g(t) - g(t_0)}{t - t_0}$$

for t in the domain of g and distinct from t_0. If $\lim_{t \to t_0} G(t)$ exists, we call its value the *derivative* of g at t_0, and denote it by $g'(t_0)$; thus

$$g'(t_0) = \lim_{t \to t_0} \frac{g(t) - g(t_0)}{t - t_0} .$$

An equivalent definition is

$$g'(t_0) = \lim_{h \to 0} \frac{g(t_0 + h) - g(t_0)}{h} .$$

Here, t_0 can be any point of the domain of g where the limit exists; hence we can omit the subscript and define the derivative g' at such points by

$$(1) \qquad g'(t) = \lim_{h \to 0} \frac{g(t + h) - g(t)}{h} .$$

Example 2.1 Let $t_0 = 2$ and $g(t) = 2t^2 + 3$; for $t \neq 2$ the difference quotient is

$$G(t) = \frac{g(t) - g(2)}{t - 2}$$

$$= \frac{2t^2 + 3 - 11}{t - 2} = \frac{2(t^2 - 4)}{t - 2}$$

$$= 2(t + 2).$$

The student should verify that

$$\lim_{t \to 2} G(t) = 8;$$

hence $g'(2) = 8$.

To compute the value of g' at any arbitrary point of its domain, we form the difference quotient on the right side of (1):

$$\frac{g(t+h) - g(t)}{h} = \frac{2(t+h)^2 + 3 - 2t^2 - 3}{h}$$

$$= 4t + 2h;$$

hence

$$g'(t) = \lim_{h \to 0} \frac{g(t+h) - g(t)}{h}$$

$$= 4t$$

for all t.

Example 2.2 Let $t_0 = 0$ and $g(t) = \sqrt{|t|}$. The difference quotient is

$$G(t) = \frac{\sqrt{|t|} - 0}{t - 0} = \pm \frac{1}{\sqrt{|t|}} \qquad (t \ne 0),$$

which does not approach a limit as t approaches zero; hence g does not have a derivative at $t_0 = 0$.

Directional Derivative

The following definition generalizes the idea of the derivative to functions whose domains are in R^n.

Definition 2.1 Let f be a real valued function defined on a subset D of R^n, let X_0 be an interior point of D, and U be a unit vector in R^n. Then the *directional derivative of f at X_0 in the direction of U* is defined by

$$(2) \qquad \frac{\partial f}{\partial U}(X_0) = \lim_{t \to 0} \frac{f(X_0 + tU) - f(X_0)}{t},$$

if the limit exists.

The problem of calculating directional derivatives quickly reduces to a problem in the ordinary calculus. Since X_0 is an interior point of D, the

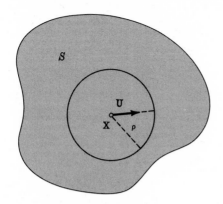

FIGURE 2.1

line segment S defined by

$$X = X_0 + tU, \qquad -\rho < t < \rho,$$

is in D for ρ sufficiently small (Fig. 2.1), and therefore

(3) $$g(t) = f(X_0 + tU)$$

is defined for $|t| < \rho$. The difference quotient on the right side of (2) can be rewritten as

$$\frac{f(X_0 + tU) - f(X_0)}{t} = \frac{g(t) - g(0)}{t};$$

letting t approach zero yields

$$\frac{\partial f}{\partial U}(X_0) = g'(0).$$

Example 2.3 Let

$$U = \begin{bmatrix} \dfrac{1}{\sqrt{2}} \\[2ex] -\dfrac{1}{\sqrt{2}} \end{bmatrix}$$

and

$$f(x, y) = x^2 + y^2 + xy.$$

In (3) take

$$g(t) = f\left(x + \frac{t}{\sqrt{2}}, y - \frac{t}{\sqrt{2}}\right)$$

$$= \left(x + \frac{t}{\sqrt{2}}\right)^2 + \left(y - \frac{t}{\sqrt{2}}\right)^2 + \left(x + \frac{t}{\sqrt{2}}\right)\left(y - \frac{t}{\sqrt{2}}\right);$$

then

$$g'(t) = \frac{2}{\sqrt{2}}\left(x + \frac{t}{\sqrt{2}}\right) - \frac{2}{\sqrt{2}}\left(y - \frac{t}{\sqrt{2}}\right) + \frac{1}{\sqrt{2}}\left(y - \frac{t}{\sqrt{2}}\right) - \frac{1}{\sqrt{2}}\left(x + \frac{t}{\sqrt{2}}\right)$$

$$= \frac{1}{\sqrt{2}}\left(x + \frac{t}{\sqrt{2}}\right) - \frac{1}{\sqrt{2}}\left(y - \frac{t}{\sqrt{2}}\right),$$

and

$$\frac{\partial f}{\partial U}(x, y) = g'(0)$$

$$= \frac{x - y}{\sqrt{2}}.$$

At

$$\mathbf{X}_0 = (1, 2),$$

$$\frac{\partial f}{\partial U}(1, 2) = -\frac{1}{\sqrt{2}}.$$

Geometrical Interpretation of Directional Derivatives

If g is a function of a single variable which has a derivative at $t = t_0$, then $g'(t_0)$ is the slope of the tangent to the curve defined by $y = g(t)$ at the point $(t_0, g(t_0))$ (Fig. 2.2); in fact, the equation of the tangent line is

$$y = g(t_0) + g'(t_0)(x - x_0),$$

or, in symmetric form,

(4)
$$x - x_0 = \frac{y - g(t_0)}{g'(t_0)}.$$

Partial derivatives can also be interpreted geometrically. Suppose S is a

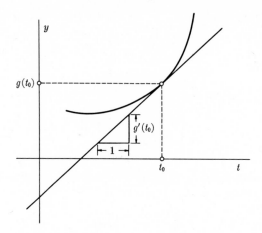

FIGURE 2.2

surface in R^3 defined by

$$z - f(x, y) = 0,$$

where f has domain D in R^2. Let (x_0, y_0) be an interior point of D, let

$$\mathbf{U} = \begin{bmatrix} u \\ v \end{bmatrix}$$

be a unit vector, and let Γ be the plane through $(x_0, y_0, 0)$ parallel to

$$\begin{bmatrix} u \\ v \\ 0 \end{bmatrix}$$

and the z-axis. Now suppose C is the trace of S on Γ (Fig. 2.3). A point (x, y, z) is on Γ if and only if $x = x_0 + tu$ and $y = y_0 + tv$; it is on S if and only if $z = f(x, y)$. Since C is the set of points which are on both Γ and S, it follows that (x, y, z) is on C if and only if

$$\begin{bmatrix} x \\ y \\ z \end{bmatrix} = \mathbf{\Phi}(t) = \begin{bmatrix} x_0 + tu \\ y_0 + tv \\ f(x_0 + tu, y_0 + tv) \end{bmatrix}.$$

This is a parametric equation for C. A tangent vector to C at

$P_0 = (x_0, y_0, z(x_0, y_0))$ is given by

$$\Phi'(0) = \begin{bmatrix} u \\ v \\ \dfrac{\partial f}{\partial \mathbf{U}}(x_0, y_0) \end{bmatrix} ;$$

consequently the tangent line T to C at P_0 (which is defined to be the line through P_0 in the direction of $\Phi'(0)$) is given by

$$\frac{x - x_0}{u} = \frac{y - y_0}{v} = \frac{z - f(x_0, y_0)}{\dfrac{\partial f}{\partial \mathbf{U}}(x_0, y_0)} ,$$

which is analogous to (4).

The tangent line is in Γ; if α is the angle between it and $\mathbf{U}$ (Fig. 2.3), then $\tan \alpha$ is the *slope of T in the plane Γ*, which is analogous to the familiar

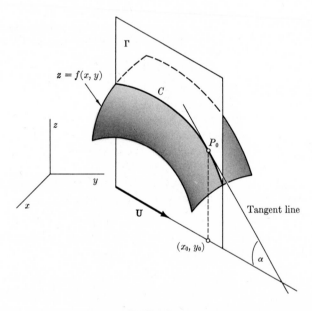

FIGURE 2.3

definition of slope for a line in R^2. Since

$$\cos \alpha = \frac{\mathbf{U} \cdot \mathbf{\Phi}'(0)}{|\mathbf{\Phi}'(0)|}$$

$$= \frac{1}{\sqrt{1 + \left(\dfrac{\partial f}{\partial \mathbf{U}} (x_0, y_0)\right)^2}},$$

it follows that

$$\sin \alpha = \frac{\dfrac{\partial f}{\partial \mathbf{U}} (x_0, y_0)}{\sqrt{1 + \left(\dfrac{\partial f}{\partial \mathbf{U}} (x_0, y_0)\right)^2}},$$

and the slope is given by

$$\tan \alpha = \frac{\partial f}{\partial \mathbf{U}} (x_0, y_0).$$

Partial Derivatives

Directional derivatives in the directions of the basis vectors

$$\mathbf{E}_1 = \begin{bmatrix} 1 \\ 0 \\ 0 \\ \vdots \\ 0 \end{bmatrix}, \quad \mathbf{E}_2 = \begin{bmatrix} 0 \\ 1 \\ 0 \\ \vdots \\ 0 \end{bmatrix}, \dots, \quad \mathbf{E}_{n-1} = \begin{bmatrix} 0 \\ \vdots \\ 0 \\ 1 \\ 0 \end{bmatrix}, \quad \mathbf{E}_n = \begin{bmatrix} 0 \\ \vdots \\ 0 \\ 0 \\ 1 \end{bmatrix}$$

occur so often that they deserve special attention. If $\mathbf{U} = \mathbf{E}_i$ we can write (2), with $t = h_i$, as

$$(5) \qquad \frac{\partial f}{\partial \mathbf{E}_i} (\mathbf{X}) = \lim_{h_i \to 0} \frac{f(\mathbf{X} + h_i \mathbf{E}_i) - f(\mathbf{X})}{h_i}$$

$$= \lim_{h_i \to 0} \frac{f(x_1, \dots, x_{i-1}, x_i + h_i, x_{i+1}, \dots, x_n) - f(x_1, \dots, x_{i-1}, x_i, x_{i+1}, \dots, x_n)}{h_i}$$

Definition 2.2 The directional derivative $(\partial f / \partial \mathbf{E}_i)(\mathbf{X})$ is called *the partial derivative of f with respect to* x_i. It is also denoted by $f_{x_i}(\mathbf{X})$ or $(\partial f / \partial x_i)(\mathbf{X})$.

If we ignore the other variables, the last member of (5) resembles the difference quotient that arises in the definition of an ordinary derivative. This leads to the following rule for calculating $f_{x_i}(\mathbf{X})$:

Treat all variables other than x_i as constants and differentiate f as if it were a function of x_i only, according to the rules of the one dimensional calculus.

Example 2.4 Let

$$f(x, y) = x^2 + y^2 + xy.$$

Considering y as a constant and differentiating with respect to x yields

$$f_x(x, y) = 2x + y;$$

considering x as a constant and differentiating with respect to y yields

$$f_y(x, y) = 2y + x.$$

The partial derivative f_{x_i} defines a new function of $\mathbf{X}$. Since it is obtained by differentiating a function of the single variable x_i according to the rules of the one-dimensional calculus, we have the following theorem.

Theorem 2.1 If $f_{x_i}(\mathbf{X})$ and $g_{x_i}(\mathbf{X})$ exist, then

$$\frac{\partial(f+g)}{\partial x_i}(\mathbf{X}) = \frac{\partial f}{\partial x_i}(\mathbf{X}) + \frac{\partial g}{\partial x_i}(\mathbf{X}),$$

$$\frac{\partial(f-g)}{\partial x_i}(\mathbf{X}) = \frac{\partial f}{\partial x_i}(\mathbf{X}) - \frac{\partial g}{\partial x_i}(\mathbf{X}),$$

$$\frac{\partial(fg)}{\partial x_i}(\mathbf{X}) = g(\mathbf{X})\frac{\partial f}{\partial x_i} + f(\mathbf{X})\frac{\partial g}{\partial x_i}(\mathbf{X})$$

and, if $g(\mathbf{X}) \neq 0$,

$$\frac{\partial}{\partial x_i}\left(\frac{f}{g}\right)(\mathbf{X}) = \frac{g(\mathbf{X})\dfrac{\partial f}{\partial x_i}(\mathbf{X}) - f(\mathbf{X})\dfrac{\partial g}{\partial x_i}(\mathbf{X})}{[g(\mathbf{X})]^2}.$$

When $n = 1$ the partial derivative f_x and the ordinary derivative $f' = df/dx$ are the same; hence in this case we shall use one of the latter symbols rather than f_x or $\partial f/\partial x$.

Higher Order Partial Derivatives

The partial derivative f_{x_i} (or, for that matter, any directional derivative) defines a function whose domain is the set of points at which it exists. We can ask the same questions about it as we did about f: Is it continuous? Does it have partial derivatives? If the partial derivative of f_{x_i} with respect to x_j exists, we denote it by

$$\frac{\partial}{\partial x_j}\left(\frac{\partial f}{\partial x_i}\right) = \frac{\partial^2 f}{\partial x_j\, \partial x_i} = f_{x_i x_j};$$

similarly

$$\frac{\partial}{\partial x_k}\left(\frac{\partial^2 f}{\partial x_j\, \partial x_i}\right) = f_{x_i x_j x_k}\, .$$

In general the function obtained by differentiating f with respect to $x_{i_1}, x_{i_2}, \ldots, x_{i_r}$ (in that order) is denoted by

$$\frac{\partial^r f}{\partial x_{i_r}\cdots \partial x_{i_2}\, \partial x_{i_1}} = f_{x_{i_1} x_{i_2}\cdots x_{i_r}};$$

here the integer r is the *order* of the partial derivative.

Example 2.5   Let $n = 3$ and

$$f(\mathbf{X}) = x^3 + 3xy + yz + 4z^2.$$

There are three first-order partial derivatives,

$$f_x(\mathbf{X}) = 3x^2 + 3y,$$
$$f_y(\mathbf{X}) = 3x + z,$$
$$f_z(\mathbf{X}) = y + 8z,$$

and nine second-order:

$$f_{xx}(\mathbf{X}) = 6x, \qquad f_{xy}(\mathbf{X}) = 3, \qquad f_{xz}(\mathbf{X}) = 0,$$
$$f_{yx}(\mathbf{X}) = 3, \qquad f_{yy}(\mathbf{X}) = 0, \qquad f_{yz}(\mathbf{X}) = 1,$$
$$f_{zx}(\mathbf{X}) = 0, \qquad f_{zy}(\mathbf{X}) = 1, \qquad f_{zz}(\mathbf{X}) = 8.$$

In general there are n^r rth-order partial derivatives of a function of n variables; however, it can be shown that the value of $f_{x_{i_1} x_{i_2}\cdots x_{i_r}}$ at a point $\mathbf{X}_0$ depends only upon which variables occur in the subscript and not on their order, provided all partial derivatives of order up to and including r are continuous at $\mathbf{X}_0$. For instance, $f_{xy} = f_{yx}$, $f_{xz} = f_{zx}$, and $f_{yz} = f_{zy}$ in Example

2.5. Whenever we speak of rth order partial derivatives, we shall assume that this condition is satisfied. Subject to this agreement we can define

$$\frac{\partial^s f}{\partial x_1^{s_1} \cdots \partial x_n^{s_n}}$$

where $s = s_1 + \cdots + s_n$, to be the function obtained by differentiating f s_1 times with respect to x_1, s_2 times with respect to $x_2, \ldots,$ s_n times with respect to x_n.

Example 2.6 If f is a function of two variables, we write

$$f_{xx} = \frac{\partial^2 f}{\partial x^2}, \qquad f_{yy} = \frac{\partial^2 f}{\partial y^2},$$

and

$$f_{xy} = f_{yx} = \frac{\partial^2 f}{\partial x \partial y}.$$

For third-order derivatives,

$$f_{xxy} = f_{xyx} = f_{yxx} = \frac{\partial^3 f}{\partial x^2 \partial y}, \quad \text{etc.}$$

EXERCISES 2.2

1. Compute $(\partial f / \partial \mathbf{U})\, (\mathbf{X}_0)$ for the following functions at $\mathbf{X}_0 = (1, 2)$ in the direction

$$\mathbf{U} = \begin{bmatrix} \dfrac{1}{\sqrt{2}} \\[2mm] \dfrac{1}{\sqrt{2}} \end{bmatrix}.$$

   (a) $f(x, y) = x^2 + y^2.$ (b) $f(x, y) = x^2 - xy + 2y^2.$

   (c) $f(x, y) = x.$ (d) $f(x, y) = \dfrac{x}{x + y}.$

2.   Repeat Exercise 1 for $\mathbf{X}_0 = (1, 2, 3)$ and

$$\mathbf{U} = \begin{bmatrix} \dfrac{1}{\sqrt{3}} \\[2mm] -\dfrac{1}{\sqrt{3}} \\[2mm] \dfrac{1}{\sqrt{3}} \end{bmatrix}.$$

(a) $f(x, y, z) = \log(x^2 + y^2 + z^2)$.

(b) $f(x, y, z) = x + y + z$.

(c) $f(x, y, z) = e^{x+y+z}$.

(d) $f(x, y, z) = \dfrac{1}{x^2 + y^2 + z^2}$.

3.   Find $\partial f/\partial x$, $\partial f/\partial y$, and $\partial f/\partial z$ for each of the following; also, calculate $f_x(\mathbf{X}_0), f_y(\mathbf{X}_0)$, and $f_z(\mathbf{X}_0)$.

(a) $f(\mathbf{X}) = 2x^2 + y^2 + z$, $\mathbf{X}_0 = (1, 1, 2)$.

(b) $f(\mathbf{X}) = e^x \cos(x^2 + y^2 + z^2)$, $\mathbf{X}_0 = \left(\sqrt{\dfrac{\pi}{4}}, \sqrt{\dfrac{\pi}{4}}, 0\right)$.

(c) $f(\mathbf{X}) = \log(1 + x^2 + y + z)$, $\mathbf{X}_0 = (0, 0, 0)$.

(d) $f(\mathbf{X}) = \exp(\log\sqrt{x^2 + y^2 + 2xy})$, $\mathbf{X}_0 = (1, 1)$.

4.   Compute f_x, f_y, and f_z for each of the following.

(a) $f(x, y, z) = xe^{yz}$.

(b) $f(x, y, z) = \dfrac{e^{xyz}}{1 + z}$.

(c) $f(x, y, z) = \log(x + y)$.

(d) $f(x, y, z) = z + \cos x^2 y$.

5.   Compute $f_{xy}(-1, 2)$ and $f_{yx}(-1, 2)$ for

(a) $f(x, y) = 3x^2 y - 2xy^2$;

(b) $f(x, y) = 2xy - y^2$.

6.   Compute all second partial derivatives for the functions in Exercises 3(a) and 4(a).

7. Compute all first partial derivatives for each of the following.

   (a) $f(x, y, z) = xye^z + 2y^2z - 3xz \cos y.$

   (b) $f(x, y) = \dfrac{2x^2 + 3y^2 - 5x^3y}{2x + y^2}.$

   (c) $f(x) = 3x + 4.$

8. Let
$$f(x, y, z) = 2x^2yz + \cos(2x - 3y + 4z).$$

   Compute

   (a) $\dfrac{\partial^2 f}{\partial x \, \partial y}$,   (b) $\dfrac{\partial^3 f}{\partial x \, \partial y^2}$,   (c) $\dfrac{\partial^3 f}{\partial x \, \partial y \, \partial z}$,

   (d) $\dfrac{\partial^4 f}{\partial x^2 \, \partial y^2}$,   (e) $\dfrac{\partial^4 f}{\partial x^2 \, \partial y \, \partial z}.$

9. If $f(x, y) = e^x \cos y$, show that $f_{xx} + f_{yy} = 0.$

10. If $f(x, y) = \cos xy$, show that $xf_x(x, y) - yf_y(x, y) = 0.$

11. Where do the following functions fail to have first partial derivatives?

   (a) $f(x, y) = \sqrt{x + y}.$

   (b) $f(x, y) = \sqrt{1 - x^2 - y^2}.$

   (c) $f(x, y) = \log(1 + y \sin x).$

THEORETICAL EXERCISES

T-1.   In R^1 there are only two unit vectors: $\mathbf{U}_1 = [1]$ and $\mathbf{U}_2 = [-1]$. Show that

$$\frac{\partial f}{\partial \mathbf{U}_1}(x) = -\frac{\partial f}{\partial \mathbf{U}_2}(x) = f'(x).$$

T-2.   If $\mathbf{V} = -\mathbf{U}$, show that

$$\frac{\partial f}{\partial \mathbf{V}}(\mathbf{X}) = -\frac{\partial f}{\partial \mathbf{U}}(\mathbf{X}).$$

T-3.   If $\mathbf{X}_0$ is an interior point of the domain of f, we say that f is *continuous at $\mathbf{X}_0$ in the direction $\mathbf{U}$* if $g(t) = f(\mathbf{X}_0 + t\mathbf{U})$ is continuous at $t = 0$.

Prove: (a) If f is continuous at $\mathbf{X}_0$, then it is continuous at $\mathbf{X}_0$ in every direction. (b) If $(\partial f/\partial \mathbf{U})(\mathbf{X}_0)$ exists, then f is continuous at $\mathbf{X}_0$ in the direction $\mathbf{U}$.

T-4. Prove that

$$\frac{\partial(f+g)}{\partial \mathbf{U}}(\mathbf{X}) = \frac{\partial f}{\partial \mathbf{U}}(\mathbf{X}) + \frac{\partial g}{\partial \mathbf{U}}(\mathbf{X}).$$

T-5. Let (x_0, y_0) be an interior point of the domain of f. Show that

$$g(y) = \frac{f(x, y_0) - f(x_0, y_0)}{x - x_0}$$

has a removable discontinuity (Exercise T-10, Section 2.1) at y_0 if and only if $f_x(x_0, y_0)$ exists.

2.3 Differentiable Functions

A function g of one variable is said to be differentiable at a point $t = t_0$ if it has a derivative there; if this is the case then

$$\lim_{t \to t_0} \frac{g(t) - g(t_0)}{t - t_0} = g'(t_0).$$

We can rewrite this as

$$\lim_{t \to t_0} \frac{g(t) - g(t_0) - g'(t_0)(t - t_0)}{t - t_0} = 0,$$

or better yet,

$$g(t) = g(t_0) + g'(t_0)(t - t_0) + \epsilon(t, t_0)(t - t_0),$$

where

$$\lim_{t \to t_0} \epsilon(t, t_0) = 0.$$

Thus, differentiability at t_0 implies that, near t_0, g can be approximated so well by

$$h(t) = g(t_0) + g'(t_0)(t - t_0)$$

that the error $g(t) - h(t)$ approaches zero faster than $t - t_0$ (Fig. 3.1). Since h is simple function of t, while g may be quite complicated, this is a desirable situation.

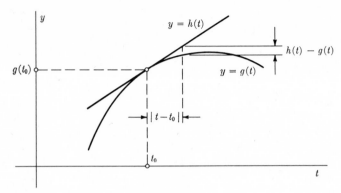

FIGURE 3.1

This interpretation of differentiability motivates the following definition for functions of n variables.

Definition 3.1 Let f be a real valued function defined on a domain D in R^n and let $\mathbf{X}_0$ be an interior point of D. We say that f is *differentiable at* $\mathbf{X}_0$ if there exist constants $a_1, a_2, \ldots, a_n$ such that

(1)
$$\lim_{\mathbf{X} \to \mathbf{X}_0} \frac{f(\mathbf{X}) - f(\mathbf{X}_0) - L(\mathbf{X} - \mathbf{X}_0)}{|\mathbf{X} - \mathbf{X}_0|} = 0,$$

where L is the function from R^n to R whose value at $\mathbf{Y} = (y_1, \ldots, y_n)$ is

$$L(\mathbf{Y}) = a_1 y_1 + a_2 y_2 + \cdots + a_n y_n.$$

A function of this form is called a *linear transformation from R^n to R.* If $\mathbf{U}$ and $\mathbf{V}$ are any two vectors in R^n then $L(\mathbf{U} + \mathbf{V}) = L(\mathbf{U}) + L(\mathbf{V})$; if c is a real number then $L(c\mathbf{U}) = cL(\mathbf{U})$.

If f is differentiable at every point of a set S, we say that f is *differentiable on S.*

Example 3.1 Let $\mathbf{X}_0 = (x_0, y_0)$ and $f(x, y) = 2x^2 + y^2$. To show that f is differentiable at $\mathbf{X}_0$, we must find constants a_1 and a_2 such that

$$\lim_{\mathbf{X} \to \mathbf{X}_0} \frac{2x^2 + y^2 - 2x_0^2 - y_0^2 - a_1(x - x_0) - a_2(y - y_0)}{|\mathbf{X} - \mathbf{X}_0|} = 0.$$

Letting $\mathbf{X}$ approach $\mathbf{X}_0$ along the line $x = x_0$, we find that

$$0 = \lim_{y \to y_0} \frac{y^2 - y_0^2 - a_2(y - y_0)}{y - y_0}$$

$$= \lim_{y \to y_0} (y + y_0 - a_2);$$

thus $a_2 = 2y_0$. Similarly, approaching $\mathbf{X}_0$ along the line $y = y_0$ leads to the conclusion that $a_1 = 4x_0$. Thus if L satisfies the requirement of Definition 3.1, then

$$L(\mathbf{X} - \mathbf{X}_0) = 4x_0(x - x_0) + 2y_0(y - y_0).$$

We have not, however, shown that f is differentiable at $\mathbf{X}_0$; to do this we must prove that

$$E(\mathbf{X}) = \frac{2x^2 + y^2 - 2x_0^2 - y_0^2 - 4x_0(x - x_0) - 2y_0(y - y_0)}{|\mathbf{X} - \mathbf{X}_0|}$$

approaches zero as $\mathbf{X}$ approaches $\mathbf{X}_0$, not only along lines parallel to the coordinate axes, but along arbitrary paths. This is accomplished by rewriting,

$$E(\mathbf{X}) = \frac{2(x - x_0)^2 + (y - y_0)^2}{|\mathbf{X} - \mathbf{X}_0|}$$

$$\leq \frac{2\,|\mathbf{X} - \mathbf{X}_0|^2}{|\mathbf{X} - \mathbf{X}_0|} = 2\,|\mathbf{X} - \mathbf{X}_0|;$$

thus, the conclusion follows.

The way in which we obtained a_1 and a_2 in this example made it clear that they are unique. This is true in general, as is shown in the following theorem.

Theorem 3.1 If f is differentiable at $\mathbf{X}_0$, then the partial derivatives $f_{x_1}(\mathbf{X}_0), \ldots, f_{x_n}(\mathbf{X}_0)$ exist, and in Definition 3.1,

$$(2) \qquad\qquad a_i = f_{x_i}(\mathbf{X}_0) \qquad (i = 1, 2, \ldots, n);$$

thus if there is a linear transformation that satisfies (1), it is unique.

Proof. In (1) let $\mathbf{X}$ approach $\mathbf{X}_0 = (c_1, \ldots, c_n)$ along the line $x_j = c_j$ $(i \neq j)$; then (1) reduces to

$$\lim_{x_i \to c_i} \frac{f(c_1, \ldots, c_{i-1}, x_i, c_{i+1}, \ldots, c_n) - f(c_1, \ldots, c_{i-1}, c_i, c_{i+1}, \ldots, c_n) - a_i(x_i - c_i)}{x_i - c_i}$$

$$= 0$$

so that (2) follows from the definition of $f_{x_i}(\mathbf{X}_0)$.

Thus, in Example 3.1, $a_1 = 4x_0 = f_x(\mathbf{X}_0)$, and $a_2 = 2y_0 = f_y(\mathbf{X}_0)$.

The Differential

Definition 3.2 If f is differentiable at $\mathbf{X}_0$, then $d_{\mathbf{X}_0} f$, the *differential of f at $\mathbf{X}_0$*, is the linear transformation from R^n to R whose value at $\mathbf{Y} = (y_1, \ldots, y_n)$ is

(3) $$(d_{\mathbf{X}_0} f)(\mathbf{Y}) = f_{x_1}(\mathbf{X}_0) y_1 + f_{x_2}(\mathbf{X}_0) y_2 + \cdots + f_{x_n}(\mathbf{X}_0) y_n.$$

When using the differential, we shall almost always write $\mathbf{Y} = \mathbf{X} - \mathbf{X}_0$ in (3); thus if $\mathbf{X}_0 = (c_1, \ldots, c_n)$,

$$(d_{\mathbf{X}_0} f)(\mathbf{X} - \mathbf{X}_0) = f_{x_1}(\mathbf{X}_0)(x_1 - c_1) + \cdots + f_{x_n}(\mathbf{X}_0)(x_n - c_n).$$

If $n = 2$, we shall usually write

$$(d_{\mathbf{X}_0} f)(\mathbf{X} - \mathbf{X}_0) = f_x(\mathbf{X}_0)(x - x_0) + f_y(\mathbf{X}_0)(y - y_0),$$

and if $n = 3$,

$$(d_{\mathbf{X}_0} f)(\mathbf{X} - \mathbf{X}_0) = f_x(\mathbf{X}_0)(x - x_0) + f_y(\mathbf{X}_0)(y - y_0) + f_z(\mathbf{X}_0)(z - z_0).$$

Example 3.2 Let $\mathbf{X}_0 = (1, 2)$ and

$$f(x, y) = 2x^2 + y^2 + xy;$$

then

$$f_x(x, y) = 4x + y,$$
$$f_y(x, y) = 2y + x,$$

and

$$(d_{\mathbf{X}_0} f)(\mathbf{X} - \mathbf{X}_0) = 6(x - 1) + 5(y - 2).$$

Also, if $\mathbf{X}_1 = (3, 5)$, then

$$(d_{\mathbf{X}_0} f)(\mathbf{X}_1 - \mathbf{X}_0) = 6(3 - 1) + 5(5 - 2) = 27.$$

Example 3.3 Let $\mathbf{X}_0 = \left(\dfrac{\pi}{8}, \dfrac{\pi}{8}, 1\right)$ and $f(x, y, z) = z \cos(x + y)$; then

$$f_x(x, y, z) = f_y(x, y, z) = -z \sin(x + y),$$
$$f_z(x, y, z) = \cos(x + y),$$

and

$$(d_{\mathbf{X}_0} f)(\mathbf{X} - \mathbf{X}_0) = -\frac{1}{\sqrt{2}}\left(x - \frac{\pi}{8}\right) - \frac{1}{\sqrt{2}}\left(y - \frac{\pi}{8}\right) + \frac{1}{\sqrt{2}}(z - 1).$$

From (1) it follows that, if f is differentiable at $\mathbf{X}_0$, the difference $f(\mathbf{X}) - f(\mathbf{X}_0)$ can be approximated so well near $\mathbf{X}_0$ by $(d_{\mathbf{X}_0} f)(\mathbf{X} - \mathbf{X}_0)$

that the error

$$f(\mathbf{X}) - f(\mathbf{X}_0) - (d_{\mathbf{X}_0} f)(\mathbf{X} - \mathbf{X}_0)$$

approaches zero faster than $|\mathbf{X} - \mathbf{X}_0|$. When we wish to use this fact in applications, we shall write simply

$$f(\mathbf{X}) - f(\mathbf{X}_0) \cong (d_{\mathbf{X}_0} f)(\mathbf{X} - \mathbf{X}_0);$$

the precise meaning of this relationship will be as given by (1); that is,

$$\lim_{\mathbf{X} \to \mathbf{X}_0} \frac{f(\mathbf{X}) - f(\mathbf{X}_0) - (d_{\mathbf{X}_0} f)(\mathbf{X} - \mathbf{X}_0)}{|\mathbf{X} - \mathbf{X}_0|} = 0.$$

Example 3.4 Suppose a box is designed to have length $l_0 = 10$ in., width $w_0 = 8$ in. and height $h_0 = 4$ in., and that the manufacturing process is such that the lengths are in error by at most one percent. To estimate an upper bound for the error in the volume V of the box we write $V = lwh$, and let $\mathbf{X} = (l, w, h)$ and $\mathbf{X}_0 = (l_0, w_0, h_0)$. Then

$$(d_{\mathbf{X}_0} V)(\mathbf{X} - \mathbf{X}_0) = \frac{\partial V}{\partial l}(l_0, w_0, h_0)(l - l_0) + \frac{\partial V}{\partial w}(l_0, w_0, h_0)(w - w_0)$$

$$+ \frac{\partial V}{\partial h}(l_0, w_0, h_0)(h - h_0)$$

$$= w_0 h_0 l_0 \left(\frac{l - l_0}{l_0} + \frac{w - w_0}{w_0} + \frac{h - h_0}{h_0} \right).$$

Since $l_0 w_0 h_0 = (10 \text{ in.})(8 \text{ in.})(4 \text{ in.}) = 320 \text{ in.}^3$ and

$$\left| \frac{l - l_0}{l_0} \right| \leq .01, \qquad \left| \frac{w - w_0}{w_0} \right| \leq .01, \quad \text{and} \quad \left| \frac{h - h_0}{h_0} \right| \leq .01,$$

it follows that

$$|(d_{\mathbf{X}_0} V)(\mathbf{X} - \mathbf{X}_0)| \leq (320 \text{ in.}^3)(0.01 + 0.01 + 0.01) = 9.6 \text{ in.}^3.$$

Thus the error in the volume, which is given by

$$\Delta V = V(\mathbf{X}) - V(\mathbf{X}_0),$$

satisfies

$$|\Delta V| \leq 9.6 \text{ in.}^3 + |E|,$$

where E is the error committed in approximating ΔV by the differential. However, $|E|$ is small compared to the terms in the differential; hence we take 9.6 in.^3 as a reasonable bound for the error.

For functions of one variable, differentiability and existence of a differential are equivalent, as we pointed out in the beginning of this section; however, if $n \geq 2$, the function may fail to be differentiable at a point even though its partial derivatives exist there.

Example 3.5   Let $\mathbf{X}_0 = (0, 0)$ and $f(x, y) = \sqrt{|xy|}$. Since $f(0, y) = f(x, 0) = 0$, it follows that $f_x(0, 0) = f_y(0, 0) = 0$, and $d_{\mathbf{X}_0} f = 0$. Hence,

$$\frac{f(x, y) - f(0, 0) - (d_{\mathbf{X}_0} f)(\mathbf{X})}{|\mathbf{X}|} = \sqrt{\frac{|xy|}{x^2 + y^2}},$$

which does not approach a limit as $\mathbf{X}$ approaches $(0, 0)$ (see Example 1.14.); thus f is not differentiable at $(0, 0)$.

Example 3.6   Let $\mathbf{X}_0 = (c_1, \ldots, c_n)$, $\mathbf{Y} = (y_1, \ldots, y_n)$, and $f_i(\mathbf{X}) = x_i$; then since

$$(d_{\mathbf{X}_0} f_i)(\mathbf{Y}) = \frac{\partial f_i}{\partial x_1}(\mathbf{X}_0) y_1 + \frac{\partial f_i}{\partial x_2}(\mathbf{X}_0) y_2 + \cdots + \frac{\partial f_i}{\partial x_n}(\mathbf{X}_0) y_n$$

and

$$\frac{\partial f_i}{\partial x_k}(\mathbf{X}_0) = \begin{cases} 0 & \text{if } k \neq i, \\ 1 & \text{if } k = i, \end{cases}$$

it follows that

$$(d_{\mathbf{X}_0} f_i)(\mathbf{Y}) = y_i$$

and

$$\frac{f_i(\mathbf{X}) - f_i(\mathbf{X}_0) - (d_{\mathbf{X}_0} f_i)(\mathbf{X} - \mathbf{X}_0)}{|\mathbf{X} - \mathbf{X}_0|} = \frac{x_i - c_i - (x_i - c_i)}{|\mathbf{X} - \mathbf{X}_0|}$$

$$= 0.$$

Thus f_i is differentiable at $\mathbf{X}_0$. Since the argument holds for all $\mathbf{X}_0$, f_i is differentiable on all of R^n.

The differential of f_i is so important that we introduce a special, simpler notation for it.

Definition 3.3   The linear transformation from R^n to R that has the value y_i at $\mathbf{Y} = (y_1, \ldots, y_n)$ is called the *differential of x_i* and is denoted by dx_i. Thus

$$dx_i(\mathbf{Y}) = y_i.$$

The linear transformation dx_i is really the differential of f_i as defined in Example 3.6, and one could correctly argue that the proper notation for it would be $d_{X_0} f_i$. We adopt the special notation because it is convenient and traditional. The student should observe that although the symbol dx_i does not identify the dimension of the space on which it is defined, this will always be clear from the context.

If $n = 2$ we shall usually write dx and dy; thus

$$dx(x - x_0, y - y_0) = x - x_0,$$
$$dy(x - x_0, y - y_0) = y - y_0;$$

hence the differential in Example 3.2 is

$$d_{X_0} f = 6\, dx + 4\, dy.$$

If $n = 3$,

$$dx(x - x_0, y - y_0, z - z_0) = x - x_0,$$
$$dy(x - x_0, y - y_0, z - z_0) = y - y_0,$$
$$dz(x - x_0, y - y_0, z - z_0) = z - z_0;$$

hence the differential in Example 3.3 is

$$d_{X_0} f = -\frac{1}{\sqrt{2}}\,(dx + dy - dz).$$

From (3), the differential of an arbitrary function can be written as

$$d_{X_0} f = f_{x_1}(\mathbf{X_0})\, dx_1 + \cdots + f_{x_n}(\mathbf{X_0})\, dx_n.$$

When we do not need to emphasize the particular point $\mathbf{X_0}$, we shall write this more simply as

$$(4) \qquad\qquad df = f_{x_1}\, dx_1 + \cdots + f_{x_n}\, dx_n,$$

which is a convenient symbolic shorthand for what we really mean:

$$d_{\mathbf{X}} f = f_{x_1}(\mathbf{X})\, dx_1 + \cdots + f_{x_n}(\mathbf{X})\, dx_n.$$

When we are dealing with the differential of a specific function, say

$$f(x, y) = x^2 + 3y^3,$$

at a general point, we shall write

$$df = 2x\, dx + 9y^2\, dy;$$

this is not entirely consistent with (4), but the inconsistency is harmless.

Example 3.7   Let $\mathbf{X_0} = (c_1, \ldots, c_n)$ and $f(\mathbf{X}) = \alpha + \beta_1 x_1 + \cdots + \beta_n x_n$, where $\alpha, \beta_1, \ldots, \beta_n$ are constants. Let $L = \beta_1\, dx_1 + \cdots + \beta_n\, dx_n$; then L is a linear transformation from R^n to R and, if $\mathbf{X_0}$ is an arbitrary point in R^n,

$$f(\mathbf{X}) - f(\mathbf{X_0}) - L(\mathbf{X} - \mathbf{X_0}) = 0.$$

Thus f is differentiable on R^n, and

$$df = \beta_1 \, dx_1 + \cdots + \beta_n \, dx_n.$$

In particular, if $\beta_1 = \cdots = \beta_n = 0$, then $df = 0$; that is, the differential of a constant function is zero.

Example 3.8   If f is a function of a single variable t, then

$$df = f' \, dt,$$

where dt is the linear transformation on R defined by $dt(y) = y$. The function df/dt is a constant with respect to y:

$$\frac{df}{dt} = f'.$$

Theorem 3.2   If f is differentiable at $\mathbf{X}_0 = (c_1, \ldots, c_n)$, then it is continuous at $\mathbf{X}_0$.

Proof.   Write

$$f(\mathbf{X}) - f(\mathbf{X}_0) = (f(\mathbf{X}) - f(\mathbf{X}_0) - (d_{\mathbf{X}_0} f)(\mathbf{X} - \mathbf{X}_0)) + (d_{\mathbf{X}_0} f)(\mathbf{X} - \mathbf{X}_0).$$

The first term approaches zero as $\mathbf{X}$ approaches $\mathbf{X}_0$, by hypothesis; the second does also, since, by Schwarz's inequality,

$$| (d_{\mathbf{X}_0} f)(\mathbf{X} - \mathbf{X}_0) | = | f_{x_1}(\mathbf{X}_0)(x_1 - c_1) + \cdots + f_{x_n}(\mathbf{X}_0)(x_n - c_n) |$$

$$\leq [f_{x_1}^2(\mathbf{X}_0) + \cdots + f_{x_n}^2(\mathbf{X}_0)]^{1/2}$$

$$\times [(x_1 - c_1)^2 + \cdots + (x_n - c_n)^2]^{1/2},$$

or

$$| (d_{\mathbf{X}_0} f)(\mathbf{X} - \mathbf{X}_0) | \leq [f_{x_1}^2(\mathbf{X}_0) + \cdots + f_{x_n}^2(\mathbf{X}_0)]^{1/2} | \mathbf{X} - \mathbf{X}_0 |.$$

We saw in Theorem 3.1 that if f is differentiable at $\mathbf{X}_0$, then $f_{x_1}(\mathbf{X}_0), \ldots, f_{x_n}(\mathbf{X}_0)$ exist. We also saw in Example 3.5 that f may fail to be differentiable at $\mathbf{X}_0$ even though the partial derivatives exist there. The next theorem gives sufficient (although not necessary) conditions for differentiability.

Theorem 3.3   Let f be a real valued function defined on a domain D of R^n, and let $\mathbf{X}_0$ be an interior point of D. If the partial derivatives f_{x_i} $(i = 1, 2, \ldots, n)$ exist in some n-ball about $\mathbf{X}_0$ and are continuous at $\mathbf{X}_0$, then f is differentiable at $\mathbf{X}_0$.

Before proving this theorem we recall, without proof, a useful result from the calculus of one variable.

Lemma 3.1 (Mean Value Theorem for Functions of One Variable) Let g be continuous in the closed interval $[a, b]$ and have a derivative at each point of the open interval (a, b). Then there is a point c, $a < c < b$, such that

$$f(b) - f(a) = f'(c)(b - a).$$

Proof of Theorem 3.3. We confine our attention to $n = 3$; the proof is easily extended to any n.

Assume that $|\mathbf{X} - \mathbf{X}_0| < \rho$ and that the ball B_ρ of radius ρ about $\mathbf{X}_0$ is contained in D. Let

$$\mathbf{X} = (x, y, z), \qquad \mathbf{X}_0 = (x_0, y_0, z_0)$$

and define

$$\mathbf{X}_1 = (x_0, y, z), \qquad \mathbf{X}_2 = (x_0, y_0, z);$$

then $\mathbf{X}_1$ and $\mathbf{X}_2$ are in B_ρ and, since B_ρ is convex (Exercise T-12, Section 2.1), so are the line segments L_1 from $\mathbf{X}$ to $\mathbf{X}_1$, L_2 from $\mathbf{X}_1$ to $\mathbf{X}_2$, and L_3 from $\mathbf{X}_2$ to $\mathbf{X}_0$ (Fig. 3.2).

Now

$$f(\mathbf{X}) - f(\mathbf{X}_0) = (f(\mathbf{X}) - f(\mathbf{X}_1)) + (f(\mathbf{X}_1) - f(\mathbf{X}_2))$$
$$+ (f(\mathbf{X}_2) - f(\mathbf{X}_0)).$$

The arguments in each of the three parentheses differ only in their first, second and third coordinates, respectively. Applying Lemma 3.1 to $f(\mathbf{X}) - f(\mathbf{X}_1)$, $f(\mathbf{X}_1) - f(\mathbf{X}_2)$, and $f(\mathbf{X}_2) - f(\mathbf{X}_0)$ as functions of single variables yields

$$(5) \quad f(\mathbf{X}) - f(\mathbf{X}_0) = f_x(\hat{\mathbf{X}}_1)(x - x_0) + f_y(\hat{\mathbf{X}}_2)(y - y_0) + f_z(\hat{\mathbf{X}}_3)(z - z_0),$$

where

$$\hat{\mathbf{X}}_1 = (\hat{x}, y, z), \qquad \hat{\mathbf{X}}_2 = (x_0, \hat{y}, z), \qquad \hat{\mathbf{X}}_3 = (x_0, y_0, \hat{z});$$

$\hat{x}$ is between x and x_0, $\hat{y}$ is between y and y_0, and $\hat{z}$ is between z and z_0. Thus $\hat{\mathbf{X}}_1$ is on L_1, $\hat{\mathbf{X}}_2$ is on L_2, and $\hat{\mathbf{X}}_3$ is on L_3.

Now subtract $(d_{\mathbf{X}_0} f)(\mathbf{X} - \mathbf{X}_0)$ from both sides of (5):

$$(6) \quad f(\mathbf{X}) - f(\mathbf{X}_0) - (d_{\mathbf{X}_0} f)(\mathbf{X} - \mathbf{X}_0)$$
$$= (f_x(\hat{\mathbf{X}}_1) - f_x(\mathbf{X}_0))(x - x_0) + (f_y(\hat{\mathbf{X}}_2) - f_y(\mathbf{X}_0))(y - y_0)$$
$$+ (f_z(\hat{\mathbf{X}}_3) - f_z(\mathbf{X}_0))(z - z_0).$$

Given an arbitrary $\epsilon > 0$, there is a $\delta > 0$ such that $|f_x(\mathbf{X}) - f_x(\mathbf{X}_0)| < \epsilon/3$, $|f_y(\mathbf{X}) - f_y(\mathbf{X}_0)| < \epsilon/3$, and $|f_z(\mathbf{X}) - f_z(\mathbf{X}_0)| < \epsilon/3$ whenever $|\mathbf{X} - \mathbf{X}_0| < \delta$, because the partial derivatives are continuous at $\mathbf{X}_0$. Since

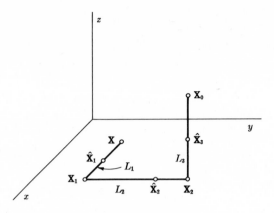

FIGURE 3.2

$|\hat{\mathbf{X}}_i - \mathbf{X}_0| < |\mathbf{X} - \mathbf{X}_0|$ $(i = 1, \; 2, \; 3)$, it follows from (6) that, if $|\mathbf{X} - \mathbf{X}_0| < \delta$, then

$$\frac{|f(\mathbf{X}) - f(\mathbf{X}_0) - (d_{\mathbf{X}_0}f)(\mathbf{X} - \mathbf{X}_0)|}{|\mathbf{X} - \mathbf{X}_0|}$$

$$< \frac{\epsilon}{3}\left(\frac{|x - x_0| + |y - y_0| + |z - z_0|}{|\mathbf{X} - \mathbf{X}_0|}\right) \le \epsilon,$$

since $|x - x_0|$, $|y - y_0|$, and $|z - z_0|$ are each $\le |\mathbf{X} - \mathbf{X}_0|$; thus f is differentiable at $\mathbf{X}_0$.

Theorem 3.4 If f and g are differentiable at $\mathbf{X}_0$ and α is a constant, then αf, $f + g$, and fg are differentiable at $\mathbf{X}_0$; the same is true of f/g provided $g(\mathbf{X}_0) \ne 0$. The differentials are given by

$$d_{\mathbf{X}_0}(\alpha f) = \alpha \, d_{\mathbf{X}_0}f,$$

$$d_{\mathbf{X}_0}(f + g) = d_{\mathbf{X}_0}f + d_{\mathbf{X}_0}g,$$

$$d_{\mathbf{X}_0}(fg) = f(\mathbf{X}_0) \, d_{\mathbf{X}_0}g + g(\mathbf{X}_0) \, d_{\mathbf{X}_0}f,$$

$$d_{\mathbf{X}_0}(f/g) = \frac{g(\mathbf{X}_0) \, d_{\mathbf{X}_0}f - f(\mathbf{X}_0) \, d_{\mathbf{X}_0}g}{[g(\mathbf{X}_0)]^2}.$$

Proof. Because it is much easier, we give the proof under the additional assumption that f and g have continuous first partial derivatives near $\mathbf{X}_0$. Then αf, $f + g$, fg, and f/g (assuming $g(\mathbf{X}_0) \ne 0$) have continuous partial derivatives near $\mathbf{X}_0$, because of Theorem 2.1, and Theorem 1.3, (applied to the partial derivatives of f and g). From Theorem 3.3 it now follows that

αf, $f + g$, fg, and f/g are differentiable at $\mathbf{X}_0$ and, for example,

$$
\begin{aligned}
d_{\mathbf{X}_0}(fg) &= (fg)_{x_1}(\mathbf{X}_0)\ dx_1 + \cdots + (fg)_{x_n}(\mathbf{X}_0)\ dx_n \\
&= (f_{x_1}(\mathbf{X}_0)g(\mathbf{X}_0) + f(\mathbf{X}_0)g_{x_1}(\mathbf{X}_0))\ dx_1 + \cdots \\
&\quad + (f_{x_n}(\mathbf{X}_0)g(\mathbf{X}_0) + f(\mathbf{X}_0)g_{x_n}(\mathbf{X}_0))\ dx_n \\
&= g(\mathbf{X}_0)(f_{x_1}(\mathbf{X}_0)\ dx_1 + \cdots + f_{x_n}(\mathbf{X}_0)\ dx_n) \\
&\quad + f(\mathbf{X}_0)(g_{x_1}(\mathbf{X}_0)\ dx_1 + \cdots + g_{x_n}(\mathbf{X}_0)\ dx_n) \\
&= g(\mathbf{X}_0)\ d_{\mathbf{X}_0}f + f(\mathbf{X}_0)\ d_{\mathbf{X}_0}g.
\end{aligned}
$$

The other assertions of the theorem can be obtained by similar manipulations.

When there is no need to emphasize $\mathbf{X}_0$ the equations of Theorem 3.4 can be written more simply as

$$d(\alpha f) = \alpha\ df,$$

$$d(f + g) = df + dg,$$

$$d(fg) = f\ dg + g\ df,$$

$$d(f/g) = \frac{g\ df - f\ dg}{g^2}.$$

Directional Derivatives from the Differential

If f is differentiable at $\mathbf{X}_0$, $\mathbf{U}$ is a unit vector, and t a real parameter, then setting $\mathbf{X} = \mathbf{X}_0 + t\mathbf{U}$ in (1) yields

$$
\lim_{t \to 0} \frac{f(\mathbf{X}_0 + t\mathbf{U}) - f(\mathbf{X}_0) - (d\mathbf{X}_0 f)(t\mathbf{U})}{t} = 0.
$$

Since $d_{\mathbf{X}_0}f$ is a linear transformation, $(d_{\mathbf{X}_0}f)(t\mathbf{U}) = t(d_{\mathbf{X}_0}f)(\mathbf{U})$; substituting this in the last equation yields

$$
\lim_{t \to 0} \frac{f(\mathbf{X}_0 + t\mathbf{U}) - f(\mathbf{X}_0)}{t} = (d_{\mathbf{X}_0}f)(\mathbf{U}),
$$

which proves the following theorem.

Theorem 3.5 If f is differentiable at a point $\mathbf{X}_0$ and

$$
\mathbf{U} = \begin{bmatrix} u_1 \\ u_2 \\ \vdots \\ u_n \end{bmatrix}
$$

is an arbitrary unit vector, then the directional derivative of f at $\mathbf{X}_0$ in the direction of $\mathbf{U}$ exists, and is given by

$$\frac{\partial f}{\partial \mathbf{U}}(\mathbf{X}_0) = (d_{\mathbf{X}_0} f)(\mathbf{U})$$

$$= f_{x_1}(\mathbf{X}_0) u_1 + f_{x_2}(\mathbf{X}_0) u_2 + \cdots + f_{x_n}(\mathbf{X}_0) u_n.$$

Example 3.9   Let $\mathbf{X}_0 = (1, 2)$,

$$\mathbf{U} = \begin{bmatrix} \dfrac{1}{\sqrt{2}} \\[2ex] -\dfrac{1}{\sqrt{2}} \end{bmatrix},$$

and

$$f(x, y) = x^2 + y^2 + xy.$$

The partial derivatives,

$$f_x(x, y) = 2x + y, \qquad f_y(x, y) = 2y + x,$$

are continuous everywhere. In particular f is differentiable at $\mathbf{X}_0$ and

$$d_{\mathbf{X}_0} f = 4 \, dx + 5 \, dy;$$

thus

$$\frac{\partial f}{\partial \mathbf{U}}(\mathbf{X}_0) = 4(dx)\left(\frac{1}{\sqrt{2}}, -\frac{1}{\sqrt{2}}\right) + 5(dy)\left(\frac{1}{\sqrt{2}}, -\frac{1}{\sqrt{2}}\right)$$

$$= 4\left(\frac{1}{\sqrt{2}}\right) + 5\left(\frac{-1}{\sqrt{2}}\right) = -\frac{1}{\sqrt{2}},$$

which agrees with the result obtained in Example 2.3.

EXERCISES 2.3

1.   Using the method of Example 3.1, find constants a and b such that

$$\lim_{(x,y)\to(x_0,y_0)} \frac{f(x, y) - f(x_0, y_0) - a(x - x_0) - b(y - y_0)}{\sqrt{(x - x_0)^2 + (y - y_0)^2}} = 0,$$

where

(a) $f(x, y) = x^2 + y^2 + 2xy + x + 1$, $(x_0, y_0) = (0, 0)$.

(b) $f(x, y) = 3x + 4y - 6$, $(x_0, y_0) = (2, 5)$.

2. Find $d_{\mathbf{X}_0} f$ and $(d_{\mathbf{X}_0} f)(\mathbf{X}_1 - \mathbf{X}_0)$:

(a) $f(\mathbf{X}) = x^2 + 3xy + y^2 + 2z$, $\mathbf{X}_0 = (1, 0, 2)$, $\mathbf{X}_1 = (2, 1, 3)$.

(b) $f(\mathbf{X}) = \log(x + y + z)$, $\mathbf{X}_0 = (1, 2, -2)$, $\mathbf{X}_1 = (3, 1, 1)$.

(c) $f(\mathbf{X}) = \sqrt{x^2 + y^2 + z^2}$, $\mathbf{X}_0 = \left(\frac{1}{\sqrt{2}}, \frac{1}{2}, -\frac{1}{2}\right)$, $\mathbf{X}_1 = \left(\frac{1}{\sqrt{2}}, 1, -1\right)$

(d) $f(\mathbf{X}) = e^x \cos(y + z)$, $\mathbf{X}_0 = \left(0, \frac{\pi}{2}, 0\right)$, $\mathbf{X}_1 = \left(0, \frac{\pi}{2}, \frac{\pi}{2}\right)$.

3. Find $d_{\mathbf{X}_0} f$ and $(d_{\mathbf{X}_0} f)(\mathbf{X}_1 - \mathbf{X}_0)$:

(a) $f(\mathbf{X}) = 3x^3 + x^2 y + x + y$, $\mathbf{X}_0 = (-1, 0, 1)$, $\mathbf{X}_1 = (3, 2, 1)$.

(b) $f(\mathbf{X}) = \log(x^2 + y^2 + z^2)$, $\mathbf{X}_0 = \left(\frac{1}{\sqrt{3}}, -\frac{1}{\sqrt{3}}, \frac{1}{\sqrt{3}}\right)$,

$$\mathbf{X}_1 = \left(\frac{2}{\sqrt{3}}, \frac{1}{\sqrt{3}}, -\frac{1}{\sqrt{3}}\right).$$

(c) $f(\mathbf{X}) = e^{\cos xyz}$, $\mathbf{X}_0 = \left(1, 1, \frac{\pi}{2}\right)$, $\mathbf{X}_1 = \left(2, 2, \frac{3\pi}{2}\right)$.

(d) $f(\mathbf{X}) = \dfrac{x^2 + y^2}{xy}$, $\mathbf{X}_0 = (1, 1, 3)$, $\mathbf{X}_1 = (2, 3, 17)$.

4. Verify (4) directly for f and $\mathbf{X}_0$ as given in Exercise 1(a), Exercise 2(a), and Exercise 3(a).

5. The area of a triangle with sides of length a and b and included angle ψ is $A = \frac{1}{2}ab \sin \psi$. Let $a = 5$, $b = 10$, and $\psi = 30$ degrees, to within errors of at most 1 percent in the lengths and 1 degree in the angle. Find an upper bound for the error in A.

6. A satellite is above a point P on a plane Γ. A radar at a point Q in Γ, at a distance $d = 300$ miles from P, measures the line of sight range from Q to the satellite and finds it to be $r = 500$ miles. From this information the height of the satellite above P is calculated. If the error in d is at most 2 miles and the error in the range is at most $\frac{1}{2}$ percent, what is the maximum error in h?

7. Find:

   (a) $dx(0, 1, 2)$,   (b) $dy(0, 1, 2)$,   (c) $dz(0, 1, 2)$,

   (d) $dx_i(2, 5, 6, 7, 9)$ $(1 \le i \le 5)$.

8. Express $d_{X_0} f$, as calculated in Exercise 2, in terms of dx, dy, and dz.

9. Express $d_{X_0} f$, as calculated in Exercise 3, in terms of dx, dy, and dz.

10. Write the general expression for df for each function in Exercise 2.

11. Write the general expression for df for each function in Exercise 3.

12. Let $\mathbf{X_0} = (-1, \ 0, \ 1)$, $f(\mathbf{X}) = x^2 + 2x^2 yz + x + 3y + 1$, and $g(\mathbf{X}) = x^2 + 3y + z$. Find $d_{X_0}(2f), d_{X_0}(f + g), d_{X_0}(fg)$, and $d_{X_0}(f/g)$.

13. Repeat Exercise 12 for $\mathbf{X_0} = (\pi/8, \pi/16, \pi/16), f(\mathbf{X}) = \sin(x + y + z)$, and $g(\mathbf{X}) = \cos(x + y + z)$.

14. Calculate the directional derivatives called for in Exercise 1, Section 2.2, by the method of Theorem 3.5.

15. Calculate the directional derivatives called for in Exercise 2, Section 2.2, by the method of Theorem 3.5.

THEORETICAL EXERCISES

T-1.   Give a direct proof, without assuming continuity of the first partial derivatives, that $f + g$ is differentiable at $\mathbf{X_0}$ if f and g are.

T-2.   If there are constants $a_1, \ldots, a_n$, not all zero, such that $a_1 f_1(\mathbf{X}) + \cdots + a_n f_n(\mathbf{X}) = 0$ for all $\mathbf{X}$ in a set S in R^n, then we say that $f_1, \ldots, f_n$ are *linearly dependent on* S. Prove: If $f_1, \ldots, f_n$ are linearly dependent on a set S and differentiable at a point $\mathbf{X_0}$ in S, then $d_{X_0} f_1, \ldots, d_{X_0} f_n$ are linearly dependent on R^n.

T-3.   Prove Theorem 3.3 for $n = 2$.

2.4   The Mean-Value Theorem

We shall now generalize the mean-value theorem (Lemma 3.1) to functions of several variables.

Theorem 4.1   Let f be continuous at distinct points $\mathbf{X_1}$ and $\mathbf{X_2}$, and differentiable on the line segment L joining them. Then there is a point

$\mathbf{X}_0$ on L such that

$$f(\mathbf{X}_2) - f(\mathbf{X}_1) = (d_{\mathbf{X}_0} f)(\mathbf{X}_2 - \mathbf{X}_1).$$

Proof.   Define the unit vector

$$\mathbf{U} = \frac{\mathbf{X}_2 - \mathbf{X}_1}{|\mathbf{X}_2 - \mathbf{X}_1|}$$

and

$$g(t) = f(\mathbf{X}_1 + t\mathbf{U}) \qquad (0 \le t \le |\mathbf{X}_2 - \mathbf{X}_1|).$$

Since f is continuous on L (including the endpoints), g is continuous on the closed interval $[0, |\mathbf{X}_2 - \mathbf{X}_1|]$. We now show that g has a derivative in the open interval $(0, |\mathbf{X}_2 - \mathbf{X}_1|)$.

Let t_0 be a fixed but arbitrary point in the open interval; then

$$\frac{g(t) - g(t_0)}{t - t_0} = \frac{f(\mathbf{X}_1 + t\mathbf{U}) - f(\mathbf{X}_1 + t_0\mathbf{U})}{t - t_0}.$$

Setting $t - t_0 = \tau$ and $\mathbf{X}_1 + t_0\mathbf{U} = \hat{\mathbf{X}}$, we rewrite this as

$$\frac{g(t) - g(t_0)}{t - t_0} = \frac{f(\hat{\mathbf{X}} + \tau\mathbf{U}) - f(\hat{\mathbf{X}})}{\tau}.$$

Since f is differentiable at $\hat{\mathbf{X}}$, the right side approaches $(\partial f/\partial \mathbf{U})(\hat{\mathbf{X}})$ as τ approaches zero; thus $g'(t_0)$ exists and

$$g'(t_0) = \lim_{t \to t_0} \frac{g(t) - g(t_0)}{t - t_0}$$

(1)
$$= \lim_{t \to t_0} \frac{f(\hat{\mathbf{X}} + \tau\mathbf{U}) - f(\hat{\mathbf{X}})}{\tau}$$

$$= \frac{\partial f}{\partial \mathbf{U}}(\hat{\mathbf{X}}) = (d_{\hat{\mathbf{X}}} f)(\mathbf{U}),$$

where the last equality follows from Theorem 3.5.

Since g has a derivative at all points of $(0, |\mathbf{X}_2 - \mathbf{X}_1|)$, the mean-value theorem for functions of one variable implies that there is a point s_0 $(0 < s_0 < |\mathbf{X}_2 - \mathbf{X}_1|)$ such that

$$g(|\mathbf{X}_2 - \mathbf{X}_1|) - g(0) = |\mathbf{X}_2 - \mathbf{X}_1| g'(s_0).$$

But $g(|\mathbf{X}_2 - \mathbf{X}_1|) = f(\mathbf{X}_2)$ and $g(0) = f(\mathbf{X}_1)$; using (1) with $t_0 = s_0$ and $\mathbf{X}_0 = \mathbf{X}_1 + s_0\mathbf{U}$, we rewrite the last equation as

$$f(\mathbf{X}_2) - f(\mathbf{X}_1) = |\mathbf{X}_2 - \mathbf{X}_1| (d_{\mathbf{X}_0} f)(\mathbf{U}) = (d_{\mathbf{X}_0} f)(\mathbf{X}_2 - \mathbf{X}_1),$$

which completes the proof, since $\mathbf{X}_0$ is on L.

If $n = 3$ the last equation can be written

$$f(\mathbf{X_2}) - f(\mathbf{X_1}) = f_x(\mathbf{X_0})(x_2 - x_1) + f_y(\mathbf{X_0})(y_2 - y_1) + f_z(\mathbf{X_0})(z_2 - z_1).$$

As an application of the mean-value theorem we now establish an analog of the chain rule of the one dimensional calculus, which says that if f is a differentiable function of x and x is a differentiable function of t, then the derivative of $h(t) = f(x(t))$ is $h'(t) = f'(x(t))x'(t)$.

Example 4.1   Let

$$f(x) = \frac{1}{1 - x^2} \qquad (|x| < 1)$$

and

$$x(t) = \sin t \qquad \left(|t| < \frac{\pi}{2}\right).$$

Then

$$h(t) = f(x(t)) = \frac{1}{(1 - \sin^2 t)^2}$$

is defined if $|t| < \pi/2$, and f and x are differentiable:

$$f'(x) = \frac{2x}{(1 - x^2)^2},$$

$$x'(t) = \cos t.$$

Thus

$$h'(t) = f'(x(t))x'(t) = \frac{2 \sin t}{(1 - \sin^2 t)^2} \cos t$$

$$= \frac{2 \sin t}{\cos^3 t}.$$

To state sufficient conditions under which an analogous rule holds for a function of several variables, we need a preliminary definition.

Definition 4.1   A function f is said to be *continuously differentiable* on a domain D of R^n if all of its first partial derivatives exist and are continuous on D.

The next theorem is an immediate consequence of this definition and Theorem 3.3.

Theorem 4.2   If f is continuously differentiable in a domain D of R^n, then f is differentiable on D.

Theorem 4.3   Let f be continuously differentiable in a domain D of R^n, let $x_1, \ldots, x_n$ be differentiable functions of t for $a < t < b$, and $X(t) = (x_1(t), \ldots, x_n(t))$ be an interior point of D for all t in (a, b). Define $h(t) = f(X(t))$; then h has a derivative in (a, b), given by

$$(2) \qquad h'(t) = f_{x_1}(X(t))x_1'(t) + \cdots + f_{x_n}(X(t))x_n'(t)$$

$$= (d_{X_0}f)(x_1'(t), \ldots, x_n'(t)).$$

Proof.   Let t_0 be fixed in (a, b); since

$$|X(t) - X(t_0)| = \sqrt{(x_1(t) - x_1(t_0))^2 + \cdots + (x_n(t) - x_n(t_0))^2}$$

and each x_i is continuous at t_0 (being differentiable there), it follows that

$$\lim_{t \to t_0} |X(t) - X(t_0)| = 0.$$

Thus, for t sufficiently near t_0, the line segment connecting $X(t)$ and $X(t_0)$ lies entirely in D (because of the convexity of the n-ball), and Theorem 4.1 implies that there is a point $\hat{X}(t)$ on this segment such that

$$f(X(t)) - f(X(t_0)) = (d_{\hat{X}(t)}f)(X(t) - X(t_0)).$$

Hence

$$(3) \quad \frac{h(t) - h(t_0)}{t - t_0} = f_{x_1}(\hat{X}(t))\frac{x_1(t) - x_1(t_0)}{t - t_0} + \cdots + f_{x_n}(\hat{X}(t))\frac{x_n(t) - x_n(t_0)}{t - t_0}.$$

Now, $|\hat{X}(t) - X(t_0)| \le |X(t) - X(t_0)|$, which implies that $\hat{X}(t)$ approaches $X(t_0)$ as t approaches t_0; hence the continuity of f_{x_i} implies that

$$\lim_{t \to t_0} f_{x_i}(\hat{X}(t)) = f_{x_i}(X(t_0)) \qquad (1 \le i \le n).$$

The differentiability of x_i implies that

$$\lim_{t \to t_0} \frac{x_i(t) - x_i(t_0)}{t - t_0} = x_i'(t_0).$$

Now let t approach t_0 in (3) to find that

$$h'(t_0) = f_{x_1}(X(t_0))x_1'(t_0) + \cdots + f_{x_n}(X(t_0))x_n'(t_0).$$

Since this argument can be carried out for each t_0 in (a, b), we can replace t_0 by t in the last equation and obtain (2). This completes the proof.

Example 4.2   Let
$$f(x, y) = 2x^2 - y^2,$$
$$x(t) = 2t^3 + t,$$
and
$$y(t) = t + 1.$$

We wish to find the derivative of $h(t) = f(x(t), y(t))$ at $t = 0$. Then $f_x(x, y) = 4x$, $f_y(x, y) = -2y$, $x'(t) = 6t^2 + 1$, and $y'(t) = 1$; thus

$$h'(0) = f_x(x(0), y(0))x'(0) + f_y(x(0), y(0))y'(0)$$
$$= f_x(0, 1)x'(0) + f_y(0, 1)y'(0)$$
$$= (0)(1) + (-2)(1)$$
$$= -2.$$

In many applications of the chain rule, $(n - 1)$ of the components of **X** are functions of the remaining one. For example, suppose $n = 3$ and y and z are functions of x, so that $h(x) = f(x, y(x), z(x))$. To apply Theorem 4.3, we introduce an intermediate variable t and write $\mathbf{X}(t) = (t, y(t), z(t))$. Then $h(t) = f(t, y(t), z(t))$ and Theorem 4.3 yields

$$h'(t) = f_x(t, y(t), z(t)) + f_y(t, y(t), z(t))y'(t)$$
$$+ f_z(t, y(t), z(t))z'(t),$$

since $x'(t) = 1$. The name of the independent variable is immaterial; hence the last equation can be rewritten

$$h'(x) = f_x(x, y(x), z(x)) + f_y(x, y(x), z(x))y'(x)$$
$$+ f_z(x, y(x), z(x))z'(x).$$

The student should recognize that the introduction of the intermediate variable t served merely as a bookkeeping device to facilitate the application of Theorem 4.2; it can be omitted when he is thoroughly familiar with this form of the chain rule.

Example 4.3   Let $f(x, y, z) = x^2 + y^2 + z^2$, $y(x) = \cos x$,   and $z(x) = e^x$; then $f_x(x, y, z) = 2x$, $f_y(x, y, z) = 2y$, $f_z(x, y, z) = 2z$, $y'(x) = -\sin x$, and $z'(x) = e^x$. Thus, if $h(x) = f(x, y(x), z(x))$, then

$$h'(x) = 2x + 2y(x)(-\sin x) + 2z(x)e^x$$
$$= 2x - 2 \sin x \cos x + 2e^{2x}.$$

The same result can be obtained by first expressing h explicitly in terms of x,

$$h(x) = x^2 + \cos^2 x + e^{2x},$$

and then differentiating with respect to x directly; however, this does not detract from the usefulness of the chain rule, which applies to cases where direct substitution is not convenient.

The chain rule will be further extended in Section 3.3.

EXERCISES 2.4

In the following exercises let $h(t) = f(\mathbf{X}(t))$.

1. Evaluate $h'(t_0)$ by the chain rule and then by substituting $\mathbf{X}(t)$ for $\mathbf{X}$ and differentiating directly with respect to t:

   (a) $f(x, y) = x^2 + y^2$; $x(t) = \cos t$, $y(t) = \sin t$; $t_0 = \dfrac{\pi}{4}$.

   (b) $f(x, y, z) = \cos xyz$; $x(t) = 1$, $y(t) = t^2$, $z(t) = t^3$; $t_0 = \sqrt[5]{\dfrac{\pi}{2}}$

   (c) $f(x, y) = \dfrac{x^2 + y^2}{xy}$; $x(t) = \sqrt{t}$, $y(t) = \sqrt[3]{t}$; $t_0 = 1$.

   (d) $f(x, y, z) = \log(x + y + z)$; $x(t) = e^t$, $y(t) = -e^t$, $z(t) = t^2$; $t_0 = 1$.

   (e) $f(x, y, z) = \tan\dfrac{xy}{z}$; $x(t) = e^t$, $y(t) = e^{-t}$, $z(t) = t$; $t_0 = \dfrac{1}{\pi}$.

2. Let $f(x, y, z) = \log(x^2 + y^2 + z^2)$, $x(t) = 1 + t/\sqrt{3}$, $y(t) = 2 - t/\sqrt{3}$, and $z(t) = 3 + t/\sqrt{3}$. Calculate $h'(0)$ and, by comparing the result with that of Exercise 2(a) of Section 2.2, show that $h'(0) = \dfrac{\partial F}{\partial \mathbf{U}}(\mathbf{X}_0)$, with $\mathbf{U}$ and $\mathbf{X}_0$ as defined there.

3. Repeat Exercise 2 for $f(x, y) = x/(x + y)$, $x(t) = 1 + t/\sqrt{2}$, and $y(t) = 2 + t/\sqrt{2}$, comparing the result with that of Exercise 1(d), Section 2.2.

4. Find $h'(0)$:

   (a) $f(x, y, z) = x^{37}y^{12}z^6$; $x(t) = \cos t$, $y(t) = \sin t$, $z(t) = e^t$.

   (b) $f(x_1, \ldots, x_n) = x_1^2 + \cdots + x_n^2$; $x_i(t) = a_i + u_i t$ $(1 \leq i \leq n)$.

5. Find $h'(1)$:

(a) $f(x,y,z) = e^{x+y^2+z}$; $x(t) = 2\log t$, $y(t) = t - 1$, $z(t) = \sin(t-1)$.

(b) $f(x, y, z) = \tan^{-1} xyz$; $x(t) = e^t$, $y(t) = e^{-t}$, $z(t) = 1$.

6. Let $f(x, y, z) = \cos^3 x + y^3 e^z$, $y(x) = \sin x$, $z(x) = x^2$, and $h(x) = f(x, y(x), z(x))$; find $h'(x)$.

7. Let $f(x, y, z) = xz \tan y$; $x(y) = \log y$, $z(y) = e^y$, and $h(y) = f(x(y), y, z(y))$; find $h'(y)$.

8. Let $f(\mathbf{X}) = x^2 + y^2 + z^2$, $\mathbf{X}_1 = (2, 0, -1)$, and $\mathbf{X}_2 = (1, -1, 1)$. Verify Theorem 4.1 by finding $\mathbf{X}_0$ on the line connecting $\mathbf{X}_2$ and $\mathbf{X}_1$ such that

$$f(\mathbf{X}_2) - f(\mathbf{X}_1) = (d_{\mathbf{X}_0} f)(\mathbf{X}_2 - \mathbf{X}_1).$$

THEORETICAL EXERCISES

T-1. Let

$$\mathbf{U} = \begin{bmatrix} u_1 \\ \vdots \\ u_n \end{bmatrix}$$

be a unit vector, $\mathbf{X}_0 = (c_1, \ldots, c_n)$, and $x_i(t) = c_i + u_i t$ $(1 \le i \le n)$. Define $h(t) = f(\mathbf{X}(t))$ and show that $h'(0) = (\partial f/\partial \mathbf{U})(\mathbf{X}_0)$.

T-2. Prove Theorem 4.2.

2.5   Graphs and Tangent Planes

The *graph* of a real-valued function f with domain D in R^n is the set of points in R^{n+1} of the form $(\mathbf{X}, f(\mathbf{X})) = (x_1, \ldots, x_n, f(\mathbf{X}))$, where $\mathbf{X}$ is in D. Thus, the graph of f consists of those points $(\mathbf{X}, x_{n+1})$ in R^{n+1} which satisfy

$$x_{n+1} = f(\mathbf{X}).$$

This equation is *the equation of the graph*.

Example 5.1   Let $n = 1$ and

$$f(x) = \sqrt{1 - x^2}.$$

FIGURE 5.1

The graph of f is a semicircle (Fig. 5.1); its equation is

$$y = \sqrt{1 - x^2}.$$

Example 5.2 Let

$$f(x, y) = \sqrt{1 - x^2 - y^2};$$

the domain of f is the disk of radius 1 about the origin, and the graph of f is a hemisphere (Fig. 5.2); its equation is

$$z = \sqrt{1 - x^2 - y^2}.$$

Example 5.3 If $n > 2$ and

$$f(\mathbf{X}) = \sqrt{1 - x_1^2 - \cdots - x_n^2},$$

then the graph of f consists of all points in R^{n+1} of the form

$$(x_1, \ldots, x_n, \sqrt{1 - x_1^2 - \cdots - x_n^2})$$

such that $x_1^2 + \cdots + x_n^2 \leq 1$; it is impossible, however, to depict it on a two-dimensional page. The equation of the graph is

$$x_{n+1} = \sqrt{1 - x_1^2 - \cdots - x_n^2}.$$

The student should recall that if f is a function of one variable, then $f'(x_0)$ is the slope of the tangent to the graph of f at the point $(x_0, f(x_0))$. We shall discuss this in detail, to motivate the generalization of this idea to functions of n variables.

FIGURE 5.2

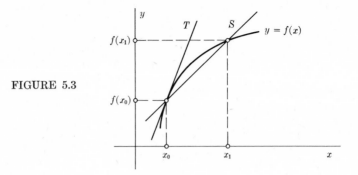

FIGURE 5.3

The point $(x_0, f(x_0))$ and any second point $(x_1, f(x_1))$ on the graph of f determine a secant line S, whose equation is

$$y = f(x_0) + \frac{f(x_1) - f(x_0)}{x_1 - x_0}(x - x_0)$$

(Fig. 5.3). As x_1 approaches x_0 the slope of the secant,

$$m = \frac{f(x_1) - f(x_0)}{x_1 - x_0},$$

approaches $f'(x_0)$, and the secant rotates into coincidence with the line T, given by

$$y = f(x_0) + f'(x_0)(x - x_0),$$

which is tangent to the graph of f at $x = x_0$.

Thus, the tangent to the graph of f at x_0 is the "limit" of the secants connecting $(x_0, f(x_0))$ and $(x_1, f(x_1))$ as x_1 approaches x_0. This limit exists if and only if f has a derivative at x_0.

Example 5.4   Let $x_0 = 1$ and $f(x) = x^2$; then $f(1) = 1$, $f'(1) = 2$, and the equation of the tangent to the graph of f at $(1, 1)$ is

$$y = 1 + 2(x - 1)$$

(Fig. 5.4).

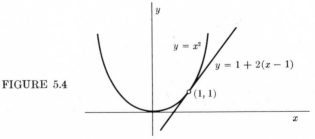

FIGURE 5.4

Example 5.5 Let $x_0 = 1$ and $f(x) = |x - 1|$. This function has no derivative at x_0; neither does its graph have a tangent at $(1, 0)$, since if $x_1 < 1$, the secant connecting $(x_1, |x_1 - 1|)$ and $(1, 0)$ has the equation

$$y = -(x - 1),$$

while if $x_1 > 1$, it has the equation

$$y = (x - 1)$$

(Fig. 5.5).

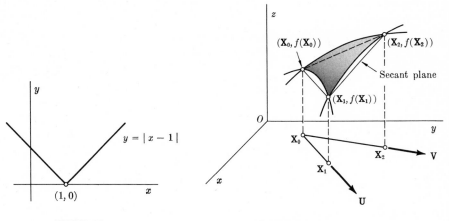

FIGURE 5.5   FIGURE 5.6

If f, a function of n (≥ 2) variables, is differentiable at $\mathbf{X}_0$, then the graph of f has a tangent plane at $(\mathbf{X}_0, f(\mathbf{X}_0))$. Before defining this plane for arbitrary n, we present an intuitive discussion for $n = 2$.

Let f be differentiable at $\mathbf{X}_0 = (x_0, y_0)$. Let $\mathbf{X}_1 = (x_1, y_1)$ and $\mathbf{X}_2 = (x_2, y_2)$ be distinct points such that the line segments from $\mathbf{X}_0$ to $\mathbf{X}_1$ and from $\mathbf{X}_0$ to $\mathbf{X}_2$ are in the domain of f, and not collinear. Then there is a unique secant plane in R^3 through $(\mathbf{X}_0, f(\mathbf{X}_0))$, $(\mathbf{X}_1, f(\mathbf{X}_1))$, and $(\mathbf{X}_2, f(\mathbf{X}_2))$ (Fig. 5.6); its equation is

$$(1) \quad z = f(\mathbf{X}_0) + \frac{(y - y_0)(x_2 - x_0) - (y_2 - y_0)(x_1 - x_0)}{(y_1 - y_0)(x_2 - x_0) - (y_2 - y_0)(x_1 - x_0)} (f(\mathbf{X}_1) - f(\mathbf{X}_0))$$

$$+ \frac{(y_1 - y_0)(x - x_0) - (y - y_0)(x_1 - x_0)}{(y_1 - y_0)(x_2 - x_0) - (y_2 - y_0)(x_1 - x_0)} (f(\mathbf{X}_2) - f(\mathbf{X}_0)).$$

(See Section 1.3, in particular, Example 3.5 and the discussion preceding it.) In allowing $\mathbf{X}_1$ and $\mathbf{X}_2$ to approach $\mathbf{X}_0$, we must restrict the method of approach so that $(\mathbf{X}_0, f(\mathbf{X}_0))$, $(\mathbf{X}_1, f(\mathbf{X}_1))$, and $(\mathbf{X}_2, f(\mathbf{X}_2))$ determine a plane; we do this by setting

(2) $\qquad\qquad \mathbf{X_1} = \mathbf{X_0} + t\mathbf{U}, \qquad \mathbf{X_2} = \mathbf{X_0} + t\mathbf{V},$

where $\mathbf{U}$ and $\mathbf{V}$ are nonparallel unit vectors:

$$\mathbf{U} = \begin{bmatrix} u_1 \\ u_2 \end{bmatrix}, \qquad \mathbf{V} = \begin{bmatrix} v_1 \\ v_2 \end{bmatrix} \qquad (u_2v_1 - u_1v_2) \neq 0.$$

We shall study the behavior of the secant plane as t approaches zero; although this is not the most general way in which the secant plane can approach its limiting position, it is sufficiently general to motivate our definition.

With $\mathbf{X_1}$ and $\mathbf{X_2}$ replaced by (2), the equation of the secant plane becomes

$$(3) \quad z = f(\mathbf{X_0}) + \frac{(y - y_0)v_1 - (x - x_0)v_2}{u_2v_1 - u_1v_2} \frac{f(\mathbf{X_0} + t\mathbf{U}) - f(\mathbf{X_0})}{t}$$

$$+ \frac{(x - x_0)u_2 - (y - y_0)u_1}{u_2v_1 - u_1v_2} \frac{f(\mathbf{X_0} + t\mathbf{V}) - f(\mathbf{X_0})}{t}.$$

In Section 2.3 we showed that

$$\lim_{t \to t_0} \frac{f(\mathbf{X_0} + t\mathbf{U}) - f(\mathbf{X_0})}{t} = \frac{\partial f}{\partial \mathbf{U}}(\mathbf{X_0}) = f_x(\mathbf{X_0})u_1 + f_y(\mathbf{X_0})u_2;$$

similarly

$$\lim_{t \to t_0} \frac{f(\mathbf{X_0} + t\mathbf{V}) - f(\mathbf{X_0})}{t} = f_x(\mathbf{X_0})v_1 + f_y(\mathbf{X_0})v_2.$$

Thus, as t approaches zero in (3), the secant plane "approaches" the *tangent plane*,

$$z = f(\mathbf{X_0}) + \frac{(y - y_0)v_1 - (x - x_0)v_2}{u_2v_1 - u_1v_2} [f_x(\mathbf{X_0})u_1 + f_y(\mathbf{X_0})u_2]$$

$$+ \frac{(x - x_0)u_2 - (y - y_0)u_1}{u_2v_1 - u_1v_2} [f_x(\mathbf{X_0})v_1 + f_y(\mathbf{X_0})v_2].$$

This is the equation of a plane through $(\mathbf{X_0}, f(\mathbf{X_0}))$. A remarkable thing about it is that it does not depend on $\mathbf{U}$ and $\mathbf{V}$, as can be seen by collecting coefficients of $f_x(\mathbf{X_0})$ and $f_y(\mathbf{X_0})$:

$$z = f(\mathbf{X_0}) + f_x(\mathbf{X_0})(x - x_0) + f_y(\mathbf{X_0})(y - y_0)$$

$$= f(\mathbf{X_0}) + (d_{\mathbf{X_0}}f)(\mathbf{X} - \mathbf{X_0})$$

(Fig. 5.7).

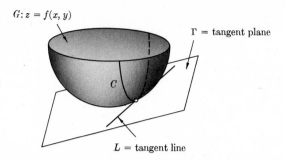

$G : z = f(x, y)$

Γ = tangent plane

C

L = tangent line

FIGURE 5.7

Definition 5.1 Let $n \geq 1$ and f be differentiable at $\mathbf{X}_0$. The *tangent plane to the graph of f at $(\mathbf{X}_0, f(\mathbf{X}_0))$* is the graph defined by

$$x_{n+1} = f(\mathbf{X}_0) + (d_{\mathbf{X}_0} f)(\mathbf{X} - \mathbf{X}_0).$$

This plane is the "limit" of secant planes to the graph of f, just as we have seen above for $n = 2$.

Thus, the tangent plane to the graph of f at $(\mathbf{X}_0, f(\mathbf{X}_0))$ is itself the graph of the function T defined by

$$T(\mathbf{X}) = f(\mathbf{X}_0) + (d_{\mathbf{X}_0} f)(\mathbf{X} - \mathbf{X}_0).$$

From the differentiability of f at $\mathbf{X}_0$ it follows that

$$\lim_{\mathbf{X} \to \mathbf{X}_0} \frac{f(\mathbf{X}) - T(\mathbf{X})}{\mathbf{X} - \mathbf{X}_0} = 0.$$

The tangent plane is the only plane with this property; that is, if $\mathbf{X}_0 = (c_1, \ldots, c_n)$ and

$$\lim_{\mathbf{X} \to \mathbf{X}_0} \frac{f(\mathbf{X}) - a - b_1(x_1 - c_1) - \cdots - b_n(x_n - c_n)}{|\mathbf{X} - \mathbf{X}_0|} = 0,$$

then the plane defined by

$$x_{n+1} = a + b_1(x_1 - c_1) + \cdots + b_n(x_n - c_n)$$

is, in fact, the tangent plane.

Example 5.6 To find the tangent plane to the graph of

$$f(x, y) = \sqrt{1 - x^2 - y^2}$$

at $(x_0, y_0, f(x_0, y_0)) = (1/2, -1/2, 1/\sqrt{2})$, compute

$$f_x\left(\frac{1}{2}, -\frac{1}{2}\right) = -f_y\left(\frac{1}{2}, -\frac{1}{2}\right) = -\frac{1}{\sqrt{2}};$$

thus the equation of the tangent plane is

$$z - \frac{1}{\sqrt{2}} = -\frac{1}{\sqrt{2}}\left(x - \frac{1}{2}\right) + \frac{1}{\sqrt{2}}\left(y + \frac{1}{2}\right),$$

or

$$z = \frac{1}{\sqrt{2}}(-x + y + 2).$$

Let G be the graph of f and Γ the tangent plane to G at $(\mathbf{X}_0, f(\mathbf{X}_0))$. If C is a curve in G passing through $(\mathbf{X}_0, f(\mathbf{X}_0))$ and L is the tangent line to C at $(\mathbf{X}_0, f(\mathbf{X}_0))$ (Fig. 5.7), then L is in Γ. To see this for $n = 2$, let C be given parametrically by $x = x(t)$, $y = y(t)$, and $z = z(t)$, where $(x(t_0), y(t_0), z(t_0)) = (x_0, y_0, f(x_0, y_0))$. Since C is in G,

$$z(t) = f(x(t), y(t));$$

from the chain rule,

$$(4) \qquad z'(t_0) = f_x(x(t_0), y(t_0))x'(t_0) + f_y(x(t_0), y(t_0))y'(t_0)$$
$$= f_x(x_0, y_0)x'(t_0) + f_y(x_0, y_0)y'(t_0).$$

If (x_1, y_1, z_1) is on L then

$$x'(t_0) = \lambda(x_1 - x_0), \qquad y'(t_0) = \lambda(y_1 - y_0),$$
$$z'(t_0) = \lambda(z_1 - f(x_0, y_0)),$$

where λ is a nonzero constant. Substituting these in (4) yields

$$z_1 - f(x_0, y_0) = f_x(x_0, y_0)(x_1 - x_0) + f_y(x_0, y_0)(y_1 - y_0);$$

hence (x_1, y_1, z_1) is in Γ, from Definition 5.1.

A similar proof holds for $n > 2$.

EXERCISES 2.5

In Exercises 1 through 7 sketch the graph of the given function.

1. $f(x) = x^2 + 1$.

2. $f(x, y) = \sqrt{36 - 9x^2 - 4y^2}$.

3. $f(x, y) = 4x^2 + 9y^2$.

4. $f(x, y) = \sqrt{36x^2 - 9y^2}$.

5. $f(x, y) = \sqrt{9x^2 - 4y^2 - 36}$.

6. $f(x, y) = \sqrt{16x^2 + 4y^2}$.

7. $f(x, y) = \sqrt{9x^2 + 4y^2 - 36}$.

8. Let $f(x, y) = x^2 + y^2$; find the equation of the secant plane to the graph of f through $(0, 2, f(0, 2))$, $(1, 1, f(1, 1))$, and $(-1, 3, f(-1, 3))$.

In Exercises 9 through 14 find the equation of the tangent plane to the graph of f at $(\mathbf{X}_0, f(\mathbf{X}_0))$.

9. $f(x, y) = \sqrt{36 - 9x^2 - 4y^2}$, $\quad \mathbf{X}_0 = (1, 2)$.

10. $f(x, y, z) = 1 + 4x^2 - y^2 + 2z^2$, $\quad \mathbf{X}_0 = (1, 1, 1)$.

11. $f(x, y) = \sqrt{4x^2 + 9y^2 - 36}$, $\quad \mathbf{X}_0 = (3, 2)$.

12. $f(x, y) = 9 - x^2$, $\quad \mathbf{X}_0 = (2, 4)$.

13. $f(x, y) = x^2 + 2y^2$, $\quad \mathbf{X}_0 = (1, -2)$.

14. Let f and $\mathbf{X}_0$ be as in Exercise 13. Find the equation of the line through $\mathbf{X}_0$ perpendicular to the tangent plane to the graph of f at $(\mathbf{X}_0, f(\mathbf{X}_0))$.

THEORETICAL EXERCISES

T-1. If f is differentiable at $\mathbf{X}_0 = (c_1, \ldots, c_n)$, show that the line through $\mathbf{X}_0$ perpendicular to the tangent plane to the graph of f at $(\mathbf{X}_0, f(\mathbf{X}_0))$ is given by

$$x_i = c_i + tf_{x_i}(\mathbf{X}_0) \qquad (1 \le i \le n),$$

$$x_{n+1} = f(\mathbf{X}_0) - t.$$

T-2. Find the equation of the tangent plane to the graph of $f(\mathbf{X}) = \sin(x_1 + \cdots + x_n)$ at $(\mathbf{X}_0, f(\mathbf{X}_0))$, where

(a) $\mathbf{X}_0 = \left(\dfrac{\pi}{n}, \ldots, \dfrac{\pi}{n} \right)$; (b) $\mathbf{X}_0 = \left(\dfrac{\pi}{2n}, \ldots, \dfrac{\pi}{2n} \right)$.

T-3. Repeat Exercise T-2 for $f(\mathbf{X}) = \log(x_1^2 + \cdots + x_n^2)$ and

(a) $\mathbf{X}_0 = \left(\dfrac{1}{\sqrt{n}}, \ldots, \dfrac{1}{\sqrt{n}} \right)$; (b) $\mathbf{X}_0 = (e, 0, \ldots, 0)$.

T-4. Verify Eq. (1) for the secant plane through $(\mathbf{X}_0, f(\mathbf{X}_0))$, $(\mathbf{X}_1, f(\mathbf{X}_1))$, and $(\mathbf{X}_2, f(\mathbf{X}_2))$.

2.6   Implicit Functions

If f is defined on a domain D in R^n, then its graph G, as defined in Section 2.5, is the set of points in R^{n+1} of the form $(\mathbf{X}, f(\mathbf{X}))$, where $\mathbf{X}$ is in D. No two points in G can differ only in their $(n+1)$th coordinates; if $(\mathbf{X}, x_{n+1})$ and $(\mathbf{X}, \hat{x}_{n+1})$ are both in G, then

$$x_{n+1} = \hat{x}_{n+1} = f(\mathbf{X}).$$

Now let us change our point of view. Suppose G is a subset of R^{n+1} with the property that no two points of G differ only in their $(n+1)$th coordinates, and let D be the set of all $\mathbf{X}$ in R^n that appear as the first n coordinates of points of G; thus for each $\mathbf{X}$ in D there is exactly one real number x_{n+1} such that $(\mathbf{X}, x_{n+1})$ is in G. Define $x_{n+1} = f(\mathbf{X})$; that is, $f(\mathbf{X})$ is the unique real number such that $(\mathbf{X}, f(\mathbf{X}))$ is in G. Then f is a function on D, and G is its graph; rather than starting with the function and arriving at the graph, we have used the graph to define the function. We say that f in this case is defined *implicitly* by G; we also say that G *defines* x_{n+1} *as a function of* $(x_1, \ldots, x_n)$.

It is convenient at this point to introduce the notation

$$S = \{\mathbf{X} \mid \cdots\}$$

to mean that S is the set of points which satisfy the conditions listed to the right of the vertical line.

Example 6.1   Let $n = 2$ and

$$(1) \qquad G = \{(x, y, z) \mid x^2 + y^2 + z^2 = 1, \quad z \geq 0\};$$

thus G is a hemisphere (Fig. 6.1), and it defines a function f with domain

$$D = \{(x, y) \mid x^2 + y^2 \leq 1\},$$

since for each (x, y) in D there is exactly one nonnegative z such that (x, y, z) is in G:

$$(2) \qquad z = f(x, y) = \sqrt{1 - x^2 - y^2}.$$

The hemisphere

$$G_1 = \{(x, y, z) \mid x^2 + y^2 + z^2 = 1, \quad z \leq 0\}$$

(Fig. 6.2) defines a function f_1 with domain D:

$$z = f_1(x, y) = -\sqrt{1 - x^2 - y^2}.$$

However, the sphere

$$S = \{(x, y, z) \mid x^2 + y^2 + z^2 = 1\}$$

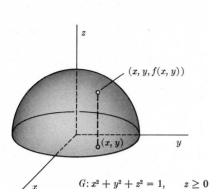

$$G: x^2 + y^2 + z^2 = 1, \quad z \geq 0$$

FIGURE 6.1

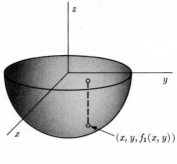

$$G_1: x^2 + y^2 + z^2 = 1, \quad z \leq 0$$

FIGURE 6.2

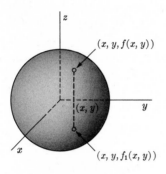

FIGURE 6.3

$$S : x^2 + y^2 + z^2 = 1$$

is not a graph, because for every $(x, y) \neq (0, 0)$ in D, the distinct points $(x, y, f(x, y))$ and $(x, y, f_1(x, y))$ are in S. (Fig. 6.3).

Sometimes it is possible to obtain an explicit expression such as (2) for an implicitly defined function, but this is not essential; one could argue that (2) does no more than provide a convenient notation for the function defined by (1). This is precisely the way in which many functions are introduced into analysis.

Example 6.2 Let $n = 1$ and

$$G = \{(x, y) \mid e^y - x = 0, \quad 0 < x < \infty \};$$

then G defines y as a function of x on $(0, \infty)$, because $e^{y_1} = e^{y_2}$ if and only if $y_1 = y_2$. This function is the natural logarithm of x;

$$y = f(x) = \log x \qquad (-\infty < x < \infty)$$

(Fig. 6.4).

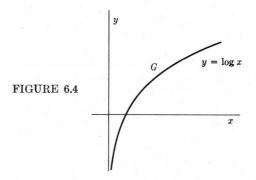

FIGURE 6.4

Example 6.3 Let $n = 1$ and

$$G = \left\{ (x, y) \mid x - \sin y = 0, \quad -\frac{\pi}{2} \le y \le \frac{\pi}{2} \right\}$$

(Fig. 6.5); again, G defines y as a function of x, since there is only one value of y in $[-\pi/2, \pi/2]$ whose sine equals a given x in $[-1, 1]$. This function is the inverse sine function; it is denoted by

$$y = f(x) = \text{Arcsin } x.$$

Placing other restrictions on y in the definition of G produces other inverse sine functions; thus

$$G_1 = \left\{ (x, y) \mid x - \sin y = 0, \quad \frac{\pi}{2} \le y \le \frac{3\pi}{2} \right\}$$

(Fig. 6.6) also defines y as a function of x.

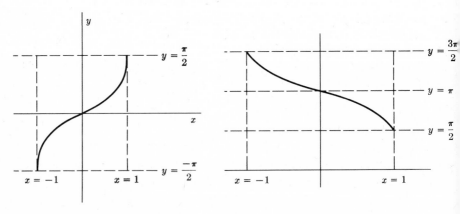

FIGURE 6.5   FIGURE 6.6

A set G in R^{n+1} may define some coordinate other than the $(n + 1)$th in terms of the remaining ones.

Example 6.4 Let

$$G = \{(1, 1, 1), (1, 1, 2), (1, 2, 3)\};$$

then G does not define z as a function of x and y because the first two points differ only in their z coordinates. However, G defines y as a function f of (x, z) with domain

$$D = \{(1, 1), (1, 2), (1, 3)\}$$

and values

$$f(1, 1) = 1, \qquad f(1, 2) = 1, \qquad f(1, 3) = 2.$$

Even though y is not the last coordinate in this example, let us agree to call G the *graph of f*, and say that G is a *graph with respect to y*.

A set G can be a graph with respect to more than one coordinate.

Example 6.5 In Example 6.4, G also defines x as a function g of (y, z) with domain

$$D_1 = \{(1, 1), \quad (1, 2), \quad (2, 3)\}$$

and values

$$g(1, 1) = g(1, 2) = g(2, 3) = 1.$$

Example 6.6 Let

$$G = \{(x_1, \ldots, x_{n+1}) \mid x_1^2 + \cdots + x_{n+1}^2 = 1, \quad x_i \geq 0 \quad (1 \leq i \leq n + 1)\};$$

then G defines each x_i as a function of the other coordinates, and is therefore a graph with respect to each coordinate.

Let F be defined on a subset of R^{n+1} and let N be the set of points at which F vanishes. We next discuss conditions under which N or certain of its subsets define x_{n+1} as a function of $\mathbf{X} = (x_1, \ldots, x_n)$. (The choice of x_{n+1} as the preferred coordinate is one of convenience; our discussion could be conducted equally well in terms of any other coordinate.)

Example 6.7 If

$$F(\mathbf{X}, x_{n+1}) = f(\mathbf{X}) - x_{n+1},$$

then N defines x_{n+1} in terms of $\mathbf{X}$ and the functional relationship is quite explicit:

$$x_{n+1} = f(\mathbf{X}).$$

In this case N is the graph of f.

Example 6.8 If

$$F(\mathbf{X}, x_{n+1}) = x_1^2 + \cdots + x_{n+1}^2 - 1,$$

then N is not a graph (because, for example $(0, \ldots, 0, 1)$ and $(0, \ldots, 0, -1)$ are both in N), but some of its subsets are. For example, the subset of N for which $x_{n+1} \geq 0$ defines x_{n+1} as a function of $\mathbf{X}$:

$$x_{n+1} = f(\mathbf{X}) = \sqrt{1 - x_1^2 - \cdots - x_n^2}.$$

The following theorem, which we state without proof, gives useful sufficient conditions which ensure that certain subsets of N define functions implicitly.

Implicit Function Theorem

Theorem 6.1 (Implicit Function Theorem)   Let F be a function of $(x_1, \ldots, x_{n+1})$. Suppose that $(\mathbf{X}_0, c)$ is an interior point of the domain of F and let $F_{x_1}, \ldots, F_{x_{n+1}}$ be continuous in an $(n+1)$-ball about $(\mathbf{X}_0, c)$. Suppose that $F(\mathbf{X}_0, c) = 0$ and $F_{x_{n+1}}(\mathbf{X}_0, c) \neq 0$. Then there is a set D in R^n with $\mathbf{X}_0$ as an interior point and a unique continuous function f with domain D, such that $f(\mathbf{X}_0) = c$ and $F(\mathbf{X}, f(\mathbf{X})) = 0$ for all $\mathbf{X}$ in D. Furthermore, f has continuous first partial derivatives on D given by

$$(3) \qquad f_{x_i}(\mathbf{X}) = \frac{-F_{x_i}(\mathbf{X}, f(\mathbf{X}))}{F_{x_{n+1}}(\mathbf{X}, f(\mathbf{X}))} \qquad (1 \leq i \leq n).$$

We say that $F(\mathbf{X}, x_{n+1}) = 0$ *defines* f *implicitly near* $\mathbf{X}_0$, or that *it defines* x_{n+1} *as a function of* $\mathbf{X}$ *near* $\mathbf{X}_0$. We write

$$x_{n+1} = f(\mathbf{X}).$$

Example 6.9   Let $(\mathbf{X}_0, c) = (0, \ldots, 0)$ and

$$F(\mathbf{X}, x_{n+1}) = a_1 x_1 + \cdots + a_{n+1} x_{n+1}.$$

Then $F_{x_i} = a_i$ $(1 \leq i \leq n+1)$ and Theorem 6.1 implies that $F(\mathbf{X}, x_{n+1}) = 0$ defines x_{n+1} as a function of $\mathbf{X}$ near $\mathbf{X}_0$, provided $a_{n+1} \neq 0$. Elementary algebra bears this out, for then

$$x_{n+1} = f(\mathbf{X}) = -\frac{1}{a_{n+1}} (a_1 x_1 + \cdots + a_n x_n)$$

satisfies $F(\mathbf{X}, f(\mathbf{X})) = 0$ for all $\mathbf{X}$, and

$$f_{x_i} = -\frac{a_i}{a_{n+1}} = \frac{-F_{x_i}}{F_{x_{n+1}}},$$

which agrees with (3).

Example 6.10 Let

$$F(x, y, z) = x^2 + y^2 + z^2 - 1$$

and

$$(x_0, y_0, z_0) = \left(\frac{1}{\sqrt{2}}, 0, -\frac{1}{\sqrt{2}} \right);$$

then

(4) $F_x(x, y, z) = 2x,$ $F_y(x, y, z) = 2y,$ and $F_z(x, y, z) = 2z.$

Since

$$F_x \left(\frac{1}{\sqrt{2}}, 0, \frac{-1}{\sqrt{2}} \right) = -F_z \left(\frac{1}{\sqrt{2}}, 0, -\frac{1}{\sqrt{2}} \right)$$

$$= \sqrt{2},$$

it follows that $F(x, y, z) = 0$ defines $z = f(x, y)$ such that $f(1/\sqrt{2}, 0) = -1/\sqrt{2}$ and $x = g(y, z)$ such that $g(0, 1/\sqrt{2}) = 1/\sqrt{2}$. These functions can also be written explicitly as

(5) $$f(x, y) = -\sqrt{1 - x^2 - y^2}$$

and

$$g(y, z) = \sqrt{1 - y^2 - z^2},$$

but this is not true for the general implicit functions, so we shall not make use of it in the following. It is possible to compute the partial derivatives of f at $(1/\sqrt{2}, 0)$ without using the explicit formula (5). From (3) and (4)

(6) $$f_x(x, y) = -\frac{x}{z}, \qquad f_y(x, y) = -\frac{y}{z};$$

hence

(7) $$f_x \left(\frac{1}{\sqrt{2}}, 0 \right) = 1, \qquad f_y \left(\frac{1}{\sqrt{2}}, 0 \right) = 0$$

(note that $z = -1/\sqrt{2}$ when $(x, y) = (1/\sqrt{2}, 0)$).
   To obtain $f_{xx}(1/\sqrt{2}, 0)$, replace z by $f(x, y)$ in (6):

$$f_x(x, y) = -\frac{x}{f(x, y)}.$$

Then

$$(8) \qquad f_{xx}(x, y) = \frac{\partial}{\partial x}\left(\frac{-x}{f(x, y)}\right)$$

$$= \frac{-f(x, y) + xf_x(x, y)}{[f(x, y)]^2}.$$

Setting $(x, y) = (1/\sqrt{2}, 0)$ and using (7) yields

$$f_{xx}\left(\frac{1}{\sqrt{2}}, 0\right) = \frac{4}{\sqrt{2}}.$$

Similar computations yield $f_{xy}(1/\sqrt{2}, 0)$ and $f_{yy}(1/\sqrt{2}, 0)$; then by differentiating (8) and using the known values for the first and second derivatives, we could calculate the third derivatives at $(1/\sqrt{2}, 0)$, etc.

Example 6.11 Let $F(x, y, z) = x^2yz + 2xy^2z^3 - 3x^3y^3z^5$ and $(x_0, y_0, z_0) = (1, 1, 1)$, so that $F(x_0, y_0, z_0) = 0$. Then

$$F_x(x, y, z) = 2xyz + 2y^2z^3 - 9x^2y^3z^5,$$

$$F_y(x, y, z) = x^2z + 4xyz^3 - 9x^3y^2z^5,$$

and

$$F_z(x, y, z) = x^2y + 6xy^2z^2 - 15x^2y^3z^4.$$

Since $F_z(x_0, y_0, z_0) = -8$, $z = f(x, y)$ on some domain D containing $(1, 1)$ as an interior point. However, it is extremely difficult to exhibit z explicitly as a function of x and y. Nevertheless, $f_x(1, 1)$ and $f_y(1, 1)$ can be computed:

$$f_x(1, 1) = \frac{-F_x(1, 1, 1)}{F_z(1, 1, 1)} = -\frac{(-5)}{(-8)} = \frac{-5}{8}$$

and

$$f_y(1, 1) = \frac{-F_y(1, 1, 1)}{F_z(1, 1, 1)} = -\frac{(-4)}{(-8)} = -\frac{1}{2}.$$

Higher partial derivatives at $(1, 1)$ can be computed as in Example 6.10.

For $n = 1$, (3) can be obtained by formally applying the chain rule. Thus if $F(x, f(x)) = 0$ for all x in some interval, differentiation with respect to x yields

$$F_x(x, f(x)) + F_y(x, f(x))f'(x) = 0;$$

hence,

$$(9) \qquad f'(x) = -\frac{F_x(x, f(x))}{F_y(x, f(x))}$$

provided $F_y(x, f(x)) \neq 0$.

The conditions of Theorem 6.1 are sufficient, but not necessary. That is, $F(\mathbf{X}, x_{n+1}) = 0$ *may* define a function f near $\mathbf{X}_0$ even though F fails to satisfy the hypotheses of Theorem 6.1 at $(\mathbf{X}_0, c)$. (See Exercises 9 and 10).

By writing

$$y = f(x), \qquad z = f(x, y), \qquad \text{and} \qquad x_{n+1} = f(x_1, \ldots, x_n)$$

in this section, we have introduced a new convention, since we are now denoting the *value* of a function at a point in R^n by the *name* of a coordinate in R^{n+1}. This is not consistent with our previous notation and can sometimes lead to ambiguity; nevertheless it is convenient for simplifying what might otherwise be extremely ugly expressions. For $n = 1$ the student should not be shocked by this notation; he must have written $y = f(x)$ hundreds of times in his first calculus course! He has also written

$$y' = \frac{dy}{dx} = f'(x), \qquad y'' = \frac{d^2y}{dx^2} = f''(x),$$

etc. We shall extend this idea and write

$$z = f(x, y),$$

$$z_x = \frac{\partial z}{\partial x} = \frac{\partial f}{\partial x}(x, y),$$

$$z_{xy} = \frac{\partial^2 z}{\partial x\, \partial y} = \frac{\partial^2 f}{\partial x\, \partial y}(x, y),$$

and so forth, when adherence to our earlier (still preferred) notation is cumbersome. Thus, we write (9) as

$$y' = \frac{-F_x(x, y)}{F_y(x, y)},$$

or, if we want an even more abbreviated expression,

$$y' = \frac{-F_x}{F_y}.$$

The student manipulates such an expression at his own risk, and should always bear in mind what it really stands for.

EXERCISES 2.6

1. Which of the following sets define $x_4 = f(x_1, x_2, x_3)$? For those which do, determine the domain and range of f.

    (a) $G_1 = \{(1, 5, 1, 1), \quad (1, 1, 2, 2), \quad (1, 1, 1, 3), \quad (1, 1, 3, 2)\}$.

(b) $G_2 = \{(1, 1, 1, 1), \quad (1, 1, 2, 2), \quad (1, 1, 1, 2), \quad (1, 1, 3, 2)\}.$

(c) $G_3 = \{(2, 7, 5, 3), \quad (1, 7, 5, 3), \quad (3, 6, 2, 3), \quad (7, 5, 3, 1)\}.$

(d) $G_4 = \{(1, 2, 3, 4), \quad (2, 3, 4, 1), \quad (3, 4, 1, 2), \quad (4, 1, 2, 3)\}.$

2. $G = \{(x, y) \mid \log y - x = 0\}$ defines $y = f(x)$. What is $f(x)$ explicitly?

3. Which of the following sets defines $y = f(x)$? For those which do, determine the domain and range of f.

(a) $G_1 = \{(x, y) \mid \cos y - x = 0\}.$

(b) $G_2 = \{(x, y) \mid \cos y - x = 0, \quad |y| < \pi/2\}.$

(c) $G_3 = \{(x, y) \mid \cos y - x = 0, \quad 0 \le y \le \pi\}.$

(d) $G_4 = \{(x, y) \mid \cos y - x = 0, \quad \pi \le y \le 2\pi\}.$

4. Which of the following sets define $z = f(x, y)$? For those which do, determine the domain and range of f.

(a) $G_1 = \{(x, y, z) \mid x^2 + y^2 - z^2 - 1 = 0, \quad z \ge 0\}.$

(b) $G_2 = \{(x, y, z) \mid x^2 + y^2 - z^2 - 1 = 0, \quad |z| < 1\}.$

(c) $G_3 = \{(x, y, z) \mid x^2 - y^2 + z^2 - 1 = 0, \quad y \ge 0\}.$

(d) $G_4 = \{(x, y, z) \mid x^2 - y^2 + z^2 - 1 = 0, \quad z \ge 0\}.$

5. Which of the sets $G_1, G_2, G_3,$ and G_4 in Exercise 1 define $x_1 = g(x_2, x_3, x_4)$? For those which do, specify the domains and ranges of the functions that they define.

6. Repeat Exercise 5 for $x = g(y)$ in Exercise 3.

7. Repeat Exercise 5 for $y = h(x, z)$ in Exercise 4.

8. Show that $1 - e^z \cos (x - y) = 0$ defines $z = f(x, y)$ near $(\pi/2, \pi/2)$. Find $f(\pi/2, \pi/2)$ and all first and second partial derivatives of f at $(\pi/2, \pi/2)$. Also, obtain explicit expressions for f_x and f_y.

9. Show that $1 - e^z \cos(x - y) = 0$ does not define $z = f(x, y)$ near $(\pi/2, 0)$. (It is *not* enough to simply state that the hypotheses of the implicit function theorem are not satisfied, since this theorem states sufficient, not necessary, conditions for the existence of an implicit function.)

10. $F(x, y) = x - y^3 = 0$ defines $y = f(x)$ near $x = 0$ (with $f(0) = 0$) even though F does not satisfy the hypothesis of Theorem 6.1 at $(0, 0)$. Does this violate Theorem 6.1?

11. Let $F(x, y) = y^2(1 - xy) - 3y[1 + \log(x + 1)] + 2e^{xy}$; then $F(x, y) = 0$ defines two functions, $y = g(x)$ and $y = h(x)$, near $x = 0$. (Does this contradict Theorem 6.1?) Find $g(0)$, $g'(0)$, $h(0)$, and $h'(0)$.

12. The equation
$$x^2 y^5 z^2 w^5 + 2xy^2 w^3 - 3x^3 z^2 w = 0$$
defines $w = f(x, y, z)$ with $f(1, 1, 1) = 1$. Find $f_x(1, 1, 1)$, $f_y(1, 1, 1)$ and $f_z(1, 1, 1)$.

13. The equation of Exercise (12) also defines $y = g(x, z, w)$ with $g(1, 1, 1) = 1$. Find $g_x(1, 1, 1)$, $g_z(1, 1, 1)$, and $g_w(1, 1, 1)$.

14. Suppose $y = f(x)$ is defined by
$$x^4 - x^3 y^2 + y^5 + 1 = 0;$$
find $f'(x)$ in terms of x and y.

THEORETICAL EXERCISES

T-1. Let $F(x, y) = 0$ define $y = f(x)$; by formally applying the chain rule to (9), show that
$$y'' = -\frac{1}{F_y}(F_{xx} + 2F_{xy}y' + F_{yy}(y')^2).$$
Compare this with the expression for $f''(x)$, completely written out in analogy with (9).

T-2. Show that the result of Exercise T-1 is consistent with the expected result if $F(x, y) = y - f(x)$ (that is, if y is defined explicitly as a function of x).

T-3. Give assumptions on F and f that make the derivation of (9) by the chain rule legitimate.

2.7 The Gradient

In Theorem 3.5, we found a useful representation for directional derivatives: if f is differentiable at $\mathbf{X}_0$ and
$$\mathbf{U} = \begin{bmatrix} u_1 \\ \vdots \\ u_n \end{bmatrix}$$

is a unit vector, then

(1) $$\frac{\partial f}{\partial \mathbf{U}}(\mathbf{X}_0) = f_{x_1}(\mathbf{X}_0)u_1 + \cdots + f_{x_n}(\mathbf{X}_0)u_n.$$

We have also seen that $(\partial f / \partial \mathbf{U})(\mathbf{X}_0)$ is the rate of change of f, at $\mathbf{X}_0$, in the direction of $\mathbf{U}$. Question: What is the maximum rate of change of f at $\mathbf{X}_0$, and in what direction is it attained?

The answer to this question is provided by Schwarz's inequality (Theorem 1.8, Section 1.1).

Definition 7.1   Let f be a function with domain D in R^n which is differentiable at $\mathbf{X}_0$. The vector

$$\nabla_{\mathbf{X}_0} f = \begin{bmatrix} f_{x_1}(\mathbf{X}_0) \\ \vdots \\ f_{x_n}(\mathbf{X}_0) \end{bmatrix}$$

is called the *gradient of f at* $\mathbf{X}_0$. If $n = 3$ we may also write

$$\nabla_{\mathbf{X}_0} f = f_x(\mathbf{X}_0)\mathbf{i} + f_y(\mathbf{X}_0)\mathbf{j} + f_z(\mathbf{X}_0)\mathbf{k}.$$

Example 7.1   Let $\mathbf{X}_0 = (1, 1, \ldots, 1)$ and

$$f(\mathbf{X}) = x_1^2 + x_2^2 + \cdots + x_n^2;$$

then

$$f_{x_i}(\mathbf{X}) = 2x_i, \qquad f_{x_i}(\mathbf{X}_0) = 2 \qquad (1 \leq i \leq n);$$

thus

$$\nabla_{\mathbf{X}_0} f = \begin{bmatrix} 2 \\ 2 \\ \vdots \\ 2 \end{bmatrix}.$$

Example 7.2   Let $\mathbf{X}_0 = (-1, 2, 1)$ and $f(x, y, z) = xy^2z^2$; then

$$f_x(x, y, z) = y^2z^2, \qquad f_y(x, y, z) = 2xyz^2, \qquad f_z(x, y, z) = 2xy^2z$$

and

$$f_x(\mathbf{X}_0) = 4, \qquad f_y(\mathbf{X}_0) = -4, \qquad f_z(\mathbf{X}_0) = -8.$$

Now (1) can be rewritten as an inner product:

$$\frac{\partial f}{\partial \mathbf{U}}(\mathbf{X}_0) = \mathbf{U} \cdot \nabla_{\mathbf{X}_0} f.$$

From Schwarz's inequality, since $|\mathbf{U}| = 1$,

$$\left| \frac{\partial f}{\partial \mathbf{U}}(\mathbf{X}_0) \right| \leq |\nabla_{\mathbf{X}_0} f|$$

and equality is attained if and only if $\mathbf{U}$ and $\nabla_{\mathbf{X}_0} f$ have the same direction. (This statement includes the case where $f_{x_1}(\mathbf{X}_0) = \cdots = f_{x_n}(\mathbf{X}_0) = 0$, if we agree that the zero vector has every direction.) This proves the following theorem.

Theorem 7.1   If f is differentiable at $\mathbf{X}_0$ and $\mathbf{U}$ is a unit vector, then

$$\frac{\partial f}{\partial \mathbf{U}}(\mathbf{X}_0) = \mathbf{U} \cdot \nabla_{\mathbf{X}_0} f.$$

The maximum value of the directional derivative of f at $\mathbf{X}_0$ is $|\nabla_{\mathbf{X}_0} f|$; it is attained in the direction of

$$\mathbf{U}_0 = \frac{\nabla_{\mathbf{X}_0} f}{|\nabla_{\mathbf{X}_0} f|},$$

and only in this direction, unless $\nabla_{\mathbf{X}_0} f = 0$, in which case $\mathbf{U}_0$ is undefined and $(\partial f / \partial \mathbf{U})(\mathbf{X}_0) = 0$ in all directions.

Example 7.3   Let $\mathbf{X}_0 = \left(\dfrac{1}{\sqrt{2}}, -\dfrac{1}{\sqrt{2}} \right)$ and $f(x, y) = x^2 + y^2$; then

$$\nabla_{\mathbf{X}_0} f = \begin{bmatrix} \sqrt{2} \\ -\sqrt{2} \end{bmatrix}$$

and, if

$$\mathbf{U} = \begin{bmatrix} u_1 \\ u_2 \end{bmatrix}$$

is an arbitrary unit vector,

$$\frac{\partial f}{\partial \mathbf{U}}(\mathbf{X}_0) = \mathbf{U} \cdot \nabla_{\mathbf{X}_0} f$$

$$= \sqrt{2}(u_1 - u_2).$$

The direction of the maximum rate of change is

$$\mathbf{U}_0 = \frac{\nabla_{\mathbf{X}_0} f}{|\nabla_{\mathbf{X}_0} f|}$$

$$= \frac{1}{2} \begin{bmatrix} \sqrt{2} \\ -\sqrt{2} \end{bmatrix} = \begin{bmatrix} \dfrac{1}{\sqrt{2}} \\ -\dfrac{1}{\sqrt{2}} \end{bmatrix}$$

and

$$\frac{\partial f}{\partial \mathbf{U}_0}(\mathbf{X}_0) = |\nabla_{\mathbf{X}_0} f| = 2.$$

Example 7.4 Let f be as in Example 7.1 and $\mathbf{X}_0 = (0, 0)$; then

$$\nabla_{\mathbf{X}_0} f = \begin{bmatrix} 0 \\ 0 \end{bmatrix} \qquad \text{and} \qquad \frac{\partial f}{\partial \mathbf{U}}(\mathbf{X}_0) = 0$$

for all $\mathbf{U}$.

The gradient provides a convenient representation for the tangent plane to the graph of an implicit function. Suppose $F(x_0, y_0, z_0) = 0$, $F_z(x_0, y_0, z_0) \neq 0$, and F satisfies the hypotheses of the implicit function theorem near $\mathbf{X}_0$. Then $F(x, y, z) = 0$ defines $z = f(x, y)$, where f is differentiable and satisfies $F(x, y, f(x, y)) = 0$ for all (x, y) in some domain D containing (x_0, y_0) in its interior; hence the graph of f has a tangent plane at $(x_0, y_0, f(x_0, y_0)) = (x_0, y_0, z_0)$ with the equation

$$(2) \qquad z = z_0 + (x - x_0)f_x(x_0, y_0) + (y - y_0)f_y(x_0, y_0).$$

From the implicit function theorem,

$$f_x(x_0, y_0) = \frac{-F_x(x_0, y_0, z_0)}{F_z(x_0, y_0, z_0)},$$

$$f_y(x_0, y_0) = \frac{-F_y(x_0, y_0, z_0)}{F_z(x_0, y_0, z_0)};$$

substituting these in (2) and rearranging terms yields

$$(3) \qquad (x - x_0)F_x(x_0, y_0, z_0) + (y - y_0)F_y(x_0, y_0, z_0)$$
$$+ (z - z_0)F_z(x_0, y_0, z_0) = 0,$$

which can be written more briefly as

$$(\mathbf{X} - \mathbf{X}_0) \cdot \nabla_{\mathbf{X}_0} F = 0.$$

Thus the gradient is normal to the tangent plane (Fig. 7.1).

Example 7.5 Let

$$F(x, y, z) = x^2 + y^2 + z^2 - 1,$$

and

$$\mathbf{X}_0 = \left(\frac{1}{\sqrt{3}}, -\frac{1}{\sqrt{6}}, \frac{1}{\sqrt{2}} \right).$$

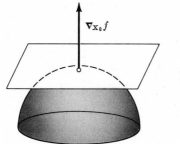

FIGURE 7.1

Then $F(x, y, z) = 0$ defines $z = f(x, y) = \sqrt{1 - x^2 - y^2}$ near $\mathbf{X}_0$ and

$$f_x(x, y) = \frac{-x}{\sqrt{1 - x^2 - y^2}},$$

$$f_y(x, y) = \frac{-y}{\sqrt{1 - x^2 - y^2}}.$$

Hence $f_x\left(\dfrac{1}{\sqrt{3}}, -\dfrac{1}{\sqrt{6}}\right) = -\sqrt{\dfrac{2}{3}}, f_y\left(\dfrac{1}{\sqrt{3}}, -\dfrac{1}{\sqrt{6}}\right) = \dfrac{1}{\sqrt{3}}$; and, from (2), the equation of the tangent plane to the graph of f at $\mathbf{X}_0$ is

$$(4) \qquad z = \frac{1}{\sqrt{2}} - \sqrt{\frac{2}{3}}\left(x - \frac{1}{\sqrt{3}}\right) + \frac{1}{\sqrt{3}}\left(y + \frac{1}{\sqrt{6}}\right).$$

On the other hand,

$$F_x(x, y, z) = 2x, \qquad F_y(x, y, z) = 2y, \qquad F_z(x, y, z) = 2z,$$

so that

$$F_x\left(\frac{1}{\sqrt{3}}, -\frac{1}{\sqrt{6}}, \frac{1}{\sqrt{2}}\right) = \frac{2}{\sqrt{3}}, \qquad F_y\left(\frac{1}{\sqrt{3}}, -\frac{1}{\sqrt{6}}, \frac{1}{\sqrt{2}}\right) = -\frac{2}{\sqrt{6}},$$

$$F_z\left(\frac{1}{\sqrt{3}}, -\frac{1}{\sqrt{6}}, \frac{1}{\sqrt{2}}\right) = \frac{2}{\sqrt{2}},$$

and the equation of the tangent plane in the form (3) is

$$(5) \qquad \frac{1}{\sqrt{3}}\left(x - \frac{1}{\sqrt{3}}\right) - \frac{1}{\sqrt{6}}\left(y + \frac{1}{\sqrt{6}}\right) + \frac{1}{\sqrt{2}}\left(z - \frac{1}{\sqrt{2}}\right) = 0,$$

which is easily shown to be equivalent to (4).

In (3) the equation of the tangent plane does not explicitly involve f;

this is an obvious advantage if it is difficult or impossible to obtain f explicitly.

Suppose we make the additional assumption that $F_x(x_0, y_0, z_0) \neq 0$; then

$$F(x, y, z) = 0$$

also defines $x = g(y, z)$ such that $g(y_0, z_0) = x_0$ and $F(g(y, z), y, z)) = 0$ at all (y, z) in some domain. The equation of the tangent plane to the graph of g at $(x_0, y_0, z_0) = (g(y_0, z_0), y_0, z_0)$ is

(6) $$x = x_0 + (y - y_0)g_y(y_0, z_0) + (z - z_0)g_z(y_0, z_0).$$

From the implicit function theorem,

$$g_y(y_0, z_0) = \frac{-F_y(x_0, y_0, z_0)}{F_x(x_0, y_0, z_0)},$$

$$g_z(y_0, z_0) = \frac{-F_z(x_0, y_0, z_0)}{F_x(x_0, y_0, z_0)},$$

which, substituted into (6), yields (3). Thus the graphs of f and g have the same tangent plane at (x_0, y_0, z_0). If $F(x, y, z) = 0$ also defined $y = h(x, z)$ near (x_0, z_0), the equation of the tangent plane to the graph of h at (x_0, y_0, z_0) would also be (3).

Example 7.6   In Example 7.5, we could have considered that

$$F(x, y, z) = 0$$

defines

$$x = g(y, z) = \sqrt{1 - y^2 - z^2};$$

then

$$g_y(y, z) = \frac{-y}{\sqrt{1 - y^2 - z^2}}$$

and

$$g_z(y, z) = \frac{-z}{\sqrt{1 - y^2 - z^2}}.$$

Hence

$$g_y\left(-\frac{1}{\sqrt{6}}, \frac{1}{\sqrt{2}}\right) = \frac{1}{\sqrt{2}},$$

$$g_z\left(-\frac{1}{\sqrt{6}}, \frac{1}{\sqrt{2}}\right) = -\sqrt{\frac{3}{2}}.$$

and, from (6), the equation of the tangent plane to the graph of g at $\mathbf{X}_0$ is

$$x = \frac{1}{\sqrt{3}} + \frac{1}{\sqrt{2}}\left(y + \frac{1}{\sqrt{6}}\right) - \sqrt{\frac{3}{2}}\left(z + \frac{1}{\sqrt{2}}\right),$$

which is equivalent to (5).

In view of this discussion it makes good sense to associate the plane (3) with F rather than with any particular implicit function defined by F. Before pursuing this further, we need two definitions.

Definition 7.2 Let $\mathbf{X}_0$ be in the domain of F. Then the *level set of F through* $\mathbf{X}_0$ is the set of points $\mathbf{X}$ such that $F(\mathbf{X}) = F(\mathbf{X}_0)$; we shall also describe it as the *level set defined by* $F(\mathbf{X}) = F(\mathbf{X}_0)$, or simply as the level set defined by $F(\mathbf{X}) = c$, where $c = F(\mathbf{X}_0)$. If $n = 2$ we shall refer to level sets as *level curves*; if $n = 3$ we shall speak of *level surfaces*.

Example 7.7 Let

$$F(x, y) = x^2 + y^2;$$

the level curve through (x_0, y_0) is the circle

$$x^2 + y^2 = x_0^2 + y_0^2.$$

The level curves through $(3, 4)$ and $(0, 4)$ are the circles defined by $F(x, y) = 5$ and $F(x, y) = 4$, respectively (Fig. 7.2).

Example 7.8 Let

$$F(x, y, z) = x + y + z;$$

the level surface through (x_0, y_0, z_0) is the plane

$$(x - x_0) + (y - y_0) + (z - z_0) = 0.$$

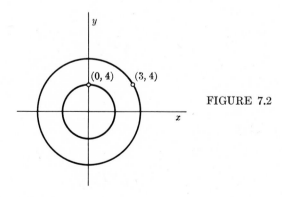

FIGURE 7.2

If $(x_0, y_0, z_0) = (1, 0, 2)$ then the level surface is the plane

$$x + y + z = 3.$$

Definition 7.3 Let F have continuous first partial derivatives in an n-ball about $\mathbf{X}_0 = (c_1, \ldots, c_n)$, and suppose $F_{x_i}(\mathbf{X}_0) \neq 0$ for some i. Let L be the level set of F through $\mathbf{X}_0$. Then the *tangent plane* (*line*, if $n = 2$) *to L at $\mathbf{X}_0$* is the plane whose equation is

$$(7) \qquad (x_1 - c_1)F_{x_1}(\mathbf{X}_0) + \cdots + (x_n - c_n)F_{x_n}(\mathbf{X}_0) = 0.$$

For each i such that $F_{x_i}(\mathbf{X}_0) \neq 0$, $F(\mathbf{X}) - F(\mathbf{X}_0) = 0$ defines x_i as a function f_i of the other coordinates (to see this take $F = F - F(\mathbf{X}_0)$); then F satisfies the hypotheses of the implicit function theorem). By the argument given earlier for $n = 3$, (7) is the tangent plane to the graph of f_i at $\mathbf{X}_0$; hence calling (7) a tangent plane in this definition is consistent with the usage in Section 2.5.

Example 7.9 To find the equation of the tangent plane to the sphere

$$x^2 + y^2 + z^2 = 1$$

at

$$\mathbf{X}_0 = \left(\frac{1}{\sqrt{3}}, \frac{-1}{\sqrt{6}}, \frac{1}{\sqrt{2}} \right),$$

take

$$F(x, y, z) = x^2 + y^2 + z^2 - 1.$$

The sphere is the level surface of F through $\mathbf{X}_0$. Since

$$F_x(\mathbf{X}_0) = \frac{2}{\sqrt{3}}, \qquad F_y(\mathbf{X}_0) = \frac{-2}{\sqrt{6}}, \qquad F_z(\mathbf{X}_0) = \frac{2}{\sqrt{2}},$$

the equation of the tangent plane is

$$\frac{1}{\sqrt{3}}\left(x - \frac{1}{\sqrt{3}} \right) - \frac{1}{\sqrt{6}}\left(y + \frac{1}{\sqrt{6}} \right) + \frac{1}{\sqrt{2}}\left(z - \frac{1}{\sqrt{2}} \right) = 0.$$

We obtained the same result in Example 7.5; the difference here is that we have dispensed with the intermediate step of introducing an implicit function.

If a level set $F(\mathbf{X}) = F(\mathbf{X}_0)$ has a tangent plane at $\mathbf{X}_0$, we define the *normal to the level set at $\mathbf{X}_0$* to be the normal to the tangent plane. Since (7) can be rewritten as

$$(\mathbf{X} - \mathbf{X}_0) \cdot \nabla_{\mathbf{X}_0} F = 0,$$

it follows that this normal is in the direction of $\nabla_{X_0}F$. More specifically, the unit vector $\mathbf{N}$ in the direction of the gradient is called the *outward normal*; $-\mathbf{N}$ is the *inward normal*.

Example 7.10 In Example 7.9

$$\nabla_{X_0}F = \begin{bmatrix} \dfrac{2}{\sqrt{3}} \\[2ex] -\dfrac{2}{\sqrt{6}} \\[2ex] \dfrac{2}{\sqrt{2}} \end{bmatrix} = 2\mathbf{X}_0$$

that is, the normal to the sphere at $\mathbf{X}_0$ is along the radius vector from the origin to $\mathbf{X}_0$, which is geometrically evident from Fig. 7.3.

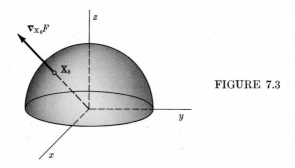

FIGURE 7.3

Example 7.11 Let $T(x, y, z)$ be the temperature at (x, y, z) in some domain D. Then the level surfaces of T are called *isothermals*. The normal to the isothermal through (x_0, y_0, z_0) is $\nabla_{X_0}T$; thus the direction of the greatest rate of temperature increase is normal to the isothermal through $\mathbf{X}_0$.

EXERCISES 2.7

1. Find $\nabla_{X_0}f$:

   (a) $f(\mathbf{X}) = a_1 x_1 + \cdots + a_n x_n$, $\mathbf{X}_0$ arbitrary.

   (b) $f(\mathbf{X}) = x^2 - 2xy + z^2$, $\mathbf{X}_0 = (1, 1, 2)$.

(c) $f(\mathbf{X}) = \log(x + y + z)$, $\mathbf{X}_0 = (1, 3, 2)$.

(d) $f(\mathbf{X}) = e^{x^2 + y^2 + z^2}$, $\mathbf{X}_0 = (1, 0, 0)$.

2. In each of the following $F(\mathbf{X}, w) = 0$ defines $w = f(\mathbf{X})$ near $\mathbf{X}_0$. Find $\nabla_{\mathbf{X}_0} f$.

(a) $F(\mathbf{X}, w) = x^2 + y^2 + z^2 - w^2 + 1$, $f(\mathbf{X}_0) = 2$,
   $\mathbf{X}_0 = (1, -1, 1)$.

(b) $F(\mathbf{X}, w) = (2x + 3y + z + 2)^2 - e^w$, $\mathbf{X}_0 = (1, -2, 3)$.

(c) $F(\mathbf{X}, w) = \tan xyz - \log w$, $\mathbf{X}_0 = (0, 1, 2)$.

(d) $F(\mathbf{X}, w) = x + 3xy + z^2 + e^{\sin w}$, $f(\mathbf{X}_0) = \pi$,
   $\mathbf{X}_0 = (-2, 0, 1)$.

3. For each function in Exercise 1 find the direction and magnitude of the greatest rate of increase of f at $\mathbf{X}_0$.

4. Find the direction in which each function in Exercise 2 decreases most rapidly at $\mathbf{X}_0$.

5. Go back to Exercise 1, (c) and (d), and find $\dfrac{\partial f}{\partial \mathbf{U}}$ $(\mathbf{X}_0)$ for

(a) $\mathbf{U} = \begin{bmatrix} \dfrac{1}{\sqrt{2}} \\ 0 \\ \dfrac{-1}{\sqrt{2}} \end{bmatrix}$,
(b) $\mathbf{U} = \begin{bmatrix} \dfrac{1}{\sqrt{3}} \\ \dfrac{-1}{\sqrt{3}} \\ \dfrac{1}{\sqrt{3}} \end{bmatrix}$.

6. Go back to Exercise 2, (a) and (b), and find $(\partial f / \partial \mathbf{U})(\mathbf{X}_0)$ for

(a) $\mathbf{U} = \begin{bmatrix} \dfrac{1}{\sqrt{2}} \\ 0 \\ \dfrac{-1}{\sqrt{2}} \end{bmatrix}$,
(b) $\mathbf{U} = \begin{bmatrix} \dfrac{1}{\sqrt{3}} \\ \dfrac{-1}{\sqrt{3}} \\ \dfrac{1}{\sqrt{3}} \end{bmatrix}$.

7. Let $F(x, y, z) = x^2 + y^2 + e^{z-2} - 3$; then $F(x, y, z) = 0$ defines $z = f(x, y)$, $y = g(x, z)$, and $x = h(y, z)$ such that the graphs of

f, g, and h all contain $(1, -1, 2)$. Show by direct computation that the three graphs have the same tangent plane at $(1, -1, 2)$.

8. Repeat Exercise 7 for $F(x, y, z) = x + y + \log(1 + x^2 + y^2) + z^2 + 2z$ and $(0, 0, 0)$.

9. What is the level surface of $g(x, y, z) = 2x + 4y - 3z$ through $\mathbf{X}_0 = (0, 1, 2)$?

10. What is the level curve of $f(x, y) = \sin xy$ through $(0, 0)$?

11. What is the level surface of $f(x, y, z) = \sin(x^2 + y^2 + z^2)$ through $(\sqrt{\pi}, 0, 0)$?

12. What is the level curve of $f(x, y) = x^2 - y^2$ through $(1, 1)$?

In Problems 13 through 17 find the tangent plane (line if $n = 2$) to the level set of F through $\mathbf{X}_0$.

13. $F(\mathbf{X}) = x^2 + 3xy + y^2 + zx$, $\quad \mathbf{X}_0 = (1, 2, -1)$.

14. $F(\mathbf{X}) = \sin(x^2 + y^2 + z^2)$, $\quad \mathbf{X}_0 = \left(\sqrt{\dfrac{\pi}{3}}, \sqrt{\dfrac{\pi}{3}}, \sqrt{\dfrac{\pi}{3}}\right)$.

15. $F(\mathbf{X}) = \sin(x^2 + y^2 + z^2)$, $\quad \mathbf{X}_0 = (\sqrt{\pi}, 0, 0)$.

16. $F(\mathbf{X}) = x^2 - y^2$, $\quad \mathbf{X}_0 = (2, 1)$.

17. $F(\mathbf{X}) = e^x + \log(1 + xy)$, $\quad \mathbf{X}_0 = (0, 0)$.

In Problems 18 through 21 find the outward normal to the given level set at $\mathbf{X}_0$.

18. $x^2 + 4y^2 = 5$, $\quad \mathbf{X}_0 = (1, 1)$.

19. $\sin(x^2 - y^2) = 0$, $\quad \mathbf{X}_0 = (1, 1)$.

20. $\sin xyz - \log(1 + x^2 + y^2 + z^2) = \log 2$; $\quad \mathbf{X}_0 = (1, 0, 0)$.

21. $\sin^2 x + \sin^2 y + \sin^2 z = 1$, $\quad \mathbf{X}_0 = \left(\dfrac{\pi}{4}, \dfrac{\pi}{4}, 0\right)$.

THEORETICAL EXERCISES

T-1. Find $(\partial F/\partial \mathbf{n})(\mathbf{X}_0)$, where $\mathbf{n}$ is a unit vector along the outward normal to the level set $F(\mathbf{X}) = F(\mathbf{X}_0)$ at $\mathbf{X}_0$.

T-2. Let f and g be differentiable at $\mathbf{X}_0$. Show that

  (a) $\nabla_{\mathbf{X}_0}(\alpha f) = \alpha \nabla_{\mathbf{X}_0} f$ $\quad (\alpha = \text{constant})$,

(b) $\nabla_{\mathbf{X}_0}(f + g) = \nabla_{\mathbf{X}_0} f + \nabla_{\mathbf{X}_0} g,$

(c) $\nabla_{\mathbf{X}_0}(fg) = f(\mathbf{X}_0) \nabla_{\mathbf{X}_0} g + g(\mathbf{X}_0) \nabla_{\mathbf{X}_0} f,$

and, if $g(\mathbf{X}_0) \neq 0,$

(d) $\nabla_{\mathbf{X}_0}(f/g) = \dfrac{g(\mathbf{X}_0) \nabla_{\mathbf{X}_0} f - f(\mathbf{X}_0) \nabla_{\mathbf{X}_0} g}{[g(\mathbf{X}_0)]^2}.$

2.8 Taylor's Theorem

By Definition 3.1 (Section 2.3), a function f which is differentiable at $\mathbf{X}_0 = (c_1, \ldots, c_n)$ can be so well approximated near $\mathbf{X}_0$ by

$$T_1(\mathbf{X}) = f(\mathbf{X}_0) + (d_{\mathbf{X}_0} f)(\mathbf{X} - \mathbf{X}_0)$$

$$= f(\mathbf{X}_0) + \sum_{r=1}^{n} \frac{\partial f}{\partial x_r}(\mathbf{X}_0)(x_r - c_r)$$

that

$$\lim_{\mathbf{X} \to \mathbf{X}_0} \frac{f(\mathbf{X}) - T_1(\mathbf{X})}{|\mathbf{X} - \mathbf{X}_0|} = 0.$$

The function T_1, or more generally, any function of the form

$$P(\mathbf{X}) = a + b_1(x_1 - c_1) + \cdots + b_n(x_n - c_n),$$

where $a, b_1, \ldots, b_n$ are constants, is a *polynomial of first degree in* $\mathbf{X} - \mathbf{X}_0$ provided at least one of $b_1, \ldots, b_n$ is nonzero. If $b_1 = \cdots = b_n = 0$ and $a \neq 0$, then P_1 is a polynomial of degree zero. The *zero polynomial*, which is the polynomial whose coefficients all vanish, is arbitrarily assigned the degree minus infinity.

In this section we consider the problem of approximating functions near a point $\mathbf{X}_0$ by means of polynomials of higher degree. Since the situation is simplest for functions of one variable, we consider that case first and in more detail.

Definition 8.1 Let g be a function of a single variable t which possesses k derivatives at $t = t_0$. Then the *k*th *Taylor polynomial of g about t_0* is defined by

$$T_k(t) = \sum_{r=0}^{k} \frac{g^{(r)}(t_0)}{r!}(t - t_0)^r$$

$$= g(t_0) + \frac{g'(t_0)}{1!}(t - t_0) + \cdots + \frac{g^{(k)}(t_0)}{k!}(t - t_0)^k.$$

Clearly the degree of T_k, which is defined to be the highest power of $(t - t_0)$ that appears in T_k with a nonzero coefficient, or minus infinity if all the coefficients vanish, does not exceed k.

Example 8.1 Let $g(t) = \log t$ and $t_0 = 1$. Then $g(1) = 0$ and $g^{(r)}(1) = (-1)^{r-1}(r-1)!$ if $r \geq 1$. Hence

$$T_0(t) = 0$$

and

$$T_k(t) = \sum_{r=1}^{k} \frac{(-1)^r}{r} (t-1)^r \qquad (k \geq 1).$$

Example 8.2 Let $g(t) = e^t$ and $t_0 = 0$. Then $g^{(r)}(0) = 1$ for all $r \geq 0$ and

$$T_k(t) = \sum_{r=0}^{k} \frac{t^r}{r!}.$$

Example 8.3 Let $g(t) = \cos t$; then $g'(t) = -\sin t$, $g''(t) = -\cos t$, $g'''(t) = \sin t$, and the cycle repeats. The derivatives of odd order vanish at $t = 0$, while those of even order alternate between $+1$ and -1. Thus the Taylor polynomials about $t = 0$ are given by

$$T_0(t) = T_1(t) \quad = 1,$$

$$T_2(t) = T_3(t) \quad = 1 - \frac{t^2}{2!},$$

$$T_4(t) = T_5(t) \quad = 1 - \frac{t^2}{2!} + \frac{t^4}{4!},$$

$$\vdots$$

$$T_{2k}(t) = T_{2k+1}(t) = \sum_{r=0}^{k} (-1)^r \frac{t^{2r}}{(2r)!}.$$

Taylor polynomials have the following remarkable property:

Theorem 8.1 (Taylor's Theorem) Let g have k continuous derivatives in an open interval J about t_0 and let T_k be its kth Taylor polynomial about t_0:

$$T_k(t) = g(t_0) + \frac{g'(t_0)}{1!}(t-t_0) + \cdots + \frac{g^{(k)}(t_0)}{k!}(t-t_0)^k.$$

Then

(1)
$$\lim_{t \to t_0} \frac{g(t) - T_k(t)}{(t - t_0)^k} = 0,$$

and T_k is the only polynomial of degree $\leq k$ with this property.

For $k = 1$ this theorem reduces to the situation discussed in the beginning of Section 2.3. We shall give a partial proof for arbitrary k, leaving the proof of uniqueness to the student (Exercise T-2).

Proof. Let $\epsilon > 0$ be given. From the definition of derivative, there is a $\delta > 0$ such that if $0 < |t - t_0| < \delta$ then

$$\left| \frac{g^{(k-1)}(t) - g^{(k-1)}(t_0)}{t - t_0} - g^{(k)}(t_0) \right| < \epsilon,$$

which can be rewritten as

$$| g^{(k-1)}(t) - g^{(k-1)}(t_0) - g^{(k)}(t_0)(t - t_0) | < \epsilon |t - t_0|.$$

Therefore, if $0 < |t - t_0| < \delta$,

$$\left| \int_{t_0}^t [g^{(k-1)}(\tau) - g^{(k-1)}(t_0) - g^{(k)}(t_0)(\tau - t_0)] \, d\tau \right| < \epsilon \left| \int_{t_0}^t (\tau - t_0) \, d\tau \right|,$$

or

$$\left| g^{(k-2)}(t) - g^{(k-2)}(t_0) - \frac{g^{(k-1)}(t_0)}{1!}(t - t_0) - \frac{g^{(k)}(t_0)}{2!}(t - t_0)^2 \right| < \epsilon \frac{|t - t_0|^2}{2!}.$$

If $k = 2$ we are finished, since ϵ is arbitrary; if $k > 2$, we integrate $k - 2$ more times to obtain

$$| g(t) - T_k(t) | < \epsilon \frac{|t - t_0|^k}{k!},$$

which completes the proof.

Example 8.4   Let $t_0 = 0$ and

$$g(t) = \frac{1}{1 - t} ;$$

then $g^{(r)}(0) = r!$ and

$$T_k(t) = 1 + t + \cdots + t^k.$$

Theorem 8.1 implies that

$$\lim_{t \to 0} t^{-k} \left(\frac{1}{1-t} - T_k(t) \right) = 0,$$

which in this case can be verified directly, since

$$\frac{1}{1-t} = 1 + t + \cdots + t^k + \frac{t^{k+1}}{1-t};$$

thus

$$t^{-k} \left(\frac{1}{1-t} - T_k(t) \right) = \frac{t}{1-t}.$$

Since T_k is the only polynomial of degree not greater than k for which (1) holds, it follows that $T_k = g$ if g is a polynomial of degree $\leq k$.

A polynomial in one variable written as a linear combination of powers of $t - t_0$ is said to be *expanded about* $t = t_0$. Theorem 8.1 provides a convenient way to find such an expansion.

Example 8.5 The polynomial

$$p(t) = 3 + 4(t-1) + (t-1)^3$$

is expanded about $t = 1$. To obtain its expansion about $t = -1$ we use the fact that $p = T_3$, where

$$T_3(t) = \sum_{r=0}^{3} \frac{p^{(r)}(-1)}{r!} (t+1)^r.$$

Thus,

$$p(-1) = -13$$

$$p'(t) = 4 + 3(t-1)^2, \qquad p'(-1) = 16;$$

$$p''(t) = 6(t-1), \qquad p''(-1) = -12;$$

$$p''' = 6,$$

and

$$p(t) = p(-1) + p'(-1)(t+1) + \frac{p''(-1)}{2!}(t+1)^2 + \frac{p'''(-1)}{3!}(t+1)^3$$

$$= -13 + 16(t+1) - 6(t+1)^2 + (t+1)^3.$$

Although Theorem 8.1 says that the error $g(t) - T_k(t)$ goes to zero faster than $(t - t_0)^k$, it does not tell us how small $| t - t_0 |$ must be to ensure that the error is below some given bound. The following theorem is useful in this connection, provided $g^{(k+1)}$ exists and is continuous in some interval about t_0.

Theorem 8.2   Let g be as in Theorem 8.1 and also let $g^{(k+1)}$ be continuous in J. For t in J let $M(t, t_0)$ be a nonnegative number such that

(2)
$$| g^{(k+1)}(x) | \leq M(t, t_0)$$

if x is in the closed interval with endpoints t and t_0. Then

(3)
$$| g(t) - T_k(t) | \leq \frac{M(t, t_0)}{(k+1)!} | t - t_0 |^{k+1}.$$

Proof.   From the fundamental theorem of the calculus,

$$g(t) = g(t_0) + \int_{t_0}^{t} g'(x) \, dx$$

$$= T_0(t) + \int_{t_0}^{t} g'(x) \, dx$$

if t is in J. Integration by parts yields

$$g(t) = g(t_0) + \frac{g'(x)}{1!} (x - t) \Big|_{t_0}^{t} - \int_{t_0}^{t} g''(x)(x - t) \, dx$$

$$= g(t_0) + \frac{g'(t_0)}{1!} (t - t_0) - \int_{t_0}^{t} g''(x)(x - t) \, dx$$

$$= T_1(t) - \int_{t_0}^{t} g''(x)(x - t) \, dx.$$

If $k > 1$, we integrate by parts again to obtain

$$f(t) = T_1(t) - g''(x) \frac{(x - t)^2}{2!} \Big|_{t_0}^{t} + \frac{1}{2!} \int_{t_0}^{t} g'''(x)(x - t)^2 \, dx$$

$$= T_1(t) + \frac{g''(t_0)}{2!} (t - t_0)^2 + \frac{1}{2!} \int_{t_0}^{t} g'''(x)(x - t)^2 \, dx$$

$$= T_2(t) + \frac{1}{2!} \int_{t_0}^{t} g'''(x)(x - t)^2 \, dx.$$

Continuing in this way we obtain

$$g(t) = T_k(t) + \frac{(-1)^k}{k!} \int_{t_0}^{t} g^{(k+1)}(x)(x-t)^k \, dx.$$

From this and (2),

$$|g(t) - T_k(t)| \le \frac{1}{k!} \left| \int_{t_0}^{t} g^{(k+1)}(x)(x-t)^k \, dx \right|$$

$$\le \frac{M(t,t_0)}{k!} \left| \int_{t_0}^{t} (x-t)^k \, dx \right|$$

$$= \frac{M(t,t_0)}{(k+1)!} |t-t_0|^{k+1},$$

which completes the proof.

Example 8.6 Let $t_0 = 0$ and $g(t) = e^t$; from Example 8.2,

$$T_k(t) = 1 + \frac{t}{1!} + \cdots + \frac{t^k}{k!}.$$

If $t \ge 0$, then $M(t,0) = e^t$ and

$$|e^t - T_k(t)| \le \frac{e^t}{(k+1)!} t^{k+1};$$

if $t \le 0$, then $M(t,0) = 1$ and

$$|e^t - T_k(t)| \le \frac{|t|^{k+1}}{(k+1)!}.$$

To choose k so that the error is less than a given $\epsilon > 0$ for all t in $[-1, 1]$ we combine these estimates to obtain

$$|e^t - T_k(t)| \le \frac{e}{(k+1)!}.$$

Thus, any k such that

$$\frac{e}{(k+1)!} < \epsilon$$

will be satisfactory.

Example 8.7 To compute $\sqrt{5}$ accurate to within an error less than 0.0005, we can use an appropriate Taylor polynomial about $t = 0$ for

$$g(t) = \sqrt{t+4},$$

since we know $g(0) = \sqrt{4} = 2$. Thus we argue that $\sqrt{5}$ is approximately equal to $T_k(1)$ for some k, and Theorem 8.2 tells us how to choose k: we have only to compute the error bound (3) at $t = 1$ for $k = 0, 1, \ldots$, until we arrive at a bound less than 0.0005. The following calculation shows that the choice $k = 3$ yields an acceptable error:

(4)
$$\begin{cases} g(t) = (t+4)^{1/2}, \quad g'(t) = \tfrac{1}{2}(t+4)^{-1/2}, \quad g''(t) = -\tfrac{1}{4}(t+4)^{-3/2}, \\ g'''(t) = \tfrac{3}{8}(t+4)^{-5/2}, \quad g^{(4)}(t) = -\tfrac{15}{16}(t+4)^{-7/2}. \end{cases}$$

The maximum of $|g^{(4)}(t)|$ in $[0, 1]$ is

$$|g^4(0)| = \frac{15}{16}\left(\frac{1}{2^7}\right),$$

and therefore, from (3), since $g(1) = \sqrt{5}$,

$$|\sqrt{5} - T_3(1)| \le \left(\frac{15}{16}\right)\left(\frac{1}{2^7}\right)\frac{(1)^4}{4!} < 0.0004.$$

From (4), $g(0) = 2$, $g'(0) = 1/4$, $g''(0) = -1/32$, and $g'''(0) = 3/256$: thus

$$T_3(t) = 2 + \frac{1}{4}t - \frac{1}{64}t^2 + \frac{1}{512}t^3$$

and

$$T_3(1) = \frac{1145}{512} \cong 2.2363 \cong \sqrt{5}.$$

From a table of square roots,

$$\sqrt{5} \cong 2.2361,$$

rounded off to four decimal places.

The following theorem is a useful by-product of the proof of Theorem 8.2.

Theorem 8.3 (Taylor's Theorem with Integral Remainder) Let $g, g', \ldots, g^{(k+1)}$ be continuous in some open interval J about t_0. Then

$$g(t) = T_k(t) + R_k(t)$$

for all t in J, where the *remainder* $R_k(t)$ is given by

$$R_k(t) = \frac{1}{k!}\int_{t_0}^{t} g^{(k+1)}(x)(t-x)^k \, dx.$$

Taylor Polynomials in *n* Variables

Taylor's theorem for functions of n variables at first appears to be quite formidable. However, when expressed in terms of a sensible notation, it can be recognized as a natural generalization of Theorem 8.1.

Definition 8.2 If all partial derivatives of the function f through order $r\ (\geq 1)$ exist and are continuous at a point $\mathbf{X}_0$ in R^n, then the polynomial

$$(5) \qquad (d^r_{\mathbf{X}_0} f)\,(\mathbf{Y}) = \sum_r \frac{r!}{r_1!\cdots r_n!}\,\frac{\partial^r f(\mathbf{X}_0)}{\partial x_1^{r_1}\cdots \partial x_n^{r_n}}\,y_1^{r_1}\cdots y_n^{r_n},$$

is called the *rth differential* of f at $\mathbf{X}_0$. Here $\sum_r$ indicates the sum over all ordered n-tuples $(r_1,\ldots,r_n)$ of nonnegative integers such that $r_1 + \cdots + r_n = r$. Since the first differential is the differential previously defined in Section 2.3, we write $d^1_{\mathbf{X}_0} f = d_{\mathbf{X}_0} f$; by convention, $d^0_{\mathbf{X}_0} f = f(\mathbf{X}_0)$ (a constant).

We observe that $d^r_{\mathbf{X}_0} f$ is a *homogeneous* polynomial of degree r (provided at least one of its coefficients is nonzero); that is, the sum of the powers of $y_i\ (1 \leq i \leq n)$ appearing in each term of (5) is r.

Example 8.8 If $n = 1$ then

$$(d^r_{x_0} f)\,(y) = f^{(r)}(x_0)\,y^r.$$

Example 8.9 If $n = 2$ then

$$(d^r_{\mathbf{X}_0} f)\,(\mathbf{Y}) = \sum_r \frac{r!}{r_1! r_2!}\,\frac{\partial^r f(\mathbf{X}_0)}{\partial x_1^{r_1}\partial x_2^{r_2}}\,y_1^{r_1} y_2^{r_2}.$$

Setting $r_1 = m$ and $r_2 = r - m$, we can write this more conveniently as

$$(d^r_{\mathbf{X}_0} f)\,(\mathbf{Y}) = \sum_{m=0}^{r} \binom{r}{m}\,\frac{\partial^r f(\mathbf{X}_0)}{\partial x_1^{m}\partial x_2^{r-m}}\,y_1^{m} y_2^{r-m},$$

where $\binom{r}{m}$ is the binomial coefficient:

$$\binom{r}{m} = \frac{r!}{m!(r - m)!}.$$

Most often in the applications we shall be interested in $(d^r_{\mathbf{X}_0} f)\,(\mathbf{X} - \mathbf{X}_0)$;

for $n = 2$ it is convenient to write this as

$$(d_{\mathbf{X}_0}^r f)(\mathbf{X} - \mathbf{X}_0) = \sum_{m=0}^{r} \binom{r}{m} \frac{\partial^r f(\mathbf{X}_0)}{\partial x^m \partial y^{r-m}} (x - x_0)^m (y - y_0)^{r-m};$$

thus,

$$(d_{\mathbf{X}_0}^0 f)(\mathbf{X} - \mathbf{X}_0) = f(\mathbf{X}_0),$$

$$(d_{\mathbf{X}_0} f)(\mathbf{X} - \mathbf{X}_0) = \frac{\partial f}{\partial x}(\mathbf{X}_0)(x - x_0) + \frac{\partial f}{\partial y}(\mathbf{X}_0)(y - y_0),$$

$$(d_{\mathbf{X}_0}^2 f)(\mathbf{X} - \mathbf{X}_0) = \frac{\partial^2 f}{\partial x^2}(\mathbf{X}_0)(x - x_0)^2 + 2\frac{\partial^2 f}{\partial x\, \partial y}(\mathbf{X}_0)(x - x_0)(y - y_0)$$

$$+ \frac{\partial^2 f}{\partial y^2}(\mathbf{X}_0)(y - y_0)^2,$$

$$(d_{\mathbf{X}_0}^3 f)(\mathbf{X} - \mathbf{X}_0) = \frac{\partial^3 f}{\partial x^3}(\mathbf{X}_0)(x - x_0)^3 + 3\frac{\partial^3 f}{\partial x^2\, \partial y}(\mathbf{X}_0)(x - x_0)^2(y - y_0)$$

$$+ 3\frac{\partial^3 f}{\partial x\, \partial y^2}(\mathbf{X}_0)(x - x_0)(y - y_0)^2$$

$$+ \frac{\partial^3 f}{\partial y^3}(\mathbf{X}_0)(y - y_0)^3, \quad \text{etc.}$$

Example 8.10 Suppose $\mathbf{X}_0 = (c_1, \ldots, c_n)$ and p is a homogeneous polynomial of degree r in $\mathbf{X} - \mathbf{X}_0$; that is

$$p(\mathbf{X}) = \sum_r a_{r_1 \ldots r_n} (x_1 - c_1)^{r_1} \cdots (x_n - c_n)^{r_n}.$$

Then all differentials of p except the rth vanish, and

$$(6) \qquad\qquad p(\mathbf{X}) = \frac{1}{r!}(d_{\mathbf{X}_0}^r p)(\mathbf{X} - \mathbf{X}_0).$$

(To see this, use the result of Exercise T-3.) For example, the polynomial

$$p(\mathbf{X}) = (x_1 + x_2 + \cdots + x_n)^r$$

is homogeneous of degree r in $\mathbf{X}$; its partial derivatives vanish at $\mathbf{X} = 0$, except those of order r, all of which equal $r!$ (Prove.) Substituting this result into (5) and using (6) yields

$$(x_1 + \cdots + x_n)^r = \sum_r \frac{r!}{r_1! \cdots r_n!} x_1^{r_1} \cdots x_n^{r_n};$$

for this reason the coefficients $\dfrac{r!}{r_1!\cdots r_n!}$, where $r_1 + \cdots + r_n = r$, are called the *multinomial coefficients*. They are generalizations of the familiar binomial coefficients.

We can now define Taylor polynomials for functions of several variables.

Definition 8.3 If all partial derivatives of f through the rth order exist and are continuous at $\mathbf{X}_0 = (c_1, \ldots, c_n)$ then the *Taylor polynomial of f about* $\mathbf{X}_0$ is defined by

$$T_k(\mathbf{X}) = \sum_{r=0}^{k} \frac{1}{r!} (d_{\mathbf{X}_0}^r f)(\mathbf{X} - \mathbf{X}_0),$$

$$= \sum_{r=0}^{k} \sum_r \frac{1}{r_1!\cdots r_n!} \frac{\partial^r f(\mathbf{X}_0)}{\partial x_1^{r_1}\cdots \partial x_n^{r_n}} (x_1 - c_1)^{r_1}\cdots (x_n - c_n)^{r_n}.$$

We point out that Example 8.8 shows that this definition is consistent with Definition 8.1 when $n = 1$.

Example 8.11 Let
$$f(\mathbf{X}) = e^{x_1 + x_2 + \cdots + x_n};$$
then all partial derivatives of f at $\mathbf{X}_0 = 0$ are equal to 1. Hence

$$T_k(\mathbf{X}) = \sum_{r=0}^{k} \sum_r \frac{x_1^{r_1}\cdots x_n^{r_n}}{r_1!\cdots r_n!}$$

Example 8.12 Let $f(x, y) = \log(1 + 2x + y)$ and $\mathbf{X}_0 = (0, 0)$. Then

$$\frac{\partial f}{\partial x}(x, y) = \frac{2}{1 + 2x + y}, \qquad \frac{\partial f}{\partial y}(x, y) = \frac{1}{1 + 2x + y};$$

$$\frac{\partial^2 f}{\partial x^2}(x, y) = \frac{-4}{(1 + 2x + y)^2}, \qquad \frac{\partial^2 f}{\partial x\, \partial y}(x, y) = \frac{-2}{(1 + 2x + y)^2},$$

$$\frac{\partial^2 f}{\partial y^2}(x, y) = \frac{-1}{(1 + 2x + y)^2};$$

in general,

$$\frac{\partial^r f(x, y)}{\partial x^m\, \partial y^{r-m}} = \frac{(-1)^{r-1}(r-1)!\, 2^m}{(1 + 2x + y)^r}.$$

Therefore,

$$T_0(x, y) = 0,$$

$$T_1(x, y) = 2x + y,$$

$$T_2(x, y) = 2x + y - 2x^2 - xy - \tfrac{1}{2}y^2,$$

$$\vdots$$

$$T_k(x, y) = \sum_{r=0}^{k} \frac{(-1)^{r-1}}{r} \sum_{m=0}^{r} \binom{r}{m} 2^m x^m y^{r-m}.$$

The next theorem plays a role in the approximation of functions of several variables similar to that of Theorem 8.1 for functions of a single variable. However, a detailed study of this question is beyond the scope of this book. Hence we shall merely state the theorem here and apply it, in Section 2.9, to the problem of finding maxima and minima of functions of several variables.

Theorem 8.4 (Taylor's Theorem for Functions of n Variables) Let f and all its partial derivatives through order k be continuous in an n-ball about $\mathbf{X}_0 = (c_1, \ldots, c_n)$, and let T_k be its kth Taylor polynomial about $\mathbf{X}_0$. Then

$$\lim_{\mathbf{X} \to \mathbf{X}_0} \frac{f(\mathbf{X}) - T_k(\mathbf{X})}{|\mathbf{X} - \mathbf{X}_0|^k} = 0,$$

and T_k is the only polynomial in $\mathbf{X} - \mathbf{X}_0$, of degree $\leq k$, with this property.

EXERCISES 2.8

1. Find the Taylor polynomial T_2 about t_0 for each of the following functions.

   (a) $f(t) = e^t;$ $t_0 = 1.$

   (b) $f(t) = \sqrt{t};$ $t_0 = 1.$

   (c) $f(t) = \sqrt{1 + t^2};$ $t_0 = 0.$

2. Find the Taylor polynomial T_3 about t_0 for each of the following functions.

   (a) $f(t) = \tan \pi t;$ $t_0 = \tfrac{1}{4}.$

    (b) $f(t) = \arcsin t;$ $t_0 = 0.$

    (c) $f(t) = e^{3t+1};$ $t_0 = 0.$

3. Expand $p(t) = 2 + 3t + 2t^2 + t^3$ in powers of $t - 1$.

4. Expand $p(t) = 2 - 5(t + 2)^2 + 3(t + 2)^3$ about $t = 3$.

5. Using Theorem 8.2 and the method of Example 8.6, find the smallest value of k such that $|T_k(t) - f(t)| \le \epsilon$ whenever $|x - x_0| \le \delta$:

    (a) $f(t) = \sin t,$ $t_0 = 0,$ $\epsilon = 0.001,$ $\delta = \dfrac{\pi}{4}.$

    (b) $f(t) = \sqrt{t},$ $t_0 = 1,$ $\epsilon = 0.0001,$ $\delta = \frac{1}{2}.$

    (c) $f(t) = e^{t^2},$ $t_0 = 0,$ $\epsilon = 0.005,$ $\delta = 1.$

    (d) $f(t) = \log t,$ $t_0 = 2,$ $\epsilon = 0.005,$ $\delta = 1.$

6. Use a Taylor polynomial for $\sqrt{1+t}$ to compute $\sqrt{1.2}$ to two decimal places.

7. Use a Taylor polynomial for e^t to compute $\sqrt{e}$ to within 0.001.

8. Find $(d_{X_0}^r f)(\mathbf{X} - \mathbf{X}_0)$ for $r = 0, 1, 2, 3$:

    (a) $f(x, y) = x + y + 2x^2 + 3xy + 4y^2 + 6x^3,$ $\mathbf{X}_0 = (0, 0).$

    (b) $f(x, y) = \sin(x + y)\pi,$ $\mathbf{X}_0 = (\frac{1}{8}, \frac{1}{8}).$

    (c) $f(x, y) = e^{x^2} + y^2,$ $\mathbf{X}_0 = (0, 0).$

9. Find $(d_{X_0}^r f)(\mathbf{X} - \mathbf{X}_0)$ for $r = 0, 1, 2$:

    (a) $f(x, y) = \log(2x + y),$ $\mathbf{X}_0 = (0, 1).$

    (b) $f(x, y) = \sin(x^2 + y^2),$ $\mathbf{X}_0 = (0, 0).$

10. Find $(d_{X_0}^r f)(\mathbf{X} - \mathbf{X}_0)$ for $r = 0, 1, 2, 3$:

    (a) $f(x, y, z) = xyz + x^3 + y,$ $\mathbf{X}_0 = (1, 2, -1).$

    (b) $f(x, y, z) = \log(x + y + z),$ $\mathbf{X}_0 = (1, 1, -1).$

11. Find $(d_{X_0}^r f)(\mathbf{X} - \mathbf{X}_0)$ for $r = 0, 1, 2$:

    (a) $f(x, y, z) = \sin(x + y + z)\pi,$ $\mathbf{X}_0 = (\frac{1}{6}, \frac{1}{6}, \frac{1}{6}).$

    (b) $f(x, y, z) = \sqrt{x^2 + y^2 + z^2},$ $\mathbf{X}_0 = (1, 0, 0).$

12. Find the Taylor polynomial T_3 about $\mathbf{X}_0$ for each function in Exercise 8.

13. Find the Taylor polynomial T_2 about $\mathbf{X}_0$ for each function in Exercise 9.

14.   Find the Taylor polynomial T_3 about $\mathbf{X}_0$ for each function in Exercise 10.

15.   Find the Taylor polynomial T_2 about $\mathbf{X}_0$ for each function in Exercise 11.

16.   Using Theorem 8.4, expand

$$p(x, y) = 3 + x + 2(y - 3) + x^2 + 4x(y - 3) + 2(y - 3)^2$$

about $\mathbf{X}_0 = (1, 2)$.

17.   Using Theorem 8.4, expand

$$p(x, y) = 2 + (x - 1) + 3(y - 2) + (x - 1)^2 - (x - 1)(y - 2)$$

about $\mathbf{X}_0 = (2, 3)$.

18.   Using Theorem 8.4, expand

$$p(x, y, z) = 4 + x + y + 2z + xz + xyz$$

about $\mathbf{X}_0 = (0, 1, -1)$.

THEORETICAL EXERCISES

T-1.   Let p be a polynomial of degree $\leq k$ such that

$$\lim_{t \to t_0} \frac{p(t)}{(t - t_0)^k} = 0.$$

Show that $p = 0$ (that is, $p(t) = 0$ for all t).

T-2.   Use Exercise T-1 to establish the uniqueness of the Taylor polynomial in Theorem 8.1.

T-3.   If $\phi(\mathbf{X}) = (x_1 - c_1)^{r_1} \cdots (x_n - c_n)^{r_n}$ show that

$$\frac{\partial^{s_1 + \cdots + s_n}}{\partial x_1^{s_1} \cdots \partial x_n^{s_n}} \phi(c_1, \ldots, c_n) = \begin{cases} r_1! \cdots r_n! & \text{if} \quad s_i = r_i \quad (1 \leq i \leq n) \\ 0 & \text{if} \quad s_j \neq r_j \quad \text{for some} \quad j. \end{cases}$$

T-4.   If F_k and G_k are the kth Taylor polynomials about $\mathbf{X}_0$ of f and g, respectively, show that $F_k + G_k$ is the kth Taylor polynomial of $f + g$ about $\mathbf{X}_0$.

T-5.   Let $n = 1$ and f, g, F_k, and G_k be as in Exercise T-2;

$$F_k(t) = a_0 + a_1(t - t_0) + \cdots + a_k(t - t_0)^k,$$
$$G_k(t) = b_0 + b_1(t - t_0) + \cdots + b_k(t - t_0)^k.$$

Let H_k be the kth Taylor polynomial of $fg = h$ about t_0:

$$H_k = c_0 + c_1(t - t_0) + \cdots + c_k(t - t_0)^k.$$

Show that

$$
\begin{aligned}
c_0 &= a_0b_0, \\
c_1 &= a_0b_1 + a_1b_0, \\
c_2 &= a_0b_2 + a_1b_1 + a_2b_0, \\
&\vdots \\
c_k &= a_0b_k + a_1b_{k-1} + \cdots + a_kb_0,
\end{aligned}
$$

which implies that H_k can be obtained by multiplying $F_k(t)$ by $G_k(t)$, collecting coefficients of like powers of $(t - t_0)$ through degree k, and ignoring higher powers of $(t - t_0)$. (*Hint:* Use Leibniz's rule for differentiating a product.)

T-6. As an application of Exercise T-5 find H_4 for $h(t) = e^{3t}$, taking $f(t) = e^t$ and $g(t) = e^{2t}$. Also, obtain H_4 directly.

T-7. Suppose f and g are functions of one variable and $h = f \circ g$ (composite function). Let $y_0 = g(t_0)$ and G_k be the kth Taylor polynomial of g about t_0:

$$G_k(t) = y_0 + b_1(t - t_0) + \cdots + b_k(t - t_0)^k.$$

Let F_k be the kth Taylor polynomial of f about y_0:

$$F_k(y) = a_0 + a_1(y - y_0) + \cdots + a_k(y - y_0)^k.$$

Then H_k, the kth Taylor polynomial of h about t_0, can be obtained by substituting $y = G_k(t)$ into F_k, collecting coefficients of like powers of $(t - t_0)$, and retaining only those powers which do not exceed k. Thus, if

$$F_k(G_k(t)) = c_0 + c_1(t - t_0) + \cdots + c_k(t - t_0)^k + \cdots,$$

then

$$H_k(t) = c_0 + c_1(t - t_0) + \cdots + c_k(t - t_0)^k.$$

We shall not prove this result. However, the student can use it to find H_4, with $t_0 = 0$, for the following functions. Also, he should find H_4 by some other means to check his result.

(a) $h(t) = e^{t^2} + 2t + 2.$ (b) $h(t) = \log e^t.$

(c) $h(t) = e^{\sin t}.$ (d) $h(t) = e^{\cos t}.$

2.9   Maxima and Minima

The problem of locating extreme values of a function of one variable is an important application of the one-dimensional calculus. We now consider this problem for functions of several variables.

Let f be defined on a domain D in R^n. Then f is said to have a *relative maximum* at $\mathbf{X}_0$ if there is an n-ball S about $\mathbf{X}_0$ such that

$$(1) \qquad\qquad f(\mathbf{X}) \le f(\mathbf{X}_0)$$

for all $\mathbf{X}$ belonging to both S and D; if (1) holds for all $\mathbf{X}$ in D, then f is said to have an *absolute maximum* at $\mathbf{X}_0$. Thus, an absolute maximum is a relative maximum, but a relative maximum need not be an absolute maximum.

Definitions of *relative minimum* and *absolute minimum* are obtained by replacing "maximum" by "minimum" in the last paragraph and replacing (1) by

$$f(\mathbf{X}) \ge f(\mathbf{X}_0).$$

An *extreme value* of f is a relative maximum or minimum, and a point where an extreme value is attained is an *extreme point*.

Example 9.1   Let

$$f(x, y) = \begin{cases} 1 - x^2 - y^2 & (x^2 + y^2 < 1), \\ 0 & (1 \le x^2 + y^2 \le 2), \\ \frac{1}{2} & ((x, y) = (3, 0)). \end{cases}$$

This function has an absolute maximum at $(0, 0)$; at every (x, y) satisfying $1 \le x^2 + y^2 \le 2$, f has an absolute minimum, and, at those for which $1 < x^2 + y^2 \le 2$, a relative maximum as well; f has both a relative maximum and minimum at $(3, 0)$, simply because there is a circle about $(3, 0)$ which contains no other point of its domain.

In the one dimensional calculus it is shown that if f is differentiable on the open interval (a, b) and has an extreme point x_0 in (a, b), then $f'(x_0) = 0$. Since $d_{x_0} f = f'(x_0)\, dx$, this can be rewritten as $d_{x_0} f = 0$; in this form the result is true for arbitrary n.

Theorem 9.1   If $\mathbf{X}_0$ is both an interior point of the domain of f and an extreme point of f, and if f is differentiable at $\mathbf{X}_0$, then $d_{\mathbf{X}_0} f = 0$.

Proof. If $d_{\mathbf{X}_0}f \neq 0$ there is a unit vector $\mathbf{U}$ such that $(d_{\mathbf{X}_0}f)(\mathbf{U}) = \sigma > 0$. From Theorem 3.5 (Section 2.3),

$$\lim_{t \to 0} \frac{f(\mathbf{X}_0 + t\mathbf{U}) - f(\mathbf{X}_0)}{t} = \sigma;$$

hence there is a $\delta > 0$ such that

$$\frac{f(\mathbf{X}_0 + t\mathbf{U}) - f(\mathbf{X}_0)}{t} > 0$$

if $0 < |t| < \delta$. Therefore

$$f(\mathbf{X}_0 + t\mathbf{U}) > f(\mathbf{X}_0) \qquad (0 < t < \delta)$$

and

$$f(\mathbf{X}_0 + t\mathbf{U}) < f(\mathbf{X}_0) \qquad (-\delta < t < 0),$$

which is impossible if $\mathbf{X}_0$ is an extreme point. Thus $d_{\mathbf{X}_0}f = 0$ if $\mathbf{X}_0$ is an extreme point.

Example 9.2 Let

$$f(x, y) = 4 - 4x^2 - 9y^2;$$

then f has an absolute maximum at $(0, 0)$ since $4x^2 + 9y^2 \geq 0$, with equality only if $x = y = 0$. The theorem asserts that $(\partial f/\partial x)(0, 0) = (\partial f/\partial y)(0, 0) = 0$, which is easily verified, since $(\partial f/\partial x)(x, y) = -8x$ and $(\partial f/\partial y)(x, y) = -18y$. Since $(0, 0)$ is the only point where the differential vanishes, f has no other extreme points.

The tangent plane to the graph of f at $(0, 0)$ is horizontal (Fig. 9.1). Theorem 9.1 says that this is always true at an extreme point where f is differentiable.

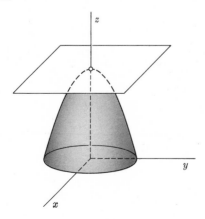

FIGURE 9.1

Differentiability of f at $\mathbf{X}_0$ requires at least that $\mathbf{X}_0$ be an interior point of the domain D of f; hence Theorem 9.1 does not apply to extreme values at boundary points (limit points that are not interior points) of D, or at isolated points such as $(3, 0)$ in Example 9.1. Obviously, Theorem 9.1 cannot be used to test for an extreme value at an interior point where f is not differentiable.

Example 9.3 Let $n = 1$ and

$$f(x) = |x| \qquad (-1 \leq x \leq 1);$$

then f has an absolute minimum at $x = 0$ and absolute maxima at $x = \pm 1$. Theorem 9.1 does not apply to them since ± 1 are not interior points of the domain of f and, although $x = 0$ is an interior point, f is not differentiable there. Since $f'(x) = 1$ if $0 < x < 1$ and $f'(x) = -1$ if $-1 < x < 0$, Theorem 9.1 implies that f has no other extreme points; this is also evident from the graph of f (Fig. 9.2).

FIGURE 9.2

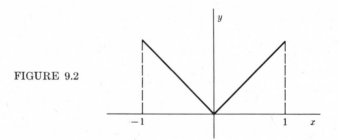

In the rest of this section we shall restrict our attention to extreme points where f is differentiable.

The converse of Theorem 9.1 is false, as is shown by the next two examples.

Example 9.4 Let $\mathbf{X}_0 = (0, 0)$ and

$$f(x, y) = x^2 - y^2;$$

then $d_{\mathbf{X}_0} f = 0$ since $f_x(0, 0) = f_y(0, 0) = 0$. Nevertheless, $(0, 0)$ is not an extreme point of f. Near $(0, 0)$ the graph of f looks like a saddle (Fig. 9.3).

Example 9.5 Let n be arbitrary, $\mathbf{X}_0 = (0, 0, \ldots, 0)$ and

$$f(\mathbf{X}) = x_1^3;$$

then $d_{\mathbf{X}_0} f = 0$ but $\mathbf{X}_0$ is not an extreme point, because $f(\mathbf{X}) < 0$ if $x_1 \leq 0$ and $f(\mathbf{X}) > 0$ if $x_1 \geq 0$. Figure 9.4 shows the graph of $f(x) = x^3$ $(n = 1)$; in this case $x = 0$ is a point of inflection.

FIGURE 9.3

Definition 9.1 A point $\mathbf{X}_0$ at which f is differentiable is called a *critical point of f* if $d_{\mathbf{X}_0} f = 0$.

We have seen that an extreme point must be a critical point, but that a critical point need not be an extreme point. Although this may seem to be only half of a result, it is nevertheless quite useful. In many problems the nature of a critical point can be deduced from the problem itself, and it is only necessary to locate it.

Example 9.6 Find the distance from the point $\mathbf{X}_1 = (0, 1, -1)$ to the plane $\Gamma: 2x + 3y + z = 1$.

Solution. The distance between $\mathbf{X}_1$ and any point $\mathbf{X}$ is

$$(2) \qquad |\mathbf{X} - \mathbf{X}_1| = \sqrt{x^2 + (y - 1)^2 + (z + 1)^2}$$

and the distance from $\mathbf{X}_1$ to Γ is the minimum value attained by $|\mathbf{X} - \mathbf{X}_1|$ as $\mathbf{X}$ varies over Γ. To be sure that we consider only points on Γ we substitute

$$z + 1 = 2 - 2x - 3y$$

in (2) to obtain

$$|\mathbf{X} - \mathbf{X}_1| = \sqrt{x^2 + (y - 1)^2 + (2 - 2x - 3y)^2}$$

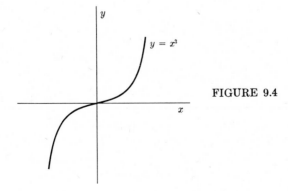

$y = x^3$

FIGURE 9.4

for **X** on P. Minimizing $|\mathbf{X} - \mathbf{X}_1|$ is equivalent to minimizing $|\mathbf{X} - \mathbf{X}_1|^2$, so our problem reduces to finding the minimum of

$$\rho(x, y) = x^2 + (y - 1)^2 + (2 - 2x - 3y)^2.$$

The partial derivative are

$$\frac{\partial \rho}{\partial x}(x, y) = 10x + 12y - 8$$

and

$$\frac{\partial \rho}{\partial y}(x, y) = 12x + 20y - 14.$$

The only point at which both derivatives vanish is

$$(x_0, y_0) = \left(\frac{-2}{14}, \frac{11}{14}\right).$$

Theorem 9.1 does not, by itself, tell us that ρ attains its maximum at (x_0, y_0), but the following argument does.

(a) It is geometrically evident that the problem has a solution; hence ρ must have an absolute minimum at some point $(\hat{x}_0, \hat{y}_0)$.

(b) Since ρ is differentiable everywhere, the differential must vanish at $(\hat{x}_0, \hat{y}_0)$, from Theorem 9.1.

(c) The differential vanishes only at (x_0, y_0); hence $(\hat{x}_0, \hat{y}_0) = (x_0, y_0)$.

The minimum value of ρ is

$$\rho(x_0, y_0) = \tfrac{1}{14}$$

and the distance from $\mathbf{X}_1$ to P is

$$d = \sqrt{\rho(x_0, y_0)} = \frac{1}{\sqrt{14}}.$$

Example 9.7 (Least Squares Curve Fitting) Suppose $t_1, \ldots, t_m$ are distinct and $y_1, \ldots, y_m$ are arbitrary real numbers $(m > 2)$. For each first degree polynomial

$$p(t) = a + bt,$$

define

$$Q(a, b) = \sum_{i=1}^{m} (y_i - a - bt_i)^2.$$

The polynomial

$$p_0(t) = a_0 + b_0 t$$

for which Q is minimized is called the *least squares linear polynomial fit* to the points $(t_1, y_1), \ldots, (t_m, y_m)$. To locate the critical points of Q we equate the first partial derivatives to zero:

$$\frac{\partial Q}{\partial a}(a_0, b_0) = -2 \sum_{i=1}^{m} (y_i - a_0 - b_0 t_i) = 0,$$

$$\frac{\partial Q}{\partial b}(a_0, b_0) = -2 \sum_{i=1}^{m} (y_i - a_0 - b_0 t_i) t_i = 0.$$

Hence (a_0, b_0) satisfies the system

(3)
$$\sum_{i=1}^{m} y_i = a_0 m + b_0 \sum_{i=1}^{m} t_i$$

$$\sum_{i=1}^{m} t_i y_i = a_0 \sum_{i=1}^{m} t_i + b_0 \sum_{i=1}^{m} t_i^2,$$

whose determinant can be shown to be

$$m \sum_{i=1}^{m} t_i^2 - \left(\sum_{i=1}^{m} t_i\right)^2 = \tfrac{1}{2} \sum_{i,j=1}^{m} (t_i - t_j)^2 > 0.$$

Therefore (3) has exactly one solution (a_0, b_0); if we assume that the curve fitting problem has a solution, then a_0 and b_0 must be its coefficients, by the reasoning used in Example 9.6.

Taylor's theorem yields a test which is sometimes useful for deciding whether a critical point is an extreme point. To establish this test, we first need a definition.

Definition 9.2   A homogeneous polynomial p in $\mathbf{X} - \mathbf{X}_0$ is said to be *positive (negative) semidefinite* if $p(\mathbf{X}) \geq 0$ $(p(\mathbf{X}) \leq 0)$ for all $\mathbf{X}$; if, in addition, $p(\mathbf{X}) = 0$ only when $\mathbf{X} = \mathbf{X}_0$, p is said to be *positive (negative) definite*. A *semidefinite (definite)* homogeneous polynomial is one that is either positive or negative semidefinite (definite).

Example 9.8   Let $n = 2$. Then

$$p_1(x, y) = (x - 1)^2 + 2(x - 1)y + 2y^2,$$

which is homogeneous in $(x - 1, y)$, is positive definite, since it can be rewritten as

$$p_1(x, y) = (x + y - 1)^2 + y^2.$$

The polynomial

$$p_2(x, y) = x^2 - 2xy + y^2 = (x - y)^2$$

is positive semidefinite. Clearly $-p_1$ and $-p_2$ are negative definite and negative semidefinite, respectively, while

$$p_3(x, y) = x^2 - y^2$$

is not semidefinite.

Theorem 9.2   Let the partial derivatives of f through order $k \geq 2$ be continuous in an n-ball about the critical point $\mathbf{X}_0$, and

(4)
$$d_{\mathbf{X}_0}^r f = 0 \qquad (1 \leq r \leq k - 1),$$
$$d_{\mathbf{X}_0}^k f \neq 0;$$

thus $d_{\mathbf{X}_0}^k f$ is the first nonvanishing higher differential of f at $\mathbf{X}_0$. Then

(a) $\mathbf{X}_0$ is not an extreme point unless $d_{\mathbf{X}_0}^k f$ is semidefinite (in particular, it is not an extreme point if k is odd);

(b) $\mathbf{X}_0$ is a relative minimum (maximum) if $d_{\mathbf{X}_0}^k f$ is positive (negative) definite;

(c)   if $d_{\mathbf{X}_0}^k f$ is semidefinite but not definite, no general conclusion can be drawn (that is, $\mathbf{X}_0$ is not necessarily an extreme point, although it may be).

Proof.   From (4) and Taylor's theorem,

(5)
$$\lim_{\mathbf{X} \to \mathbf{X}_0} \frac{f(\mathbf{X}) - f(\mathbf{X}_0) - \dfrac{(d_{\mathbf{X}_0}^k f)(\mathbf{X} - \mathbf{X}_0)}{k!}}{|\mathbf{X} - \mathbf{X}_0|^k} = 0,$$

which implies that

(6)
$$\lim_{t \to 0} \frac{f(\mathbf{X}_0 + t\mathbf{U}) - f(\mathbf{X}_0) - \dfrac{t^k}{k!}(d_{\mathbf{X}_0}^k f)(\mathbf{U})}{t^k} = 0,$$

where $\mathbf{U}$ is any unit vector and t is a real variable. For (a), if $d_{\mathbf{X}_0}^k f$ is not semidefinite, there are unit vectors $\mathbf{U}_1$ and $\mathbf{U}_2$ such that

$$(d_{\mathbf{X}_0}^k f)(\mathbf{U}_1) = \sigma_1 > 0 \qquad \text{and} \qquad (d_{\mathbf{X}_0}^k f)(\mathbf{U}_2) = -\sigma_2 < 0,$$

and from (6) there are positive numbers δ_1 and δ_2 such that

$$\frac{f(\mathbf{X}_0 + t\mathbf{U}_1) - f(\mathbf{X}_0)}{t^k} > 0$$

if $0 < |t| < \delta_1$, and

$$\frac{f(\mathbf{X}_0 + t\mathbf{U}_2) - f(\mathbf{X}_0)}{t^k} < 0$$

if $0 < |t| < \delta_2$. Thus if $0 < t < \min(\delta_1, \delta_2)$,

$$f(\mathbf{X}_0 + t\mathbf{U}_1) > f(\mathbf{X}_0)$$

and

$$f(\mathbf{X}_0 + t\mathbf{U}_2) < f(\mathbf{X}_0);$$

hence $\mathbf{X}_0$ is not an extreme point of f.

If k is odd and $d_{\mathbf{X}_0}^k f \neq 0$, it cannot be semidefinite (Exercise T-1). This completes the proof of (a).

For (b), assume that $d_{\mathbf{X}_0}^k f$ is positive definite; then it can be shown that there is a number $\rho > 0$ such that

$$(7) \qquad \frac{(d_{\mathbf{X}_0}^k f)\,(\mathbf{X} - \mathbf{X}_0)}{k!} > \rho\,|\,\mathbf{X} - \mathbf{X}_0\,|^k$$

for all $\mathbf{X}$. From (5) and the definition of limit, the length of

$$\frac{f(\mathbf{X}) - f(\mathbf{X}_0) - \dfrac{(d_{\mathbf{X}_0}^k f)\,(\mathbf{X} - \mathbf{X}_0)}{k!}}{|\,\mathbf{X} - \mathbf{X}_0\,|^k}$$

can be made as small as we wish by choosing $|\,\mathbf{X} - \mathbf{X}_0\,|$ sufficiently small (but not zero); in particular, it can be made less than $\rho/2$. Thus

$$f(\mathbf{X}) - f(\mathbf{X}_0) - \frac{(d_{\mathbf{X}_0}^k f)\,(\mathbf{X} - \mathbf{X}_0)}{k!} > -\frac{\rho}{2}\,|\,\mathbf{X} - \mathbf{X}_0\,|^k$$

if $\mathbf{X}$ is sufficiently close to $\mathbf{X}_0$; from this and (7),

$$f(\mathbf{X}) - f(\mathbf{X}_0) > \frac{\rho}{2}\,|\,\mathbf{X} - \mathbf{X}_0\,|^k$$

for such $\mathbf{X}$. Therefore f has a relative minimum at $\mathbf{X}_0$.

The proof of the other half of (b) is similar to this and we leave it to the student (Exercise T-2). Part (c) is proved in Example 9.9.

Corollary 1   Let f, a function of a single variable, have $k \geq 2$ continuous derivatives in an open interval containing a critical point x_0 and suppose that

$$f^{(r)}(x_0) = 0 \qquad (1 \leq r \leq k - 1)$$

while

$$f^{(k)}(x_0) \neq 0.$$

Then x_0 is a point of inflection if k is odd; if k is even and $f^{(k)}(x_0) > 0$ $(f^{(k)}(x_0) < 0)$ then x_0 is a relative minimum (maximum).

For the definition of a point of inflection, see Example 9.5. If $k = 2$ this corollary reduces to the second derivative test.

Example 9.9 The following polynomials are homogeneous in $\mathbf{X}$; $\mathbf{X}_0 = (0, 0)$ is a critical point, $p(0, 0) = 0$, and $k = 2$ in all cases.

(a) $p(x, y) = x^2 - y^2 + x^4 + y^4$; then $(d^2_{\mathbf{X}_0}p)(x, y) = 2(x^2 - y^2)$ is not semidefinite. As Theorem 9.2 asserts, $(0, 0)$ is not an extreme point, since $p(t, 2t) = -t^2(3 - 17t^2)$, which is negative if $0 < t\sqrt{3/17}$, while $p(2t, t) = t^2(3 + 17t^2)$, which is positive for all $t \neq 0$.

(b) $p(x, y) = x^2 + y^2 - x^4 - y^4$; then $(d^2_{\mathbf{X}_0}p)(x, y) = 2(x^2 + y^2)$, which is positive definite. Since $p(x, y) = x^2(1 - x^2) + y^2(1 - y^2)$, it is clear that $p(x, y) \geq 0$ if $-1 < x, y < 1$; hence $\mathbf{X}_0$ is a relative minimum, as asserted by Theorem 9.2.

(c) $p(x, y) = x^2 - 2xy + y^2 + x^4 + y^4$; then $(d^2_{\mathbf{X}_0}p)(x, y) = 2(x^2 - 2xy + y^2) = -2(x - y)^2$ is positive semidefinite, but not positive definite; however $\mathbf{X}_0$ is relative minimum because for all (x, y), $p(x, y) = (x - y)^2 + x^4 + y^4 \geq 0$.

(d) $p(x, y) = x^2 - 2xy + y^2 - x^4 - y^4$; again $(d^2_{\mathbf{X}_0}p)(x, y) = 2(x - y)^2$ is positive semidefinite, but not positive definite. In this case, however, $\mathbf{X}_0$ is not an extreme point, since $p(t, t) = -2t^4$ is negative for all $t \neq 0$, while $p(t, -t) = 2t^2(2 - t^2)$ is positive if $0 < |t| < \sqrt{2}$.

The following consequence of Theorem 9.2 is of special interest.

Theorem 9.3 Let f be a function of (x, y) with continuous second partial derivatives f_{xx}, f_{xy}, and f_{yy} in a disk about a critical point $\mathbf{X}_0 = (x_0, y_0)$, and define

(8) $$D = f_{xx}(\mathbf{X}_0)f_{yy}(\mathbf{X}_0) - f^2_{xy}(\mathbf{X}_0).$$

(a) If $D > 0$, $\mathbf{X}_0$ is an extreme point. It is a relative minimum if $f_{xx}(\mathbf{X}_0) > 0$, a relative maximum if $f_{xx}(\mathbf{X}_0) < 0$.

(b) If $D < 0$, $\mathbf{X}_0$ is not an extreme point.

Proof. Let $\mathbf{X} - \mathbf{X}_0 = (u, v)$ and write

$$(d^2_{\mathbf{X}_0}f)(u, v) = p(u, v) = Au^2 + 2Buv + Cv^2$$

where $A = f_{xx}(\mathbf{X}_0)$, $B = f_{xy}(\mathbf{X}_0)$, $C = f_{yy}(\mathbf{X}_0)$, and $D = AC - B^2$. If $D > 0$, then $A \neq 0$, and we can write

$$p(u, v) = A\left(u^2 + \frac{2B}{A}uv + \frac{B^2}{A^2}v^2\right) + \left(C - \frac{B^2}{A}\right)v^2$$

$$= A\left(u + \frac{B}{A}v\right)^2 + \frac{D}{A}v^2,$$

which vanishes only if $(u, v) = (0, 0)$. Hence $d_{X_0}^2 f$ is positive definite if $A > 0$ or negative definite if $A < 0$, and (a) follows from (b) of Theorem 9.2.

If $D < 0$ there are three possibilities:

(i) $A \neq 0$; then $p(1, 0) = A$ and $p\left(\dfrac{-B}{A}, 1\right) = \dfrac{D}{A}$;

(ii) $C \neq 0$; then $p(0, 1) = C$ and $p\left(1, \dfrac{-B}{C}\right) = \dfrac{D}{C}$;

(iii) $A = C = 0$; then $B \neq 0$ and $p(1, 1) = -p(1, -1) = 2B$.

The two values of p in each of these cases are of opposite sign; hence if $D < 0$, $d_{X_0}^2 f$ is not semidefinite and X_0 is not an extreme point, from (a) of Theorem 9.2.

Example 9.10   Consider the polynomials of Example 9.9, taking $X_0 = (0, 0)$ in all cases.

(a) $f_{xx}(0, 0) = -f_{yy}(0, 0) = 2$, and $f_{xy}(0, 0) = 0$; hence $D = -4$, and $(0, 0)$ is not an extreme point.

(b) $f_{xx}(0, 0) = f_{yy}(0, 0) = 2$, and $f_{xy}(0, 0) = 0$; hence $D = 4$, and $(0, 0)$ is a relative minimum.

(c) and (d) $f_{xx}(0, 0) = f_{yy}(0, 0) = -f_{xy}(0, 0) = 2$; hence $D = 0$ and the test yields no information.

EXERCISES 2.9

1.   Find the extreme values of f, with domain D:

(a) $f(x) = 2x^3 - 9x^2 + 12x + 6$; $D = \{x \mid |x| \leq \tfrac{3}{2}\}$.

(b) $f(x) = x$; $D = \{x \mid 0 < x < 1\}$.

(c) $f(x) = \sqrt{1 - x^2}$; $D = \{x \mid |x| \leq 1\}$.

(d) $f(x) = |x^2 - 1|$; $D = R^1$.

(e) $f(x, y) = e^{|x+y|}$; $D = R^2$.

(f) $f(x, y) = \sqrt{|\sin xy|}$; $D = R^2$.

2.   Find all critical points:

(a) $f(x, y) = x^3 + y^3 - x - y$.

(b) $f(x, y, z) = \cos(x^2 + y^2 + z^2)$.

(c) $f(x, y) = 2x^2 + 2xy - 2x + 3y^2 + 2y + 2$.

(d) $f(x, y) = (x^2 + y^2)e^{-2xy}$.

(e) $f(x, y) = (xy)^2 e^{-(x^2+y^2)}$

3. Find all critical points:

(a) $f(x, y) = \dfrac{x^3}{3} + x^2y - xy^2 + \dfrac{y^3}{3}$.

(b) $f(x, y) = x^2 e^{-x^2} + y^2 e^{-y^2}$.

(c) $f(x, y) = x \log x + y \log y$.

(d) $f(x, y) = e^{x^2 - y^2}$.

4. Show that among all rectangles with a given perimeter P, the square has the largest area.

5. Find the distance from $(2, -1, 3)$ to the plane $2x + y - 2z = 4$.

6. Find the least squares linear polynomial fit to the points $(2, 1)$, $(1, 3)$, and $(3, 4)$.

7. A rectangular box with no top is to have a given volume V. Find dimensions for the box so as to minimize its surface area.

8. A rectangular box with no top is to have a given surface area S. Find dimensions for the box so as to maximize its volume.

9. Find the distance from the point $(1, 2, -1)$ to the line $x = 2 - t$, $y = 3 + t, z = 1 - 2t$.

10. Find the distance between the lines $x = 2 - t, y = 3 + t, z = 1 - 2t$ and $x = 1 - s, y = 2 - s, z = 3 + s$.

11. Find the dimensions of the largest rectangular box with three faces in the coordinate planes and a vertex in the plane

$$\frac{x}{a} + \frac{y}{b} + \frac{z}{c} = 1.$$

12. Find the distance between the line $x = 1 - 2t, y = 3 + t, z = 1 - t$ and the plane $x + 2y - z = 3$.

13. Classify the critical points of the functions in Exercise 3; that is, state whether they are maxima, minima, or neither.

14. Repeat Exercise 13 for the functions in Exercise 2.

15. Find the volume of the largest rectangular parallelepiped that can be inscribed in the ellipsoid

$$\frac{x^2}{a^2} + \frac{y^2}{b^2} + \frac{z^2}{c^2} = 1.$$

16. Classify the critical points:

(a) $f(x, y) = \cos(x^2 + y^2)$.   (b) $f(x, y, z) = e^{x^3+y^3+z^3}$.

(c) $f(x, y) = e^{x^2+y^2}$.   (d) $f(x, y, z) = e^{x^4+y^3+z^3}$.

17. Classify the critical points:

(a) $f(x, y) = \sin(x^2 + y^2)$.   (b) $f(x, y, z) = e^{x^2-y^2+z^2}$.

(c) $f(x, y, z) = e^{x^4+y^4+z^4}$.   (d) $f(x, y) = e^{x^4+y^3}$.

(e) $f(x, y) = e^{x^4+y^6}$.

THEORETICAL EXERCISES

T-1. Show that the only semidefinite homogeneous polynomial of odd degree is $p = 0$.

T-2. Prove (b) of Theorem 9.2 if $d_{\mathbf{x}_0}^k f$ is negative definite.

T-3. Minimize $f(x_1, \ldots, x_n) = x_1^2 + \cdots + x_n^2$ subject to $x_1 + \cdots + x_n = 1$.

T-4. Let $t_1, \ldots, t_n$ be distinct, $y_1, \ldots, y_n$ be arbitrary, and $m \geq k + 1$. For each polynomial of degree not exceeding k,

$$p(t) = a_0 + a_1 t + \cdots + a_k t^k,$$

define

$$Q(a_0, \ldots, a_k) = \sum_{i=1}^{m} (y_i - a_0 - a_1 t_i - \cdots - a_k t_i^k)^2.$$

The polynomial

$$p_0(t) = \alpha_0 + \alpha_1 t + \cdots + \alpha_k t^k$$

whose coefficients minimize Q is called the *least squares polynomial fit of degree k to the points* $(t_1, y_1), \ldots, (t_n, y_n)$. Show that $\alpha_0, \ldots, \alpha_k$

satisfy the *normal equations*:

$$\sum_{j=0}^{k} s_{i+j}\alpha_j = d_i \qquad (0 \le i \le k),$$

where

$$s_r = \sum_{j=1}^{m} t_j^r \qquad \text{and} \qquad d_r = \sum_{j=1}^{m} y_j t_j^r.$$

T-5.   Using Theorem 9.3 show that $f(x, y) = ax^2 + 2bxy + cy^2$ is positive definite if $a > 0$ and

$$\begin{vmatrix} a & b \\ b & c \end{vmatrix} > 0.$$

T-6.   If f is differentiable on R^n, how are the critical points of f related to those of f^2?

2.10   The Method of Lagrange Multipliers

Suppose f and $g_1, \ldots, g_k$ are defined on a subset D of R^n and that

(1) $$g_1(\mathbf{X}) = \cdots = g_m(\mathbf{X}) = 0$$

for all $\mathbf{X}$ in some nonempty subset D_1 of D. Then $\mathbf{X}_0$ is said to be a *relative maximum point of f subject to the constraints (1)* if there is an n-ball S about $\mathbf{X}_0$ such that

$$f(\mathbf{X}) \le f(\mathbf{X}_0)$$

for every $\mathbf{X}$ which is in both S and D_1. If

$$f(\mathbf{X}) \ge f(\mathbf{X}_0)$$

under the same assumptions on $\mathbf{X}$ and $\mathbf{X}_0$, then $\mathbf{X}_0$ is a *relative minimum point of f subject to the constraints* (1). In either case we also say that $\mathbf{X}_0$ is a *relative extreme point of f subject to the constraints* (1). We also use the terms *constrained minimum point, constrained maximum point,* and *constrained extreme point*.

Example 9.6, Section 2.9 involved a constrained minimum point, since we wished to find the point in the plane

$$2x + 3y + z = 1$$

that is closest to $(0, 1, -1)$. To solve this problem we must minimize the function

$$f(x, y, z) = x^2 + (y - 1)^2 + (z + 1)^2$$

subject to the constraint

(2) $$g(x, y, z) = 2x + 3y + z - 1 = 0.$$

We accomplished this by solving (2) for z and minimizing

$$h(x, y) = x^2 + (y - 1)^2 + (2 - 2x - 3y)^2.$$

In many applications it is inconvenient to solve the constraint equations as we did there. Fortunately the method of Lagrange multipliers makes this unnecessary.

We shall derive the method only for the simplest case, and state it for the general case. We assume that all functions in this section are continuously differentiable on their domains.

Two Variables Subject to One Constraint

Suppose we wish to extremize a function f of two variables (x, y) subject to the constraint

(3) $$g(x, y) = 0.$$

Let $\mathbf{X}_0 = (x_0, y_0)$ be a constrained extreme point and suppose that $g_y(\mathbf{X}_0) \neq 0$. (An argument similar to the following could be given under the assumption that $g_x(\mathbf{X}_0) \neq 0$.) Then the implicit function theorem implies that (3) defines y as a differentiable function of x is an open interval J about x_0; that is, there is a function $y = y(x)$ such that $y(x_0) = y_0$ and

(4) $$g(x, y(x)) = 0$$

for all x in J. Now define

$$h(x) = f(x, y(x));$$

since $(x, y(x))$ satisfies the constraint equation for all x in J and $\mathbf{X}_0$ is a relative extreme point of f subject to the constraint, it follows that x_0 must be a relative extreme point of h with respect to x. Therefore

(5) $$h'(x_0) = f_x(\mathbf{X}_0) + f_y(\mathbf{X}_0) y'(x_0) = 0.$$

Differentiating both sides of (4) and setting $x = x_0$ yields

(6) $$g_x(\mathbf{X}_0) + g_y(\mathbf{X}_0) y'(x_0) = 0.$$

From (5) and (6), the gradients

$$\nabla_{\mathbf{X}_0} f = \begin{bmatrix} f_x(\mathbf{X}_0) \\ f_y(\mathbf{X}_0) \end{bmatrix} \quad \text{and} \quad \nabla_{\mathbf{X}_0} g = \begin{bmatrix} g_x(\mathbf{X}_0) \\ g_y(\mathbf{X}_0) \end{bmatrix}$$

are both perpendicular to the vector $\begin{bmatrix} 1 \\ y'(x_0) \end{bmatrix}$. Hence they are parallel, and there is a constant λ such that

$$(7) \qquad\qquad \nabla_{\mathbf{X}_0} f = \lambda \nabla_{\mathbf{X}_0} g;$$

therefore

$$(8) \qquad f_x(\mathbf{X}_0) - \lambda g_x(\mathbf{X}_0) = f_y(\mathbf{X}_0) - \lambda g_y(\mathbf{X}_0) = 0.$$

Geometrically, (7) means that at $\mathbf{X}_0$ the normals to the level curves $g(x, y) = 0$ and $f(x, y) = f(\mathbf{X}_0)$ are parallel; consequently, the two curves have a common tangent. We say in this case that the curves are tangent (Fig. 10.1). In general, two surfaces in R^n are *tangent at a point* $\mathbf{X}_0$ if they have the same tangent plane at $\mathbf{X}_0$.

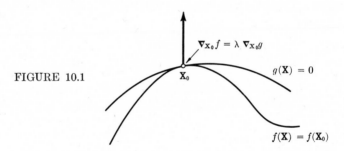

FIGURE 10.1

Analytically (8) means that any extreme point of f subject to the constraint (3) is a critical point of

$$(9) \qquad\qquad F = f - \lambda g;$$

for some value of λ. This is the basis of the method of Lagrange multipliers for finding constrained extreme points, which we now state for a function of two variables subject to one constraint:

(a) Find all critical points of the *auxiliary function* (9), treating λ as a fixed but unspecified constant.

(b) Determine which of the points obtained in (a) satisfy the constraint (3). Any such point is called a *critical point of f subject to the constraint*, or simply a *constrained critical point of f.*

(c)   Determine which of the constrained critical points are actually constrained extreme points.

In general, not all of the constrained critical points discovered in part (b) are actually constrained extreme points; however, no point which is not a constrained critical point can be a constrained extreme point.

The parameter λ introduced in the auxiliary function is called a *Lagrange multiplier*. The power of the method of Lagrange multipliers is due to the fact that it does not require solving the constraint equation for one of the variables as a function of the other.

Example 10.1   Suppose we wish to choose the dimensions of a rectangle with given perimeter p so that its area A is a maximum. We must maximize

$$A(x, y) = xy$$

subject to the constraint

(10) $2x + 2y - p = 0.$

The auxiliary function is

$$F(x, y) = xy - \lambda(2x + 2y - p),$$

which has critical points satisfying

$$\frac{\partial F}{\partial x}(x_0, y_0) = y_0 - 2\lambda = 0,$$

$$\frac{\partial F}{\partial y}(x_0, y_0) = x_0 - 2\lambda = 0.$$

Thus any constrained critical point of A must satisfy $x_0 = y_0$ and (10), which implies that $x_0 = y_0 = p/4$. Since we know on geometrical grounds that the problem has a solution, we conclude that, of all rectangles with given perimeter p, the one with the largest area is the square with sides of length $p/4$.

In this example we could just as well have taken

$$F_1(x, y) = xy - \lambda(2x + 2y)$$

to be the auxiliary function, since it differs from F only by a constant, and therefore has the same critical points. We could also have replaced λ by $\lambda/2$, and considered the auxiliary function

$$F_2(x, y) = xy - \lambda(x + y).$$

In general it is legitimate to add an arbitrary constant to an auxiliary function and replace λ by $c\lambda$, where c is any convenient constant. This allows us to dispense with "nuisance constants" in our calculations.

Example 10.2   To find the points on the circle $x^2 + y^2 = 80$ which are closest to and farthest from $(1, 2)$, we must extremize

$$f(x, y) = (x - 1)^2 + (y - 2)^2$$

subject to

(11) $$g(x, y) = x^2 + y^2 - 80 = 0.$$

Define the auxiliary function

$$F(x, y) = (x - 1)^2 + (y - 2)^2 - \lambda(x^2 + y^2);$$

its critical points satisfy

$$\frac{\partial F}{\partial x}(x_0, y_0) = 2(x_0 - 1) - 2\lambda x_0 = 0,$$

$$\frac{\partial F}{\partial y}(x_0, y_0) = 2(y_0 - 2) - 2\lambda y_0 = 0.$$

Therefore

$$\frac{x_0 - 1}{x_0} = \frac{y_0 - 2}{y_0} = \lambda,$$

which implies that $y_0 = 2x_0$. Substituting this in (11) yields

$$5x_0^2 = 80$$

or

$$x_0 = \pm 4.$$

Therefore there are two constrained critical points: $(4, 8)$ and $(-4, -8)$. Since $f(4, 8) = 45$ and $f(-4, -8) = 125$, it follows that $(4, 8)$ is the point on the circle closest to $(1, 2)$ and $(-4, -8)$ is the farthest (Fig. 10.2).

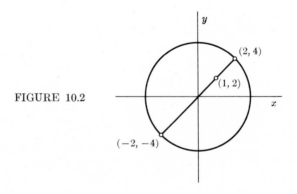

FIGURE 10.2

The General Case

The method of Lagrange multipliers for extremizing a function of n variables subject to m constraints is based on the following theorem, which we state without proof.

Theorem 10.1 Let f and $g_1, \ldots, g_m$ be continuously differentiable on a domain D of R^n, where $n > m \geq 1$, and suppose that at every $\mathbf{X}$ in D it is possible to construct a nonzero $m \times m$ determinant by choosing some m columns from the matrix

$$
\begin{bmatrix}
\dfrac{\partial g_1}{\partial x_1}(\mathbf{X}) & \dfrac{\partial g_1}{\partial x_2}(\mathbf{X}) & \cdots & \dfrac{\partial g_1}{\partial x_n}(\mathbf{X}) \\[2ex]
\dfrac{\partial g_2}{\partial x_1}(\mathbf{X}) & \dfrac{\partial g_2}{\partial x_2}(\mathbf{X}) & \cdots & \dfrac{\partial g_2}{\partial x_n}(\mathbf{X}) \\[1ex]
\vdots & \vdots & & \vdots \\[1ex]
\dfrac{\partial g_m}{\partial x_1}(\mathbf{X}) & \dfrac{\partial g_m}{\partial x_2}(\mathbf{X}) & \cdots & \dfrac{\partial g_m}{\partial x_n}(\mathbf{X})
\end{bmatrix}.
$$

Then at any point $\mathbf{X}_0$ in D where f attains an extreme value subject to the constraints

$$(12) \qquad\qquad g_1(\mathbf{X}) = \cdots = g_m(\mathbf{X}) = 0,$$

the gradient of f is a linear combination of the gradients of $g_1, \ldots, g_m$; that is,

$$(13) \qquad\qquad \nabla_{\mathbf{X}_0} f = \lambda_1 \nabla_{\mathbf{X}_0} g_1 + \cdots + \lambda_m \nabla_{\mathbf{X}_0} g_m.$$

A proof of this theorem is sketched in Exercises T-2 and T-3, Section 3.5.

We can rewrite (13) in terms of components as

$$
\frac{\partial f}{\partial x_i}(\mathbf{X}_0) - \lambda_1 \frac{\partial g_1}{\partial x_i}(\mathbf{X}_0) - \cdots - \lambda_m \frac{\partial g_m}{\partial x_i}(\mathbf{X}_0) = 0 \qquad (1 \leq i \leq m);
$$

thus, if we define the auxiliary function

$$(14) \qquad\qquad F = f - \lambda_1 g_1 - \cdots - \lambda_m g_m,$$

it follows that

$$
F_{x_1}(\mathbf{X}_0) = F_{x_2}(\mathbf{X}_0) = \cdots = F_{x_n}(\mathbf{X}_0) = 0.
$$

We therefore conclude that any extremum of f subject to (12) is a critical point of F, for some choice of the constraints $\lambda_1, \ldots, \lambda_m$. This is the basis for the general method of Lagrange multipliers for locating constrained extreme points, which consists of the following three steps.

(a)   Find all critical points of the auxiliary function (14), treating the *Lagrange multipliers* $\lambda_1, \ldots, \lambda_m$ as fixed but arbitrary constants.

(b)   Determine which of the points obtained in (a) satisfy the constraints (12). These are the *constrained critical points* of f.

(c)   Determine which of the constrained critical points are actually constrained extreme points.

We shall now consider several examples of the method of Lagrange multipliers. We observe again that, in forming the auxiliary function (14), it is permissible to omit additive constants and replace $\lambda_1, \ldots, \lambda_m$ by $c_1\lambda_1, \ldots, c_m\lambda_m$, where $c_1, \ldots, c_m$ are constants.

Example 10.3   Suppose we wish to design a rectangular box, open at the top, with maximum volume V and given surface area A. Let x and y be the dimensions of the base and let z be the height (Fig. 10.3). We must maximize

$$V(x, y, z) = xyz$$

subject to the constraint

(15) $$xy + 2yz + 2xz = A.$$

Define the auxiliary function

$$F(x, y, z) = xyz - \lambda(xy + 2yz + 2xz);$$

its critical points satisfy

(16) $$y_0z_0 - \lambda(y_0 + 2z_0) = 0,$$

(17) $$x_0z_0 - \lambda(x_0 + 2z_0) = 0,$$

(18) $$x_0y_0 - \lambda(2y_0 + 2x_0) = 0.$$

We need only consider critical points of F with positive coordinates. For such points (16) and (17) imply that

$$\frac{y_0 + 2z_0}{y_0z_0} = \frac{x_0 + 2z_0}{x_0z_0},$$

from which it follows that $x_0 = y_0$. Now (18) yields $\lambda = x_0/4$. Substituting this in (17) and solving for z_0 yields $z_0 = x_0/2$.

Thus the length, width, and height of the box are x_0, x_0, and $x_0/2$, respectively.

FIGURE 10.3

From (15),

$$3x_0^2 = A;$$

thus

$$x_0 = \sqrt{\frac{A}{3}}$$

and the maximum volume is

$$V = \frac{x_0^3}{2} = \frac{A^{3/2}}{6\sqrt{3}}.$$

In the remaining examples in this section we shall, for simplicity of notation, omit the zero subscript which we have previously attached to the coordinates of constrained critical points.

Example 10.4 By the distance d between the sphere

$$x^2 + y^2 + z^2 = 1$$

and the plane

$$x + y + z = 2\sqrt{3}$$

we mean the minimum value of

$$f(\mathbf{X}_1, \mathbf{X}_2) = \sqrt{(x_1 - x_2)^2 + (y_1 - y_2)^2 + (z_1 - z_2)^2}$$

as $\mathbf{X}_1 = (x_1, y_1, z_1)$ and $\mathbf{X}_2 = (x_2, y_2, z_2)$ vary over the sphere and plane, respectively. To find d we must minimize f subject to the constraints

$$(19) \qquad x_1^2 + y_1^2 + z_1^2 = 1$$

and

$$(20) \qquad x_2 + y_2 + z_2 = 2\sqrt{3}.$$

Clearly we can minimize f^2 to simplify the calculations. Consider the auxiliary function of six variables,

$$\begin{aligned}
F(\mathbf{X}_1, \mathbf{X}_2) &= f^2(\mathbf{X}_1, \mathbf{X}_2) - \lambda_1(x_1^2 + y_1^2 + z_1^2) - 2\lambda_2(x_2 + y_2 + z_2) \\
&= (x_1 - x_2)^2 + (y_1 - y_2)^2 + (z_1 - z_2)^2 \\
&\quad - \lambda_1(x_1^2 + y_1^2 + z_1^2) - 2\lambda_2(x_2 + y_2 + z_2).
\end{aligned}$$

Equating the partial derivatives of F to zero yields

(21) $\qquad x_1 - x_2 - \lambda_1 x_1 = y_1 - y_2 - \lambda_1 y_1 = z_1 - z_2 - \lambda_1 z_1 = 0$

and

(22) $\qquad x_1 - x_2 + \lambda_2 = y_1 - y_2 + \lambda_2 = z_1 - z_2 + \lambda_2 = 0.$

From (22),

(23) $\qquad\qquad\qquad x_1 - x_2 = y_1 - y_2 = z_1 - z_2;$

since their common value cannot be zero (the plane and the sphere do not intersect), it follows that $\lambda_1 \neq 0$ in (21), and therefore

$$x_1 = y_1 = z_1.$$

Therefore, from (19),

(24) $\qquad\qquad\qquad x_1 = y_1 = z_1 = \pm \dfrac{1}{\sqrt{3}}.$

Now it follows from (20) and (23) that

$$x_2 = y_2 = z_2 = \frac{2}{\sqrt{3}}.$$

Clearly we must choose the plus sign in the last member of (24), since

$$(1/\sqrt{3}, \quad 1/\sqrt{3}, \quad 1/\sqrt{3})$$

is closer to

$$(2/\sqrt{3}, \quad 2/\sqrt{3}, \quad 2/\sqrt{3})$$

than is

$$(-1/\sqrt{3}, \quad -1/\sqrt{3}, \quad -1/\sqrt{3}).$$

Thus the distance from the plane to the sphere is

$$d = \sqrt{\left(\frac{2}{\sqrt{3}} - \frac{1}{\sqrt{3}}\right)^2 + \left(\frac{2}{\sqrt{3}} - \frac{1}{\sqrt{3}}\right)^2 + \left(\frac{2}{\sqrt{3}} - \frac{1}{\sqrt{3}}\right)^2} = 1.$$

Example 10.5   Let

$$\mathbf{A} = \begin{bmatrix} a_{11} & a_{12} & \cdots & a_{1n} \\ a_{21} & a_{22} & \cdots & a_{2n} \\ \vdots & \vdots & & \vdots \\ a_{n1} & a_{n2} & \cdots & a_{nn} \end{bmatrix}$$

be a symmetric matrix; that is,

(25) $\qquad\qquad\qquad a_{ij} = a_{ji}.$

Let Q be the associated quadratic form:

(26)
$$Q(\mathbf{X}) = \mathbf{X}^{\mathrm{T}} \cdot \mathbf{AX} = \sum_{i,j=1}^{n} a_{ij} x_i x_j.$$

(Here we think of $\mathbf{X}$ as a column vector.) Suppose we wish to maximize Q subject to the requirement

(27)
$$|\mathbf{X}| = 1.$$

Using (25) and equating the partial derivatives of

(28)
$$F(\mathbf{X}) = Q(\mathbf{X}) - \lambda |\mathbf{X}|^2.$$

to zero yields

$$\sum_{j=1}^{n} a_{ij} x_j = \lambda x_i \qquad (1 \le i \le n)$$

(verify!), or, in matrix notation,

$$\mathbf{AX} = \lambda \mathbf{X},$$

which means that $\mathbf{X}$ is an eigenvector and λ an eigenvalue of $\mathbf{A}$. If $\mathbf{X}$ satisfies this equation then, from (26) and (27),

$$Q(\mathbf{X}) = \mathbf{X} \cdot (\lambda \mathbf{X}) = \lambda |\mathbf{X}|^2 = \lambda.$$

Therefore, the maximum value of Q subject to (27) is equal to the largest eigenvalue of $\mathbf{A}$, and it is attained when $\mathbf{X}$ is an associated eigenvector. Similarly, the minimum value of Q subject to (27) is the smallest eigenvalue of Q.

Example 10.6 Let us minimize

$$Q(\mathbf{X}) = \sum_{r=1}^{n} x_r^2$$

subject to the constraints

(29)
$$\sum_{r=1}^{n} x_r = 1, \qquad \sum_{r=1}^{n} r x_r = 0.$$

Define the auxiliary function

$$F(\mathbf{X}) = \sum_{r=1}^{n} x_r^2 - 2\lambda_0 \sum_{r=1}^{n} x_r - 2\lambda_1 \sum_{r=1}^{n} r x_r;$$

equating its partial derivatives to zero yields

$$x_r = \lambda_0 + \lambda_1 r \qquad (1 \le r \le n).$$

To determine λ_0 and λ_1 we substitute in (29) to find that

$$\sum_{r=1}^{n} (\lambda_0 + \lambda_1 r) = 1,$$

$$\sum_{r=1}^{n} (\lambda_0 + \lambda_1 r) r = 0;$$

thus λ_0 and λ_1 satisfy the simultaneous equations

(30)
$$s_0 \lambda_0 + s_1 \lambda_1 = 1,$$
$$s_1 \lambda_0 + s_2 \lambda_1 = 0,$$

where

$$s_0 = \sum_1^n 1,$$

$$s_1 = \sum_1^n r = \frac{n(n+1)}{2},$$

$$s_2 = \sum_1^n r^2 = \frac{n(n+1)(2n+1)}{6}.$$

Solving (30) yields

$$\lambda_0 = \frac{2(2n+1)}{n(n-1)},$$

$$\lambda_1 = -\frac{6}{n(n-1)}.$$

Thus the only extreme point of Q subject to (29) is $\hat{\mathbf{X}} = (\hat{x}_1, \ldots, \hat{x}_n)$ where

$$\hat{x}_r = \frac{2(2n+1) - 6r}{n(n-1)} \qquad (1 \leq r \leq n).$$

To see that $\hat{\mathbf{X}}$ is a constrained minimum, suppose $\mathbf{X} = (x_1, \ldots, x_n)$ is any point which satisfies the constraints; then

$$\sum_{r=1}^{n} (x_r - \hat{x}_r) = \sum_{r=1}^{n} (x_r - \hat{x}_r) r = 0,$$

and therefore

$$\sum_{r=1}^{n} \hat{x}_r (x_r - \hat{x}_r) = 0.$$

Hence

$$\sum_{r=1}^{n} x_r^2 = \sum_{r=1}^{n} [\hat{x}_r + (x_r - \hat{x}_r)]^2$$

$$= \sum_{r=1}^{n} \hat{x}_r^2 + \sum_{r=1}^{n} (x_r - \hat{x}_r)^2 \geq \sum_{r=1}^{n} \hat{x}_r^2,$$

and equality occurs only if $x_r - \hat{x}_r = 0 \ (1 \leq r \leq n)$.

EXERCISES 2.10

Use the method of Lagrange multipliers in the following problems.

1. Find the dimensions of the rectangle with minimum perimeter subject to the requirement that its area be A.

2. A rectangular box with no top is to have given volume V. Find dimensions for the box so as to minimize its surface area.

3. Find the extreme points of $f(x, y) = 2x + y$ subject to $x^2 + y^2 = 4$.

4. Find the distance from $(1, 2, -3)$ to the plane $x + 2y - z = 4$.

5. Find the extreme points of $f(x, y, z) = 2x + y + 2z$ subject to $x^2 + y^2 = 4$ and $x + z = 2$.

6. Find the maximum and minimum of $f(x, y, z) = x + y^2 + 2z$ subject to $4x^2 + 9y^2 - 36z^2 = 36$.

7. Find the dimensions of the largest rectangular box with three faces in the coordinate planes and a vertex in the plane

$$\frac{x}{a} + \frac{y}{b} + \frac{z}{c} = 1.$$

8. Find the distance from the point $(-1, 2, 3)$ to the line of intersection of the planes $x + 2y - 3z = 4$ and $2x - y + 2z = 5$.

9. Find the points on the ellipse $4x^2 + 9y^2 = 36$ which are closest to and farthest from $(0, 0)$.

10. Find the extreme value of xy on the line $3x + 2y = 6$.

11. Find the volume of the largest rectangular parallelepiped that can be inscribed in the ellipsoid

$$\frac{x^2}{a^2} + \frac{y^2}{b^2} + \frac{z^2}{c^2} = 1.$$

12. Find the points on the ellipsoid

$$4x^2 + 9y^2 + 36z^2 = 1.$$

whose distances from the origin are relative extremes.

13. Extremize $f(x, y, z) = x^2 + y^2 + z^2$ subject to $2x + 3y - 4z = 5$.

14. Find the points on the curve $x^2 + 4xy + y^2 = 36$ which are closest to the origin.

15. Find the distance between the circle $x^2 + y^2 = 1$ and the part of the hyperbola $xy = 1$ which is in the first quadrant.

In Exercises 16, 17, and 18 let

$$\mathbf{X} = \begin{bmatrix} x_1 \\ x_2 \\ x_3 \\ x_4 \end{bmatrix}, \qquad \mathbf{A} = \begin{bmatrix} 0 & 0 & -2 & 0 \\ 0 & -2 & 0 & 0 \\ -2 & 0 & 3 & 0 \\ 0 & 0 & 0 & 1 \end{bmatrix}$$

and $Q(\mathbf{X}) = \mathbf{X}^{\mathrm{T}} \cdot \mathbf{A} \mathbf{X}$.

16. Find the maximum value of $Q(\mathbf{X})$ subject to the constraint $x_1^2 + x_2^2 + x_3^2 + x_4^2 = 1$.

17. Find the maximum value of $Q(\mathbf{X})$ subject to the constraint in Exercise 16, and the additional constraint $x_1 - 2x_3 = 0$.

18. Find the maximum value of $Q(\mathbf{X})$ subject to the constraints of Exercise 17 and the additional constraint $x_4 = 0$.

THEORETICAL EXERCISES

T-1. Prove the Schwarz inequality by extremizing $F(\mathbf{U}, \mathbf{V}) = \mathbf{U} \cdot \mathbf{V}$ where $\mathbf{U}$ and $\mathbf{V}$ are n-vectors) subject to appropriate constraints.

T-2. What is wrong with defining the auxiliary function in Example 10.5 as $F(\mathbf{X}) = Q(\mathbf{X}) - \lambda \mid \mathbf{X} \mid$?

T-3.   Let α be an eigenvalue of the symmetric matrix $\mathbf{A}$ of Example 10.5 and let $\mathbf{U}$ be the associated eigenvector. Show that the extreme values of $Q(\mathbf{X})$ subject to the constraints $|\mathbf{X}| = 1$ and $\mathbf{X} \cdot \mathbf{U} = 0$ are eigenvalues of $\mathbf{A}$, and that they are attained when $\mathbf{X}$, considered as a column vector, is an associated eigenvector.

T-4.   Let $n \geq k + 1$ and $c_0, \ldots, c_k$ be given constants. Show that the minimum value of

$$Q(\mathbf{X}) = \sum_{r=1}^{n} x_r^2$$

subject to the constraints

$$\sum_{r=1}^{n} x_r r^i = c_i \qquad (0 \leq i \leq k)$$

is attained when

$$x_r = \lambda_0 + \lambda_1 r + \cdots + \lambda_k r^k,$$

where

$$\sum_{j=0}^{k} \sigma_{i+j} \lambda_j = c_i \qquad (0 \leq i \leq k)$$

and

$$\sigma_i = \sum_{r=1}^{n} r^i.$$

T-5.   Suppose $\nabla_{\mathbf{X}} g$ is continuous and nonzero for all $\mathbf{X}$ in R^n, let S be the level set defined by $g(\mathbf{X}) = 0$, and let $g(\mathbf{X}_1) \neq 0$. *Prove*: If $\mathbf{X}_0$ is the point on S closest to $\mathbf{X}_1$, then the vector $\mathbf{X}_1 - \mathbf{X}_0$ is normal to the tangent plane to S at $\mathbf{X}_0$.

Chapter 3

Differential Calculus of Vector-Valued Functions

3.1 Functions, Limits, and Continuity

If $f_1, \ldots, f_m$ are real valued functions defined on a set D in R^n, then

$$\mathbf{f} = \begin{bmatrix} f_1 \\ \vdots \\ f_m \end{bmatrix}$$

is the function that assigns to every $\mathbf{X}$ in D the m-vector

(1)
$$\mathbf{f(X)} = \begin{bmatrix} f_1(\mathbf{X}) \\ \vdots \\ f_m(\mathbf{X}) \end{bmatrix}.$$

We say that $\mathbf{f}$ is *vector valued*, and $f_1, \ldots, f_m$ are called the *coordinate functions* or *components* of $\mathbf{f}$. The *range* (or *image*) of $\mathbf{f}$ is the set of m-vectors obtained from (1) as $\mathbf{X}$ varies over D, which is the *domain* of $\mathbf{f}$.

As a convenient method for stating the dimensions of the spaces containing the domain and range of $\mathbf{f}$, we shall write

(2)
$$\mathbf{f}: \quad R^n \to R^m$$

when we mean that D is a subset of R^n, and that the range of $\mathbf{f}$ is a subset of R^m. Many authors use this notation to mean that the domain of $\mathbf{f}$ is *all* of R^n. It should be emphasized that when we use this notation we mean only that the domain of $\mathbf{f}$ is a subset D of R^n; of course this includes the possibility that $D = R^n$.

If f is a real valued function such as those studied in Chapter 2, we shall write

$$f: \quad R^n \to R^1$$

or simply

$$f: \quad R^n \to R;$$

when $m = 1$, it is to be understood that $\mathbf{f} = f_1$ in (1).

When we write (2) without further definition of the domain of $\mathbf{f}$, we shall mean that D is the largest subset of R^n common to the domains of $f_1, \ldots, f_m$.

Example 1.1 Let $\mathbf{f}: R^2 \to R^3$ have components

$$f_1(x, y) = x + y,$$

$$f_2(x, y) = x - y,$$

$$f_3(x, y) = x^2 + y^2.$$

Thus

$$\mathbf{f} = \begin{bmatrix} f_1 \\ f_2 \\ f_3 \end{bmatrix} \quad \text{and} \quad \mathbf{f}(x, y) = \begin{bmatrix} x + y \\ x - y \\ x^2 + y^2 \end{bmatrix}.$$

The domain of $\mathbf{f}$ is all of R^2 and its range is a subset of R^3.

Example 1.2 Let $\mathbf{f}: R^2 \to R^2$ be defined by

$$\mathbf{f}(x, y) = \begin{bmatrix} \dfrac{1}{x - y} \\ x^2 + y^2 \end{bmatrix};$$

the components of $\mathbf{f}$ are

$$f_1(x, y) = \frac{1}{x - y},$$

$$f_2(x, y) = x^2 + y^2.$$

The domain of $\mathbf{f}$ is all of R^2 except points of the form (x, x), which are not in the domain of f_1. The range of $\mathbf{f}$ is a subset of R^2.

Example 1.3   Let

$$\mathbf{f}(x, y) = \begin{bmatrix} x \\ y \\ \sqrt{1 - x^2 - y^2} \end{bmatrix}.$$

The domain of $\mathbf{f}$ is the set of (x, y) such that $x^2 + y^2 \leq 1$; its range is a subset of R^3.

If D is the largest subset of R^n on which both

$$\mathbf{f} = \begin{bmatrix} f_1 \\ \vdots \\ f_m \end{bmatrix} \quad \text{and} \quad \mathbf{g} = \begin{bmatrix} g_1 \\ \vdots \\ g_m \end{bmatrix}$$

are defined, then we define

$$\mathbf{f} + \mathbf{g} = \begin{bmatrix} f_1 + g_1 \\ \vdots \\ f_m + g_m \end{bmatrix} \quad \text{and} \quad \mathbf{f} - \mathbf{g} = \begin{bmatrix} f_1 - g_1 \\ \vdots \\ f_m - g_m \end{bmatrix}$$

on D. If $\mathbf{f} \colon R^n \to R^m$ and $\mathbf{g} \colon R^{n_1} \to R^{m_1}$, then $\mathbf{f} + \mathbf{g}$ and $\mathbf{f} - \mathbf{g}$ are not defined unless $n = n_1$ and $m = m_1$.

If c is a constant then

$$c\mathbf{f} = \begin{bmatrix} cf_1 \\ \vdots \\ cf_m \end{bmatrix}.$$

Example 1.4   Let

$$\mathbf{f}(x, y) = \begin{bmatrix} x \\ \sqrt{1 - x^2 - y^2} \\ y \end{bmatrix}$$

and

$$\mathbf{g}(x, y) = \begin{bmatrix} x \\ x\sqrt{1 - 2x^2 - 2y^2} \\ y \end{bmatrix};$$

then

$$(\mathbf{f} + \mathbf{g})(x, y) = \begin{bmatrix} 2x \\ \sqrt{1 - x^2 - y^2} + x\sqrt{1 - 2x^2 - 2y^2} \\ 2y \end{bmatrix}$$

$$(x^2 + y^2 \le 1/2).$$

Example 1.5 Let

$$\mathbf{f}(x, y) = \begin{bmatrix} x \\ y \\ x^2 - y^2 \end{bmatrix}$$

and $c = -3$. Then $-3\mathbf{f}$ has values

$$(-3\mathbf{f})(x, y) = \begin{bmatrix} -3x \\ -3y \\ -3x^2 + 3y^2 \end{bmatrix}.$$

Definition 1.1 Let $\mathbf{f}: R^n \to R^m$, $\mathbf{g}: R^m \to R^p$, and let D be the subset of the domain of $\mathbf{f}$ consisting of those n-vectors $\mathbf{X}$ such that the m-vector $\mathbf{f}(\mathbf{X})$ is in the domain of $\mathbf{g}$. Then the *composite function* $\mathbf{g} \circ \mathbf{f}$ is defined for $\mathbf{X}$ in D by

$$(\mathbf{g} \circ \mathbf{f})(\mathbf{X}) = \mathbf{g}(\mathbf{f}(\mathbf{X})).$$

Thus $\mathbf{g} \circ \mathbf{f}: R^n \to R^p$.

If $m = p = 1$, this reduces to Definition 1.2, Section 2.1.

Composite functions are so important that we consider them in more detail. Let

$$\mathbf{f}(\mathbf{X}) = \begin{bmatrix} f_1(\mathbf{X}) \\ \vdots \\ f_m(\mathbf{X}) \end{bmatrix}$$

and

(3) $$\mathbf{g}(u_1, \ldots, u_m) = \begin{bmatrix} g_1(u_1, \ldots, u_m) \\ \vdots \\ g_p(u_1, \ldots, u_m) \end{bmatrix};$$

then $(\mathbf{g} \circ \mathbf{f})(\mathbf{X})$ is obtained by setting $u_1 = f_1(\mathbf{X}), \ldots, u_m = f_m(\mathbf{X})$ in (3):

$$(4) \qquad (\mathbf{g} \circ \mathbf{f})(\mathbf{X}) = \begin{bmatrix} g_1(f_1(\mathbf{X}), \ldots, f_m(\mathbf{X})) \\ \vdots \\ g_p(f_1(\mathbf{X}), \ldots, f_m(\mathbf{X})) \end{bmatrix}.$$

However, according to Definition 1.1 with $\mathbf{g}$ replaced by g_i and p $= 1$,

$$g_i(f_1(\mathbf{X}), \ldots, f_m(\mathbf{X})) = (g_i \circ f)(\mathbf{X});$$

thus, from (4),

$$\mathbf{g} \circ \mathbf{f} = \begin{bmatrix} g_1 \circ \mathbf{f} \\ \vdots \\ g_p \circ \mathbf{f} \end{bmatrix};$$

that is, $(g \circ \mathbf{f})_i = g_i \circ \mathbf{f}$.

Example 1.6 A *linear transformation from R^n to R^m* is a function $\mathbf{f}: R^n \to R^m$ whose components are linear transformations from R^n to R, as defined in Section 2.3. Thus

$$\mathbf{f} = \begin{bmatrix} f_1 \\ \vdots \\ f_m \end{bmatrix}$$

is a linear transformation from R^n to R^m if

$$f_r(\mathbf{X}) = \sum_{j=1}^{n} a_{rj} x_j \qquad (1 \le r \le m)$$

where the a_{rj} are constants.

Now suppose $\mathbf{g}$ is a linear transformation from R^m to R^p; thus if $\mathbf{U} = (u_1, \ldots, u_m)$ is a point in R^m then

$$\mathbf{g} = \begin{bmatrix} g_1 \\ \vdots \\ g_p \end{bmatrix}$$

where

$$g_i(\mathbf{U}) = \sum_{s=1}^{m} b_{is} u_s \qquad (1 \le i \le p).$$

Since the range of $\mathbf{f}$ is contained in the domain of $\mathbf{g}$, we can define the composite function $\mathbf{g} \circ \mathbf{f}$. To obtain $(\mathbf{g} \circ \mathbf{f})_i$, we take

$$u_s = f_s(\mathbf{X}) = \sum_{j=1}^{n} a_{sj} x_j;$$

then

$$(\mathbf{g} \circ \mathbf{f})_i(\mathbf{X}) = (g_i \circ \mathbf{f})(\mathbf{X})$$

$$= \sum_{s=1}^{m} b_{is} \sum_{j=1}^{n} a_{sj} x_j$$

$$= \sum_{j=1}^{n} (\sum_{s=1}^{m} b_{is} a_{sj}) x_j.$$

Thus $\mathbf{g} \circ \mathbf{f}$ is a linear transformation from R^n to R^p:

$$(\mathbf{g} \circ \mathbf{f})_i(\mathbf{X}) = \sum_{j=1}^{n} c_{ij} x_j \qquad (1 \le i \le p),$$

where

$$c_{ij} = \sum_{s=1}^{m} b_{is} a_{sj} \qquad (1 \le i \le p, \ \ 1 \le j \le m).$$

In terms of the matrices $\mathbf{M_f} = (a_{ij})$, $\mathbf{M_g} = (b_{ij})$, and $\mathbf{M_{g \circ f}} = (c_{ij})$ we have

$$\mathbf{M_{g \circ f}} = \mathbf{M_g} \mathbf{M_f}.$$

Example 1.7 Let

$$\mathbf{f}(x, y) = \begin{bmatrix} x \\ x \cos y \\ x \sin y \end{bmatrix} \qquad (\mathbf{f}: \quad R^2 \to R^3)$$

and

$$\mathbf{g}(u, v, w) = \begin{bmatrix} uv \\ vw \\ uw \\ \sqrt{1 - u^2 - v^2 - w^2} \end{bmatrix} \qquad (\mathbf{g}: \quad R^3 \to R^4);$$

then

$$(5) \qquad (\mathbf{g} \circ \mathbf{f})\,(x,\, y) = \begin{bmatrix} x^2 \cos y \\[6pt] x^2 \sin y \cos y \\[6pt] x^2 \sin y \\[6pt] \sqrt{1 - 2x^2} \end{bmatrix} \qquad (\mathbf{g} \circ \mathbf{f}\!: \ R^2 \to R^4),$$

and the domain of $\mathbf{g} \circ \mathbf{f}$ is the set of points $(x,\, y)$ such that $|\,x\,| \leq 1/\sqrt{2}$.

Definition 1.2 If $\mathbf{X}_0$ is a limit point of the domain of

$$\mathbf{f} = \begin{bmatrix} f_1 \\ \vdots \\ f_m \end{bmatrix} \qquad \text{and} \qquad \mathbf{U}_0 = \begin{bmatrix} b_1 \\ \vdots \\ b_m \end{bmatrix},$$

we say that $\lim\limits_{\mathbf{X} \to \mathbf{X}_0} \mathbf{f}(\mathbf{X}) = \mathbf{U}_0$ if $\lim\limits_{\mathbf{X} \to \mathbf{X}_0} f_i(\mathbf{X}) = b_i \ (1 \leq i \leq m)$.

Example 1.8 Let $\mathbf{X}_0 = (-1,\, 2)$ and

$$\mathbf{f}(x,\, y) = \begin{bmatrix} 2x + 3y \\[6pt] x + y \\[6pt] x - y \end{bmatrix};$$

then $\lim\limits_{\mathbf{X} \to \mathbf{X}_0} f_1(x,\, y) = 4$, $\lim\limits_{\mathbf{X} \to \mathbf{X}_0} f_2(x,\, y) = 1$, and $\lim\limits_{\mathbf{X} \to \mathbf{X}_0} f_3(x,\, y) = -3$. Hence

$$\lim_{\mathbf{X} \to \mathbf{X}_0} \mathbf{f}(x,\, y) = \begin{bmatrix} 4 \\[6pt] 1 \\[6pt] -3 \end{bmatrix}.$$

Example 1.9 Let $\mathbf{X}_0 = (0,\, 0)$ and

$$\mathbf{f}(x,\, y) = \begin{bmatrix} x + y \\[6pt] \dfrac{2xy}{x^2 + y^2} \end{bmatrix}.$$

Then $\lim\limits_{X \to X_0} f_1(x, y) = 0$, but f_2 has no limit as (x, y) approaches $(0, 0)$. Hence

$\lim\limits_{X \to X_0} \mathbf{f}(x, y)$ does not exist.

Theorem 1.1 $\lim\limits_{X \to X_0} \mathbf{f}(\mathbf{X}) = \mathbf{U}_0$ if and only if $\lim\limits_{X \to X_0} |\mathbf{f}(X) - \mathbf{U}_0| = 0$.

Proof. Write

$$|\mathbf{f}(X) - \mathbf{U}_0| = \sqrt{(f_1(\mathbf{X}) - b_1)^2 + \cdots + (f_m(\mathbf{X}) - b_m)^2}.$$

The sum of squares on the right approaches zero as $\mathbf{X}$ approaches $\mathbf{X}_0$ if and only if the individual terms do; that is, if and only if

$$\lim\limits_{X \to X_0} f_i(\mathbf{X}) = b_i \qquad (1 \leq i \leq m).$$

The following corollary gives a useful interpretation of Theorem 1.1.

Corollary 1.1 If $\mathbf{f}: R^n \to R^m$, then $\lim\limits_{X \to X_0} \mathbf{f}(\mathbf{X}) = \mathbf{U}_0$ if and only if, for

every $\epsilon > 0$, there is a $\delta > 0$ such that $|\mathbf{f}(\mathbf{X}) - \mathbf{U}_0| < \epsilon$ whenever $\mathbf{X}$ is in the domain of $\mathbf{f}$ and $0 < |\mathbf{X} - \mathbf{X}_0| < \delta$.

We leave the proof to the student.

From Definition 1.2 it follows that all questions about limits of a vector valued function can be answered by examining its components. Thus, limits are unique, and the following theorems can be proved by applying Theorem 1.2 of Section 2.1 to the components of $\mathbf{f}$ and $\mathbf{g}$.

Theorem 1.2 Let $\mathbf{X}_0$ be a limit point of the set of points common to the domains of $\mathbf{f}$ and $\mathbf{g}$. If $\lim\limits_{X \to X_0} \mathbf{f}(\mathbf{X}) = \mathbf{U}_0$ and $\lim\limits_{X \to X_0} \mathbf{g}(\mathbf{X}) = \mathbf{U}_1$, then

$$\lim\limits_{X \to X_0} (\mathbf{f} + \mathbf{g})(\mathbf{X}) = \mathbf{U}_0 + \mathbf{U}_1$$

and

$$\lim\limits_{X \to X_0} (\mathbf{f} - \mathbf{g})(\mathbf{X}) = \mathbf{U}_0 - \mathbf{U}_1.$$

Theorem 1.3 Let c be a constant and $\lim\limits_{X \to X_0} \mathbf{f}(\mathbf{X}) = \mathbf{U}_0$. Then

$$\lim\limits_{X \to X_0} (c\mathbf{f})(\mathbf{X}) = c\mathbf{U}_0.$$

Definition 1.3 A vector valued function is said to be *continuous* at a point $\mathbf{X}_0$ if each of its components is continuous at $\mathbf{X}_0$.

Example 1.10 Let $\mathbf{X}_0 = (0, 0)$ and

$$\mathbf{f}(x, y) = \begin{bmatrix} \cos(x + y) \\ e^{x+y} \\ 2x^2 - y^2 + 1 \end{bmatrix};$$

then $\mathbf{f}$ is continuous at $\mathbf{X}_0$, since each of its components is continuous at $\mathbf{X}_0$. Thus

$$\lim_{\mathbf{X} \to \mathbf{X}_0} \mathbf{f}(\mathbf{X}) = \begin{bmatrix} 1 \\ 1 \\ 1 \end{bmatrix} = \mathbf{f}(\mathbf{X}_0).$$

Example 1.11 Let $\mathbf{X}_0 = (1, 1)$ and

$$\mathbf{f}(x, y) = \begin{cases} \begin{bmatrix} x - y \\ \dfrac{x^2 - y^2}{x - y} \end{bmatrix} & \text{if} \quad x \neq y, \\[4ex] \begin{bmatrix} 0 \\ 0 \end{bmatrix} & \text{if} \quad x = y. \end{cases}$$

Then f_1 is continuous at $\mathbf{X}_0$, but f_2 is not, because

$$\lim_{\mathbf{X} \to \mathbf{X}_0} f_2(x, y) = \lim_{\mathbf{X} \to \mathbf{X}_0} (x + y)$$

$$= 2 \neq 0 = f_2(\mathbf{X}_0);$$

hence $\mathbf{f}$ is not continuous at $\mathbf{X}_0$.

Example 1.12 A linear transformation $\mathbf{L} : R^n \to R^m$

$$\mathbf{L} = \begin{pmatrix} L_1 \\ \vdots \\ L_m \end{pmatrix}$$

is continuous on R^n, since the real valued linear transformations $L_1, \ldots, L_m$ are continuous on R^n.

The following theorem can be obtained by applying Theorem 1.3 of Section 2.1 to the component functions of **f** and **g**.

Theorem 1.4   (a) If a vector valued function **f** is continuous at $\mathbf{X}_0$ and c is a constant, then $c\mathbf{f}$ is continuous at $\mathbf{X}_0$. (b) Let D be the set of points common to the domains of **f** and **g** (both $R^n \to R^m$). Let $\mathbf{X}_0$ be in D and be a limit point of D. Then, if **f** and **g** are continuous at $\mathbf{X}_0$, so are $\mathbf{f} + \mathbf{g}$ and $\mathbf{f} - \mathbf{g}$.

A continuous function of a continuous function is continuous; this is stated more precisely in the following theorem, which generalizes Theorem 1.4 of Section 2.1.

Theorem 1.5   Let **f** be continuous at $\mathbf{X}_0$, and $\mathbf{U}_0 = \mathbf{f}(\mathbf{X}_0)$ be an interior point of the domain of **g**, at which **g** is continuous. Then the composite function $\mathbf{g} \circ \mathbf{f}$ is continuous at $\mathbf{X}_0$.

Example 1.13   Let **f** and **g** be as in Example 1.7, and let $\mathbf{X}_0 = (1/2, \pi/2)$. Then **f** is continuous at $\mathbf{X}_0$ and

$$\mathbf{U}_0 = \mathbf{f}(\mathbf{X}_0) = \begin{bmatrix} \frac{1}{2} \\ 0 \\ \frac{1}{2} \end{bmatrix}$$

is an interior point of the domain of **g**; furthermore, **g** is continuous at $\mathbf{U}_0$. Theorem 1.5 implies that $\mathbf{g} \circ \mathbf{f}$ is continuous at $\mathbf{X}_0$, which can be verified from the explicit expression (5) for $(\mathbf{g} \circ \mathbf{f})(\mathbf{X})$.

EXERCISES 3.1

1. Find the coordinate functions and domains of the following functions.

(a) $\mathbf{f}(x, y) = \begin{bmatrix} \cos xy \\ \sqrt{x^2 + y^2 - 4} \\ \dfrac{1}{xy} \end{bmatrix}$.   (b) $\mathbf{f}(x, y) = \begin{bmatrix} e^{x^2 + y^2} \\ \dfrac{x}{y^2 - 1} \end{bmatrix}$.

(c) $f(x, y, z) = \begin{bmatrix} x^2 + y^2 + z^2 \\ x^2 \\ -y^2 \end{bmatrix}$.   (d) $f(x, y) = \begin{bmatrix} \sqrt{4 - x^2} \\ \sqrt{4 - y^2} \\ x + y \end{bmatrix}$.

2.   Repeat Exercise 1 for the following functions.

(a) $f(x, y, z) = \begin{bmatrix} x + \dfrac{1}{z} \\ \dfrac{z}{y^2 - 9} \end{bmatrix}$.   (b) $f(x, y) = \begin{bmatrix} \log(x - y) \\ e^{x-y} \end{bmatrix}$.

(c) $f(x, y, z) = \begin{bmatrix} \dfrac{1}{x - y} \\ \dfrac{1}{y^2 - z^2} \\ \dfrac{1}{x - z} \end{bmatrix}$.   (d) $f(x, y) = \begin{bmatrix} \dfrac{1}{\cos(x^2 - y^2)} \\ e^{x^2 - y^2} \\ \sqrt{x^2 - y^2} \end{bmatrix}$.

3.   For each function in Exercise 1, fill in m and n in "$f: R^n \to R^m$."

4.   For each function in Exercise 2, fill in m and n in "$f: R^n \to R^m$."

5.   Let

$$f(x, y) = \begin{bmatrix} x^2 + y^2 \\ \dfrac{1}{xy} \\ y \end{bmatrix}, \qquad g(x, y) = \begin{bmatrix} \dfrac{1}{xy} \\ \sqrt{2 - x^2 - y^2} \\ x \end{bmatrix}.$$

(a)   What are the domains of $f + g$ and $f - g$?

(b)   Find $(f + g)\left(\frac{1}{2}, \frac{1}{2}\right)$ and $(f - g)\left(\frac{1}{2}, -\frac{1}{2}\right)$.

(c)   Find $(3f + 2g)(1, 1)$.

6. Let

$$\mathbf{f}(x, y, z) = \begin{bmatrix} x^2 + y^2 - z^2 \\ 1 \\ \dfrac{1}{x^2 - y^2} \end{bmatrix}, \qquad \mathbf{g}(x, y, z) = \begin{bmatrix} x + y - z \\ \sqrt{x^2 + y^2} \end{bmatrix}.$$

(a) What are the domains of $\mathbf{f} + \mathbf{g}$ and $\mathbf{f} - \mathbf{g}$?

(b) Find $(\mathbf{f} + \mathbf{g})\,(1, 0, 1)$ and $(\mathbf{f} - \mathbf{g})\,(1, 0, -1)$.

(c) Find $(2\mathbf{f} + 4\mathbf{g})\,(0, 1, 0)$.

7. Let

$$\mathbf{f}(x, y) = \begin{bmatrix} 2x + y \\ 3x - 2y \\ -2x + 3y \end{bmatrix}, \qquad \mathbf{g}(u, z, w) = \begin{bmatrix} u - v + w \\ 2u + v + 2w \\ 3u - v \end{bmatrix}.$$

(a) Find $(\mathbf{g} \circ \mathbf{f})\,(2, -1)$.

(b) Find $(\mathbf{g} \circ \mathbf{f})\,(x, y)$.

(c) Verify that the matrices of $\mathbf{f}$, $\mathbf{g}$, and $\mathbf{g} \circ \mathbf{f}$ satisfy

$$\mathbf{M}_{g \circ f} = \mathbf{M}_g \mathbf{M}_f.$$

(d) Is $\mathbf{f} \circ \mathbf{g}$ defined?

8. Let

$$\mathbf{f}(x, y) = \begin{bmatrix} x + 2y \\ 2x - y \end{bmatrix}, \qquad \mathbf{g}(u, v) = \begin{bmatrix} 3u + v \\ 2u - v \end{bmatrix}.$$

(a) Find $(\mathbf{g} \circ \mathbf{f})\,(x, y)$ and $(\mathbf{f} \circ \mathbf{g})\,(u, v)$.

(b) Find $\mathbf{M}_{g \circ f}$ and $\mathbf{M}_{f \circ g}$.

9. Let

$$\mathbf{f}(x, y, z) = \begin{bmatrix} \log(x - y) \\ \log z \end{bmatrix}, \qquad \mathbf{g}(u, v) = \begin{bmatrix} e^{u-v} \\ e^{v/2} \end{bmatrix}.$$

(a) Find $(\mathbf{g} \circ \mathbf{f})\,(2, 0, 4)$.

(b) Find $(\mathbf{g} \circ \mathbf{f}) (x, y, z)$.

(c) What is the domain of $\mathbf{g} \circ \mathbf{f}$?

10. Let

$$\mathbf{f}(x, y, z) = \begin{bmatrix} x + y + z \\ x - y \end{bmatrix}, \qquad \mathbf{g}(u, v) = \begin{bmatrix} u \\ u + v \\ v \end{bmatrix}.$$

(a) Find $(\mathbf{g} \circ \mathbf{f}) (x, y, z)$.

(b) Find $(\mathbf{f} \circ \mathbf{g}) (u, v, w)$.

11. Let

$$\mathbf{f}(x, y) = \begin{bmatrix} x \tan y \\ x \end{bmatrix}, \qquad \mathbf{g}(u, z) = \begin{bmatrix} \dfrac{1}{u^2 + v^2} \\ v \end{bmatrix}.$$

(a) Find $(\mathbf{g} \circ \mathbf{f}) (1, \pi/4)$.

(b) Write $(\mathbf{g} \circ \mathbf{f}) (x, y)$ in its simplest form.

12. Find $\lim_{\mathbf{X} \to \mathbf{X}_0} \mathbf{f}(\mathbf{X})$, if it exists:

(a) $\mathbf{f}(\mathbf{X}) = \begin{bmatrix} \dfrac{x^2 - y^2}{x + y} \\ x - y \end{bmatrix}, \qquad \mathbf{X}_0 = (1, -1).$

(b) $\mathbf{f}(\mathbf{X}) = \begin{bmatrix} \dfrac{x^2 - y^2}{x + y} \\ \dfrac{x - y}{x^2 - y^2} \end{bmatrix}, \qquad \mathbf{X}_0 = (1, -1).$

(c) $\mathbf{f}(\mathbf{X}) = \begin{bmatrix} \log \sqrt{1 - x^2 - y^2 - z^2} \\ e^{x^2 + y^2 + z^2} \end{bmatrix}, \qquad \mathbf{X}_0 = (0, 0, 0).$

13. Find $\lim_{X \to X_0} f(X)$, if it exists:

(a) $f(X) = \begin{bmatrix} \dfrac{x}{y^2 - 1} \\[2ex] 2x - 3y \\[2ex] \dfrac{xy}{\sqrt{x^2 + y^2}} \end{bmatrix}$, $X_0 = (3, 4)$.

(b) $f(X) = \begin{bmatrix} \cos \dfrac{\pi}{4 - 2xyz} \\[2ex] \sin \dfrac{2\pi}{2x + y + z} \\[2ex] \dfrac{x}{2x - y - z} \end{bmatrix}$, $X_0 = (1, 1, 1)$.

(c) $f(X) = \begin{bmatrix} \dfrac{xy}{\sqrt{x^2 + y^2}} \\[2ex] \dfrac{x^2 + y^2 - 2xy}{x - y} \\[2ex] \sqrt{x^2 + y^2} \end{bmatrix}$, $X_0 = (0, 0)$.

14. Let $X_0 = (1, 2)$ and

$$f(X) = \begin{bmatrix} 2x + 3 \\ x + y \\ y + 1 \end{bmatrix}.$$

(a) Find $L = \lim_{X \to X_0} f(X)$.

(b) Find $|f(X) - L|$.

(c) Show that $\lim_{X \to X_0} |f(X) - L| = 0$.

15.   Repeat Exercise 14 for $\mathbf{X}_0 = (1, 0, -1)$ and

$$\mathbf{f}(\mathbf{X}) = \begin{bmatrix} x + y + z + 2 \\ x - y + z - 3 \end{bmatrix}.$$

16.   Let $\mathbf{f}$ and $\mathbf{g}$ be as in Exercise 10, $\mathbf{h} = \mathbf{g} \circ \mathbf{f}$ and $\mathbf{X}_0 = (1, 0, -1)$.

(a)   Find $\mathbf{h}$ explicitly in terms of (x, y, z) and evaluate

$$\lim_{\mathbf{X} \to \mathbf{X}_0} \mathbf{h}(x, y, z).$$

(b)   Find $\lim_{\mathbf{X} \to \mathbf{X}_0} \mathbf{h}(x, y, z)$ by means of Theorem 1.4.

17.   Repeat Exercise 16 for $\mathbf{f}$ and $\mathbf{g}$ as defined in Exercise 9, and $\mathbf{X}_0 = (4, 2, 1)$.

18.   Where are the functions of Exercise 1 continuous?

19.   Where are the functions of Exercise 2 continuous?

20.   Where are the following functions continuous?

(a) $\mathbf{f}(x, y) = \begin{cases} \begin{bmatrix} \dfrac{x^2 - y^2}{x + y} \\ x - y \end{bmatrix} & \text{if } x \neq -y \\[2em] \begin{bmatrix} x - y \\ x - y \end{bmatrix} & \text{if } x = -y. \end{cases}$

(b) $\mathbf{f}(x, y) = \begin{cases} \begin{bmatrix} \dfrac{x^2 - y^2}{x + y} \\ x - y \end{bmatrix} & \text{if } x \neq -y \\[2em] \begin{bmatrix} 0 \\ x - y \end{bmatrix} & \text{if } x = -y. \end{cases}$

(c) $\quad f(x, y) = \begin{cases} \begin{bmatrix} x + y \\ xy \end{bmatrix} & \text{if} \quad x \neq y \\ \\ \begin{bmatrix} 2x \\ 4 \end{bmatrix} & \text{if} \quad x = y. \end{cases}$

21. Let f and g be as defined in Exercise 5. Where are $f + g$ and $f - g$ continuous?

22. Let f and g be as defined in Exercise 6. Where are $f + g$ and $f - g$ continuous?

23. Where is $g \circ f$ continuous, if f and g are as defined in Exercise 7?

24. Where is $g \circ f$ continuous, if f and g are as defined in Exercise 9?

25. Where are $g \circ f$ and $f \circ g$ continuous, if f and g are as defined in Exercise 11?

THEORETICAL EXERCISES

T-1. Let $f: R^n \to R^m$ and $g: R^p \to R^q$. If both $f \circ g$ and $g \circ f$ are defined, what relations must hold between m, n, p, and q?

T-2. Show that Corollary 1.1 follows from Theorem 1.1.

T-3. Using Theorem 1.1, prove Theorems 1.2 and 1.3.

T-4. Prove Theorem 1.5.

T-5. Let $f: R^n \to R^m$, $h: R^n \to R$, and define $F = hf$ by

$$F(X) = \begin{bmatrix} h(X)f_1(X) \\ \vdots \\ h(X)f_m(X) \end{bmatrix}.$$

Prove: If h and f are continuous at X_0, so is F.

3.2   Differentiable Functions

In this section we generalize the idea of differentiability, previously defined for real valued functions, to vector valued functions. Looking back at the corresponding generalization of continuity, the student might guess that a differentiable vector valued function is nothing more than an ordered m-tuple of differentiable real valued functions; this is exactly the case, as we shall now see.

Definition 2.1   A vector valued function

$$\mathbf{f} = \begin{bmatrix} f_1 \\ \vdots \\ f_m \end{bmatrix}$$

is said to be *differentiable* at $\mathbf{X}_0$ if its components are differentiable at $\mathbf{X}_0$. If $\mathbf{f}$ is differentiable at every point of a set S, we say that $\mathbf{f}$ is *differentiable on S*.

Example 2.1   Let

$$\mathbf{f}(x, y) = \begin{bmatrix} 2x^2y + y^2x \\ x^2 - y^2 \end{bmatrix}.$$

The components of $\mathbf{f}$ have continuous partial derivatives everywhere; hence according to Theorem 3.3, Section 2.3, they are differentiable on R^2 and so is $\mathbf{f}$, by definition.

If $\mathbf{f}$ is differentiable at $\mathbf{X}_0$ then

$$(1) \qquad \lim_{\mathbf{X} \to \mathbf{X}_0} \frac{f_i(\mathbf{X}) - f_i(\mathbf{X}_0) - (d_{\mathbf{X}_0} f_i)(\mathbf{X} - \mathbf{X}_0)}{|\mathbf{X} - \mathbf{X}_0|} = 0$$

for $i = 1, \ldots, m$, and the definition of limit for vector valued functions implies that

$$(2) \qquad \lim_{\mathbf{X} \to \mathbf{X}_0} \frac{\mathbf{f}(\mathbf{X}) - \mathbf{f}(\mathbf{X}_0) - \mathbf{L}(\mathbf{X} - \mathbf{X}_0)}{|\mathbf{X} - \mathbf{X}_0|} = \mathbf{0},$$

where $\mathbf{L}$ is the linear transformation

$$\mathbf{L} = \begin{bmatrix} d_{\mathbf{X}_0} f_1 \\ \vdots \\ d_{\mathbf{X}_0} f_m \end{bmatrix}.$$

This motivates the following definition.

Definition 2.2 If $f: R^n \to R^m$ is differentiable at $\mathbf{X}_0$, then the *differential of f at $\mathbf{X}_0$* is the linear transformation from R^n to R^m given by

$$(3) \qquad \mathbf{d}_{\mathbf{X}_0}\mathbf{f} = \begin{bmatrix} d_{\mathbf{X}_0}f_1 \\ \vdots \\ d_{\mathbf{X}_0}f_m \end{bmatrix}.$$

Let $\mathbf{Y} = (y_1, \ldots, y_n)$ and $\mathbf{U} = (\mathbf{d}_{\mathbf{X}_0}\mathbf{f})(\mathbf{Y})$; then the components of $\mathbf{U}$ are

$$u_1 = \frac{\partial f_1}{\partial x_1}(\mathbf{X}_0)\, y_1 + \frac{\partial f_1}{\partial x_2}(\mathbf{X}_0)\, y_2 + \cdots + \frac{\partial f_1}{\partial x_n}(\mathbf{X}_0)\, y_n,$$

$$u_2 = \frac{\partial f_2}{\partial x_1}(\mathbf{X}_0)\, y_1 + \frac{\partial f_2}{\partial x_2}(\mathbf{X}_0)\, y_2 + \cdots + \frac{\partial f_2}{\partial x_n}(\mathbf{X}_0)\, y_n,$$

$$\vdots$$

$$u_m = \frac{\partial f_m}{\partial x_1}(\mathbf{X}_0)\, y_1 + \frac{\partial f_m}{\partial x_2}(\mathbf{X}_0)\, y_2 + \cdots + \frac{\partial f_m}{\partial x_n}(\mathbf{X}_0)\, y_n.$$

Definition 2.3 If $\mathbf{f}$ is as in Definition 2.2, then the matrix of $\mathbf{d}_{\mathbf{X}_0}\mathbf{f}$ is called the *Jacobian matrix of f at $\mathbf{X}_0$* and denoted by $\mathbf{J}_{\mathbf{X}_0}\mathbf{f}$; thus

$$\mathbf{J}_{\mathbf{X}_0}\mathbf{f} = \begin{bmatrix} \dfrac{\partial f_1}{\partial x_1}(\mathbf{X}_0) & \dfrac{\partial f_1}{\partial x_2}(\mathbf{X}_0) & \cdots & \dfrac{\partial f_1}{\partial x_n}(\mathbf{X}_0) \\[2ex] \dfrac{\partial f_2}{\partial x_1}(\mathbf{X}_0) & \dfrac{\partial f_2}{\partial x_2}(\mathbf{X}_0) & \cdots & \dfrac{\partial f_2}{\partial x_n}(\mathbf{X}_0) \\[2ex] \vdots & \vdots & & \vdots \\[2ex] \dfrac{\partial f_m}{\partial x_1}(\mathbf{X}_0) & \dfrac{\partial f_m}{\partial x_2}(\mathbf{X}_0) & \cdots & \dfrac{\partial f_m}{\partial x_n}(\mathbf{X}_0) \end{bmatrix}.$$

In matrix notation, $(\mathbf{d}_{\mathbf{X}_0}\mathbf{f})(\mathbf{Y}) = (\mathbf{J}_{\mathbf{X}_0}\mathbf{f})\mathbf{Y}$, where we consider $\mathbf{Y}$ as a column vector on the right side.

Just as we did when discussing real valued functions, we shall write simply

$$\mathbf{df} = \begin{bmatrix} df_1 \\ df_2 \\ \vdots \\ df_m \end{bmatrix}$$

when it is not important to emphasize $\mathbf{X}_0$; in the same situation we shall write

$$\mathbf{Jf} = \begin{bmatrix} \dfrac{\partial f_1}{\partial x_1} & \cdots & \dfrac{\partial f_1}{\partial x_n} \\ \vdots & & \vdots \\ \dfrac{\partial f_m}{\partial x_1} & \cdots & \dfrac{\partial f_m}{\partial x_n} \end{bmatrix}.$$

Example 2.2 Let $\mathbf{X}_0 = (0, 1, 2)$ and

$$\mathbf{f}(x, y, z) = \begin{bmatrix} 2x^2 + y^2 + z \\ \\ x^2 - y^2 \end{bmatrix};$$

then

$$\mathbf{Jf} = \begin{bmatrix} \dfrac{\partial f_1}{\partial x} & \dfrac{\partial f_1}{\partial y} & \dfrac{\partial f_1}{\partial z} \\ \\ \dfrac{\partial f_2}{\partial x} & \dfrac{\partial f_2}{\partial y} & \dfrac{\partial f_2}{\partial z} \end{bmatrix} = \begin{bmatrix} 4x & 2y & 1 \\ 2x & -2y & 0 \end{bmatrix}.$$

In this equation we have a harmless inconsistency: the entries in the first matrix are functions, while those in the second are their values at the point (x, y, z). We shall not hesitate to write the Jacobian matrices of particular functions in this way when it does not lead to confusion.

Now

$$\mathbf{df} = \begin{bmatrix} df_1 \\ df_2 \end{bmatrix} = \begin{bmatrix} 4x\,dx + 2y\,dy + dz \\ \\ 2x\,dx - 2y\,dy \end{bmatrix},$$

$$\mathbf{d_{X_0}f} = \begin{bmatrix} d_{X_0} f_1 \\ d_{X_0} f_2 \end{bmatrix} = \begin{bmatrix} 2dy + dz \\ \\ -2dy \end{bmatrix}$$

and

(4) $$(\mathbf{d_{X_0}f})\,(\mathbf{X} - \mathbf{X}_0) = (\mathbf{d_{X_0}f})\,(x, y - 1, z - 2)$$

$$= \begin{bmatrix} 2y + z - 4 \\ \\ -2y + 2 \end{bmatrix}.$$

Example 2.3 If **f** is a vector valued function of one variable,

$$\mathbf{f}(t) = \begin{bmatrix} f_1(t) \\ \vdots \\ f_m(t) \end{bmatrix},$$

then the Jacobian matrix is a column vector,

$$\mathbf{Jf} = \begin{bmatrix} f_1' \\ \vdots \\ f_m' \end{bmatrix},$$

and

$$\mathbf{df} = \begin{bmatrix} f_1' \, dt \\ \vdots \\ f_m' \, dt \end{bmatrix}.$$

For instance, let $t_0 = 0$ and

$$\mathbf{f}(t) = \begin{bmatrix} e^t \\ \sin t \\ \cos t \end{bmatrix};$$

then

$$\mathbf{Jf} = \begin{bmatrix} e^t \\ \cos t \\ -\sin t \end{bmatrix}, \qquad \mathbf{df} = \begin{bmatrix} e^t \, dt \\ \cos t \, dt \\ -\sin t \, dt \end{bmatrix},$$

$$\mathbf{d}_{t_0}\mathbf{f} = \begin{bmatrix} dt \\ dt \\ 0 \end{bmatrix}, \qquad (\mathbf{d}_{t_0}\mathbf{f})\,(t - t_0) = \begin{bmatrix} t \\ t \\ 0 \end{bmatrix}.$$

If a vector valued function **f** is differentiable at $\mathbf{X}_0$ then (2) can be rewritten as

(5) $$\lim_{\mathbf{X} \to \mathbf{X}_0} \frac{\mathbf{f}(\mathbf{X}) - \mathbf{f}(\mathbf{X}_0) - (\mathbf{d}_{\mathbf{X}_0}\mathbf{f})\,(\mathbf{X} - \mathbf{X}_0)}{|\mathbf{X} - \mathbf{X}_0|} = \mathbf{0},$$

which is identical in appearance with the definition of differentiability for real valued functions; the difference is that (5) is a statement about vectors rather than real numbers. It says that

(6) $$\mathbf{f}(\mathbf{X}) \cong \mathbf{f}(\mathbf{X}_0) + (\mathbf{d}_{\mathbf{X}_0}\mathbf{f})\,(\mathbf{X} - \mathbf{X}_0),$$

where the magnitude of the error vector approaches zero "very fast" as $\mathbf{X}$ approaches $\mathbf{X}_0$; that is, the ratio of the length of the error vector to $|\mathbf{X} - \mathbf{X}_0|$ approaches zero as $|\mathbf{X} - \mathbf{X}_0|$ approaches zero.

The function $\mathbf{g}$ defined by

$$\mathbf{g}(\mathbf{X}) = \mathbf{f}(\mathbf{X}_0) + (\mathbf{d}_{\mathbf{X}_0}\mathbf{f})(\mathbf{X} - \mathbf{X}_0)$$

is the sum of a constant vector and a linear transformation. A function of this kind is called an *affine function*. That is, an affine function $\mathbf{g}: R^n \to R^m$ is defined by

$$\mathbf{g} = \begin{bmatrix} g_1 \\ \vdots \\ g_m \end{bmatrix},$$

where

$$
\begin{aligned}
g_1(\mathbf{X}) &= b_1 + a_{11}(x_1 - c_1) + \cdots + a_{1n}(x_n - c_n), \\
g_2(\mathbf{X}) &= b_2 + a_{21}(x_1 - c_1) + \cdots + a_{2n}(x_n - c_n), \\
&\vdots \\
g_m(\mathbf{X}) &= b_m + a_{m1}(x_1 - c_1) + \cdots + a_{mn}(x_n - c_n),
\end{aligned}
$$

(7)

and all quantities on the right side except $(x_1, \ldots, x_n)$ are constants. We can write (7) in matrix form as

$$\mathbf{g}(\mathbf{X}) = \mathbf{B} + \mathbf{A}(\mathbf{X} - \mathbf{X}_0)$$

where

$$\mathbf{B} = \begin{bmatrix} b_1 \\ \vdots \\ b_m \end{bmatrix},$$

$$\mathbf{A} = \begin{bmatrix} a_{11} & a_{12} & \cdots & a_{1n} \\ a_{21} & a_{22} & \cdots & a_{2n} \\ \vdots & \vdots & & \vdots \\ a_{m1} & a_{m2} & & a_{mn} \end{bmatrix}$$

and

$$\mathbf{X} - \mathbf{X}_0 = \begin{bmatrix} x_1 - c_1 \\ \vdots \\ x_n - c_n \end{bmatrix}.$$

Equation (6) says that a differentiable vector valued function can be closely approximated by an affine function. This is analogous to the statement that a differentiable real valued function can be closely approximated by a polynomial of degree ≤ 1, to which (5) reduces when $m = 1$.

Example 2.4 Let f and X_0 be as defined in Example 2.2. Since

$$f(X_0) = \begin{bmatrix} 3 \\ -1 \end{bmatrix},$$

it follows from (4) that f is closely approximated near X_0 by the affine function

$$g(X) = \begin{bmatrix} 3 \\ -1 \end{bmatrix} + \begin{bmatrix} 2y + z - 4 \\ -2y + 2 \end{bmatrix}$$

$$= \begin{bmatrix} 2y + z - 1 \\ -2y + 1 \end{bmatrix}.$$

The next theorem is obtained by applying the corresponding results for real valued functions to the components of f and g.

Theorem 2.1 If f is differentiable at X_0, then it is continuous there. Also cf, where c is a constant, is differentiable at X_0; furthermore

$$d_{X_0}(cf) = c\, d_{X_0}f.$$

If g is also differentiable at X_0, the same is true of $f + g$ and $f - g$, and

$$d_{X_0}(f + g) = d_{X_0}f + d_{X_0}g \quad \text{and} \quad d_{X_0}(f - g) = d_{X_0}f - d_{X_0}g.$$

A vector valued function may fail to be differentiable at a point X_0 even though the components have partial derivatives there, as is shown by the following example.

Example 2.5 Let $X_0 = (0, 0)$ and

$$f(x, y) = \begin{bmatrix} \sqrt{|xy|} \\ x \end{bmatrix};$$

then

$$\frac{\partial f_1}{\partial x}(X_0) = \frac{\partial f_1}{\partial y}(X_0) = \frac{\partial f_2}{\partial y}(X_0) = 0$$

and

$$\frac{\partial f_2}{\partial x}(X_0) = 1.$$

However, f_1 is not differentiable at $\mathbf{X}_0$ (Example 3.5, Section 2.3). Hence neither is $\mathbf{f}$.

The following theorem gives sufficient conditions for differentiability at a point.

Theorem 2.2   A vector valued function is differentiable at $\mathbf{X}_0$ if all first partial derivatives of its components exist in some n-ball about $\mathbf{X}_0$, and are continuous at $\mathbf{X}_0$.

Proof.   The hypothesis implies that the components are differentiable at $\mathbf{X}_0$ (Theorem 3.3, Section 2.3); hence $\mathbf{f}$ is differentiable by definition.

Example 2.6   As previously noted in Example 2.1,

$$\mathbf{f}(x, y) = \begin{bmatrix} 2x^2y + y^2x \\ x^2 - y^2 \end{bmatrix}$$

is differentiable everywhere, and

$$\mathbf{df} = \begin{bmatrix} (4xy + y^2) \, dx + (2x^2 + 2xy) \, dy \\ 2x \, dx - 2y \, dy \end{bmatrix}.$$

Example 2.7   If $\mathbf{g}$ is the affine function (7), then $\mathbf{g}$ is differentiable on R^n, because the partial derivatives of $g_1, \cdots, g_m$, being constants, are continuous. By evaluating the partial derivatives, we find that

$$\mathbf{Jg} = \begin{bmatrix} a_{11} & a_{12} & \cdots & a_{1n} \\ a_{21} & a_{22} & \cdots & a_{2n} \\ \vdots & \vdots & & \vdots \\ a_{m1} & a_{m2} & \cdots & a_{mn} \end{bmatrix}.$$

For any $\mathbf{X}_0$,

$$\mathbf{g}(\mathbf{X}) - \mathbf{g}(\mathbf{X}_0) - (\mathbf{d}_{\mathbf{X}_0}\mathbf{g})(\mathbf{X} - \mathbf{X}_0) = 0$$

(verify!), a stronger result than (5).

Our view has been that a differentiable vector valued function is simply an ordered m-tuple of differentiable real valued functions; thus we started from (1) and deduced (2). An alternate approach would be to take (2) as the starting point; that is, define $\mathbf{f} \colon R^n \to R^m$ to be differentiable if there is a linear transformation

$$\mathbf{L} = \begin{bmatrix} L_1 \\ \vdots \\ L_m \end{bmatrix}$$

that satisfies (2). This definition, however, is equivalent to ours, since (2) and the definition of limit for vector valued functions imply that

$$\lim_{\mathbf{X} \to \mathbf{X}_0} \frac{f_i(\mathbf{X}) - f_i(\mathbf{X}_0) - L_i(\mathbf{X} - \mathbf{X}_0)}{|\mathbf{X} - \mathbf{X}_0|} = 0 \qquad (1 \le i \le m),$$

which in turn implies that f_i is differentiable at $\mathbf{X}_0$ and $L_i = d_{\mathbf{X}_0} f_i$. Hence

$$\mathbf{L} = \begin{bmatrix} d_{\mathbf{X}_0} f_1 \\ \vdots \\ d_{\mathbf{X}_0} f_m \end{bmatrix} = d_{\mathbf{X}_0} \mathbf{f},$$

as defined by (3).

EXERCISES 3.2

1. Find $d_{\mathbf{X}_0} \mathbf{f}$ and $(d_{\mathbf{X}_0} \mathbf{f})(\mathbf{X}_1 - \mathbf{X}_0)$:

(a) $\mathbf{f}(\mathbf{X}) = \begin{bmatrix} \sqrt{x^2 + y^2 + z^2} \\ 3x + 2y + 4z + 1 \end{bmatrix}$; $\quad \begin{array}{l} \mathbf{X}_0 = (1, 2, -2), \\ \mathbf{X}_1 = (1, 1, 1). \end{array}$

(b) $\mathbf{f}(\mathbf{X}) = \begin{bmatrix} \log(x - y) \\ x^3 + y^2 \\ 2x^2 + y \end{bmatrix}$; $\quad \mathbf{X}_0 = (2, 1), \qquad \mathbf{X}_1 = (3, 2).$

(c) $\mathbf{f}(\mathbf{X}) = \begin{bmatrix} ze^{x+y} \\ \sqrt{x + y} \\ (xyz)^{3/2} \end{bmatrix}$; $\quad \mathbf{X}_0 = (1, 1, 1), \qquad \mathbf{X}_1 = (1, 1, 2).$

2. Find $\mathbf{J}_{\mathbf{X}_0} \mathbf{f}$ for each $\mathbf{f}$ in Exercise 1.

3. Find $d_{\mathbf{X}_0} \mathbf{f}$ and $(d_{\mathbf{X}_0} \mathbf{f})(\mathbf{X}_1 - \mathbf{X}_0)$:

(a) $\mathbf{f}(\mathbf{X}) = \begin{bmatrix} x + y + z \\ x^2 + y^2 + z^2 \end{bmatrix}$; $\quad \mathbf{X}_0 = (1, 1, 2), \qquad \mathbf{X}_1 = (1, 2, 3).$

(b) $f(\mathbf{X}) = \begin{bmatrix} \sin(x+y) \\ \tan(x+y) \\ \cos(x+y) \end{bmatrix}$; $\mathbf{X}_0 = \left(\dfrac{\pi}{8}, \dfrac{\pi}{8}\right)$, $\mathbf{X}_1 = \left(\dfrac{\pi}{4}, -\dfrac{\pi}{4}\right)$.

(c) $f(\mathbf{X}) = \begin{bmatrix} e^x \\ x^2 \\ x+1 \end{bmatrix}$; $x_0 = 0$, $x_1 = 1$.

4.   Find $J_{\mathbf{X}_0}f$ for each f in Exercise 3.

5.   Find df:

(a) $f(x, y) = \begin{bmatrix} \cos(x+y) \\ \sin(x+y) \end{bmatrix}$.

(b) $f(x) = \begin{bmatrix} x \\ x^2 \\ x^3 \end{bmatrix}$.

(c) $f(x, y, z) = \begin{bmatrix} e^x \sin yz \\ e^y \sin xz \\ e^z \sin xy \end{bmatrix}$.

6.   Find Jf for each f in Exercise 5.

7.   Find df:

(a) $f(x, y, z, w) = \begin{bmatrix} e^x(y+z+w) \\ \sin(x+y+z+w) \\ xyz \end{bmatrix}$.

(b) $f(x, y) = \begin{bmatrix} \sin(1-x-y) \\ 2x+xy \\ x^3+y^2 \end{bmatrix}$.

(c) $\qquad f(x) = \begin{bmatrix} e^x \sin x \\ e^x \cos x \end{bmatrix}.$

8. Find $\mathbf{Jf}$ for each $\mathbf{f}$ in Exercise 7.

9. Let $\mathbf{X}_0 = (1, 2)$,

$$f(x, y) = \begin{bmatrix} x^2 + 2y^2 \\ x^2 + y \end{bmatrix} \quad \text{and} \quad g(x, y) = \begin{bmatrix} 2x - y \\ x^3 + y^2 \end{bmatrix}.$$

Find

(a) $d_{\mathbf{X}_0}(3\mathbf{f})$, (b) $d_{\mathbf{X}_0}(\mathbf{f} + \mathbf{g})$, (c) $d_{\mathbf{X}_0}(2\mathbf{f} - 3\mathbf{g})$.

10. For $\mathbf{X}_0$, $\mathbf{f}$, and $\mathbf{g}$ as defined in Exercise 9, compute

(a) $\mathbf{J}_{\mathbf{X}_0}\mathbf{f}$ and $\mathbf{J}_{\mathbf{X}_0}\mathbf{g}$;

(b) $\mathbf{J}_{\mathbf{X}_0}(3\mathbf{f})$ and $3\,\mathbf{J}_{\mathbf{X}_0}\mathbf{f}$;

(c) $\mathbf{J}_{\mathbf{X}_0}(\mathbf{f} + \mathbf{g})$ and $\mathbf{J}_{\mathbf{X}_0}\mathbf{f} + \mathbf{J}_{\mathbf{X}_0}\mathbf{g}$.

11. For each $\mathbf{f}$ and corresponding $\mathbf{X}_0$ in Exercise 1 find an affine transformation $\mathbf{g}$ such that

$$\lim_{\mathbf{X} \to \mathbf{X}_0} \frac{\mathbf{f}(\mathbf{X}) - \mathbf{g}(\mathbf{X})}{|\mathbf{X} - \mathbf{X}_0|} = \mathbf{0}.$$

12. Repeat Exercise 11 for each $\mathbf{f}$ in Exercise 3.

THEORETICAL EXERCISES

T-1. If f is a real valued function, show that $\nabla_{\mathbf{X}_0} f$ is the transpose of $\mathbf{J}_{\mathbf{X}_0} f$.

T-2. Prove Theorem 2.1.

T-3. Prove that $\mathbf{J}(c\mathbf{f}) = c\,\mathbf{Jf}$.

T-4. Prove that $\mathbf{J}(\mathbf{f} + \mathbf{g}) = \mathbf{Jf} + \mathbf{Jg}$.

T-5. What is $\mathbf{dg}$ in Example 2.7?

3.3 The Chain Rule

In this section we generalize the chain rule of Section 2.4, Chapter 2.

Theorem 3.1 Let $\mathbf{f}: R^n \to R^m$ be differentiable at $\mathbf{X}_0$ and $\mathbf{g}: R^m \to R^p$ be differentiable at $\mathbf{U}_0 = \mathbf{f}(\mathbf{X}_0)$. Then $\mathbf{h} = \mathbf{g} \circ \mathbf{f}$ is differentiable at $\mathbf{X}_0$ and

$$(1) \qquad d_{\mathbf{X}_0}\mathbf{h} = d_{\mathbf{U}_0}\mathbf{g} \circ d_{\mathbf{X}_0}\mathbf{f};$$

thus, the differential of the composite function is the composition of the differentials.

Proof. We first observe that $\mathbf{X}_0$ is an interior point of the domain of $\mathbf{h}$ (Exercise T-1), so that it is leigitimate to ask if $\mathbf{h}$ is differentiable at $\mathbf{X}_0$.

From the differentiability of $\mathbf{f}$ and $\mathbf{g}$,

$$(2) \qquad \mathbf{g}(\mathbf{U}) - \mathbf{g}(\mathbf{U}_0) = (d_{\mathbf{U}_0}\mathbf{g})(\mathbf{U} - \mathbf{U}_0) + |\mathbf{U} - \mathbf{U}_0|\,\mathbf{E}_1(\mathbf{U})$$

and

$$(3) \qquad \mathbf{f}(\mathbf{X}) - \mathbf{f}(\mathbf{X}_0) = (d_{\mathbf{X}_0}\mathbf{f})(\mathbf{X} - \mathbf{X}_0) + |\mathbf{X} - \mathbf{X}_0|\,\mathbf{E}_2(\mathbf{X}),$$

where

$$(4) \qquad \begin{aligned} &\lim_{\mathbf{U}\to\mathbf{U}_0} \mathbf{E}_1(\mathbf{U}) = \mathbf{0}, \\ &\lim_{\mathbf{X}\to\mathbf{X}_0} \mathbf{E}_2(\mathbf{X}) = \mathbf{0}. \end{aligned}$$

(Here the zeroes stand for zero vectors of dimensions p and m, respectively.) Substituting $\mathbf{U} = \mathbf{f}(\mathbf{X})$ and $\mathbf{U}_0 = \mathbf{f}(\mathbf{X}_0)$ in (2) gives

$$\mathbf{g}(\mathbf{f}(\mathbf{X})) - \mathbf{g}(\mathbf{f}(\mathbf{X}_0)) = (d_{\mathbf{U}_0}\mathbf{g})(\mathbf{f}(\mathbf{X}) - \mathbf{f}(\mathbf{X}_0)) + |\mathbf{f}(\mathbf{X}) - \mathbf{f}(\mathbf{X}_0)|\,\mathbf{E}_1(\mathbf{f}(\mathbf{X})),$$

and substituting (3) into this yields

$$\mathbf{g}(\mathbf{f}(\mathbf{X})) - \mathbf{g}(\mathbf{f}(\mathbf{X}_0)) = (d_{\mathbf{U}_0}\mathbf{g})((d_{\mathbf{X}_0}\mathbf{f})(\mathbf{X} - \mathbf{X}_0) + |\mathbf{X} - \mathbf{X}_0|\,\mathbf{E}_2(\mathbf{X}))$$
$$+ |(d_{\mathbf{X}_0}\mathbf{f})(\mathbf{X} - \mathbf{X}_0) + |(\mathbf{X} - \mathbf{X}_0)|\,\mathbf{E}_2(\mathbf{X})|\,\mathbf{E}_1(\mathbf{f}(\mathbf{X})).$$

Defining

$$\mathbf{E}_3(\mathbf{X}) = \mathbf{g}(\mathbf{f}(\mathbf{X})) - \mathbf{g}(\mathbf{f}(\mathbf{X}_0)) - (d_{\mathbf{U}_0}\mathbf{g})((d_{\mathbf{X}_0}\mathbf{f})(\mathbf{X} - \mathbf{X}_0)),$$

and using the linearity of $d_{\mathbf{U}_0}\mathbf{g}$, we find that

$$(5) \qquad \mathbf{E}_3(\mathbf{X}) = |\mathbf{X} - \mathbf{X}_0|\,(d_{\mathbf{U}_0}\mathbf{g})(\mathbf{E}_2(\mathbf{X}))$$
$$+ |(d_{\mathbf{X}_0}\mathbf{f})(\mathbf{X} - \mathbf{X}_0) + |\mathbf{X} - \mathbf{X}_0|\,\mathbf{E}_2(\mathbf{X})|\,\mathbf{E}_1(\mathbf{f}(\mathbf{X})).$$

There is a constant M such that

$$| (d_{\mathbf{X}_0}\mathbf{f}) (\mathbf{X} - \mathbf{X}_0) | < M | \mathbf{X} - \mathbf{X}_0 |$$

(Exercise T-2); hence from (5),

(6) $$\frac{| \mathbf{E}_3(\mathbf{X}) |}{| \mathbf{X} - \mathbf{X}_0 |} \leq | (d_{\mathbf{U}_0}\mathbf{g}) (\mathbf{E}_2(\mathbf{X})) | + \left| M + | \mathbf{E}_2(\mathbf{X}) | \right| | \mathbf{E}_1(\mathbf{f}(\mathbf{X})) |.$$

The right side approaches zero as $\mathbf{X}$ approaches $\mathbf{X}_0$ (Exercise T-3); hence

$$\lim_{\mathbf{X} \to \mathbf{X}_0} \frac{\mathbf{g} \circ \mathbf{f}(\mathbf{X}) - \mathbf{g} \circ \mathbf{f}(\mathbf{X}_0) - (d_{\mathbf{U}_0}\mathbf{g} \circ d_{\mathbf{X}_0}\mathbf{f}) (\mathbf{X} - \mathbf{X}_0)}{| \mathbf{X} - \mathbf{X}_0 |} = \mathbf{0}.$$

Since $d_{\mathbf{U}_0}\mathbf{g} \circ d_{\mathbf{X}_0}\mathbf{f}$ is a linear transformation, $\mathbf{h}$ is differentiable at $\mathbf{X}_0$ and (1) follows. This completes the proof.

In Example 1.6, we saw that the matrix of the composition of two linear transformations is the product of their matrices; hence from (1)

(7) $$\mathbf{J}_{\mathbf{X}_0}\mathbf{h} = (\mathbf{J}_{\mathbf{U}_0}\mathbf{g}) (\mathbf{J}_{\mathbf{X}_0}\mathbf{f}),$$

so that, from the definition of matrix multiplication,

(8) $$\frac{\partial h_i}{\partial x_j} (\mathbf{X}_0) = \sum_{r=1}^{m} \frac{\partial g_i}{\partial u_r} (\mathbf{f}(\mathbf{X}_0)) \frac{\partial f_r}{\partial x_j} (\mathbf{X}_0).$$

When it is not important to emphasize a particular point $\mathbf{X}_0$, we shall simplify the notation by rewriting the last two equations as

$$\mathbf{J}\mathbf{h} = (\mathbf{J}\mathbf{g}) (\mathbf{J}\mathbf{f})$$

and

$$\frac{\partial h_i}{\partial x_j} = \sum_{r=1}^{m} \frac{\partial g_i}{\partial u_r} \frac{\partial f_r}{\partial x_j} ;$$

however, it must be remembered, when calculating $(\partial h_i/\partial x_j) (\mathbf{X})$, that the argument of $\partial g_i/\partial u_r$ is the vector $\mathbf{f}(\mathbf{X})$. To further simplify the formula for $\partial h_i/\partial x_j$, we shall often dispense with the symbol $\mathbf{f}$ entirely and write $\mathbf{f} = \mathbf{U} = \mathbf{U}(\mathbf{X})$ (a natural extension of the informal notation introduced in Section 2.6); we then have

$$\frac{\partial h_i}{\partial x_j} (\mathbf{X}) = \sum_{r=1}^{m} \frac{\partial g_i}{\partial u_r} (\mathbf{U}(\mathbf{X})) \frac{\partial u_r}{\partial x_j} (\mathbf{X}),$$

or simply

$$\frac{\partial h_i}{\partial x_j} = \sum_{r=1}^{m} \frac{\partial g_i}{\partial u_r} \frac{\partial u_r}{\partial x_j}.$$

Then (1) becomes

$$d_{X_0}h = d_{U_0}g \circ d_{X_0}U,$$

(where $U_0 = U(X_0)$), or simply

$$dh = dg \circ dU.$$

The simplified form of (7) is

$$Jh = (Jg)(JU).$$

Example 3.1   Let $X_0 = (1, 0, -1)$,

$$U = \begin{bmatrix} u \\ v \end{bmatrix} = \begin{bmatrix} \sqrt{x^2 + y^2 + z^2} \\ \sqrt{x^2 - y^2} \end{bmatrix}$$

and

$$g(u, v) = \begin{bmatrix} u^2 \\ u^2 + v^2 \\ v^2 \end{bmatrix}.$$

In this case h can be conveniently expressed directly in terms of (x, y, z):

$$h(x, y, z) = g(u(x, y, z), v(x, y, z))$$

$$= \begin{bmatrix} x^2 + y^2 + z^2 \\ 2x^2 + z^2 \\ x^2 - y^2 \end{bmatrix},$$

and

$$Jh = \begin{bmatrix} \dfrac{\partial h_1}{\partial x} & \dfrac{\partial h_1}{\partial y} & \dfrac{\partial h_1}{\partial z} \\[2mm] \dfrac{\partial h_2}{\partial x} & \dfrac{\partial h_2}{\partial y} & \dfrac{\partial h_2}{\partial z} \\[2mm] \dfrac{\partial h_3}{\partial x} & \dfrac{\partial h_3}{\partial y} & \dfrac{\partial h_3}{\partial z} \end{bmatrix} = \begin{bmatrix} 2x & 2y & 2z \\ 4x & 0 & 2z \\ 2x & -2y & 0 \end{bmatrix}.$$

Hence

$$\mathbf{J}_{\mathbf{x}_0}\mathbf{h} = \begin{bmatrix} 2 & 0 & -2 \\ 4 & 0 & -2 \\ 2 & 0 & 0 \end{bmatrix}.$$

Now let us compute $\mathbf{Jh}$ by the chain rule:

$$\mathbf{JU} = \begin{bmatrix} \dfrac{\partial u}{\partial x} & \dfrac{\partial u}{\partial y} & \dfrac{\partial u}{\partial z} \\[2mm] \dfrac{\partial v}{\partial x} & \dfrac{\partial v}{\partial y} & \dfrac{\partial v}{\partial z} \end{bmatrix}$$

$$= \begin{bmatrix} \dfrac{x}{\sqrt{x^2 + y^2 + z^2}} & \dfrac{y}{\sqrt{x^2 + y^2 + z^2}} & \dfrac{z}{\sqrt{x^2 + y^2 + z^2}} \\[4mm] \dfrac{x}{\sqrt{x^2 + y^2}} & \dfrac{-y}{\sqrt{x^2 - y^2}} & 0 \end{bmatrix}$$

and

$$\mathbf{Jg} = \begin{bmatrix} \dfrac{\partial g_1}{\partial u} & \dfrac{\partial g_1}{\partial v} \\[2mm] \dfrac{\partial g_2}{\partial u} & \dfrac{\partial g_2}{\partial v} \\[2mm] \dfrac{\partial g_3}{\partial u} & \dfrac{\partial g_3}{\partial v} \end{bmatrix} = \begin{bmatrix} 2u & 0 \\ 2u & 2v \\ 0 & 2v \end{bmatrix}.$$

Now, $\mathbf{U}_0 = (\sqrt{2}, 1)$ and

$$\mathbf{J}_{\mathbf{x}_0}\mathbf{h} = \mathbf{J}_{\mathbf{U}_0}\mathbf{g}\,\mathbf{J}_{\mathbf{x}_0}\mathbf{U} = \begin{bmatrix} 2\sqrt{2} & 0 \\ 2\sqrt{2} & 2 \\ 0 & 2 \end{bmatrix} \begin{bmatrix} \dfrac{1}{\sqrt{2}} & 0 & -\dfrac{1}{\sqrt{2}} \\ 1 & 0 & 0 \end{bmatrix} = \begin{bmatrix} 2 & 0 & -2 \\ 4 & 0 & -2 \\ 2 & 0 & 0 \end{bmatrix},$$

which agrees with the previous result.

The differential of $\mathbf{h}$ at $\mathbf{X}_0$ is

$$
\mathbf{d}_{\mathbf{X}_0}\mathbf{h} = \begin{bmatrix} 2\,dx - 2\,dz \\ 4\,dx - 2\,dz \\ 2\,dx \end{bmatrix}.
$$

For simplicity we take g to be real valued in the following examples. The results are easily generalized to vector valued functions.

Example 3.2 Let g be a real valued function of $\mathbf{U} = (u_1,\ldots,u_m)$, which is in turn a function of a single variable x, and

$$
h(x) = g(u_1(x),\ldots,u_m(x)).
$$

Then

$$
J\mathbf{U} = \begin{bmatrix} \dfrac{du_1}{dx} \\[2mm] \dfrac{du_2}{dx} \\ \vdots \\ \dfrac{du_m}{dx} \end{bmatrix}
\qquad \text{and} \qquad
J\mathbf{g} = \begin{bmatrix} \dfrac{\partial g}{\partial u_1} & \dfrac{\partial g}{\partial u_2} & \cdots & \dfrac{\partial g}{\partial u_m} \end{bmatrix};
$$

hence

$$
J\mathbf{h} = \frac{dh}{dx} = \begin{bmatrix} \dfrac{\partial g}{\partial u_1} & \dfrac{\partial g}{\partial u_2} & \cdots & \dfrac{\partial g}{\partial u_m} \end{bmatrix} \begin{bmatrix} \dfrac{du_1}{dx} \\[2mm] \dfrac{du_2}{dx} \\ \vdots \\ \dfrac{du_m}{dx} \end{bmatrix}
$$

$$
= \frac{\partial g_1}{\partial u_1}\frac{du_1}{dx} + \cdots + \frac{\partial g_m}{\partial u_m}\frac{du_m}{dx},
$$

which, aside from differences in notation, agrees with the result of Theorem 4.3, Section 2.4.

Example 3.3   Let g be a real valued function of (x, y).

$$\begin{bmatrix} x \\ y \end{bmatrix} = \begin{bmatrix} x(u, v) \\ y(u, v) \end{bmatrix}$$

and

$$h(u, v) = g(x(u, v), y(u, v)).$$

(Note that x and y are now the intermediate variables, rather than u and v). Then

$$\begin{bmatrix} \dfrac{\partial h}{\partial u} & \dfrac{\partial h}{\partial v} \end{bmatrix} = \begin{bmatrix} \dfrac{\partial g}{\partial x} & \dfrac{\partial g}{\partial y} \end{bmatrix} \begin{bmatrix} \dfrac{\partial x}{\partial u} & \dfrac{\partial x}{\partial v} \\[2ex] \dfrac{\partial y}{\partial u} & \dfrac{\partial y}{\partial v} \end{bmatrix} ;$$

hence

$$\frac{\partial h}{\partial u} = \frac{\partial g}{\partial x}\frac{\partial x}{\partial u} + \frac{\partial g}{\partial y}\frac{\partial y}{\partial u} ,$$

$$\frac{\partial h}{\partial v} = \frac{\partial g}{\partial x}\frac{\partial x}{\partial v} + \frac{\partial g}{\partial y}\frac{\partial y}{\partial v} .$$

Example 3.4   Let (x, y) and (r, θ) be cartesian and polar coordinates in the plane;

$$\begin{bmatrix} x \\ y \end{bmatrix} = \begin{bmatrix} r \cos \theta \\ r \sin \theta \end{bmatrix} .$$

If g is a real valued function of (x, y) and $h(r, \theta) = g(r \cos \theta, r \sin \theta)$, then

$$\begin{bmatrix} \dfrac{\partial h}{\partial r} & \dfrac{\partial h}{\partial \theta} \end{bmatrix} = \begin{bmatrix} \dfrac{\partial g}{\partial x} & \dfrac{\partial g}{\partial y} \end{bmatrix} \begin{bmatrix} \dfrac{\partial x}{\partial r} & \dfrac{\partial x}{\partial \theta} \\[2ex] \dfrac{\partial y}{\partial r} & \dfrac{\partial y}{\partial \theta} \end{bmatrix}$$

$$= \begin{bmatrix} \dfrac{\partial g}{\partial x} & \dfrac{\partial g}{\partial y} \end{bmatrix} \begin{bmatrix} \cos \theta & -r \sin \theta \\[1ex] \sin \theta & r \cos \theta \end{bmatrix} .$$

Hence

$$\frac{\partial h}{\partial r} = \cos\theta\,\frac{\partial g}{\partial x} + \sin\theta\,\frac{\partial g}{\partial y},$$

$$\frac{\partial h}{\partial \theta} = -r\sin\theta\,\frac{\partial g}{\partial x} + r\cos\theta\,\frac{\partial g}{\partial y}.$$

Example 3.5 Let $(x,\,y,\,z)$ and $(\rho,\,\theta,\,z)$ be cartesian and cylindrical coordinates in R^3:

$$\begin{bmatrix} x \\ y \\ z \end{bmatrix} = \begin{bmatrix} \rho\cos\theta \\ \rho\sin\theta \\ z \end{bmatrix}.$$

If g is a real valued function of $(x,\,y,\,z)$ and

$$h(\rho,\,\theta,\,z) = g(\rho\cos\theta,\,\rho\sin\theta,\,z),$$

then

$$\begin{bmatrix} \dfrac{\partial h}{\partial \rho} & \dfrac{\partial h}{\partial \theta} & \dfrac{\partial h}{\partial z} \end{bmatrix} = \begin{bmatrix} \dfrac{\partial g}{\partial x} & \dfrac{\partial g}{\partial y} & \dfrac{\partial g}{\partial z} \end{bmatrix} \begin{bmatrix} \dfrac{\partial x}{\partial \rho} & \dfrac{\partial x}{\partial \theta} & \dfrac{\partial x}{\partial z} \\[2mm] \dfrac{\partial y}{\partial \rho} & \dfrac{\partial y}{\partial \theta} & \dfrac{\partial y}{\partial z} \\[2mm] \dfrac{\partial z}{\partial \rho} & \dfrac{\partial z}{\partial \theta} & \dfrac{\partial z}{\partial z} \end{bmatrix}$$

$$= \begin{bmatrix} \dfrac{\partial g}{\partial x} & \dfrac{\partial g}{\partial y} & \dfrac{\partial g}{\partial z} \end{bmatrix} \begin{bmatrix} \cos\theta & -\rho\sin\theta & 0 \\ \sin\theta & \rho\cos\theta & 0 \\ 0 & 0 & 1 \end{bmatrix};$$

hence

$$\frac{\partial h}{\partial r} = \cos\theta\,\frac{\partial g}{\partial x} + \sin\theta\,\frac{\partial g}{\partial y},$$

$$\frac{\partial h}{\partial \theta} = -\rho\sin\theta\,\frac{\partial g}{\partial x} + \rho\cos\theta\,\frac{\partial g}{\partial y},$$

$$\frac{\partial h}{\partial z} = \frac{\partial g}{\partial z}.$$

Example 3.6 Let (r, θ, ϕ) be spherical coordinates in R^3; then

$$\begin{bmatrix} x \\ y \\ z \end{bmatrix} = \begin{bmatrix} r \cos \theta \sin \phi \\ r \sin \theta \sin \phi \\ r \cos \phi \end{bmatrix}$$

If g is a real valued function of (x, y, z) and $h(r, \theta, \phi) = g(r \cos \theta \sin \phi, r \sin \theta \sin \phi, r \cos \phi)$, then

$$\begin{bmatrix} \dfrac{\partial h}{\partial r} & \dfrac{\partial h}{\partial \theta} & \dfrac{\partial h}{\partial \phi} \end{bmatrix} = \begin{bmatrix} \dfrac{\partial g}{\partial x} & \dfrac{\partial g}{\partial y} & \dfrac{\partial g}{\partial z} \end{bmatrix} \begin{bmatrix} \cos \theta \sin \phi & -r \sin \theta \sin \phi & r \cos \theta \cos \phi \\ \sin \theta \sin \phi & r \cos \theta \sin \phi & r \cos \theta \cos \phi \\ \cos \phi & 0 & -r \sin \phi \end{bmatrix}.$$

It may happen that some of the variables $x_1, \ldots, x_n$ are functions of the others, as in Example 4.3 of Section 2.4 and the following example.

Example 3.7 Let g be a real valued function of (x, y, z, w),

$$\begin{bmatrix} x \\ y \end{bmatrix} = \begin{bmatrix} x(z, w) \\ y(z, w) \end{bmatrix}$$

and

$$h(z, w) = g(x(z, w), y(z, w), z, w);$$

then

$$\mathbf{U} = \begin{bmatrix} x(z, w) \\ y(z, w) \\ z \\ w \end{bmatrix}$$

and

$$\begin{bmatrix} \dfrac{\partial h}{\partial z} & \dfrac{\partial h}{\partial w} \end{bmatrix} = \begin{bmatrix} \dfrac{\partial g}{\partial x} & \dfrac{\partial g}{\partial y} & \dfrac{\partial g}{\partial z} & \dfrac{\partial g}{\partial w} \end{bmatrix} \begin{bmatrix} \dfrac{\partial x}{\partial z} & \dfrac{\partial x}{\partial w} \\ \dfrac{\partial y}{\partial z} & \dfrac{\partial y}{\partial w} \\ 1 & 0 \\ 0 & 1 \end{bmatrix}.$$

Thus

$$\frac{\partial h}{\partial z} = \frac{\partial g}{\partial x}\frac{\partial x}{\partial z} + \frac{\partial g}{\partial y}\frac{\partial y}{\partial z} + \frac{\partial g}{\partial z},$$

$$\frac{\partial h}{\partial w} = \frac{\partial g}{\partial x}\frac{\partial x}{\partial w} + \frac{\partial g}{\partial y}\frac{\partial y}{\partial w} + \frac{\partial g}{\partial w}.$$

Second derivatives of composite functions can be computed by repetition of the chain rule, as in the following examples.

Example 3.8 In Example 3.3 we found that

$$\frac{\partial h}{\partial u} = \frac{\partial g}{\partial x}\frac{\partial x}{\partial u} + \frac{\partial g}{\partial y}\frac{\partial y}{\partial u};$$

hence

$$\frac{\partial^2 h}{\partial u^2} = \frac{\partial}{\partial u}\left(\frac{\partial h}{\partial u}\right) = \frac{\partial}{\partial u}\left(\frac{\partial g}{\partial x}\frac{\partial x}{\partial u}\right) + \frac{\partial}{\partial u}\left(\frac{\partial g}{\partial y}\frac{\partial y}{\partial u}\right)$$

$$= \frac{\partial x}{\partial u}\frac{\partial}{\partial u}\left(\frac{\partial g}{\partial x}\right) + \frac{\partial g}{\partial x}\frac{\partial^2 x}{\partial u^2} + \frac{\partial y}{\partial u}\frac{\partial}{\partial u}\left(\frac{\partial g}{\partial y}\right) + \frac{\partial g}{\partial y}\frac{\partial^2 y}{\partial u^2}.$$

In calculating $(\partial/\partial u)(\partial g/\partial x)$ we must recall that we are actually after

$$\frac{\partial}{\partial u}\left(\frac{\partial g}{\partial x}\left[(x(u,v), y(u,z))\right]\right);$$

the safest procedure is to define

$$g_1(x, y) = \frac{\partial g}{\partial x}(x, y) \text{ and } h_1(u, v) = g_1(x(u,v), y(u,v)),$$

and apply the chain rule to h_1:

$$\frac{\partial h_1}{\partial u} = \frac{\partial g_1}{\partial x}\frac{\partial x}{\partial u} + \frac{\partial g_1}{\partial y}\frac{\partial y}{\partial u}.$$

But

$$\frac{\partial h_1}{\partial u} = \frac{\partial}{\partial u}\left(\frac{\partial g}{\partial x}\right), \qquad \frac{\partial g_1}{\partial x} = \frac{\partial^2 g}{\partial x^2}, \qquad \text{and} \qquad \frac{\partial g_1}{\partial y} = \frac{\partial^2 g}{\partial x \partial y};$$

hence

$$\frac{\partial}{\partial u}\left(\frac{\partial g}{\partial x}\right) = \frac{\partial^2 g}{\partial x^2}\frac{\partial x}{\partial u} + \frac{\partial^2 g}{\partial x \partial y}\frac{\partial y}{\partial u}.$$

Similarly,

$$\frac{\partial}{\partial u}\left(\frac{\partial g}{\partial y}\right) = \frac{\partial^2 g}{\partial x\, \partial y}\frac{\partial x}{\partial u} + \frac{\partial^2 g}{\partial y^2}\frac{\partial y}{\partial u};$$

hence

$$\frac{\partial^2 h}{\partial u^2} = \frac{\partial^2 g}{\partial x^2}\left(\frac{\partial x}{\partial u}\right)^2 + 2\frac{\partial^2 g}{\partial x\, \partial y}\frac{\partial x}{\partial u}\frac{\partial x}{\partial v} + \frac{\partial^2 g}{\partial y^2}\left(\frac{\partial y}{\partial u}\right)^2 + \frac{\partial g}{\partial x}\frac{\partial^2 x}{\partial u^2} + \frac{\partial g}{\partial y}\frac{\partial^2 y}{\partial u^2}.$$

Again we emphasize that all partial derivatives of g are to be evaluated at $(x, y) = (x(u, v), y(u, v))$.

Example 3.9 In Example 3.4,

$$(9) \quad \frac{\partial^2 h}{\partial \theta\, \partial r} = \frac{\partial}{\partial \theta}\left(\frac{\partial h}{\partial r}\right) = \frac{\partial}{\partial \theta}\left(\cos \theta\, \frac{\partial g}{\partial x} + \sin \theta\, \frac{\partial g}{\partial y}\right)$$

$$= -\sin \theta\, \frac{\partial g}{\partial x} + \cos \theta\, \frac{\partial g}{\partial y} + \cos \theta\, \frac{\partial}{\partial \theta}\left(\frac{\partial g}{\partial x}\right) + \sin \theta\, \frac{\partial}{\partial \theta}\left(\frac{\partial g}{\partial y}\right).$$

But

$$\frac{\partial}{\partial \theta}\left(\frac{\partial g}{\partial x}\right) = \frac{\partial^2 g}{\partial x^2}\frac{\partial x}{\partial \theta} + \frac{\partial^2 g}{\partial x\, \partial y}\frac{\partial y}{\partial \theta}$$

$$= -r \sin \theta\, \frac{\partial^2 g}{\partial x^2} + r \cos \theta\, \frac{\partial^2 g}{\partial x\, \partial y}$$

and

$$\frac{\partial}{\partial \theta}\left(\frac{\partial g}{\partial y}\right) = \frac{\partial^2 g}{\partial x\, \partial y}\frac{\partial x}{\partial \theta} + \frac{\partial^2 g}{\partial y^2}\frac{\partial y}{\partial \theta}$$

$$= -r \sin \theta\, \frac{\partial^2 g}{\partial x\, \partial y} + r \cos \theta\, \frac{\partial^2 g}{\partial y^2}.$$

Substituting these in (9) yields

$$\frac{\partial^2 h}{\partial \theta\, \partial r} = -\sin \theta\, \frac{\partial g}{\partial x} + \cos \theta\, \frac{\partial g}{\partial y} + r(\cos^2 \theta - \sin^2 \theta)\frac{\partial^2 g}{\partial x\, \partial y}$$

$$+ r \sin \theta \cos \theta \left(\frac{\partial^2 g}{\partial y^2} - \frac{\partial^2 g}{\partial x^2}\right).$$

Example 3.10 In Example 3.7,

$$(10) \qquad \frac{\partial^2 h}{\partial z \, \partial w} = \frac{\partial}{\partial z}\left(\frac{\partial h}{\partial w}\right) = \frac{\partial}{\partial z}\left(\frac{\partial g}{\partial x}\frac{\partial x}{\partial w} + \frac{\partial g}{\partial y}\frac{\partial y}{\partial w} + \frac{\partial g}{\partial w}\right)$$

$$= \frac{\partial x}{\partial w}\frac{\partial}{\partial z}\left(\frac{\partial g}{\partial x}\right) + \frac{\partial g}{\partial x}\frac{\partial^2 x}{\partial z \, \partial w}$$

$$+ \frac{\partial y}{\partial w}\frac{\partial}{\partial z}\left(\frac{\partial g}{\partial y}\right) + \frac{\partial g}{\partial y}\frac{\partial^2 y}{\partial z \, \partial w} + \frac{\partial}{\partial z}\left(\frac{\partial g}{\partial w}\right).$$

To obtain $(\partial/\partial z)(\partial g/\partial x)$, we first define $g_1(x, y) = (\partial g/\partial x)(x, y)$ and $h_1(z, w) = g_1(x(z, w), y(z, w), z, w)$; then

$$\frac{\partial}{\partial z}\left(\frac{\partial g}{\partial x}\right) = \frac{\partial h_1}{\partial z} = \frac{\partial g_1}{\partial x}\frac{\partial x}{\partial z} + \frac{\partial g_1}{\partial y}\frac{\partial y}{\partial z} + \frac{\partial g_1}{\partial z}$$

$$= \frac{\partial^2 g}{\partial x^2}\frac{\partial x}{\partial z} + \frac{\partial^2 g}{\partial x \, \partial y}\frac{\partial y}{\partial z} + \frac{\partial^2 g}{\partial x \, \partial z}.$$

Similarly,

$$\frac{\partial}{\partial z}\left(\frac{\partial g}{\partial y}\right) = \frac{\partial^2 g}{\partial x \, \partial y}\frac{\partial x}{\partial z} + \frac{\partial^2 g}{\partial y^2}\frac{\partial y}{\partial z} + \frac{\partial^2 g}{\partial y \, \partial z}$$

and

$$\frac{\partial}{\partial z}\left(\frac{\partial g}{\partial w}\right) = \frac{\partial^2 g}{\partial x \, \partial w}\frac{\partial x}{\partial z} + \frac{\partial^2 g}{\partial y \, \partial w}\frac{\partial y}{\partial z} + \frac{\partial^2 g}{\partial z \, \partial w}.$$

Substitution of these last three expressions into (10) yields $\partial^2 h/\partial z \, \partial w$.

After attaining proficiency with the chain rule, the student will probably prefer to dispense with the explicit introduction of intermediate functions such as g_1 in the last example.

EXERCISES 3.3

1. Find $\mathbf{h} = \mathbf{g} \circ \mathbf{f}$ explicitly in terms of (x, y, z) and compute $\mathbf{d_{X_0}h}$; then find $\mathbf{d_{X_0}h}$ by the chain rule.

$$\text{(a)} \quad \mathbf{f}(x, y, z) = \begin{bmatrix} 2x - y + z \\ \\ e^{x^2 - y^2} \end{bmatrix}, \quad \mathbf{g}(u, v) = \begin{bmatrix} u + v \\ \\ u - v \end{bmatrix};$$

$$\mathbf{X_0} = (1, 1, -2).$$

(b) $\quad \mathbf{f}(x, y) = \begin{bmatrix} e^x \cos y \\ e^x \sin y \end{bmatrix}, \quad \mathbf{g}(u, v) = \begin{bmatrix} u^2 + v^2 \\ u^2 - v^2 \end{bmatrix};$

$$\mathbf{X}_0 = (0, 0).$$

2. Repeat Exercise 1 for the following functions.

(a) $\quad \mathbf{f}(x, y) = \begin{bmatrix} x + 2y \\ 2x - y^2 \end{bmatrix}, \quad \mathbf{g}(u, v) = \begin{bmatrix} u + 2v \\ u - v^2 \\ u^2 + v \end{bmatrix};$

$$\mathbf{X}_0 = (1, -2).$$

(b) $\quad \mathbf{f}(x, y) = \begin{bmatrix} \log(x + y) \\ \log(x - y) \end{bmatrix}, \quad \mathbf{g}(u, v) = \begin{bmatrix} e^{2v} \\ e^{u+v} \end{bmatrix};$

$$\mathbf{X}_0 = (2, 1).$$

3. Find $\mathbf{J}_{\mathbf{X}_0}\mathbf{f}$, $\mathbf{J}_{\mathbf{f}(\mathbf{X}_0)}\mathbf{g}$, $\mathbf{J}_{\mathbf{f}(\mathbf{X}_0)}\mathbf{g} \cdot \mathbf{J}_{\mathbf{X}_0}\mathbf{f}$ and $\mathbf{J}_{\mathbf{X}_0}\mathbf{h}$ in Exercise 1.

4. Repeat Exercise 3 for the functions in Exercise 2.

5. Let

$$\mathbf{U} = \begin{bmatrix} u \\ v \\ w \end{bmatrix} = \begin{bmatrix} \sqrt{x^2 + y^2 + z^2} \\ \tan^{-1} \dfrac{y}{x} \\ \tan^{-1} \dfrac{z}{\sqrt{x^2 + y^2}} \end{bmatrix} \quad \left(|z| < \frac{\pi}{2}, |w| < \frac{\pi}{2} \right),$$

$$\mathbf{g}(u, v, w) = \begin{bmatrix} uv \\ vw \\ uw \end{bmatrix},$$

and $\mathbf{h} = \mathbf{g} \circ \mathbf{U}$.

(a) Find $\mathbf{JU}$ and $\mathbf{Jg}$.

(b) Find $\mathbf{J}_{\mathbf{X}_0}\mathbf{U}$, $\mathbf{J}_{\mathbf{U}_0}\mathbf{g}$, and $\mathbf{J}_{\mathbf{X}_0}\mathbf{h}$, with $\mathbf{X}_0 = (1/\sqrt{2}, -1/\sqrt{2}, 0)$.

(c) Find $\mathbf{d}_{\mathbf{X}_0}\mathbf{U}$, $\mathbf{d}_{\mathbf{U}_0}\mathbf{g}$ and $\mathbf{d}_{\mathbf{X}_0}\mathbf{h}$.

6. Repeat Exercise 5 with $\mathbf{X}_0 = (2, 1, 2)$,

$$
\mathbf{U} = \begin{bmatrix} x^2 - y^2 \\ 2xy \\ z \end{bmatrix}, \quad \text{and} \quad \mathbf{g}(u, v, w) = \begin{bmatrix} u + w^2 \\ \dfrac{u}{w} \end{bmatrix}.
$$

7. Let $h(t) = f(\mathbf{X}(t))$ and find $h'(t_0)$ and $d_{t_0}h$:

(a) $f(x, y, z) = \sqrt{x^2 + y + z^4}$, $\quad \mathbf{X}(t) = \begin{bmatrix} t \\ t^2 \\ \sqrt{t} \end{bmatrix}$ $\quad (t_0 = 1)$.

(b) $f(x, y, z) = xye^z + 2y^2z - 3xz \cos y$, $\quad \mathbf{X}(t) = \begin{bmatrix} e^t \\ \sin t \\ 1 - \cos t \end{bmatrix}$

$(t_0 = 0)$.

(c) $f(x, y, z) = \log(x + y + z)$, $\quad \mathbf{X}(t) = \begin{bmatrix} t + 2 \\ t^2 \\ -2 \sin \pi t \end{bmatrix}$

$(t_0 = 1)$.

8. If $h(r, \theta) = g(r \cos \theta, r \sin \theta)$, show that

$$
\left(\frac{\partial h}{\partial r}\right)^2 + \frac{1}{r^2}\left(\frac{\partial h}{\partial \theta}\right)^2 = \left(\frac{\partial g}{\partial x}\right)^2 + \left(\frac{\partial g}{\partial y}\right)^2.
$$

9. Let $\mathbf{h}(r, \theta) = \mathbf{g}(r \cos \theta, r \sin \theta)$ and $\mathbf{U}_0 = (r_0, \theta_0) = (2, \pi/4)$. Find $d_{\mathbf{U}_0}\mathbf{h}$:

(a) $g(x, y) = x^2 + y$. $\quad$ (b) $\mathbf{g}(x, y) = \begin{bmatrix} x - y^2 \\ 2x + y \end{bmatrix}$.

(c) $\mathbf{g}(x, y) = \begin{bmatrix} (x + y)^2 \\ x^2 + y^2 \end{bmatrix}$.

10. Let f and g be functions of a single variable and $F(x, y) = f(y + ax) + g(y - ax)$. Show that $F_{xx} = a^2 F_{yy}$.

11. Let $\mathbf{h}(r, \theta, z) = \mathbf{g}(r \cos \theta, r \sin \theta, z)$ and $\mathbf{U}_0 = (r_0, \theta_0, z_0) = (1, \pi/2, -1)$. Find $d_{\mathbf{U}_0}\mathbf{h}$:

(a) $g(x, y, z) = ze^{xy}$. (b) $\mathbf{g}(x, y, z) = \begin{bmatrix} x^3 + y^3 + 2yz \\ x^2 + y^2 + z^2 \end{bmatrix}$.

(c) $\mathbf{g}(x, y, z) = \begin{bmatrix} \dfrac{xy}{z} \\ \dfrac{xz}{y} \end{bmatrix}$.

12. If f is a function of one variable and $h(u, v) = f(u/v)$, show that

$$v \frac{\partial h}{\partial v} + u \frac{\partial h}{\partial u} = 0.$$

13. Let $\mathbf{h}(r, \theta, \phi) = \mathbf{g}(r \cos \theta \sin \phi, r \sin \theta \sin \phi, r \cos \phi)$ and $\mathbf{U}_0 = (r_0, \theta_0, \phi_0) = (1, \pi/4, -\pi/4)$. Find $d_{\mathbf{U}_0}\mathbf{h}$:

(a) $\mathbf{g}(x, y, z) = \begin{bmatrix} \dfrac{x^2 - y^2}{z^2} \\ x^2 + y^2 + z^2 \end{bmatrix}$. (b) $\mathbf{g}(x, y, z) = \begin{bmatrix} x^2 \\ y^2 \\ z^2 \end{bmatrix}$.

(c) $\mathbf{g}(x, y, z) = \dfrac{x}{x + y}$.

14. If $h(x, y) = f(ax + by)$, show that

$$a \frac{\partial h}{\partial y} - b \frac{\partial h}{\partial x} = 0.$$

15. Let $g(x, y, z, w) = \log(1 + x + y + z + w)$,

$$\begin{bmatrix} z \\ w \end{bmatrix} = \begin{bmatrix} x^2 - y^2 \\ 2xy \end{bmatrix}$$

and $h(x, y) = g(x, y, z(x, y), w(x, y))$. Find $d_{(1,1)}h$.

16. Let $g(x, y, z, w) = e^{x+y+z+w}$, $z = z(x, y)$, $w = w(x)$, and $h(x, y) = g(x, y, z(x, y), w(x))$. Find dg.

17. Find $(\partial^2 h/\partial x^2)(1, 1)$, $(\partial^2 h/\partial x\, \partial y)(1, 1)$, and $(\partial^2 h/\partial y^2)(1, 1)$ in Exercise 15.

18. If f is a function of (u, v) and $h(x, y) = f(x + y, x - y)$, show that

$$\frac{\partial h}{\partial x}\frac{\partial h}{\partial y} = \left(\frac{\partial f}{\partial u}\right)^2 - \left(\frac{\partial f}{\partial v}\right)^2.$$

19. Calculate $\partial^2 h/\partial u\, \partial v$ and $\partial^2 h/\partial v^2$ in Example 3.3. (*Hint:* See Example 3.8.)

20. If U is a real valued function of (x, y) and $h(r, \theta) = U(r \cos\theta, r \sin\theta)$, show that

$$U_{xx} + U_{yy} = h_{rr} + \frac{1}{r}h_r + \frac{1}{r^2}h_{\theta\theta}.$$

THEORETICAL EXERCISES

T-1. Prove that $\mathbf{X}_0$ is an interior point of the domain of $\mathbf{g} \circ \mathbf{f}$ in Theorem 3.1. (*Hint:* From the definition of differentiability, $\mathbf{X}_0$ and $\mathbf{U}_0$ are interior points of the domains of $\mathbf{f}$ and $\mathbf{g}$, respectively; also $\mathbf{f}$, being differentiable at $\mathbf{X}_0$, is continuous there.)

T-2. If $\mathbf{L}$ is a linear transformation,

$$\mathbf{L}(\mathbf{X}) = \begin{bmatrix} a_{11}x_1 + \cdots + a_{1n}x_n \\ \vdots \\ a_{m1}x_1 + \cdots + a_{mn}x_n \end{bmatrix},$$

show that $|\, \mathbf{L}(\mathbf{X})\, | < mn\, K\, |\, \mathbf{X}\, |$, where $K = \max |\, a_{ij}\, |$.

T-3. Show that the right side of (6) approaches zero as $\mathbf{X}$ approaches $\mathbf{X}_0$. Use (4), the continuity of $\mathbf{f}$ at $\mathbf{X}_0$ and Theorem 1.5.

T-4. Prove (8) directly, under the assumption that $\mathbf{f}$ has first partial derivatives at $\mathbf{X}_0$ and that $\mathbf{g}$ has continuous first partial derivatives near $\mathbf{U}_0 = \mathbf{f}(\mathbf{X}_0)$. (*Hint:* See the proof of Theorem 4.3, Section 2.4.)

3.4 Vector and Scalar Fields

Vector functions of the form $\mathbf{F}: R^n \to R^n$ occur in so many applications that we give them a special name.

Definition 4.1 A *vector field* is a vector valued function **F** whose domain and range are both subsets of R^n.

Example 4.1 Consider a fluid flowing in a channel in R^3 with velocity at each point **X** depending on **X** alone, and not on time. (We say that the flow is *steady*.) With each point **X** of the channel we associate the velocity **V(X)** of the fluid at that point; the vector function so defined is a vector field. We can depict the vector field by drawing the velocity vector **V(X)** at each **X** in the channel (Fig. 4.1).

FIGURE 4.1

Example 4.2 Let a particle P_0 of mass m_0 be fixed at the origin of R^3. According to Newton's law of gravitation the force exerted by P_0 on a unit mass at an arbitrary point $\mathbf{X} \neq \mathbf{0}$ is given by

$$\mathbf{F(X)} = -\frac{Gm_0}{|\mathbf{X}|^3}\mathbf{X},$$

where G is a universal constant. Thus **F** is a continuous vector field for all $\mathbf{X} \neq \mathbf{0}$. The direction of **F(X)** is toward the origin and its magnitude varies inversely with $|\mathbf{X}|^2$; **F** is an *inverse square law* force field.

The following definition introduces a convenient terminology for a familiar concept.

Definition 4.2 .A real valued function $f: R^n \to R$ is called a *scalar field*.

Example 4.3 In connection with the inverse square law force field of Example 4.2, we may associate with each $\mathbf{X} \neq \mathbf{0}$ the real number

$$f(\mathbf{X}) = |\mathbf{F(X)}| = \frac{Gm_0}{|\mathbf{X}|^2};$$

then f is a scalar field.

The Gradient as a Vector Field

In Section 2.5 we defined the gradient of a real valued function $f \colon R^n \to R$ at a point $\mathbf{X}_0$ to be the vector

$$
\nabla_{\mathbf{X}_0} f =
\begin{bmatrix}
\dfrac{\partial f}{\partial x_1}(\mathbf{X}_0) \\[2ex]
\dfrac{\partial f}{\partial x_2}(\mathbf{X}_0) \\[2ex]
\vdots \\[2ex]
\dfrac{\partial f}{\partial x_n}(\mathbf{X}_0)
\end{bmatrix},
$$

or, if $n = 3$,

$$
\nabla_{\mathbf{X}_0} f = \frac{\partial f}{\partial x}(\mathbf{X}_0)\,\mathbf{i} + \frac{\partial f}{\partial y}(\mathbf{X}_0)\,\mathbf{j} + \frac{\partial f}{\partial z}(\mathbf{X}_0)\,\mathbf{k},
$$

provided the indicated partial derivatives exist. If f is a scalar field which has partial derivatives at every point $\mathbf{X}$ of a domain D we can define the vector field ∇f, called the *gradient of f*, to be the vector field whose value at each $\mathbf{X}$ in D is $\nabla_{\mathbf{X}} f$; thus

$$
\nabla f =
\begin{bmatrix}
\dfrac{\partial f}{\partial x_1} \\[2ex]
\dfrac{\partial f}{\partial x_2} \\[2ex]
\vdots \\[2ex]
\dfrac{\partial f}{\partial x_n}
\end{bmatrix}
$$

or, if $n = 3$,

$$
\nabla f = \frac{\partial f}{\partial x}\mathbf{i} + \frac{\partial f}{\partial y}\mathbf{j} + \frac{\partial f}{\partial z}\mathbf{k}. \tag{1}
$$

Example 4.4 Let f be the scalar field defined by $f(x,\,y,\,z) = 3x^2yz$. Then

$$
\nabla f = 6xyz\mathbf{i} + 3x^2z\mathbf{j} + 3x^2y\mathbf{k}.
$$

Example 4.5 The scalar field f defined by

$$f(\mathbf{X}) = \frac{Gm_0}{|\mathbf{X}|} = \frac{Gm_0}{\sqrt{x^2 + y^2 + z^2}}$$

is defined for all $\mathbf{X} \neq \mathbf{0}$ in R^3. Since

$$f_x(\mathbf{X}) = -\frac{Gm_0 x}{|\mathbf{X}|^3}, \qquad f_y(\mathbf{X}) = -\frac{Gm_0 y}{|\mathbf{X}|^3}, \qquad f_z(\mathbf{X}) = -\frac{Gm_0 z}{|\mathbf{X}|^3},$$

it follows that the inverse square law force field of Example 4.2 is the gradient of f.

If a vector field $\mathbf{F}$ can be expressed as the gradient of a scalar field f,

$$\mathbf{F} = \nabla f,$$

then $\mathbf{F}$ is said to be *conservative*, and $-f$ is called the *potential function* of $\mathbf{F}$. We shall encounter these ideas again in Section 4.5.

The next theorem follows from Exercise T-2, Section 2.7.

Theorem 4.1 Let f and g be differentiable scalar fields in a domain D in R^n and let a be a real number. Then

$$\text{(a)} \qquad \nabla(af) = a \nabla f,$$

$$\text{(b)} \quad \nabla(f + g) = \nabla f + \nabla g,$$

and

$$\text{(c)} \qquad \nabla(fg) = f \nabla g + g \nabla f$$

on D. Moreover $\nabla(f/g)$ is defined on the subset D' of D where $g(\mathbf{X}) \neq 0$, and

$$\text{(d)} \qquad \nabla(f/g) = \frac{g \nabla f - f \nabla g}{g^2}.$$

If $n = 3$ we may view ∇f as the vector field obtained by applying the operator

$$\nabla = \frac{\partial}{\partial x}\mathbf{i} + \frac{\partial}{\partial y}\mathbf{j} + \frac{\partial}{\partial z}\mathbf{k}$$

to f. Thus we write (1) as

$$\nabla f = \left(\frac{\partial}{\partial x}\mathbf{i} + \frac{\partial}{\partial y}\mathbf{j} + \frac{\partial}{\partial z}\mathbf{k}\right)f,$$

which suggests that we are multiplying the "vector" ∇ by a scalar f. Of course this is not actually the case, but it is sometimes useful to think of ∇ as a vector (which it is not!), especially in connection with the divergence and curl, defined below. The operator ∇ is called the "del" operator.

Divergence of a Vector Field

Throughout the rest of this section we consider only scalar and vector fields in R^3, except where it is specifically stated otherwise.

Definition 4.3 Let $\mathbf{F} = F_1\mathbf{i} + F_2\mathbf{j} + F_3\mathbf{k}$ be a differentiable vector field defined on a domain D in R^3. The *divergence of* $\mathbf{F}$, denoted by div $\mathbf{F}$, is the scalar field defined by

$$(2) \qquad\qquad \text{div } \mathbf{F} = \frac{\partial F_1}{\partial x} + \frac{\partial F_2}{\partial y} + \frac{\partial F_3}{\partial z};$$

its value at a point $\mathbf{X}_0$ in D is

$$\text{div } \mathbf{F}(\mathbf{X}_0) = \frac{\partial F_1}{\partial x}(\mathbf{X}_0) + \frac{\partial F_2}{\partial y}(\mathbf{X}_0) + \frac{\partial F_3}{\partial z}(\mathbf{X}_0).$$

Example 4.6 If

$$\mathbf{F}(\mathbf{X}) = x\mathbf{i} + (xz + 1)\mathbf{j} + (yz + x)\mathbf{k}$$

then

$$\text{div } \mathbf{F}(\mathbf{X}) = (1) + (0) + (y) = 1 + y;$$

thus if $\mathbf{X}_0 = (2, -2, 3)$, then

$$\text{div } \mathbf{F}(\mathbf{X}_0) = 1 - 2 = -1.$$

It is sometimes useful to view the right side of (2) as the "inner product" of ∇ and $\mathbf{F}$; thus

$$\nabla \cdot \mathbf{F} = \left(\frac{\partial}{\partial x}\mathbf{i} + \frac{\partial}{\partial y}\mathbf{j} + \frac{\partial}{\partial z}\mathbf{k}\right) \cdot (F_1\mathbf{i} + F_2\mathbf{j} + F_3\mathbf{k}) = \text{div } \mathbf{F}.$$

However, caution is in order here. The inner product of two vectors is commutative ($\mathbf{U} \cdot \mathbf{V} = \mathbf{V} \cdot \mathbf{U}$), but $\nabla \cdot \mathbf{F}$ and $\mathbf{F} \cdot \nabla$ are different: $\nabla \cdot \mathbf{F}$ is a scalar field, while $\mathbf{F} \cdot \nabla$ is the operator

$$\mathbf{F} \cdot \nabla = F_1\frac{\partial}{\partial x} + F_2\frac{\partial}{\partial y} + F_3\frac{\partial}{\partial z}.$$

Definition 4.3 appears to be quite arbitrary. Nevertheless, the divergence of a vector field has important geometrical and physical applications which we shall explore further in Chapter 4. For now we shall simply show that although div $\mathbf{F}$ is defined in terms of a particular coordinate system, its value at a given point is determined by the point itself, and not by the particular rectangular coordinate system chosen to identify points.

To see this, let us recall the viewpoint discussed in Section 1.6, where we distinguished between a point P in R^3 and its coordinates (x, y, z) with respect to a particular rectangular coordinate system. Suppose (x_1, x_2, x_3) and (y_1, y_2, y_3) are the coordinates of the same point P with respect to two rectangular systems. Denote the unit vectors along the (x_1, x_2, x_3) axes by $\mathbf{i}_1$, $\mathbf{i}_2$, and $\mathbf{i}_3$, and those along the (y_1, y_2, y_3) axes by $\mathbf{j}_1$, $\mathbf{j}_2$, and $\mathbf{j}_3$. Then we can write

$$(3) \qquad \mathbf{i}_r = \sum_{s=1}^{3} l_{sr}\mathbf{j}_s,$$

where the matrix

$$\mathbf{M} = \begin{bmatrix} l_{11} & l_{12} & l_{13} \\ l_{21} & l_{22} & l_{23} \\ l_{31} & l_{32} & l_{33} \end{bmatrix}$$

is orthogonal; that is

$$(4) \qquad \mathbf{M}^{\mathrm{T}} = \mathbf{M}^{-1}.$$

This result was derived, in different notation, in Section 1.6. It was also shown there that the coordinates of a given point P with respect to the two systems are related by

$$(5) \qquad \begin{bmatrix} y_1 \\ y_2 \\ y_3 \end{bmatrix} = \mathbf{M} \begin{bmatrix} x_1 - x_1^0 \\ x_2 - x_2^0 \\ x_3 - x_3^0 \end{bmatrix},$$

where (x_1^0, x_2^0, x_3^0) are the coordinates of the origin of the (y_1, y_2, y_3) system with respect to the (x_1, x_2, x_3) system.

To emphasize that $\mathbf{F}$ is a function of P let us temporarily denote its values by $\mathbf{F}(P)$ instead of the usual $\mathbf{F}(\mathbf{X})$. Then we can express $\mathbf{F}(P)$ as a linear combination of $\mathbf{i}_1$, $\mathbf{i}_2$, and $\mathbf{i}_3$, and also as a linear combination of $\mathbf{j}_1$, $\mathbf{j}_2$, and $\mathbf{j}_3$:

$$(6) \qquad \mathbf{F}(P) = F_1(x_1, x_2, x_3)\mathbf{i}_1 + F_2(x_1, x_2, x_3)\mathbf{i}_2 + F_3(x_1, x_2, x_3)\mathbf{i}_3$$

$$= G_1(y_1, y_2, y_3)\mathbf{j}_1 + G_2(y_1, y_2, y_3)\mathbf{j}_2 + G_3(y_1, y_2, y_3)\mathbf{j}_3 .$$

By substituting (3) into (6) and equating coefficients of j_1, j_2, and j_3 on both sides of the second equality, we obtain

$$\begin{bmatrix} G_1(y_1,\, y_2,\, y_3) \\[6pt] G_2(y_1,\, y_2,\, y_3) \\[6pt] G_3(y_1,\, y_2,\, y_3) \end{bmatrix} = \mathbf{M} \begin{bmatrix} F_1(x_1,\, x_2,\, x_3) \\[6pt] F_2(x_1,\, x_2,\, x_3) \\[6pt] F_3(x_1,\, x_2,\, x_3) \end{bmatrix},$$

or, in terms of components,

$$(7) \qquad G_r(y_1,\, y_2,\, y_3) = \sum_{s=1}^{3} l_{rs} F_s(x_1,\, x_2,\, x_3) \qquad (r = 1,\, 2,\, 3)$$

(Exercise T-1). Differentiating both sides with respect to y_r yields

$$(8) \qquad \frac{\partial G_r}{\partial y_r} = \sum_{s=1}^{3} l_{rs} \sum_{q=1}^{3} \frac{\partial F_s}{\partial x_q} \frac{\partial x_q}{\partial y_r} \qquad (r = 1,\, 2,\, 3),$$

where the partial derivatives of G_r are to be evaluated at $(y_1,\, y_2,\, y_3)$ and those of F_s at $(x_1,\, x_2,\, x_3)$.

Now (4) and (5) imply that

$$\begin{bmatrix} x_1 - x_1^0 \\[6pt] x_2 - x_2^0 \\[6pt] x_3 - x_3^0 \end{bmatrix} = M^{\mathrm{T}} \begin{bmatrix} y_1 \\[6pt] y_2 \\[6pt] y_3 \end{bmatrix},$$

from which it follows that

$$\frac{\partial x_q}{\partial y_r} = l_{rq}$$

(Exercise T-2). Substituting this in (8) yields

$$\frac{\partial G_r}{\partial y_r} = \sum_{s=1}^{3} l_{rs} \sum_{q=1}^{3} l_{rq} \frac{\partial F_s}{\partial x_q} \qquad (r = 1,\, 2,\, 3).$$

Adding these equations for $r = 1$, 2, and 3 yields

$$\sum_{r=1}^{3} \frac{\partial G_r}{\partial y_r} = \sum_{r=1}^{3} \sum_{s=1}^{3} l_{rs} \sum_{q=1}^{3} l_{rq} \frac{\partial F_s}{\partial x_q}.$$

It is convenient to rewrite the sum on the right so that the summation on r is performed first; thus

$$(9) \qquad \sum_{r=1}^{3} \frac{\partial G_r}{\partial x_r} = \sum_{s=1}^{3} \sum_{q=1}^{3} \left(\sum_{r=1}^{3} l_{rs} l_{rq} \right) \frac{\partial F_s}{\partial x_q}.$$

According to (4), $\mathbf{MM}^{\mathrm{T}} = \mathbf{I}$; thus

$$\sum_{r=1}^{3} l_{rs} l_{rq} = \begin{cases} 0 & \text{if} \quad q \neq s, \\ 1 & \text{if} \quad q = s \end{cases}$$

(Exercise T-3). Therefore (9) reduces to

(10)
$$\sum_{r=1}^{3} \frac{\partial G_r}{\partial x_r} (y_1, y_2, y_3) = \sum_{s=1}^{3} \frac{\partial F_s}{\partial x_s} (x_1, x_2, x_3).$$

We have therefore proved the following theorem.

Theorem 4.2 Let $\mathbf{F}$ be a vector field defined at all points P in a domain D, which is expressed by

(11) $\mathbf{F}(P) = F_1(x_1, x_2, x_3)\mathbf{i}_1 + F_2(x_1, x_2, x_3)\mathbf{i}_2 + F_3(x_1, x_2, x_3)\mathbf{i}_3$

with respect to a given rectangular coordinate system in R^3, where F_1, F_2, and F_3 are differentiable functions of (x_1, x_2, x_3). Furthermore suppose

(12) $\mathbf{F}(P) = G_1(y_1, y_2, y_3)\mathbf{j}_i + G_2(y_1, y_2, y_3)\mathbf{j}_2 + G_3(y_1, y_2, y_3)\mathbf{j}_3$

is a representation of the same vector field with respect to a second rectangular coordinate system. Then div $\mathbf{F}(P)$ can be computed in terms of either coordinate system, since (10) holds whenever (x_1, x_2, x_3) and (y_1, y_2, y_3) are the coordinates of the same point P with respect to the two coordinate systems.

We leave the verification of the following theorem to the student (Exercise T-4).

Theorem 4.3 If a and b are constants, $\mathbf{F}$ and $\mathbf{G}$ differentiable vector fields, and g a differentiable scalar field, then

(a)   div$(a\mathbf{F} + b\mathbf{G}) = a$ div $\mathbf{F} + b$ div $\mathbf{G}$;
(b)   div$(g\mathbf{F}) = g$ div $\mathbf{F} + (\nabla g) \cdot \mathbf{F}$.

The divergence of a vector field

$$\mathbf{F} = \begin{bmatrix} F_1 \\ F_2 \\ \vdots \\ F_n \end{bmatrix} \quad (R^n \to R^n)$$

can be defined for arbitrary $n > 1$ by

$$\text{div } \mathbf{F} = \sum_{i=1}^{n} \frac{\partial F_i}{\partial x_i}.$$

The Laplacian of a Scalar Field

If f is a scalar field with second partial derivatives f_{xx}, f_{yy}, and f_{zz} in a domain D in R^3, then the divergence of ∇f is given by

$$\mathbf{\nabla} \cdot \mathbf{\nabla} f = \left(\frac{\partial}{\partial x} \mathbf{i} + \frac{\partial}{\partial y} \mathbf{j} + \frac{\partial}{\partial z} \mathbf{k} \right) \cdot \left(\frac{\partial f}{\partial x} \mathbf{i} + \frac{\partial f}{\partial y} \mathbf{j} + \frac{\partial f}{\partial z} \mathbf{k} \right)$$

$$= \frac{\partial^2 f}{\partial x^2} + \frac{\partial^2 f}{\partial y^2} + \frac{\partial^2 f}{\partial z^2}.$$

The expression on the right is called the *Laplacian of f* and is denoted more briefly by $\nabla^2 f$. The operator

$$\nabla^2 = \frac{\partial^2}{\partial x^2} + \frac{\partial^2}{\partial y^2} + \frac{\partial^2}{\partial z^2}$$

is called the *Laplacian operator*.

Example 4.7   If

$$f(x, y, z) = x^3 + 2xy^2 + yz^3$$

then

$$\nabla^2 f = 6x + 4x + 6yz.$$

A scalar field f is said to be *harmonic* in a domain D if

(13) $$\nabla^2 f = 0$$

in D; (13) is called *Laplace's equation*. Harmonic functions have many applications in the natural sciences. We shall study them further in Chapter 4.

Example 4.8   If

$$f(x, y, z) = x^2 + y^2 - 2z^2$$

then

$$\nabla^2 f = 2 + 2 - 2(2) = 0;$$

hence f is harmonic.

The Laplacian $\nabla^2 f$ can be shown to be independent of the particular rectangular coordinate system chosen for R^3 (Exercise T-5). It can also be defined for scalar fields in R^n for any $n > 1$ by

$$\nabla^2 f = \sum_{r=1}^{n} \frac{\partial^2 f}{\partial x_r^2}.$$

Curl of a Vector Field

The *curl* of a vector field $\mathbf{F} = F_1 \mathbf{i} + F_2 \mathbf{j} + F_3 \mathbf{k}$ is defined by

$$(14) \quad \text{curl } \mathbf{F} = \left(\frac{\partial F_3}{\partial y} - \frac{\partial F_2}{\partial z} \right) \mathbf{i} + \left(\frac{\partial F_1}{\partial z} - \frac{\partial F_3}{\partial x} \right) \mathbf{j} + \left(\frac{\partial F_2}{\partial x} - \frac{\partial F_1}{\partial y} \right) \mathbf{k}.$$

The value of **curl F** at a point $\mathbf{X}_0$ is denoted by **curl** $\mathbf{F}(\mathbf{X}_0)$. This definition is easily remembered if we write

$$\text{curl } \mathbf{F} = \nabla \times \mathbf{F} = \begin{vmatrix} \mathbf{i} & \mathbf{j} & \mathbf{k} \\ \dfrac{\partial}{\partial x} & \dfrac{\partial}{\partial y} & \dfrac{\partial}{\partial z} \\ F_1 & F_2 & F_3 \end{vmatrix}.$$

Formal expansion of this determinant yields (14).

Example 4.9 If

$$\mathbf{F}(\mathbf{X}) = 3x^2 \mathbf{i} + xy \mathbf{j} + z \mathbf{k}$$

then

$$\text{curl } \mathbf{F} = \begin{vmatrix} \mathbf{i} & \mathbf{j} & \mathbf{k} \\ \dfrac{\partial}{\partial x} & \dfrac{\partial}{\partial y} & \dfrac{\partial}{\partial z} \\ 3x^2 & xy & z \end{vmatrix}$$

$$= \left(\frac{\partial}{\partial y}(z) - \frac{\partial}{\partial z}(xy) \right) \mathbf{i} + \left(\frac{\partial}{\partial z}(3x^2) - \frac{\partial}{\partial x}(z) \right) \mathbf{j}$$

$$+ \left(\frac{\partial}{\partial x}(xy) - \frac{\partial}{\partial y}(3x^2) \right) \mathbf{k} = y\mathbf{k}.$$

Thus if $\mathbf{X}_0 = (2, 3, -4)$ then

$$\mathbf{curl\ F}(\mathbf{X}_0) = -4\mathbf{k}.$$

Example 4.10 If f is a real valued function with continuous second partial derivatives, then

$$(15) \qquad \mathbf{curl}\ (\nabla f) = \begin{vmatrix} \mathbf{i} & \mathbf{j} & \mathbf{k} \\ \dfrac{\partial}{\partial x} & \dfrac{\partial}{\partial y} & \dfrac{\partial}{\partial z} \\ f_x & f_y & f_z \end{vmatrix}$$

$$= (f_{zy} - f_{yz})\mathbf{i} + (f_{xz} - f_{zx})\mathbf{j} + (f_{yx} - f_{xy})\mathbf{k}$$

$$= \mathbf{0},$$

since the mixed derivatives are equal. A vector field $\mathbf{F}$ for which $\mathbf{curl\ F} = \mathbf{0}$ is said to be *irrotational*. Thus we have shown that any vector field which is the gradient of a scalar field with continuous second partial derivatives is irrotational. Notice that (15) can be written formally as

$$\nabla \times (\nabla f) = \mathbf{0}.$$

Example 4.11 If $\mathbf{F}$ has continuous second partial derivatives then

$$(16) \quad \mathrm{div}(\mathbf{curl\ F}) = \frac{\partial}{\partial x}\left(\frac{\partial F_3}{\partial y} - \frac{\partial F_2}{\partial z}\right) + \frac{\partial}{\partial y}\left(\frac{\partial F_1}{\partial z} - \frac{\partial F_3}{\partial x}\right)$$

$$+ \frac{\partial}{\partial z}\left(\frac{\partial F_2}{\partial x} - \frac{\partial F_1}{\partial y}\right)$$

$$= \frac{\partial^2 F_3}{\partial x \partial y} - \frac{\partial^2 F_2}{\partial x \partial z} + \frac{\partial^2 F_1}{\partial y \partial z} - \frac{\partial^2 F_3}{\partial y \partial z} + \frac{\partial^2 F_2}{\partial z \partial x} - \frac{\partial^2 F_1}{\partial z \partial y}$$

$$= 0.$$

Notice that (16) can be written as

$$\nabla \cdot (\nabla \times \mathbf{F}) = 0.$$

The curl of a vector field is independent of the rectangular coordinate system used in its definition (Exercise T-7).

Theorem 4.4   If a and b are constants, **F** and **G** differentiable vector fields, and g a differentiable scalar field, then

(a) $\operatorname{curl}(a\mathbf{F} + b\mathbf{G}) = a\,\operatorname{curl}\mathbf{F} + b\,\operatorname{curl}\mathbf{G}$;

(b) $\operatorname{curl}(g\mathbf{F}) = \operatorname{curl}\mathbf{F} + \nabla g \times \mathbf{F}$.

We leave the proof of this theorem to the student (Exercise T-8).

EXERCISES 3.4

1.   Compute ∇f.

(a) $f(\mathbf{X}) = e^{xy}\sin z$.

(b) $f(\mathbf{X}) = \log |\,\mathbf{X}\,|$.

(c) $f(\mathbf{X}) = \dfrac{x^2 z}{2} + xz$.

2.   Compute ∇f.

(a) $f(\mathbf{X}) = 4x - 2y + z$.

(b) $f(\mathbf{X}) = x \log x + y \log y + z \log z$.

(c) $f(\mathbf{X}) = \cos xyz$.

3.   Verify Theorem 4.1 for $a = 2$, $f(x, y, z) = xyz^2$, and $g(x, y, z) = yz^2$.

4.   Compute $(\operatorname{div}\mathbf{F})(\mathbf{X})$:

(a) $\mathbf{F}(\mathbf{X}) = x^2 y\mathbf{i} + xyz\mathbf{j} + yz^2\mathbf{k}$;

(b) $\mathbf{F}(\mathbf{X}) = e^{xy}\sin z\,\mathbf{i} + e^{xz}\cos y\,\mathbf{j} + e^{yz}\sin x\,\mathbf{k}$;

(c) $\mathbf{F}(\mathbf{X}) = (y + z)x\mathbf{i} + (x - z^2)y^2\mathbf{j} + (y^2 - z^2)\mathbf{k}$;

(d) $\mathbf{F}(\mathbf{X}) = \nabla f$,   where $f(x, y, z) = ye^x - xe^y$;

(e) $\mathbf{F}(\mathbf{X}) = \nabla f$,   where $f(x, y, z) = 4x - 2y + z$.

5.   Compute $(\operatorname{div}\mathbf{F})(\mathbf{X})$:

(a) $\mathbf{F}(\mathbf{X}) = y \log x\mathbf{i} + x \log y\mathbf{j} + xy \log z\mathbf{k}$;

(b) $\mathbf{F}(\mathbf{X}) = 2\mathbf{i} + xy\mathbf{j} + 3\mathbf{k}$;

(c) $\mathbf{F}(\mathbf{X}) = x^3\mathbf{i} + y^2\mathbf{j} + z^3\mathbf{k}$;

(d) $\mathbf{F}(\mathbf{X}) = \nabla f$, where $f(x, y, z) = x^3 + y^3 + z^3$;

(e) $\mathbf{F}(\mathbf{X}) = \nabla f$, where $f(x, y, z) = \cos xyz$.

6. Verify Theorem 4.3 for $a = 2, b = -3$,

$$\mathbf{F}(\mathbf{X}) = x^2\mathbf{i} + y^2 z\mathbf{j} + (y^2 - z^2)\mathbf{k},$$

and

$$\mathbf{G}(\mathbf{X}) = x^3 y\mathbf{i} + (y^2 + x)\mathbf{j} + (y^2 - z^2)\mathbf{k}.$$

7. Compute $\nabla^2 f$.

(a) $f(\mathbf{X}) = ze^{xy}$.

(b) $f(\mathbf{X}) = e^{yz} \cos x$.

(c) $f(\mathbf{X}) = \log(1 + x^2 + y^2 + z^2)$.

(d) $f(\mathbf{X}) = x^2 - y^2 + 2z^2$.

(e) $f(\mathbf{X}) = \sin xyz$.

8. Compute $\nabla^2 f$.

(a) $f(\mathbf{X}) = (x^2 + y^2)e^{xyz}$.

(b) $f(\mathbf{X}) = e^{-(x^2+y^2+z^2)}$.

(c) $f(\mathbf{X}) = \log(\cos x + \sin y)$.

(d) $f(\mathbf{X}) = x^2 - 3xyz + 2xz$.

(e) $f(\mathbf{X}) = y \log x + z \log y$.

9. Show that f is harmonic.

(a) $f(\mathbf{X}) = \dfrac{1}{\sqrt{x^2 + y^2 + z^2}}$.

(b) $f(\mathbf{X}) = x^2 - y^2 + 2z$.

(c) $f(\mathbf{X}) = \sinh 2x \cos y \sin \sqrt{3}\, z$.

(d) $f(\mathbf{X}) = \sin \dfrac{x}{\sqrt{2}} \cos \dfrac{y}{\sqrt{2}} \cosh z.$

(e) $f(\mathbf{X}) = \sinh x \cos \sqrt{2y} \cosh y.$

10.   Compute **curl F**.

(a) $\mathbf{F}(\mathbf{X}) = (x^2 + y^3)\mathbf{i} + (xz^3 + x^2 y)\mathbf{j} + (x^2 + xyz)\mathbf{k}.$

(b) $\mathbf{F}(\mathbf{X}) = \left(x^2 + \dfrac{y^2}{2}\right)\mathbf{i} + \left(xy + \dfrac{z^2}{2}\right)\mathbf{j} + (yz)\mathbf{k}.$

(c) $\mathbf{F}(\mathbf{X}) = e^x y\mathbf{i} + e^y z\mathbf{j} + e^z xy\mathbf{k}.$

(d) $\mathbf{F}(\mathbf{X}) = y \sin x\, \mathbf{i} + z \cos y\, \mathbf{j} + xy \sin z\, \mathbf{k}.$

11.   Compute **curl F**.

(a) $\mathbf{F}(\mathbf{X}) = (xyz + x)\mathbf{i} + \left(\dfrac{x^2 z}{2} + y\right)\mathbf{j} + (xz)\mathbf{k}.$

(b) $\mathbf{F}(\mathbf{X}) = (2x + y + z)\mathbf{i} + (x - y - z)\mathbf{j} + (y - z)\mathbf{k}.$

(c) $\mathbf{F}(\mathbf{X}) = x^2 y e^z \mathbf{i} + xe^y z^2 \mathbf{j} + e^x y^2 z\mathbf{k}.$

(d) $\mathbf{F}(\mathbf{X}) = e^x \sin y\, \mathbf{i} + e^y \cos x\, \mathbf{j} + e^z \cos z\, \mathbf{k}.$

12.   Verify Theorem 4.3 for a, b, **F**, **G**, and g as defined in Example 4.6.

THEORETICAL EXERCISES

T-1.   Verify Eq. (7).

T-2.   If (x_1, x_2, x_3) and (y_1, y_2, y_3) are related by (5), show that

$$\frac{\partial x_q}{\partial y_r} = l_{rq}.$$

T-3.   If

$$\mathbf{M} = \begin{bmatrix} l_{11} & l_{12} & l_{13} \\ l_{21} & l_{22} & l_{23} \\ l_{31} & l_{32} & l_{33} \end{bmatrix}$$

is an orthogonal matrix ($\mathbf{M}^{-1} = \mathbf{M}^{\mathbf{T}}$) show that

$$\sum_{j=1}^{3} l_{jr}l_{js} = \sum_{j=1}^{3} l_{rj}l_{sj} = \begin{cases} 0 & \text{if } r \neq s, \\ 1 & \text{if } r = s. \end{cases}$$

T-4. Prove Theorem 4.3.

T-5. Let f and g be real valued functions of three variables which satisfy

$$f(x_1, x_2, x_3) = g(y_1, y_2, y_3)$$

whenever (x_1, x_2, x_3) and (y_1, y_2, y_3) are related by (5), where $\mathbf{M}$ is an orthogonal matrix. Let f and g have continuous second partial derivatives. Show that

$$\sum_{r=1}^{3} \frac{\partial^2 f}{\partial x_r^2} (x_1, x_2, x_3) = \sum_{s=1}^{3} \frac{\partial^2 g}{\partial y_s^2} (y_1, y_2, y_3).$$

T-6. Under the assumption of Exercise T-5 (except that we need only assume that f and g are differentiable), show that

$$\sum_{r=1}^{3} \frac{\partial f}{\partial x_r} (x_1, x_2, x_3) \, \mathbf{i}_r = \sum_{s=1}^{3} \frac{\partial g}{\partial y_s} (y_1, y_2, y_3) \, \mathbf{j}_s.$$

(*Hint*: Use Eq. (3) of Section 3.4). Use this result to give a precise meaning to the statement: The gradient of a vector field is independent of the particular rectangular coordinate system chosen for R^3.

T-7. If the vector field $\mathbf{F} = \mathbf{F}(P)$ is defined with respect to two rectangular coordinate systems by (11) and (12), show that

$$\begin{vmatrix} \mathbf{i}_1 & \mathbf{i}_2 & \mathbf{i}_3 \\ \dfrac{\partial}{\partial x_1} & \dfrac{\partial}{\partial x_2} & \dfrac{\partial}{\partial x_3} \\ F_1 & F_2 & F_3 \end{vmatrix} = \begin{vmatrix} \mathbf{j}_1 & \mathbf{j}_2 & \mathbf{j}_3 \\ \dfrac{\partial}{\partial y_1} & \dfrac{\partial}{\partial y_2} & \dfrac{\partial}{\partial y_3} \\ G_1 & G_2 & G_3 \end{vmatrix}.$$

(*Hint*: Refer to the proof of Theorem 4.2.) Use this result to give a precise meaning to the statement: The curl of a vector field is independent of the particular rectangular coordinate system chosen for R^3.

T-8. Prove Theorem 4.4.

T-9.   Let $\mathbf{F}$ be defined by

$$\mathbf{F}(\mathbf{X}) \;=\; -\,\frac{Gm_0}{|\mathbf{X}|^3}\,\mathbf{X}.$$

Show that div $\mathbf{F} = 0$ and curl $\mathbf{F} = \mathbf{0}$.

T-10. Show that $\operatorname{div}(\mathbf{F} \times \mathbf{G}) = \mathbf{G}\cdot\operatorname{curl}\mathbf{F} - \mathbf{F}\cdot\operatorname{curl}\mathbf{G}$.

T-11. Show that $\operatorname{div}(\nabla f \times \nabla g) = 0$.

T-12. Show that div $u\nabla v = \nabla u\cdot\nabla v + u\nabla^2 v$.

T-13. The *Laplacian of a vector field* $\mathbf{F}$ is defined by

$$\nabla^2\mathbf{F} \;=\; \nabla^2 F_1\,\mathbf{i} + \nabla^2 F_2\,\mathbf{j} + \nabla^2 F_3\,\mathbf{k}.$$

Show that

$$\operatorname{curl}(\operatorname{curl}\mathbf{F}) \;=\; \nabla(\operatorname{div}\mathbf{F}) - \nabla^2\mathbf{F}.$$

3.5   Implicit Functions

In this section we study the implicit functions theorem for vector valued functions, a generalization of Theorem 6.1 of Section 2.6.

It will be convenient to denote points in R^{n+m} by $(\mathbf{X},\ \mathbf{U})$, where $\mathbf{X} = (x_1, \ldots, x_n)$ and $\mathbf{U} = (u_1, \ldots, u_m)$.

Let $\mathbf{F}\colon R^{n+m} \to R^m$. For a given $\mathbf{X}$ the vector equation

(1) $$\mathbf{F}(\mathbf{X}, \mathbf{U}) = \mathbf{0},$$

which is equivalent to the system

$$F_1(\mathbf{X}, u_1, \ldots, u_m) = 0,$$
$$\vdots$$
$$F_m(\mathbf{X}, u_1, \ldots, u_m) = 0,$$

may have no solution, one solution, or more than one solution for the un-known $\mathbf{U}$.

Definition 5.1 A rule which assigns to each $\mathbf{X}$ in a domain D exactly one solution $\mathbf{U}$ of (1) defines a function $\mathbf{f}\colon R^n \to R^m$ on D such that $\mathbf{F}(\mathbf{X}, \mathbf{f}(\mathbf{X})) = \mathbf{0}$ for all $\mathbf{X}$ in D. We say that $\mathbf{F}(\mathbf{X}, \mathbf{U}) = \mathbf{0}$ *defines* $\mathbf{f}$ *implicitly*, and that it defines $\mathbf{U} = \mathbf{f}(\mathbf{X})$ on D. If $\mathbf{X}_0$ is an interior point of D, we say that $\mathbf{F}(\mathbf{X}, \mathbf{U}) = \mathbf{0}$ *defines* $\mathbf{f}$ *implicitly near* $\mathbf{X}_0$.

Example 5.1   If F is real valued and satisfies the hypotheses of Theorem 6.1, Section 2.6, at $(\mathbf{X}_0, c)$, then $F(\mathbf{X}_0, u) = 0$ defines near $\mathbf{X}_0$ the unique continuous function $\mathbf{U} = f(\mathbf{X})$ such that $f(\mathbf{X}_0) = c$.

Example 5.2   If $\mathbf{F}(\mathbf{X}, \mathbf{U}) = \mathbf{U} - \mathbf{f}(\mathbf{X})$ where $\mathbf{f}: R^n \to R^m$ has domain D, then $\mathbf{F}(\mathbf{X}, \mathbf{U}) = \mathbf{0}$ defines $\mathbf{f}$ on D.

Example 5.3   Let $\mathbf{F}$ be the linear transformation

$$\mathbf{F}(\mathbf{X}, \mathbf{U}) = \begin{bmatrix} a_{11}x_1 + \cdots + a_{1n}x_n + b_{11}u_1 + \cdots + b_{1m}u_m \\ a_{21}x_1 + \cdots + a_{2n}x_n + b_{21}u_1 + \cdots + b_{2m}u_m \\ \vdots \\ a_{m1}x_1 + \cdots + a_{mn}x_n + b_{m1}u_1 + \cdots + b_{mm}u_m \end{bmatrix}$$

$$= \mathbf{AX} + \mathbf{BU}.$$

If $\mathbf{B}$ is nonsingular (that is, there exists a matrix $\mathbf{B}^{-1}$ such that $\mathbf{BB}^{-1} = \mathbf{B}^{-1}\mathbf{B} = \mathbf{I}$, the $m \times m$ identity matrix), then $\mathbf{F}(\mathbf{X}, \mathbf{U}) = \mathbf{0}$ defines

$$\mathbf{U} = \mathbf{f}(\mathbf{X}) = -\mathbf{B}^{-1}\mathbf{AX}$$

on R^n.

The general problem of existence of implicit functions defined by $\mathbf{F}(\mathbf{X}, \mathbf{U}) = \mathbf{0}$ is too broad, so we shall consider a more specific question: If $\mathbf{F}(\mathbf{X}_0, \mathbf{U}_0) = \mathbf{0}$, what conditions on $\mathbf{F}$ are sufficient to guarantee that there is a unique *differentiable* function $\mathbf{f}$, defined on a domain D including $\mathbf{X}_0$, such that $\mathbf{f}(\mathbf{X}_0) = \mathbf{U}_0$ and $\mathbf{F}(\mathbf{X}, \mathbf{f}(\mathbf{X})) = \mathbf{0}$ for all $\mathbf{X}$ in D?

In Example 5.3 we saw that a linear transformation from R^{n+m} to R^m defines an implicit function on R^n if a certain $m \times m$ submatrix of its matrix is nonsingular. The fact that the matrix of a linear transformation is its Jacobian matrix is the clue to a generalization of this result.

Let $\mathbf{F}: R^{n+m} \to R^m$ (not necessarily a linear transformation) and divide its Jacobian matrix into two parts:

$$\mathbf{JF} = \begin{bmatrix} \dfrac{\partial F_1}{\partial x_1} & \cdots & \dfrac{\partial F_1}{\partial x_n} & \vdots & \dfrac{\partial F_1}{\partial u_1} & \cdots & \dfrac{\partial F_1}{\partial u_m} \\ \vdots & & \vdots & \vdots & \vdots & & \vdots \\ \dfrac{\partial F_m}{\partial x_1} & \cdots & \dfrac{\partial F_m}{\partial x_n} & \vdots & \dfrac{\partial F_m}{\partial u_1} & \cdots & \dfrac{\partial F_m}{\partial u_m} \end{bmatrix}.$$

Generalizing the symbol for partial derivatives, we denote the matrices to the left and right of the dashed line by $\mathbf{F}_X$ and $\mathbf{F}_U$, respectively. Then in Example 5.3, $\mathbf{F_X} = \mathbf{A}$ and $\mathbf{F_U} = \mathbf{B}$ are constant matrices, and $\mathbf{F(X, U)} = \mathbf{0}$ defines $\mathbf{U} = \mathbf{f(X)}$ if $\mathbf{F_U}$ is nonsingular. For more general functions the result is not as simple as this, but the following generalization holds.

Theorem 5.1 (The Implicit Function Theorem) Let $\mathbf{F}: R^{n+m} \to R^m$ be a function of $(x_1, \ldots, x_n, u_1, \ldots, u_m)$ with continuous partial derivatives $\mathbf{F}_{x_1}, \ldots, \mathbf{F}_{x_n}, \mathbf{F}_{u_1}, \ldots, \mathbf{F}_{u_m}$ in an $(n + m)$-ball about $(\mathbf{X_0, U_0})$. Suppose that $\mathbf{F(X_0, U_0)} = \mathbf{0}$ and $\mathbf{F_U(X_0, U_0)}$ is nonsingular. Then there is a set D in R^n with $\mathbf{X_0}$ as an interior point, and a unique continuous function $\mathbf{f}: R^n \to R^m$ with domain D, such that $\mathbf{f(X_0)} = \mathbf{U_0}$ and $\mathbf{F(X, f(X))} = \mathbf{0}$. Furthermore, $\mathbf{F_U(X, f(X))}$ is nonsingular, $\mathbf{f}$ has continuous partial derivatives, and

$$(2) \qquad \mathbf{J_X f} = -[\mathbf{F_U(X, f(X))}]^{-1} \mathbf{F_X(X, f(X))}$$

for all $\mathbf{X}$ in D.

This theorem reduces to Theorem 6.1, Section 2.6, when $m = 1$. Its proof is beyond the scope of this book; however, given that $\mathbf{f}$ is differentiable and $\mathbf{F(X, f(X))} = \mathbf{0}$, (2) follows from the chain rule. In terms of the components of $\mathbf{F}$,

$$F_i(\mathbf{X}, f_i(\mathbf{X}), \ldots, f_m(\mathbf{X})) = 0 \qquad (1 \le i \le m);$$

differentiating both sides with respect to x_j yields

$$\frac{\partial F_i}{\partial x_j} + \frac{\partial F_i}{\partial u_1}\frac{\partial f_1}{\partial x_j} + \frac{\partial F_i}{\partial u_2}\frac{\partial f_2}{\partial x_j} + \cdots + \frac{\partial F_i}{\partial u_m}\frac{\partial f_m}{\partial x_j} = 0 \qquad (1 \le i, j \le m)$$

for all $\mathbf{X}$ in D, where the partial derivatives of $\mathbf{f}$ and $\mathbf{F}$ are evaluated at $\mathbf{X}$ and $(\mathbf{X, f(X)})$, respectively. In vector form,

$$\begin{bmatrix} \dfrac{\partial F_1}{\partial x_j} \\ \vdots \\ \dfrac{\partial F_m}{\partial x_j} \end{bmatrix} + \mathbf{F_U} \begin{bmatrix} \dfrac{\partial f_1}{\partial x_j} \\ \vdots \\ \dfrac{\partial f_m}{\partial x_j} \end{bmatrix} = \mathbf{0} \qquad (1 \le j \le m).$$

The vectors here are the columns of $\mathbf{F_X}$ and $\mathbf{Jf}$; hence these m equations can be combined to yield

$$\mathbf{F_X} + \mathbf{F_U Jf} = \mathbf{0},$$

from which (2) follows.

In the informal notation, where we write $\mathbf{U} = \mathbf{f}$, (2) becomes

$$
(3) \quad
\begin{bmatrix}
\dfrac{\partial u_1}{\partial x_1} & \dfrac{\partial u_1}{\partial x_2} & \cdots & \dfrac{\partial u_1}{\partial x_n} \\[2ex]
\dfrac{\partial u_2}{\partial x_1} & \dfrac{\partial u_2}{\partial x_2} & \cdots & \dfrac{\partial u_2}{\partial x_n} \\[2ex]
\vdots & \vdots & & \vdots \\[2ex]
\dfrac{\partial u_m}{\partial x_1} & \dfrac{\partial u_m}{\partial x_2} & \cdots & \dfrac{\partial u_m}{\partial x_n}
\end{bmatrix}
$$

$$
= -
\begin{bmatrix}
\dfrac{\partial F_1}{\partial u_1} & \dfrac{\partial F_1}{\partial u_2} & \cdots & \dfrac{\partial F_1}{\partial u_m} \\[2ex]
\dfrac{\partial F_2}{\partial u_1} & \dfrac{\partial F_2}{\partial u_2} & \cdots & \dfrac{\partial F_2}{\partial u_m} \\[2ex]
\vdots & \vdots & & \vdots \\[2ex]
\dfrac{\partial F_m}{\partial u_1} & \dfrac{\partial F_m}{\partial u_2} & \cdots & \dfrac{\partial F_m}{\partial u_m}
\end{bmatrix}^{-1}
\begin{bmatrix}
\dfrac{\partial F_1}{\partial x_1} & \dfrac{\partial F_1}{\partial x_2} & \cdots & \dfrac{\partial F_1}{\partial x_n} \\[2ex]
\dfrac{\partial F_2}{\partial x_1} & \dfrac{\partial F_2}{\partial x_2} & \cdots & \dfrac{\partial F_2}{\partial x_n} \\[2ex]
\vdots & \vdots & & \vdots \\[2ex]
\dfrac{\partial F_m}{\partial x_1} & \dfrac{\partial F_m}{\partial x_2} & \cdots & \dfrac{\partial F_m}{\partial x_n}
\end{bmatrix} .
$$

Theorem 5.1 is an existence theorem; it says that under certain conditions (1) has a differentiable solution for all $\mathbf{X}$ in some domain D, but does not tell how to determine D or the solution. It is nevertheless a useful theorem, for often it is enough just to know that a solution exists in some domain. Furthermore, even if $\mathbf{f}$ cannot be obtained explicitly, its partial derivatives can be evaluated at any $\mathbf{X}$ in D where $\mathbf{f}(\mathbf{X})$ is known, by means of (2).

Example 5.4   Let

$$
\mathbf{F}(\mathbf{X}, \mathbf{U}) =
\begin{bmatrix}
x^2 + y^2 + z^2 - u^2 - v^2 \\[1.5ex]
(x - y)^2 + u + v
\end{bmatrix} ;
$$

thus $m = 2$ and $n = 3$. Then

$$
\mathbf{F_X}(\mathbf{U}) =
\begin{bmatrix}
2x & 2y & 2z \\[1.5ex]
2x - 2y & 2y - 2x & 0
\end{bmatrix}
$$

and

$$
\mathbf{F_U}(\mathbf{X}, \mathbf{U}) =
\begin{bmatrix}
-2u & -2v \\[1.5ex]
1 & 1
\end{bmatrix} .
$$

If

$$\mathbf{X}_0 = \left(\frac{1}{\sqrt{3}}, \frac{1}{\sqrt{3}}, -\frac{1}{\sqrt{3}} \right) \quad \text{and} \quad \mathbf{U}_0 = \left(\frac{1}{\sqrt{2}}, -\frac{1}{\sqrt{2}} \right),$$

then $\mathbf{F}(\mathbf{X}_0, \mathbf{U}_0) = \mathbf{0}$ and

$$[\mathbf{F}_\mathbf{U}(\mathbf{X}_0, \mathbf{U}_0)]^{-1} = \begin{bmatrix} -\sqrt{2} & \sqrt{2} \\ 1 & 1 \end{bmatrix}^{-1}$$

$$= \begin{bmatrix} -\dfrac{1}{2\sqrt{2}} & \dfrac{1}{2} \\[2mm] \dfrac{1}{2\sqrt{2}} & \dfrac{1}{2} \end{bmatrix};$$

hence $\mathbf{F}(\mathbf{X}, \mathbf{U}) = \mathbf{0}$ defines $\mathbf{U} = \mathbf{U}(\mathbf{X})$ near $\mathbf{X}_0$ and

$$\begin{bmatrix} u_x(\mathbf{X}_0) & u_y(\mathbf{X}_0) & u_z(\mathbf{X}_0) \\ v_x(\mathbf{X}_0) & v_y(\mathbf{X}_0) & v_z(\mathbf{X}_0) \end{bmatrix} = -(\mathbf{F}_\mathbf{U}(\mathbf{X}_0, \mathbf{U}_0))^{-1}\, \mathbf{F}_\mathbf{X}(\mathbf{X}_0, \mathbf{U}_0)$$

$$= -\begin{bmatrix} -\dfrac{1}{2\sqrt{2}} & \dfrac{1}{2} \\[2mm] \dfrac{1}{2\sqrt{2}} & \dfrac{1}{2} \end{bmatrix} \begin{bmatrix} \dfrac{2}{\sqrt{3}} & \dfrac{2}{\sqrt{3}} & -\dfrac{2}{\sqrt{3}} \\[2mm] 0 & 0 & 0 \end{bmatrix}$$

$$= \begin{bmatrix} \dfrac{1}{\sqrt{6}} & \dfrac{1}{\sqrt{6}} & -\dfrac{1}{\sqrt{6}} \\[2mm] -\dfrac{1}{\sqrt{6}} & -\dfrac{1}{\sqrt{6}} & \dfrac{1}{\sqrt{6}} \end{bmatrix}.$$

It is not necessary to memorize (2) or (3), since the partial derivatives of an implicit function can be calculated from the chain rule and Cramer's rule, as in the next example.

Example 5.5 Suppose we wish to find only $u_x(\mathbf{X}_0)$ in Example 5.4. Since

$$x^2 + y^2 + z^2 - u^2 - v^2 = 0,$$

$$(x - y)^2 + u + v = 0,$$

differentiation with respect to x yields

$$2x - 2uu_x - 2vv_x = 0,$$

$$2(x - y) + u_x + v_x = 0.$$

From Cramer's rule,

$$u_x = \frac{\begin{vmatrix} 2x & 2v \\ 2(x-y) & -1 \end{vmatrix}}{\begin{vmatrix} 2u & 2v \\ -1 & -1 \end{vmatrix}}$$

and

$$u_x(\mathbf{X}_0) = \frac{\begin{vmatrix} \dfrac{2}{\sqrt{3}} & -\sqrt{2} \\ 0 & -1 \end{vmatrix}}{\begin{vmatrix} \sqrt{2} & -\sqrt{2} \\ -1 & -1 \end{vmatrix}} = \frac{1}{\sqrt{6}}.$$

Example 5.6   Let $(x_0,\, y_0,\, z_0) = (1,\, 0,\, 0)$ and

$$\mathbf{F}(x,\, y,\, z) = \begin{bmatrix} x^2 + y^2 + z^2 + y - z \\ xyz + e^x \sin(y + z) \end{bmatrix};$$

then

$$\mathbf{F}_{(y,\,z)} = \begin{bmatrix} \dfrac{\partial F_1}{\partial y} & \dfrac{\partial F_1}{\partial z} \\ \dfrac{\partial F_2}{\partial y} & \dfrac{\partial F_2}{\partial z} \end{bmatrix} = \begin{bmatrix} 2y + 1 & 2z - 1 \\ xz + e^x \cos(y + z) & xy + e^x \cos(y + z) \end{bmatrix},$$

and

$$\mathbf{F}_{(y,\,z)}(1,\, 0,\, 0) = \begin{bmatrix} 1 & -1 \\ e & e \end{bmatrix}$$

is nonsingular. Therefore $\mathbf{F}(x, y, z) = \mathbf{0}$ defines

$$\begin{bmatrix} y \\ z \end{bmatrix} = \begin{bmatrix} y(x) \\ z(x) \end{bmatrix}$$

near $x_0 = 1$, with

$$\begin{bmatrix} y(0) \\ z(0) \end{bmatrix} = \begin{bmatrix} 0 \\ 0 \end{bmatrix}.$$

The derivatives satisfy

$$2x + (2y + 1)y' + (2z - 1)z' = 0$$

$$yz + e^x \sin(y + z) + [xz + e^x \cos(y + z)]y'$$

$$+ [xy + e^x \cos(y + z)]z' = 0;$$

hence

$$2 + y'(0) - z'(0) = 0,$$

$$y'(0) + z'(0) = 0,$$

which has the solution $y'(0) = -1$, $z'(0) = 1$.

The positions of $u_1, \ldots, u_m$ among the $(n + m)$ coordinates of a point in R^{n+m} are not important; they do not have to be last as we have assumed until now. Any m variables for which the inverse matrix on the right of (3) exists are defined implicitly by $\mathbf{F} = \mathbf{0}$, with derivatives given by (3).

Example 5.7 In Example 5.4,

$$\mathbf{F}_{(x, u)} = \begin{bmatrix} \dfrac{\partial F_1}{\partial x} & \dfrac{\partial F_1}{\partial u} \\ \\ \dfrac{\partial F_2}{\partial x} & \dfrac{\partial F_2}{\partial u} \end{bmatrix} = \begin{bmatrix} 2x & -2u \\ 2(x - y) & 1 \end{bmatrix},$$

and

$$\mathbf{F}_{(x, u)}(x_0, u_0) = \begin{bmatrix} \dfrac{2}{\sqrt{3}} & -\sqrt{2} \\ \\ 0 & 1 \end{bmatrix}$$

is nonsingular. Hence $\mathbf{F}(x, y, z, u, v) = \mathbf{0}$ also defines

$$\begin{bmatrix} x \\ u \end{bmatrix} = \begin{bmatrix} x(y, z, v) \\ u(y, z, v) \end{bmatrix}$$

such that

$$\begin{bmatrix} x(\mathbf{V}_0) \\ u(\mathbf{V}_0) \end{bmatrix} = \begin{bmatrix} \dfrac{1}{\sqrt{3}} \\ \dfrac{1}{\sqrt{2}} \end{bmatrix},$$

where $\mathbf{V}_0 = (1/\sqrt{3}, -1/\sqrt{3}, 1/\sqrt{2})$. Furthermore, for $\mathbf{V} = (y, z, v)$ near $\mathbf{V}_0$,

$$\begin{bmatrix} \dfrac{\partial x}{\partial y} & \dfrac{\partial x}{\partial z} & \dfrac{\partial x}{\partial v} \\ \\ \dfrac{\partial u}{\partial y} & \dfrac{\partial u}{\partial z} & \dfrac{\partial u}{\partial v} \end{bmatrix} = - \begin{bmatrix} \dfrac{\partial F_1}{\partial x} & \dfrac{\partial F_1}{\partial u} \\ \\ \dfrac{\partial F_2}{\partial x} & \dfrac{\partial F_2}{\partial u} \end{bmatrix}^{-1} \begin{bmatrix} \dfrac{\partial F_1}{\partial y} & \dfrac{\partial F_1}{\partial z} & \dfrac{\partial F_1}{\partial v} \\ \\ \dfrac{\partial F_2}{\partial y} & \dfrac{\partial F_2}{\partial z} & \dfrac{\partial F_2}{\partial v} \end{bmatrix}$$

$$= - \begin{bmatrix} 2x & -2u \\ 2(x - y) & 1 \end{bmatrix}^{-1} \begin{bmatrix} 2y & 2z & -2v \\ 2(y - x) & 0 & 1 \end{bmatrix}.$$

EXERCISES 3.5

1. Let

$$\mathbf{F}(x, y, u, v) = \begin{bmatrix} x + 2y - u + v \\ -2x + y + 2u + 2v \end{bmatrix};$$

then $\mathbf{F}(x, y, u, v) = \mathbf{0}$ defines $\begin{bmatrix} u \\ v \end{bmatrix} = \mathbf{f}(x, y)$ on R^2. Find $\mathbf{f}$ explicitly;

also find $\mathbf{Jf}$.

2. Repeat Exercise 1 for

$$\mathbf{F}(x, y, u, v) = \begin{bmatrix} 3x + 2y - 3u + 2v \\ 5x - 2y + u - 2v \end{bmatrix}.$$

3. Let $\mathbf{X} = (x, y)$ and $\mathbf{U} = (u, v)$ in (1) and (2). Find $\mathbf{F_X}$ and $\mathbf{F_U}$ for $\mathbf{F}$ in Exercise 1, and calculate $\mathbf{Jf}$ by means of (2).

4. Repeat Exercise 3 for $\mathbf{F}$ as defined in Exercise 2.

5. Let $\mathbf{X_0} = (x_0, y_0) = (1, 1)$ and

$$\mathbf{F}(x, y, z, w) = \begin{bmatrix} xyz - 2yz + xw \\ 2xy - y^2 - w^2 \end{bmatrix}.$$

Then $\mathbf{F}(x, y, z, w) = \mathbf{0}$ defines near $\mathbf{X_0}$ two differentiable functions,

$$\begin{bmatrix} z \\ w \end{bmatrix} = \mathbf{f}(x, y) \qquad \text{and} \qquad \begin{bmatrix} z \\ w \end{bmatrix} = \mathbf{g}(x, y).$$

(Does this contradict Theorem 5.1?) Find $\mathbf{f}(\mathbf{X_0})$, $\mathbf{g}(\mathbf{X_0})$, $\mathbf{J_{X_0}f}$, $\mathbf{J_{X_0}g}$, $\mathbf{d_{X_0}f}$, and $\mathbf{d_{X_0}g}$.

6. Let

$$\mathbf{F}(x, y, z) = \begin{bmatrix} e^x \cos z - y \\ e^x \sin z - y + 1 \end{bmatrix};$$

then $\mathbf{F}(x, y, z) = \mathbf{0}$ defines exactly one continuous function

$$\begin{bmatrix} x \\ y \end{bmatrix} = \begin{bmatrix} x(z) \\ y(z) \end{bmatrix}$$

near $z_0 = 0$. Find $x(0)$, $y(0)$, $x'(0)$, and $y'(0)$.

7. Let

$$\mathbf{F}(x, y, z, w) = \begin{bmatrix} x^2y + xy^2 + z^2 - w^2 \\ e^{x+y} - w \end{bmatrix};$$

then $\mathbf{F}(x, y, z, w) = \mathbf{0}$ defines

$$\begin{bmatrix} z \\ w \end{bmatrix} = \begin{bmatrix} z(x, y) \\ w(x, y) \end{bmatrix}$$

near $(x_0, y_0) = (0, 0)$ such that

$$\begin{bmatrix} z(0, 0) \\ w(0, 0) \end{bmatrix} = \begin{bmatrix} 1 \\ 1 \end{bmatrix}.$$

Using the chain rule, find $\dfrac{\partial z}{\partial y}(0, 0)$.

8. Let $\mathbf{F}$ be as in Exercise 7; then $\mathbf{F}(x, y, z, w) = \mathbf{0}$ defines

$$\begin{bmatrix} x \\ z \end{bmatrix} = \begin{bmatrix} x(y, w) \\ z(y, w) \end{bmatrix}$$

near $(y_0, w_0) = (0, 1)$ such that

$$\begin{bmatrix} x(0, 1) \\ z(0, 1) \end{bmatrix} = \begin{bmatrix} 0 \\ 1 \end{bmatrix}.$$

Using the chain rule, find $(\partial z/\partial y)(0, 1)$.

9. Let

$$\mathbf{F}(x, y, z, u, v, w) = \begin{bmatrix} e^x \cos y + e^z \cos u + e^v \cos w + x - 3 \\ e^x \sin y + e^z \sin u + e^v \cos w - 1 \\ e^x \tan y + e^z \tan u + e^v \tan w + z \end{bmatrix}.$$

(a) Find $\mathbf{F}_{(x,y,z)}(0, 0, 0)$ and $\mathbf{F}_{(u,v,w)}(0, 0, 0)$.

(b) Let

$$\begin{bmatrix} u \\ v \\ w \end{bmatrix} = \mathbf{f}(x, y, z)$$

be defined by $\mathbf{F}(x, y, z, u, v, w) = \mathbf{0}$ near $(x_0, y_0, z_0) = (0, 0, 0)$ such that $\mathbf{f}(0, 0, 0) = \mathbf{0}$. Find $\mathbf{J}_{(0,0,0)}\mathbf{f}$.

10. In Exercise 1, which of the pairs (x, y), (x, u), (x, v), (y, u), (y, v), and (u, v) are defined as functions of the remaining variables by $\mathbf{F}(x, y, u, v) = \mathbf{0}$?

11. Repeat Exercise 10 for $\mathbf{F}$ as defined in Exercise 2.

12. Let $\mathbf{F}$ be as in Exercise 5. Which of the pairs (x, y), (x, z), (x, w), (y, z), (y, w), and (z, w) are, according to the implicit function theorem, defined as functions of the remaining variables by $\mathbf{F}(x, y, z, w) = \mathbf{0}$, such that $(x_0, y_0, z_0, w_0) = (1, 1, 1, 1)$?

THEORETICAL EXERCISES

T-1. Suppose $n = m$ in Theorem 5.1 and $\mathbf{F}_{\mathbf{X}}(\mathbf{X}_0, \mathbf{U}_0)$ is also nonsingular. Let $\mathbf{g}$ be the differentiable function defined near $\mathbf{U}_0$ such that $\mathbf{g}(\mathbf{U}_0) = \mathbf{X}_0$ and $\mathbf{F}(\mathbf{g}(\mathbf{U}), \mathbf{U}) = \mathbf{0}$. Show that $\mathbf{J}_{\mathbf{X}}\mathbf{f} = (\mathbf{J}_{\mathbf{U}}\mathbf{g})^{-1}$.

T-2. Let $1 \leq m < n$ and suppose f and $g_1, \ldots, g_m$ are continuously differentiable real valued functions on a domain D in R^n. Let $\mathbf{X}_0 = (c_1, c_2, \ldots, c_n)$ be an interior point of D at which f assumes an extreme value subject to the constraints

(a) $$g_1(\mathbf{X}) = g_2(\mathbf{X}) = \cdots = g_m(\mathbf{X}) = 0,$$

and suppose that the matrix

$$\mathbf{A} = \begin{bmatrix} \dfrac{\partial g_1}{\partial x_1}(\mathbf{X}_0) & \dfrac{\partial g_1}{\partial x_2}(\mathbf{X}_0) & \cdots & \dfrac{\partial g_1}{\partial x_m}(\mathbf{X}_0) \\[2ex] \dfrac{\partial g_2}{\partial x_1}(\mathbf{X}_0) & \dfrac{\partial g_2}{\partial x_2}(\mathbf{X}_0) & \cdots & \dfrac{\partial g_2}{\partial x_m}(\mathbf{X}_0) \\[2ex] \vdots & \vdots & & \vdots \\[2ex] \dfrac{\partial g_m}{\partial x_1}(\mathbf{X}_0) & \dfrac{\partial g_m}{\partial x_2}(\mathbf{X}_0) & \cdots & \dfrac{\partial g_m}{\partial x_m}(\mathbf{X}_0) \end{bmatrix}$$

is nonsingular. Show that, for $m + 1 \leq i \leq n$,

(b) $$\begin{bmatrix} \dfrac{\partial f}{\partial x_1}(\mathbf{X}_0) \\[2ex] \dfrac{\partial f}{\partial x_2}(\mathbf{X}_0) \\[1ex] \vdots \\[1ex] \dfrac{\partial f}{\partial x_m}(\mathbf{X}_0) \\[2ex] \dfrac{\partial f}{\partial x_i}(\mathbf{X}_0) \end{bmatrix} = \sum_{j=1}^{m} \lambda_{ij} \begin{bmatrix} \dfrac{\partial g_j}{\partial x_1}(\mathbf{X}_0) \\[2ex] \dfrac{\partial g_j}{\partial x_2}(\mathbf{X}_0) \\[1ex] \vdots \\[1ex] \dfrac{\partial g_j}{\partial x_m}(\mathbf{X}_0) \\[2ex] \dfrac{\partial g_j}{\partial x_i}(\mathbf{X}_0) \end{bmatrix}.$$

Hint: From the implicit function theorem, (a) defines $x_1, \ldots, x_m$ as functions of $x_{m+1}, \ldots, x_n$ near $(x_{m+1}, \ldots, x_n) = (c_{m+1}, \ldots, c_n)$. Therefore the function h defined by

$$h(x_{m+1}, \ldots, x_n) = f(x_1(x_{m+1}, \ldots, x_n), \ldots, x_m(x_{m+1}, \ldots, x_n), x_{m+1}, \ldots, x_n)$$

assumes a relative extreme value at $(c_{m+1}, \ldots, c_n)$. Show that this implies that the vectors on both sides of (b) are all orthogonal to

$$\mathbf{N}_i = \begin{bmatrix} \dfrac{\partial x_1}{\partial x_i}(c_{m+1}, \ldots, c_n) \\[2ex] \dfrac{\partial x_2}{\partial x_i}(c_{m+1}, \ldots, c_n) \\[2ex] \vdots \\[2ex] \dfrac{\partial x_m}{\partial x_i}(c_{m+1}, \ldots, c_n) \\[2ex] 1 \end{bmatrix}$$

and are therefore linearly dependent. Since the vectors on the right side of (b) are linearly independent (verify this from the fact that the columns of a nonsingular matrix are linearly independent), (b) now follows.

T-3. Show that the nonsingularity of the matrix $\mathbf{A}$ in Exercise T-2 implies that $\lambda_{i1}, \ldots, \lambda_{im}$ are in fact independent of i, and conclude that $\nabla_{\mathbf{X}_0} f$ is a linear combination of $\nabla_{\mathbf{X}_0} g_1, \ldots, \nabla_{\mathbf{X}_0} g_m$. This result is the basis for the method of Lagrange multipliers for locating extreme values of a function of n variables subject to m constraints (Section 2.10).

3.6 Inverse Functions and Coordinate Transformations

In Example 5.2 of the last section we observed that

(1) $$\mathbf{F}(\mathbf{X}, \mathbf{U}) = \mathbf{U} - \mathbf{f}(\mathbf{X}) = 0$$

defines $\mathbf{U} = \mathbf{f}(\mathbf{X})$ on the domain of $\mathbf{f}$. We now consider a less trivial question: Under what conditions on $\mathbf{f}$ does (1) define $\mathbf{X}$ as a function of $\mathbf{U}$? This involves the idea of an inverse function, which we now develop.

In the following D is the domain of $\mathbf{f}$ and S is a subset of D.

Definition 6.1   A function f is said to be *one-to-one on* S if $f(\mathbf{X}_1)$ and $f(\mathbf{X}_2)$ are distinct whenever $\mathbf{X}_1$ and $\mathbf{X}_2$ are distinct points of S; thus, $f(\mathbf{X}_1) = f(\mathbf{X}_2)$ implies $\mathbf{X}_1 = \mathbf{X}_2$. If $S = D$ we say simply that f *is one-to-one.*

Example 6.1   Let

$$f(x, y) = \begin{bmatrix} x + y \\ x - y \end{bmatrix};$$

then f is one-to-one, because if

$$\begin{bmatrix} x_1 + y_1 \\ x_1 - y_1 \end{bmatrix} = \begin{bmatrix} x_2 + y_2 \\ x_2 - y_2 \end{bmatrix},$$

then

$$(x_1 - x_2) + (y_1 - y_2) = 0,$$
$$(x_1 - x_2) - (y_1 - y_2) = 0,$$

from which it follows that $x_1 - x_2 = y_1 - y_2 = 0$.

Definition 6.2   *The restriction of* f *to a subset* S *of its domain* D is denoted by f_S and defined by $f_S(\mathbf{X}) = f(\mathbf{X})$ for $\mathbf{X}$ in S; it is undefined if $\mathbf{X}$ is not in S. If $S = D$ we write simply $f_S = f$.

From Definitions 6.1 and 6.2, f is one-to-one on S if and only f_S is one-to-one.

Example 6.2   Let

$$f(x, y) = \begin{bmatrix} x^2 + y^2 \\ x^2 - y^2 \end{bmatrix};$$

then f is not one-to-one on any set which contains, along with a point $(x_0, y_0) \neq (0, 0)$, one of the points $(-x_0, y_0)$, $(x_0, -y_0)$, and $(-x_0, -y_0)$ which is distinct from (x_0, y_0). In particular, f is not one-to-one on any set containing points on the coordinate axes in its interior. It is one-to-one on any set S contained within one of the quadrants of R^2 (Exercise T-1). In particular, if S is the first quadrant,

$$S = \{(x, y) \mid x \geq 0, \quad y \geq 0\},$$

then f_S is one-to-one.

Definition 6.3   Let $f \colon R^n \to R^m$ be one-to-one on S and let $f(S)$ be the set of values attained by f as $\mathbf{X}$ varies over S. For each $\mathbf{U}$ in $f(S)$ let $f_S^{-1}(\mathbf{U})$ be the unique $\mathbf{X}$ in S such that $\mathbf{U} = f(\mathbf{X})$ (Fig. 6.1); then f_S^{-1} is a function

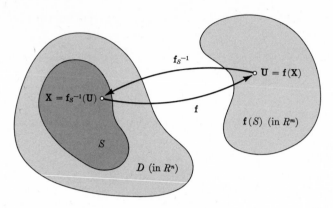

FIGURE 6.1

with domain $f(S)$, which we call the *inverse of* f *on* S, or simply the *inverse of* f if $S = D$. We say that f_S^{-1} is the inverse of f near X_0 if X_0 is in the interior of S.

This terminology is somewhat misleading. It does not mean that X_0 is in the domain of f_S^{-1} (in fact, it is in the range); it means that the restriction of f to some set S containing X_0 in its interior is one-to-one.

Example 6.3 We saw in Example 6.1 that if

$$f(x, y) = \begin{bmatrix} x + y \\ x - y \end{bmatrix}$$

then $S = R^2$. The image T is also R^2, since for (u, v) in R^2 there is a unique (x, y) such that

$$\begin{bmatrix} u \\ v \end{bmatrix} = \begin{bmatrix} x + y \\ x - y \end{bmatrix};$$

explicitly,

$$\begin{bmatrix} x \\ y \end{bmatrix} = f^{-1}(u, v) = \begin{bmatrix} \dfrac{u + v}{2} \\ \dfrac{u - v}{2} \end{bmatrix}.$$

Example 6.4 Let

$$f(x, y) = \begin{bmatrix} x^2 + y^2 \\ x^2 - y^2 \end{bmatrix}$$

as in Example 6.2, and let S be the first quadrant of R^2. Then $\mathbf{f}_S$ is one-to-one (Exercise T-1); to find $T = \mathbf{f}(S)$, we observe that if

$$\begin{bmatrix} u \\ v \end{bmatrix} = \begin{bmatrix} x^2 + y^2 \\ x^2 - y^2 \end{bmatrix}$$

then

$$u + v = 2x^2,$$

$$u - v = 2y^2;$$

hence $u + v \geq 0$ and $u - v \geq 0$ if (u, v) is in T. Conversely, any point which satisfies these inequalities is the image of

$$\begin{bmatrix} x \\ y \end{bmatrix} = \begin{bmatrix} \sqrt{\dfrac{u+v}{2}} \\ \\ \sqrt{\dfrac{u-v}{2}} \end{bmatrix},$$

(2)

which is in S. Therefore

$$\mathbf{f}(S) = \{(u, v) \mid u - v \geq 0, \quad u + v \geq 0\}$$

is the shaded area shown in Fig. 6.2. The inverse of $\mathbf{f}_S$ is given by

$$\mathbf{f}_S^{-1}(u, v) = \begin{bmatrix} \sqrt{\dfrac{u+v}{2}} \\ \\ \sqrt{\dfrac{u-v}{2}} \end{bmatrix}.$$

A similar argument shows that if S_1 is the second quadrant then $\mathbf{f}(S_1) = \mathbf{f}(S)$ and

$$\mathbf{f}_{S_1}^{-1}(u, v) = \begin{bmatrix} -\sqrt{\dfrac{u+v}{2}} \\ \\ \sqrt{\dfrac{u-v}{2}} \end{bmatrix}.$$

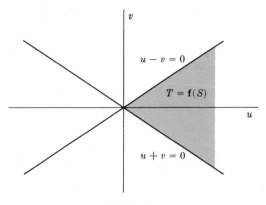

FIGURE 6.2

Theorem 6.1 Let S be a subset of the domain D of $\mathbf{f}$, and suppose $\mathbf{f}_S$ is one-to-one. Then $\mathbf{f}_{\overline{S}}^{1}$ is one-to-one on $T = \mathbf{f}(S)$ and $\mathbf{f}_S$ is the inverse of $\mathbf{f}_{\overline{S}}^{1}$. Furthermore $\mathbf{f}_{\overline{S}}^{1}(\mathbf{f}_S(\mathbf{X})) = \mathbf{X}$ for all $\mathbf{X}$ in S and $\mathbf{f}_S(\mathbf{f}_{\overline{S}}^{1}(\mathbf{U})) = \mathbf{U}$ for all $\mathbf{U}$ in T.

We leave the proof of this theorem to the student (Exercise T-2).

We are primarily interested in the existence of inverse functions when the range and domain of $\mathbf{f}$ are both in R^n. The following theorem gives sufficient conditions under which such a function has a differentiable inverse near $\mathbf{X}_0$.

Theorem 6.2 (The Inverse Function Theorem)   Let $\mathbf{f}: R^n \to R^n$ have continuous first partial derivatives on D and let $\mathbf{X}_0$ be an interior point of D. If $\mathbf{J}_{\mathbf{X}_0}\mathbf{f}$ is nonsingular then there is an n-ball C about $\mathbf{X}_0$ such that

(a) $\mathbf{J}_{\mathbf{X}}\mathbf{f}$ is nonsingular for all $\mathbf{X}$ in C;

(b) $\mathbf{f}$ is one-to-one on C and $\mathbf{f}_{\overline{C}}^{1}$ exists on $\mathbf{f}(C)$.

(c)   every point of $\mathbf{f}(C)$ is an interior point and $\mathbf{f}_{\overline{C}}^{1}$ has continuous first partial derivatives on $\mathbf{f}(C)$;

(d)   if $\mathbf{U}$ is in $\mathbf{f}(C)$ then

(3) $$\mathbf{J}_{\mathbf{U}}\mathbf{f}_{\overline{C}}^{1} = (\mathbf{J}_{\mathbf{X}}\mathbf{f})^{-1},$$

where $\mathbf{X}$ is the unique point in C such that $\mathbf{U} = \mathbf{f}(\mathbf{X})$.

We omit the proof of this theorem, which is quite formidable. However, (3) can be obtained formally from Eq. (2) of Section 3.4 (Exercise T-3).

Example 6.5 Suppose f is a linear transformation,

$$f(X) = \begin{bmatrix} a_{11}x_1 + a_{12}x_2 + \cdots + a_{1n}x_n \\ \vdots \\ a_{n1}x_1 + a_{n2}x_2 + \cdots + a_{nn}x_n \end{bmatrix} = AX,$$

where A^{-1} exists. Since $J_{X_0}f = A$ for all X, Theorem 6.2 implies that f has an inverse near every point X_0. However, in this case we already know from linear algebra that f has an inverse on all of R^n, and that

$$f^{-1}(U) = A^{-1}U.$$

Furthermore, if $U = f(X)$ then

$$J_U f^{-1} = A^{-1} = (J_X f)^{-1},$$

which is consistent with (3).

Example 6.6 Let $X_0 = (1, 2)$ and

$$f(x, y) = \begin{bmatrix} x^2 + y^2 \\ x^2 - y^2 \end{bmatrix};$$

then

$$Jf = \begin{bmatrix} 2x & 2y \\ 2x & -2y \end{bmatrix},$$

and

$$J_{X_0}f = \begin{bmatrix} 2 & 2 \\ 2 & -4 \end{bmatrix}$$

is nonsingular. Theorem 6.1 implies that f has an inverse near X_0, and that

(4) $$J_U f_S^{-1} = (J_X f)^{-1} = \begin{bmatrix} \dfrac{1}{4x} & \dfrac{1}{4x} \\[2ex] \dfrac{1}{4y} & -\dfrac{1}{4y} \end{bmatrix}$$

for all X in some set S containing X_0, but the theorem does not tell us exactly what S is. However, we know from Example 6.4 that S is the first quadrant and that f^{-1} is given explicitly by (2), from which (4) can be verified directly.

Example 6.7 Let $X_0 = (1, -1)$ and

$$f(x, y) = \begin{bmatrix} x^4 y^5 - 4x \\ x^3 y^2 - 3y \end{bmatrix};$$

then

$$J_{X_0} f = \begin{bmatrix} -8 & 5 \\ 3 & -5 \end{bmatrix}$$

is nonsingular and Theorem 6.2 implies that f has an inverse near X_0. This time it is difficult to obtain f^{-1} and $J_U f$ explicitly; however, (3) yields

$$(J_{U_0} f_c^{-1}) = (J_{X_0} f)^{-1} = \begin{bmatrix} -\dfrac{5}{25} & -\dfrac{5}{25} \\[2ex] -\dfrac{3}{25} & -\dfrac{8}{25} \end{bmatrix},$$

where $U_0 = (-5, 4)$. Moreover,

$$d_{U_0} f^{-1} = \begin{bmatrix} -\dfrac{5}{25} du - \dfrac{5}{25} dv \\[2ex] -\dfrac{3}{25} du - \dfrac{8}{25} dv \end{bmatrix}.$$

If $f: R^n \to R^n$ is one-to-one on S then we can identify a point X_0 in S by the n-vector $U_0 = f(X_0)$ that f assigns to X_0. We say that $U = f(X)$ is a *coordinate transformation near* X_0, thus considering f not as a mapping from S to $f(S)$, but as a way of giving new names to the points of S.

Example 6.8 We saw in Example 6.4 that

$$\begin{bmatrix} u \\ v \end{bmatrix} = f(x, y) = \begin{bmatrix} x^2 + y^2 \\ x^2 - y^2 \end{bmatrix}$$

defines a coordinate transformation in the first quadrant of R^2. The point whose new coordinates are u_0 and v_0 is the intersection of the circle $x^2 + y^2 = u_0$ and the branch of the hyperbola $x^2 - y^2 = v_0$ that lies in the first quadrant (Fig. 6.3).

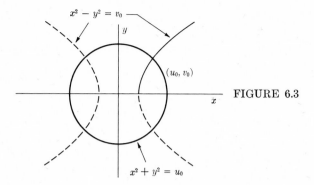

FIGURE 6.3

Definition 6.4   Let $\mathbf{f}: R^n \to R^n$ be differentiable on a set S. The determinant of $\mathbf{Jf}$ is called the *Jacobian of* $\mathbf{f}$, and is denoted by

$$(5) \qquad \frac{\partial(f_1, \ldots, f_n)}{\partial(x_1, \ldots, x_n)} ;$$

its value at a particular point $\mathbf{X}_0$ is denoted by

$$(6) \qquad \frac{\partial(f_1, \ldots, f_n)}{\partial(x_1, \ldots, x_n)} \bigg|_{\mathbf{X}_0}.$$

Consistent with the informal notation $\mathbf{U} = \mathbf{f}(\mathbf{X})$, we also write (5) and (6) as

$$\frac{\partial(u_1, \ldots, u_n)}{\partial(x_1, \ldots, x_n)}$$

and

$$\frac{\partial(u_1, \ldots, u_n)}{\partial(x_1, \ldots, x_n)} \bigg|_{\mathbf{X}_0}.$$

Theorem 6.3   If $\mathbf{f}: R^n \to R^n$ has continuous first derivatives near $\mathbf{X}_0$ and

$$\frac{\partial(u_1, \ldots, u_n)}{\partial(x_1, \ldots, x_n)} \bigg|_{\mathbf{X}_0} \neq 0,$$

then $\mathbf{U} = \mathbf{f}(\mathbf{X})$ defines a coordinate transformation on a set S containing $\mathbf{X}_0$ in its interior and $\mathbf{X} = \mathbf{f}_S^{-1}(\mathbf{U})$ defines a coordinate transformation near $\mathbf{U}_0 = \mathbf{f}(\mathbf{X}_0)$. Furthermore,

$$(7) \qquad \frac{\partial(u_1, \ldots, u_n)}{\partial(x_1, \ldots, x_n)} \frac{\partial(x_1, \ldots, x_n)}{\partial(u_1, \ldots, u_n)} = 1$$

for $\mathbf{X}$ sufficiently near $\mathbf{X}_0$.

We leave the proof, which follows quickly from Theorem 6.2, to the student (Exercise T-4).

Example 6.9 Let $\mathbf{f}$ be the transformation from rectangular to polar coordinates in R^2:

$$\mathbf{f}(x, y) = \begin{bmatrix} r(x, y) \\ \theta(x, y) \end{bmatrix},$$

where

$$r = \sqrt{x^2 + y^2},$$

$$\theta = \cos^{-1} \frac{x}{\sqrt{x^2 + y^2}}$$

$$= \sin^{-1} \frac{y}{\sqrt{x^2 + y^2}} \qquad (0 \leq \theta < 2\pi).$$

(Thus, θ is the unique angle in $(0, 2\pi)$ such that $\cos \theta = x/\sqrt{x^2 + y^2}$ and $\sin \theta = y/\sqrt{x^2 + y^2}$.) The domain of $\mathbf{f}$ is all of R^2 except $(0, 0)$, where θ is undefined, and

$$\begin{bmatrix} \dfrac{\partial r}{\partial x} & \dfrac{\partial r}{\partial y} \\[2ex] \dfrac{\partial \theta}{\partial x} & \dfrac{\partial \theta}{\partial y} \end{bmatrix} = \begin{bmatrix} \dfrac{x}{\sqrt{x^2 + y^2}} & \dfrac{y}{\sqrt{x^2 + y^2}} \\[2ex] -\dfrac{y}{x^2 + y^2} & \dfrac{x}{x^2 + y^2} \end{bmatrix};$$

hence

(8)
$$\frac{\partial(r, \theta)}{\partial(x, y)} = \frac{1}{\sqrt{x^2 + y^2}} = \frac{1}{r}.$$

The inverse function theorem implies that

$$\begin{bmatrix} r \\ \theta \end{bmatrix} = \mathbf{f}(x, y)$$

defines a coordinate transformation near every $\mathbf{X}_0 \neq (0, 0)$. However, we know more: $\mathbf{f}$ is a coordinate transformation on its entire domain, with inverse

$$\begin{bmatrix} x \\ y \end{bmatrix} = \mathbf{f}^{-1}(r, \theta) = \begin{bmatrix} r \cos \theta \\ r \sin \theta \end{bmatrix}.$$

By direct calculation,

$$\frac{\partial(x, y)}{\partial(r, \theta)} = \begin{bmatrix} \cos\theta & -r\sin\theta \\ \sin\theta & r\cos\theta \end{bmatrix} = r,$$

so that

$$\frac{\partial(r, \theta)}{\partial(x, y)} \frac{\partial(x, y)}{\partial(r, \theta)} = 1,$$

as is implied by (7) and (8).

In this example (and in Example 6.5), f has an inverse on its entire domain D. A hasty reading of Theorem 6.2 might lead the student to conclude that this is so because $\partial(u, v)/\partial(x, y)$ does not vanish in D. However, this would be faulty reasoning. The inverse function theorem is a "local" theorem. It says only that if f satisfies certain hypotheses near X_0, then f has an inverse near X_0; however, f may have an inverse near every point of D, but fail to have an inverse on D, as in the following example.

Example 6.10   Let

$$f(x, y) = \begin{bmatrix} e^x \cos y \\ e^x \sin y \end{bmatrix};$$

then

$$\frac{\partial(u, v)}{\partial(x, y)} = \begin{vmatrix} e^x \cos y & -e^x \sin y \\ e^x \sin y & e^x \cos y \end{vmatrix} = e^{2x}$$

never vanishes, and Theorem 6.2 implies that f has inverse near every point of R^2. Nevertheless, f is not one-to-one on R^2, since

$$f(x, y + 2\pi) = f(x, y),$$

and therefore f does not have an inverse on R^2.

We shall meet coordinate transformations and Jacobians again when we discuss change of variables in multiple integrals (Section 4.3).

EXERCISES 3.6

1.   Show that

$$f(x, y) = \begin{bmatrix} x + y \\ 2x + 3y \end{bmatrix}$$

is one-to-one and that its range is R^2. Find f^{-1} and Jf^{-1}.

2. Repeat Exercise 1 for

$$f(x, y) = \begin{bmatrix} x - y \\ x + 4y \end{bmatrix}.$$

3. Let

$$f(x, y) = \begin{bmatrix} x^2 + 2y^2 \\ 2x^2 + y^2 \end{bmatrix}.$$

(a)   Show that f is one-to-one in any quadrant of R^2.

(b)   Find f_S^{-1}, where S is the first quadrant. What is the domain of f_S^{-1}?

(c)   Find f_T^{-1}, where T is the second quadrant. What is the domain of f_T^{-1}?

(d)   Does f have an inverse near $(0, 1)$?

4. Repeat Exercise 3 for

$$f(x, y) = \begin{bmatrix} 3x^2 - 2y^2 \\ x^2 + y^2 \end{bmatrix}.$$

5. Let f be as in Exercise 3, $\mathbf{X}_0 = (1, 2)$ and $\mathbf{U}_0 = f(\mathbf{X}_0)$. Using the inverse function theorem, find $\mathbf{J}_{U_0} f_S^{-1}$, where S is a small disk about $\mathbf{X}_0$.

6. Repeat Exercise 5 with f as defined in Exercise 4.

7. Let

$$f(x, y) = \begin{bmatrix} x^2 y + yx \\ 2xy + xy^2 \end{bmatrix}.$$

(a)   Show that f satisfies the hypotheses of the inverse function theorem at $\mathbf{X}_0 = (1, 1)$.

(b)   Find an affine transformation $\mathbf{L}$ that approximates f_S^{-1} for $\mathbf{U}$ near $\mathbf{U}_0 = f(\mathbf{X}_0)$, where S is a small disk about $\mathbf{X}_0$.

8. Repeat Exercise 7 for

$$f(x, y) = \begin{bmatrix} x^2 - y^2 \\ 2xy \end{bmatrix}$$

and $\mathbf{X}_0 = (1, -1)$.

9. Find $\partial(r, \theta, \phi)/\partial(x, y, z)$ and $\partial(x, y, z)/\partial(r, \theta, \phi)$ if

$$\begin{bmatrix} x \\ y \\ z \end{bmatrix} = \begin{bmatrix} r \cos \theta \sin \phi \\ r \sin \theta \sin \phi \\ r \cos \phi \end{bmatrix}.$$

10. Find $\partial(r, \theta, z)/\partial(x, y, z)$ and $\partial(x, y, z)/\partial(r, \theta, z)$ if

$$\begin{bmatrix} x \\ y \\ z \end{bmatrix} = \begin{bmatrix} r \cos \theta \\ r \sin \theta \\ z \end{bmatrix}.$$

11. Find:

(a) $\dfrac{\partial(x, y, z)}{\partial(x, y, z)}$, (b) $\dfrac{\partial(x, z, y)}{\partial(x, y, z)}$, (c) $\dfrac{\partial(x, y, z)}{\partial(z, y, x)}$.

THEORETICAL EXERCISES

T-1. Prove the assertions made about

$$\mathbf{f}(x, y) = \begin{bmatrix} x^2 + y^2 \\ x^2 - y^2 \end{bmatrix}$$

in Example 6.2.

T-2. Prove Theorem 6.1.

T-3. Formally derive (4) from Eq. (2), Section 3.4.

T-4. Use Theorem 6.2 to prove (7).

T-5. Set $n = 1$ in Theorem 6.2 and show that if f has a continuous derivative near x_0 and $f'(x_0) \neq 0$, then f has an inverse near x_0. Find an equation relating the derivatives of f and f^{-1}.

Integration

4.1 Multiple Integrals

In this section we define the integral of a real valued function of n variables. First we recall the definition for $n = 1$. Let f be defined on $[a, b]$ and Δ be a partition of $[a, b]$, defined by points $x_0, x_1, \ldots, x_m$ such that

$$a = x_0 < x_1 < \cdots < x_m = b.$$

The largest of the lengths of the subintervals so defined is the *norm* of Δ; it is denoted by

$$\| \Delta \| = \max_{1 \leq i \leq m} (x_i - x_{i-1}).$$

A sum of the form

$$\sigma = \sum_{i=1}^{m} f(\xi_i)(x_i - x_{i-1}),$$

where ξ_i is a fixed but arbitrary point in $[x_i, x_{i-1}]$, is called a *Riemann sum*. If there is a number I such that $|\sigma - I|$ can be made as small as we wish by taking $\| \Delta \|$ sufficiently small—regardless of how $\xi_1, \ldots, \xi_n$ are chosen in their respective intervals—then we say that f is *integrable* on $[a, b]$ and define I to be its *integral*; we write

(1)
$$I = \int_a^b f(x)\, dx.$$

Thus if f is integrable on $[a, b]$ there corresponds to each $\epsilon > 0$ a $\delta > 0$ such that

$$\left| \sigma - \int_a^b f(x)\, dx \right| < \epsilon$$

provided $|| \Delta || < \delta$ and $x_{i-1} \leq \xi_i \leq x_i$ $(1 \leq i \leq m)$. We shall call an integral of this kind an *ordinary* integral to distinguish it from the other integrals defined in this chapter.

The student should recall that if f is integrable and nonnegative on $[a, b]$, then the integral (1) is the area of the two-dimensional region T bounded by the curve $y = f(x)$, the x-axis and the lines $x = a$ and $x = b$:

$$T = \{ (x, y) \mid 0 \leq y \leq f(x), \quad a \leq x \leq b \}$$

(Fig. 1.1).

FIGURE 1.1

The most general conditions under which f is integrable on $[a, b]$ cannot be given here. A useful sufficient condition is that f be continuous on $[a, b]$, or that there exist points $a = \alpha_0 < \alpha_1 < \cdots < \alpha_k = b$ such that f is continuous in (α_{i-1}, α_i) and the right-hand limit $f(\alpha_{i-1}+)$ and the left-hand limit $f(\alpha_i-)$ exist for $1 \leq i \leq k$. (For definitions of right- and left-hand limits see Exercise T-5). In this case f is said to be *piecewise continuous* in $[a, b]$. A continuous function is piecewise continuous, but the converse is false. The graph of a piecewise continuous function is shown in Fig. 1.2.

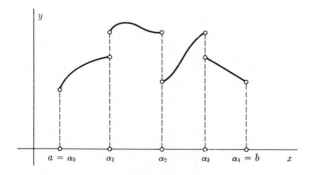

FIGURE 1.2

Double Integrals

Now suppose f is defined on a rectangle

$$S = \{(x, y) \mid a \le x \le b, \ c \le y \le d\}.$$

Let

$$a = x_0 < x_1 < \cdots < x_m = b,$$

$$c = y_0 < y_1 < \cdots < y_n = d,$$

and construct a grid on S by drawing vertical line segments through $x = x_i$ from $y = c$ to $y = d$ $(1 \le i \le m)$ and horizontal lines through $y = y_j$ from $x = a$ to $x = b$ $(1 \le j \le n)$. Let Δ be the resulting partition of S into mn subrectangles

$$S_{ij} = \{(x, y) \mid x_{i-1} \le x \le x_i, \ y_{j-1} \le y \le y_j\} \quad (1 \le i \le m, \ 1 \le j \le n)$$

(Fig. 1.3), and let $\| \Delta \|$, the *norm* of Δ, be the length of the longest side appearing in any subrectangle.

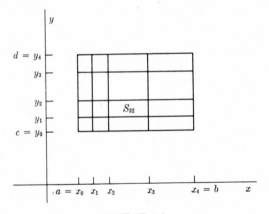

FIGURE 1.3

The sum

(2)
$$\sigma = \sum_{i=1}^{m} \sum_{j=1}^{n} f(\mathbf{X}_{ij}) (x_i - x_{i-1}) (y_j - y_{j-1}),$$

where $\mathbf{X}_{ij}$ is an arbitrary point in S_{ij}, is called a *Riemann sum*. If there is a number I such that $| \sigma - I |$ can be made as small as we wish by taking $\| \Delta \|$ sufficiently small, no matter how $\mathbf{X}_{ij}$ is chosen in S_{ij}, then we say that f is *integrable on* S. We call I the *double integral of f over S* and write

(3)
$$I = \int_{S} f \, dx \, dy;$$

when dealing with a specific function, say $f(x, y) = x^2 + 2xy$, we write

$$I = \int_S (x^2 + 2xy)\, dx\, dy.$$

Another common notation for the double integral is

$$I = \iint_S f\, dx\, dy;$$

however we shall use (3).

Example 1.1   Suppose $f(x, y) = 1$; then

$$\sigma = \sum_{i=1}^{m} \sum_{j=1}^{n} (x_i - x_{i-1})(y_j - y_{j-1})$$

$$= \left(\sum_{i=1}^{m} (x_i - x_{i-1})\right)\left(\sum_{j=1}^{n} (y_j - y_{j-1})\right)$$

$$= (x_m - x_0)(y_n - y_0)$$

$$= (b - a)(d - c).$$

Thus

$$\int_S 1\, dx\, dy = \int_S dx\, dy$$

is the area of S. We can also interpret the integral as the volume of a rectangular box with base S and unit height.

The following theorem, whose proof we omit, gives a useful sufficient condition for the existence of the double integral.

Theorem 1.1   If $f: R^2 \to R$ is continuous on a rectangle S then

$$\int_S f\, dx\, dy$$

exists.

The next example illustrates how the continuity of f leads to the existence of the integral.

Example 1.2   Let

$$S = \{(x, y) \mid -1 \le x \le 1,\quad 0 \le y \le 2\}$$

(Fig. 1.4) and consider

$$\int_S (x + 3y)\, dx\, dy.$$

A typical Riemann sum is

$$\sigma = \sum_{i=1}^{m} \sum_{j=1}^{n} (\xi_i + 3\eta_j)\, (x_i - x_{i-1})\, (y_j - y_{j-1}),$$

which can be rewritten

$$\sigma = \sigma_0 + (\sigma - \sigma_0)$$

where

$$\sigma_0 = \sum_{i=1}^{m} \sum_{j=1}^{n} \left(\frac{x_i + x_{i-1}}{2} + 3\frac{y_j + y_{j-1}}{2} \right) (x_i - x_{i-1})\, (y_j - y_{j-1})$$

and

$$(4) \quad \sigma - \sigma_0 = \sum_{i=1}^{m} \sum_{j=1}^{n} \left[\xi_i + 3\eta_j - \frac{x_i + x_{i-1}}{2} - 3\frac{y_j + y_{j-1}}{2} \right]$$

$$\times (x_i - x_{i-1})\, (y_j - y_{j-1}).$$

Since $x_{i-1} < \xi_i < x_i$,

$$\left| \xi_i - \frac{x_i + x_{i-1}}{2} \right| \le \frac{x_i - x_{i-1}}{2};$$

similarly

$$\left| \eta_j - \frac{y_j + y_{j-1}}{2} \right| \le \frac{y_j - y_{j-1}}{2}.$$

Hence

$$\left| \xi_i + 3\eta_j - \frac{x_i + x_{i-1}}{2} - 3\frac{y_j + y_{j-1}}{2} \right| \le 2\,\|\,\Delta\,\|$$

and

$$|\sigma - \sigma_0| \le 2\,\|\,\Delta\,\| \sum_{i=1}^{m} \sum_{j=1}^{n} (x_i - x_{i-1})\, (y_j - y_{j-1})$$

$$= 2\,\|\,\Delta\,\| \,(2)\,(2) = 8\,\|\,\Delta\,\|,$$

from (4) and Example 1.1, with $a = x_0 = -1$, $b = x_m = 1$, $c = y_0 = 0$,

$d = y_n = 2$. Now

$$\sigma_0 = \tfrac{1}{2} \sum_{i=1}^{m} \sum_{j=1}^{n} (x_i^2 - x_{i-1}^2)\,(y_j - y_{j-1}) + \tfrac{3}{2} \sum_{i=1}^{m} \sum_{j=1}^{n} (x_i - x_{i-1})\,(y_j^2 - y_{j-1}^2)$$

$$= \tfrac{1}{2} \Big(\sum_{i=1}^{m} (x_i^2 - x_{i-1}^2) \Big) \Big(\sum_{j=1}^{n} (y_j - y_{j-1}) \Big)$$

$$+ \tfrac{3}{2} \Big(\sum_{i=1}^{n} (x_i - x_{i-1}) \Big) \Big(\sum_{j=1}^{m} (y_j^2 - y_{j-1}^2) \Big)$$

$$= \tfrac{1}{2} (x_n^2 - x_0^2)\,(y_m - y_0) + \tfrac{3}{2} (x_n - x_0)\,(y_m^2 - y_0^2)$$

$$= 12.$$

Hence

$$|\, \sigma - 12 \,| \leq 8\, ||\, \Delta \,||$$

and therefore

$$\int_S (x + 3y)\, dx\, dy = 12.$$

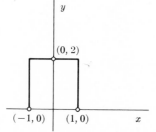

FIGURE 1.4

If f is integrable over S and $f(x, y) \geq 0$ in S, then the volume of the three dimensional region

$$T = \{ (x, y, z) \mid a \leq x \leq b, \quad c \leq y \leq d, \quad 0 \leq z \leq f(x, y) \}$$

(Fig. 1.5) is defined by

$$V(T) = \int_S f\, dx\, dy.$$

This definition is motivated by the interpretation of a typical term in the Riemann sum (2),

$$f(\mathbf{X}_{ij})\,(x_i - x_{i-1})\,(y_j - y_{j-1}),$$

as the volume of a rectangular parallelepiped with base S_{ij} and height $f(\mathbf{X}_{ij})$. Geometrically, the collection of these parallelepipeds approximates T increasingly well as $\| \Delta \|$ is made small; therefore it is reasonable to define $V(T)$ so that it is closely approximated by the sum of the volumes of the parallelepipeds.

If $f = 1$, then T is the rectangular box of Example 1.1, where we saw that

$$\int_S dx\,dy = (b - a)(d - c),$$

which is the volume of the box; thus the new definition of volume agrees with the old in this case. This is also true for other three-dimensional geometrical shapes whose volumes have been previously defined without recourse to the double integral, as we shall see in later sections after developing more practical ways to evaluate double integrals.

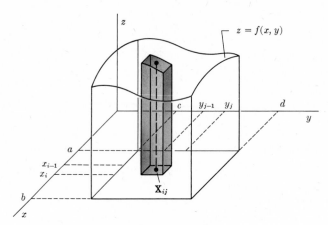

FIGURE 1.5

If f is integrable and nonpositive on S, then we define the volume of

$$T = \{ (x, y, z) \mid a \le x \le b, \quad c \le y \le d, \quad f(x, y) \le z \le 0 \}$$

to be

$$V(T) = -\int_S f\,dx\,dy.$$

If f is an arbitrary integrable function on S then we define

$$f^+(x, y) = \begin{cases} f(x, y) & \text{if} \quad f(x, y) \ge 0, \\ 0 & \text{if} \quad f(x, y) < 0; \end{cases}$$

and

$$f^-(x, y) = \begin{cases} f(x, y) & \text{if} \quad f(x, y) \leq 0, \\ 0 & \text{if} \quad f(x, y) > 0. \end{cases}$$

It can be shown that f^+ and f^- are integrable on S, and the volume of

$$T = \{ (x, y, z) \mid a \leq x \leq b,\ c \leq y \leq d,\ \min(0, f(x, y)) \leq z \leq \max(0, f(x, y)) \}$$

is defined to be

$$V(T) = \int_S f^+ \, dx \, dy - \int_S f^- \, dx \, dy.$$

Example 1.3   Let $S = \{ 0 \leq x \leq 1,\ 0 \leq y \leq 1 \}$ and $f(x, y) = x - y$; then T is shown in Fig. 1.6 and

$$f^+(x, y) = \begin{cases} x - y & (0 \leq y \leq x \leq 1), \\ 0 & (0 \leq x < y \leq 1); \end{cases}$$

$$f^-(x, y) = \begin{cases} x - y & (0 \leq x \leq y \leq 1), \\ 0 & (0 \leq y < x \leq 1). \end{cases}$$

The volume of T is given by

$$V(T) = \int_S f^+ \, dx \, dy - \int_S f^- \, dx \, dy,$$

which is calculated in Exercise 38, Section 4.2.

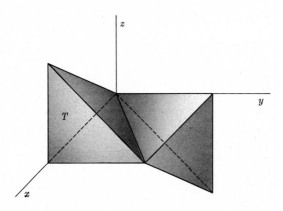

FIGURE 1.6

The assumption that the domain of f is rectangular was introduced merely for convenience. If f is defined on any bounded set R in the plane, then let S be a rectangle which contains R and define F on S by

$$F(x, y) = \begin{cases} f(x, y) & ((x, y) \text{ in } R), \\ 0 & ((x, y) \text{ in } S \text{ but not in } R). \end{cases}$$

Then define

$$\int_R f \, dx \, dy = \int_S F \, dx \, dy$$

if the integral on the right exists.

It can be shown that this definition is independent of the particular rectangle S chosen to enclose R; in fact the definition can be recast so that it does not involve an enclosing rectangle at all.

Example 1.4 Let g_1 and g_2 be continuous and $g_1(x) \leq g_2(x)$ in $[a, b]$. Then

$$R = \{(x, y) \mid a \leq x \leq b, \quad g_1(x) \leq y \leq g_2(x)\}$$

(Fig. 1.7) is an example of a more general type of domain which occurs in applications of the double integral.

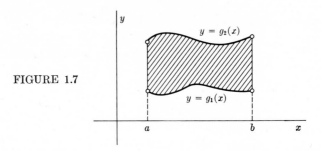

FIGURE 1.7

Example 1.5 Let h_1 and h_2 be continuous and $h_1(y) \leq h_2(y)$ in $[c, d]$. Then

$$R = \{(x, y) \mid h_1(y) \leq x \leq h_2(y), \quad c \leq y \leq d\}$$

(Fig. 1.8) is an example of another type of domain which occurs in applications of the double integral.

The following theorem, whose proof we omit, gives sufficient conditions for the existence of the double integral which are more general than those given in Theorem 1.1.

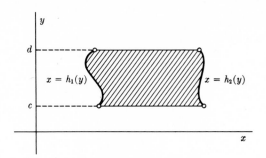

FIGURE 1.8

Theorem 1.2   If R consists of a piecewise smooth closed curve (Definition 5.4, Section 1.5) and the bounded region enclosed by it, and if f is continuous in R except possibly on a finite number of segments of smooth curves, then

$$\int_R f \, dx \, dy$$

exists.

The regions depicted in Figs. 1.7 and 1.8 have the properties required in this theorem, provided g_1, g_2, h_1, and h_2 are piecewise smooth.

Example 1.6   Let R be the shaded region bounded by the curves $y = -x^2 + 9$ and $y = x^2 - 1$ (Fig. 1.9), and let

$$f(x, y) = \begin{cases} x^2 + y^2 & (x \le 0), \\ x^2 + y^2 + 1 & (x > 0). \end{cases}$$

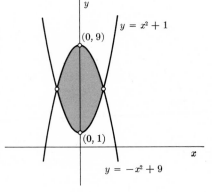

FIGURE 1.9

Then f is continuous in R except on the smooth curve defined by $x = 0$, $1 \leq y \leq 9$; hence f is integrable over R.

Theorem 1.2 tells us only that the integral exists under certain conditions; it does not provide a way to find its value (except, of course, by considering the Riemann sums). We shall consider this important matter in the next section.

If f is nonnegative and integrable over R then $\displaystyle\int_R f\, dx\, dy$ is the *volume* of

the three-dimensional region

$$T = \{ (x, y, z) \mid 0 \leq z \leq f(x, y), \quad (x, y) \text{ in } R \};$$

the motivation for this definition is the same as that given earlier for the case where R is a rectangle. If $f = 1$, then

$$\int_R f\, dx\, dy = \int_R dx\, dy$$

is the area of R. We have already seen (Example 1.1) that this is consistent with the usual definition if R is a rectangle. The theory of ordinary integrals also provides a definition of area for a more general two dimensional region. We leave it to the student to define the volume of

(5) $T = \{ (x, y, z) \mid \min(f(x, y), 0) \leq z \leq \max(f(x, y), 0) \}$

if f has negative values in R (Exercise T-1).

The following theorem, which we state without proof, shows that the double integral has properties already familiar to the student from his study of ordinary integrals.

Theorem 1.3   Let f and g be integrable over a domain R. Then
(a) $f + g$ is integrable over R and

$$\int_R (f + g)\, dx\, dy = \int_R f\, dx\, dy + \int_R g\, dx\, dy;$$

(b)   if k is a constant then kf is integrable over R and

$$\int_R (kf)\, dx\, dy = k \int_R f\, dx\, dy;$$

(c)   if $f(x, y) \geq 0$ for all (x, y) in R, then

$$\int_R f\, dx\, dy \geq 0,$$

and

(d)   If R consists of regions R_1 and R_2 with no points except possibly on their bounding curves in common, then

$$\int_R f \, dx \, dy = \int_{R_1} f \, dx \, dy + \int_{R_2} f \, dx \, dy.$$

Parts (a), (b), and (c) can be established by considering the Riemann sums that approximate the various integrals (Exercise T-2). Figure 1.10 illustrates two kinds of domains to which (d) applies; in both cases R comprises R_1 and R_2.

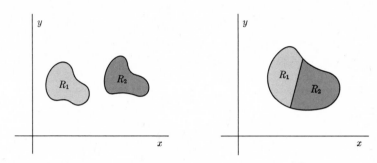

FIGURE 1.10

Corollary 1.1   If in addition to the assumptions of Theorem 1.2, $g(x, y) \leq f(x, y)$ in R, then

$$\int_R g \, dx \, dy \leq \int_R f \, dx \, dy.$$

Proof.   From (a) and (b), $F = f - g$ is integrable over R and

$$\int_R F \, dx \, dy = \int_R f \, dx \, dy - \int_R g \, dx \, dy;$$

from (c)

$$\int_R F \, dx \, dy \geq 0,$$

which yields the corollary.

Corollary 1.2   If g and $|\,g\,|$ are integrable over R then

(6)
$$\left| \int_R g \, dx \, dy \right| \leq \int_R |\,g\,| \, dx \, dy.$$

Proof. Since $g \leq |g|$ and $-g \leq |g|$, it follows from Corollary 1.1 that

$$\int_R g \, dx \, dy \leq \int_R |g| \, dx \, dy$$

and

$$-\int_R g \, dx \, dy \leq \int_R |g| \, dx \, dy.$$

These two inequalities together imply (6).

Multiple Integrals

We now give the general definition for the integral of a real valued function of n variables. A *rectangular parallelepiped* in R^n is a set of the form

(7) $$S = \{\mathbf{X} \mid a_i \leq x_i \leq b_i, \quad i = 1, \ldots, n\};$$

the sides of S have length $b_1 - a_1, \ldots, b_n - a_n$ and its *volume* is

$$V(S) = (b_1 - a_1) \cdots (b_n - a_n).$$

Example 1.7 If $n = 1$, S is a line segment, and $V(S)$ is its length; if $n = 2$, S is a rectangle and $V(S)$ its area, and if $n = 3$, S is a rectangular box and $V(S)$ its familiar volume.

If for each $i = 1, \ldots, n$ we cut through S with planes

$$x_i = x_{i0}, \qquad x_i = x_{i1}, \ldots, \qquad x_i = x_{in_i},$$

where

$$a_i = x_{i0} < x_{i1} < \cdots < x_{in_i} = b_i,$$

then we obtain a partition Δ of S into subparallelepipeds $S_1, \ldots, S_k$; furthermore

$$V(S) = V(S_1) + \cdots + V(S_k).$$

(Fig. 1.11 depicts the situation for $n = 3$.) The order in which the subparallelepipeds are numbered is immaterial.

FIGURE 1.11

The *norm* of Δ, denoted by $\| \Delta \|$, is the length of the largest side appearing on $S_1, \ldots, S_k$. Now suppose f is defined on S; in each S_i choose an arbitrary point $\mathbf{X}_i$ and form the Riemann sum,

$$\sigma = \sum_{i=1}^{k} f(\mathbf{X}_i) V(S_i).$$

We say that f *is integrable over* S if there is a number I such that $| \sigma - I |$ can be made as small as we wish by taking $\| \Delta \|$ sufficiently small, independently of how $\mathbf{X}_i$ is chosen in S_i; I is called the *n-fold multiple integral of f over S*, and we write

$$I = \int_S f \, dx_1 \cdots dx_n.$$

If f is defined on an arbitrary bounded set R in R^n, then let S be a rectangular parallelepiped of the form (7) which contains S, and define

$$F(\mathbf{X}) = \begin{cases} f(\mathbf{X}) & (\mathbf{X} \text{ in } R), \\ 0 & (\mathbf{X} \text{ in } S \text{ but not in } R) \end{cases}$$

and

$$\int_R f \, dx_1 \cdots dx_n = \int_S F \, dx_1 \cdots dx_n$$

if the integral on the right exists. If $f = 1$, then

$$\int_R dx_1 \cdots dx_n$$

is the *n-dimensional volume* of T.

We shall not state general conditions under which n-fold multiple integrals exist. In succeeding sections there are many examples of multiple integrals whose existence will be demonstrated by direct calculation.

Theorem 1.1 and its corollaries, with $dx \, dy$ replaced by $dx_1 \cdots dx_n$, also hold for n-fold multiple integrals.

The student has no doubt noticed that $dx_1, \ldots, dx_n$ as used in this section do not stand for linear transformations, as they did in Chapter 2. They are nothing more than bookkeeping devices to identify the variables of integration; indeed many writers simply omit them and write, for instance,

$$\int_R f \, dx \, dy \, dz = \int_R f.$$

It is possible, however, to ascribe a fictitious geometric meaning to these symbols which aids in understanding multiple integrals. Consider $dx_1, \ldots, dx_n$ to be small, or *differential*, lengths measured parallel to the $x_1, \ldots, x_n$ axes, respectively. Then

$$dV_n = dx_1 \cdots dx_n$$

can be viewed as a *differential* volume (strictly an intuitive notion, which we shall not attempt to define precisely), and we can write an n-fold multiple integral as

(8) $$\int_R f(\mathbf{X}) dx_1 \cdots dx_n = \int_R f(\mathbf{X}) \, dV_n.$$

If $n = 1, 2,$ or 3, we write simply

$$dV_1 = dx \qquad \text{(differential length)},$$
$$dV_2 = dA \qquad \text{(differential area)},$$

and

$$dV_3 = dV.$$

We now view the multiple integral (8) as an integral with respect to n-dimensional volume. This idea will be exploited in the following sections.

EXERCISES 4.1

In Exercises 1 through 12 let

$$S = \{(x, y) \mid -1 \le x \le 1, \quad 0 \le y \le 2\}$$

and $f(x, y) = x + 3y$. Compute the Riemann sum associated with the given partition Δ and the given choice of $\mathbf{X}_{ij}$.

1. Δ: $x_0 = -1 < 0 < 1 = x_2; \ y_0 = 0 < 1 < 2 = y_2.$

$$\mathbf{X}_{ij} = \left(\frac{x_{i-1} + x_i}{2}, \frac{y_{j-1} + y_j}{2} \right).$$

2. Δ: $x_0 = -1 < 0 < 1 = x_2, \ y_0 = 0 < \frac{1}{2} < 1 < \frac{3}{2} < 2 = y_4.$

$$\mathbf{X}_{ij} = \left(\frac{x_{i-1} - x_i}{2}, \frac{y_{j-1} + y_j}{2} \right).$$

3. Δ: $x_0 = -1 < -\frac{1}{2} < 0 < \frac{1}{2} < 1 = x_4, \ y_0 = 0 < \frac{1}{2} < 1 < \frac{3}{2} < 2 = y_4.$

$$\mathbf{X}_{ij} = \left(\frac{x_{i-1} + x_i}{2}, \frac{y_{j-1} + y_j}{2} \right).$$

4. Let Δ be as in Exercise 3 and $\mathbf{X}_{ij} = (x_{i-1}, y_{j-1})$.

5. Let Δ be as in Exercise 3 and $\mathbf{X}_{ij} = (x_i, y_{j-1})$.

6. Let Δ be as in Exercise 3 and $\mathbf{X}_{ij} = (x_{i-1}, y_j)$.

7. Let Δ be as in Exercise 3 and $\mathbf{X}_{ij} = (x_i, y_j)$.

In Exercises 8 through 12 show that the Riemann sum can be made to be as close to 12 (see Example 1.2) as we wish, by taking m and n sufficiently large. Use the formulas

$$\sum_{r=1}^{k} r = \frac{k(k+1)}{2}, \qquad \sum_{r=1}^{k} r^2 = \frac{k(k+1)(2k+1)}{6}.$$

8. Δ: $x_i = -1 + 2i/n \ (0 \le i \le n), \ y_j = 2j/m \ (0 \le j \le m)$;

$$\mathbf{X}_{ij} = \left(\frac{x_i + x_{i-1}}{2}, \frac{y_j + y_{j-1}}{2} \right).$$

9. Let Δ be as in Exercise 8 and $\mathbf{X}_{ij} = (x_{i-1}, y_{j-1})$.

10. Let Δ be as in Exercise 8 and $\mathbf{X}_{ij} = (x_i, y_{j-1})$.

11. Let Δ be as in Exercise 8 and $\mathbf{X}_{ij} = (x_{i-1}, y_j)$.

12. Let Δ be as in Exercise 8 and $\mathbf{X}_{ij} = (x_i, y_j)$.

13. Use the method of Example 1.2 to compute

$$\int_S (2x + 3y) \, dx \, dy, \quad \text{where } S = \{ (x, y) \mid 2 \le x \le 4, \ 1 \le y \le 3 \}.$$

14. Use the method of Example 1.2 to compute

$$\int_S (3x + y) \, dx \, dy, \quad \text{where } S = \{ (x, y) \mid 1 \le x \le 4, \ 2 \le y \le 6 \}.$$

15. Use the result of Exercise 13 to obtain the volume of

$$T = \{ (x, y, z) \mid 2 \le x \le 4, \ 1 \le y \le 3, \ 0 \le z \le 2x + 3y \}.$$

16. Use the result of Exercise 14 to obtain the volume of

$$T = \{ (x, y, z) \mid 1 \le x \le 4, \ 2 \le y \le 6, \ 0 \le z \le 3x + y \}.$$

17. Find the volume of

$$T = \{ (x, y, z) \mid -0 \le x \le 1, \ 0 \le y \le 1, \ x + y \le z \le 0 \}.$$

THEORETICAL EXERCISES

T-1. Define the volume of
$$T = \{(x, y, z) \mid (x, y) \text{ in } R, \ \min(f(x,y), 0) \le z \le \max(f(x, y), 0)\},$$
where R is a region in R^2.

T-2. Prove (a), (b), and (c) of Theorem 1.3.

T-3. Suppose f is integrable and $m \le f(x, y) \le M$ on a bounded set S. Show that

$$mA(S) \le \int_S f(x, y) \ dx \ dy \le MA(S).$$

T-4. Let S be the subset of R^n defined by
$$S = \{\mathbf{X} \mid a_i \le x_i \le b_i \quad (i = 1, \ldots, n)\}.$$

Suppose f_i is a continuous real valued function of the single variable x_i on $[a_i, b_i]$. Show directly from the definition of the integral that

$$\int_S f_1 f_2 \cdots f_n \ dV = \left(\int_{a_1}^{b_1} f_1(x_1) \ dx_1\right)\left(\int_{a_2}^{b_2} f_2(x_2) \ dx_2\right) \cdots \left(\int_{a_n}^{b_n} f_n(x_n) \ dx_n\right).$$

T-5. Suppose f is defined on (a, b) and $a < \alpha < b$. Then we define the *left-hand limit* $f(\alpha-)$ to be $\lim\limits_{x \to \alpha} f_1(x)$, where f_1 is defined by $f_1(x) = f(x)$ if $a \le x < \alpha$, and undefined elsewhere (See also Definition 1.7, Section 2.1).

(a) Give an analogous definition for the *right-hand limit* $f(\alpha+)$.

(b) Show that $\lim\limits_{x \to \alpha} f(x)$ exists if and only if $f(\alpha+) = f(\alpha-)$. In this case, what is $\lim\limits_{x \to \alpha} f(x)$?

4.2 Iterated Integrals

It is theoretically possible to evaluate a multiple integral from its Riemann sums, as we did in Example 1.2. However, this is impractical for all but the simplest integrals. In this section we show how to evaluate an n-fold multiple integral by considering n successive integrals, each with respect to a single variable.

If f is continuous on the rectangle

$$S = \{(x, y) \mid a \leq x \leq b, \quad c \leq y \leq d\}$$

we can define a function F on $[c, d]$ by

$$F(y) = \int_a^b f(x, y)\, dx,$$

where y is considered to be constant as we integrate with respect to x. It can be shown that F is continuous, and therefore integrable, on $[c, d]$; then

$$J_1 = \int_c^d F(y)\, dy = \int_c^d \left(\int_a^b f(x, y)\, dx \right) dy$$

is an *iterated integral* of f over S. Another iterated integral can be obtained by integrating with respect to y first,

$$G(x) = \int_c^d f(x, y)\, dy,$$

and then with respect to x:

$$J_2 = \int_a^b G(x)\, dx = \int_a^b \left(\int_c^d f(x, y)\, dy \right) dx.$$

Example 2.1   Let $f(x, y) = x + 3y$ and

$$S = \{(x, y) \mid -1 \leq x \leq 1, \quad 0 \leq y \leq 2\};$$

then

$$F(y) = \int_{-1}^1 (x + 3y)\, dx$$

$$= \left(\frac{x^2}{2} + 3xy \right) \bigg|_{x=-1}^1 = (\tfrac{1}{2} + 3y) - (\tfrac{1}{2} - 3y) = 6y.$$

Hence

$$J_1 = \int_0^2 F(y)\, dy$$

$$= \int_0^2 6y\, dy = 3y^2 \bigg|_0^2 = 12.$$

Also,

$$G(x) = \int_0^2 (x + 3y) \, dx$$

$$= (xy + \tfrac{3}{2}y^2) \Big|_{y=0}^{2} = 2x + 6$$

and

$$J_2 = \int_{-1}^{1} (2x + 6) \, dx$$

$$= (x^2 + 6x) \Big|_{-1}^{1} = 12.$$

We have shown previously (Example 1.2), that

$$\int_S (x + 3y) \, dx \, dy = 12;$$

thus, in this example the two iterated integrals are equal and their common value is the double integral. That this is no accident is indicated by the following theorem.

Theorem 2.1 If the double integral of f over the rectangle S exists, and if either of the iterated integrals J_1 and J_2 exist, then both exist and

$$\int_S f(x, y) \, dA = \int_a^b \left(\int_c^d f(x, y) \, dy \right) dx$$

$$= \int_c^d \left(\int_a^b f(x, y) \, dx \right) dy.$$

We shall not prove this theorem, nor other similar results that are given below. Rather, we ask the student to accept it on the basis of the following geometric evidence.

If $f(x, y) \geq 0$ and T is the three-dimensional region bounded above by $z = f(x, y)$ and below by S, then

$$F(\eta) = \int_a^b f(x, \eta) \, dx$$

is the area of the cross section cut out of T by the plane $y = \eta$ (Fig. 2.1).

If $y_j \leq \eta_j \leq y_{j-1}$, then

$$F(\eta_j)(y_j - y_{j-1})$$

approximates the volume of that part of T between the planes $y = y_{j-1}$ and $y = y_j$; hence the Riemann sum

$$\sum_{j=1}^{n} F(\eta_j)(y_j - y_{j-1}),$$

which approximates

$$\int_c^d \left(\int_a^b f(x, y) \, dx \right) dy,$$

also approximates the volume of T, which is given by the double integral. Hence it is to be expected that $J_1 = I$.

By considering cross sections of T cut out by planes $x = $ constant, we also find that $J_2 = I$.

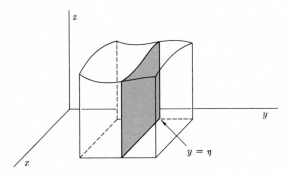

FIGURE 2.1

We shall adopt the notation

$$\int_c^d \left(\int_a^b f(x, y) \, dx \right) dy = \int_c^d dy \int_a^b f(x, y) \, dx,$$

and

$$\int_a^b \left(\int_c^d f(x, y) \, dy \right) dx = \int_a^b dx \int_c^d f(x, y) \, dy,$$

which exhibits the order of integration without cumbersome parentheses.

Suppose g_1 and g_2 are continuous and $g_1(x) \leq g_2(x)$ in $[a, b]$, and let f be defined on

$$R = \{ (x, y) \mid a \leq x \leq b, \quad g_1(x) \leq y \leq g_2(x) \}$$

(Fig. 2.2). For each fixed x in $[a, b]$ let

$$G(x) = \int_{g_1(x)}^{g_2(x)} f(x, y) \, dy;$$

then

$$\int_a^b G(x) \, dx = \int_a^b \left(\int_{g_1(x)}^{g_2(x)} f(x, y) \, dy \right) dx = \int_a^b dx \int_{g_1(x)}^{g_2(x)} f(x, y) \, dy$$

is also an iterated integral. It can be shown that if this iterated integral and the double integral of f over R both exist, then they are equal:

(1) $$\int_R f(x, y) \, dA = \int_a^b dx \int_{g_1(x)}^{g_2(x)} f(x, y) \, dy.$$

Example 2.2   Let $f = 1$, then

$$\int_a^b dx \int_{g_1(x)}^{g_2(x)} dy = \int_a^b (g_2(x) - g_1(x)) \, dx = \int_a^b g_2(x) \, dx - \int_a^b g_1(x) \, dx.$$

The first integral is the area under the curve $y = g_2(x)$ and the second is the area under the curve $y = g_1(x)$ $(a \leq x \leq b)$; the difference is the area of R, which is $\int_R dA$ (Fig. 2.2).

To convert a double integral into an iterated integral, one must specify the limits of integration in the latter. This problem can be solved intuitively by considering the multiple integral as a sum of symbolic terms of the form

$$f(x, y) \, dx \, dy,$$

and viewing the iterated integral as a way of computing this sum by means of a specific order of summation. Thus,

(2) $$dx \int_{g_1(x)}^{g_2(x)} f(x, y) \, dy$$

is the sum over the shaded strip in Fig. 2.2, and the sum of all terms of the form (2) is the double integral, which leads to (1).

Example 2.3   Let it be required to find

$$I = \int_R (x + y) \, dA,$$

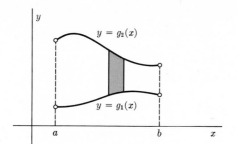

FIGURE 2.2

where R is the domain bounded by the parabola $y = x^2$ and the line $y = 2x$. We first sum $(x + y) \, dx \, dy$ over the vertical strip shown in Fig. 2.3 to obtain

$$dx \int_{x^2}^{2x} (x + y) \, dy,$$

and then over all such strips $(0 \le x \le 2)$ to obtain

$$I = \int_0^2 dx \int_{x^2}^{2x} (x + y) \, dy$$

$$= \int_0^2 \left[\left(xy + \frac{y^2}{2} \right) \Big|_{y=x^2}^{2x} \right] dx$$

$$= \int_0^2 \left(4x^2 - x^3 - \frac{x^4}{2} \right) dx$$

$$= \left(\frac{4}{3} x^3 - \frac{x^4}{4} - \frac{x^5}{10} \right) \Big|_0^2 = \frac{52}{15}.$$

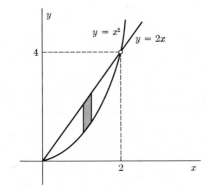

FIGURE 2.3

Example 2.4   The double integral of Example 2.3 can also be calculated by means of a second iterated integral, since the equations of the boundaries of R can also be written as $x = y/2$ and $x = \sqrt{y}$ $(0 \leq y \leq 4)$. Summing $(x + y)\, dx\, dy$ over the horizontal strip in Fig. 2.4 yields

$$dy \int_{y/2}^{\sqrt{y}} (x + y)\, dx$$

and then summing over all such horizontal strips yields

$$I = \int_0^4 dy \int_{y/2}^{\sqrt{y}} (x + y)\, dx$$

$$= \int_0^4 \left[\left(\frac{x^2}{2} + xy \right) \bigg|_{x=y/2}^{\sqrt{y}} \right] dy$$

$$= \int_0^4 \left(\frac{y}{2} + y^{3/2} - \frac{5}{8} y^2 \right) dy$$

$$= \left(\frac{y^2}{4} + \frac{2}{5} y^{5/2} - \frac{5}{24} y^3 \right) \bigg|_0^4 = \frac{52}{15}.$$

FIGURE 2.4

Examples 2.3 and 2.4 show that I can be evaluated by either of the iterated integrals, and, since they are roughly of the same difficulty, neither has an advantage over the other. The next example depicts a situation in which one of the iterated integrals is definitely simpler than the other.

Example 2.5   Find the area A of the region R bounded by $y = x^2 + 1$ and $y = 9 - x^2$ (Fig. 2.5).

Summing over the vertical strip in Fig. 2.5 yields

$$dx \int_{x^2+1}^{9-x^2} dy;$$

hence

$$A = \int_{-2}^{2} dx \int_{x^2+1}^{9-x^2} dy$$

$$= \int_{-2}^{2} \left[(9 - x^2) - (x^2 + 1) \right] dx$$

$$= 2 \int_{0}^{2} (8 - 2x^2) \, dx$$

$$= 2 \left(8x - \frac{2}{3} x^3 \right) \Big|_{0}^{2} = \frac{64}{3}.$$

To find A by means of an iterated integral with the opposite order of integration requires extra care. Summing over a horizontal strip yields

$$dy \int_{-\sqrt{9-y}}^{\sqrt{9-y}} dx$$

if $1 \leq y \leq 5$ and

$$dy \int_{-\sqrt{y-1}}^{\sqrt{y-1}} dx$$

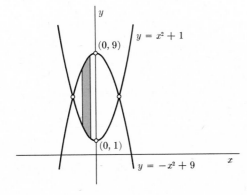

$y = x^2 + 1$

$(0, 9)$

$(0, 1)$

$y = -x^2 + 9$

FIGURE 2.5

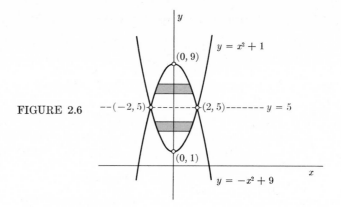

FIGURE 2.6

if $5 \leq y \leq 9$ (Fig. 2.6); thus

$$A = \int_1^5 dy \int_{-\sqrt{9-y}}^{\sqrt{9-y}} dx + \int_5^9 dy \int_{-\sqrt{y-1}}^{\sqrt{y-1}} dx$$

$$= 2 \int_1^5 \sqrt{9-y}\, dx + 2 \int_5^9 \sqrt{y-1}\, dy$$

$$= -\frac{4}{3}(9-y)^{3/2} \Big|_1^5 + \frac{4}{3}(y-1)^{3/2} \Big|_5^9 = \frac{64}{3}.$$

It is sometimes easier to evaluate an iterated integral by replacing it with an equivalent iterated integral with the opposite order of integration. The determination of the limits of integration of the latter is a geometrical problem.

Example 2.6   Let

$$J = \int_{1/2}^1 dy \int_{-\sqrt{1-y^2}}^{\sqrt{1-y^2}} f(x, y)\, dx,$$

and suppose we wish to find an equivalent iterated integral with the order of integration reversed. For each y in $[\frac{1}{2}, 1]$,

$$dy \int_{-\sqrt{1-y^2}}^{\sqrt{1-y^2}} f(x, y)\, dx$$

is a symbolic sum over a horizontal strip from $(-\sqrt{1-y^2}, y)$ to $(\sqrt{1-y^2}, y)$ (Fig. 2.7a). As y varies from $\frac{1}{2}$ to 1, these strips cover the part of the unit circle for which $\frac{1}{2} \leq y \leq 1$ (Fig. 2.7b). The same area can be covered first be summing over vertical strips from $(x, \frac{1}{2})$ to $(x, \sqrt{1-x^2})$

to obtain

$$dx \int_{1/2}^{\sqrt{1-x^2}} f(x, y) \; dy \qquad \left(-\frac{\sqrt{3}}{2} \le x \le \frac{\sqrt{3}}{2}\right)$$

(Fig. 2.7c), and then summing over all such strips to obtain

$$J = \int_{-\sqrt{3}/2}^{\sqrt{3}/2} dx \int_{1/2}^{\sqrt{1-x^2}} f(x, y) \; dy.$$

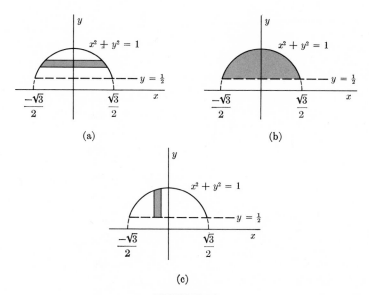

(a)

(b)

(c)

FIGURE 2.7

Iterated Integrals in n Dimensions

An iterated integral in n dimensions is of the form

$$J = \int_{a}^{b} dx_1 \int_{\eta_1(x_1)}^{\eta_2(x_1)} dx_2 \cdots \int_{\psi_1(x_1,\ldots,x_{n-2})}^{\psi_2(x_1,\ldots,x_{n-2})} dx_{n-1} \int_{\varphi_1(x_1,\ldots,x_{n-1})}^{\varphi_2(x_1,\ldots,x_{n-1})} f(x_1,\ldots,x_n) \; dx_n;$$

here the integration is with respect to x_n first, then $x_{n-1},\ldots,$ then x_1. Iterated integrals in n variables can be written with $n!$ different orders of integration.

Example 2.7 Let $n = 3$ and

$$R = \{(x, y, z) \mid x^2 + y^2 + z^2 \le 1, \quad x \ge 0, \quad y \ge 0, \quad z \ge 0\};$$

thus R is the part of the unit sphere that lies in the first octant (Fig. 2.8a).
Suppose we wish to express

$$I = \int_R f(x, y, z) \, dx \, dy \, dz$$

as an iterated integral where the integrations are with respect to x, y, and
z, in that order. First, summing from $x = 0$ to $x = \sqrt{1 - y^2 - z^2}$ gives

$$dz \, dy \int_0^{\sqrt{1-y^2-z^2}} f(x, y, z) \, dx,$$

the sum over the parallelepiped shown in Fig. 2.8b; summing this from
$y = 0$ to $y = \sqrt{1 - z^2}$ yields

(3) $$dz \int_0^{\sqrt{1-z^2}} dy \int_0^{\sqrt{1-y^2-z^2}} f(x, y, z) \, dx,$$

the sum over the slab in Fig. 2.8c. Finally, summing (3) from $z = 0$ to

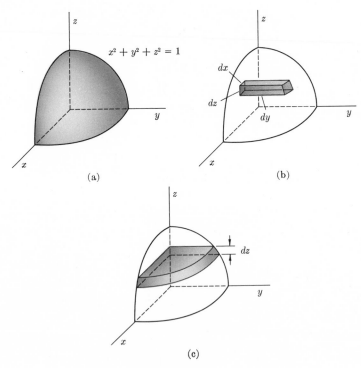

FIGURE 2.8

$z = 1$ yields

$$I = \int_0^1 dz \int_0^{\sqrt{1-z^2}} dy \int_0^{\sqrt{1-y^2-z^2}} f(x, y, z)\ dx.$$

Example 2.8   There are six iterated integrals by which

$$I = \int_R f(x, y, z)\ dx\ dy\ dz$$

can be computed. We leave it to the student to show that if R is the region bounded by the coordinate planes and the plane

$$x + 2y + 3z = 1$$

(Fig. 2.9), then

$$I = \int_0^{1/3} dz \int_0^{(1-3z)/2} dy \int_0^{1-2y-3z} f(x, y, z)\ dx$$

$$= \int_0^{1/2} dy \int_0^{(1-2y)/3} dz \int_0^{1-2y-3z} f(x, y, z)\ dx$$

$$= \int_0^1 dx \int_0^{(1-x)/3} dz \int_0^{(1-x-3z)/2} f(x, y, z)\ dy$$

$$= \int_0^{1/3} dz \int_0^{1-3z} dx \int_0^{(1-x-3z)/2} f(x, y, z)\ dy$$

$$= \int_0^1 dx \int_0^{(1-x)/2} dy \int_0^{(1-x-2y)/3} f(x, y, z)\ dz$$

$$= \int_0^{1/2} dy \int_0^{1-2y} dx \int_0^{(1-x-2y)/3} f(x, y, z)\ dz.$$

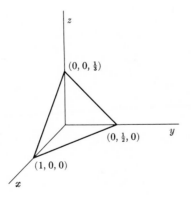

FIGURE 2.9

It is convenient and legitimate to rearrange the integrand of an iterated integral so that factors which do not depend upon the earlier variables of integration appear as far to the left as possible. Thus, we would write

$$\int_a^b dy \int_{\phi_1(y)}^{\phi_2(y)} y^2 \sin xy \, dx = \int_a^b y^2 \, dy \int_{\phi_1(y)}^{\phi_2(y)} \sin xy \, dx,$$

since y^2 can be treated as a constant while integrating with respect to x. As another example,

$$\int_a^b dz \int_{\phi_1(z)}^{\phi_2(z)} dy \int_{\psi_1(y,z)}^{\psi_2(y,z)} z^2 e^{yz} x^2 \, dx = \int_a^b z^2 \, dz \int_{\phi_1(z)}^{\phi_2(z)} e^{yz} \, dy \int_{\psi_1(y,z)}^{\psi_2(y,z)} x^2 \, dx$$

since z^2 can be treated as constant while integrating with respect to x and y, and e^{yz} can be treated as a constant while integrating with respect to x.

EXERCISES 4.2

In Exercises 1 through 9 evaluate the given iterated integrals and sketch the region of integration.

1. $\displaystyle\int_0^3 dy \int_{-1}^2 (2xy^2 + 3x^2y) \, dx.$

2. $\displaystyle\int_{-2}^5 dx \int_2^3 (3x^2 + y^4) \, dy.$

3. $\displaystyle\int_{-1}^3 dx \int_{-2x+1}^{-x^2+4} xy \, dy.$

4. $\displaystyle\int_{-\sqrt{2}}^{\sqrt{2}} dx \int_{x^2}^{-x^2+4} (x + 2y) \, dy.$

5. $\displaystyle\int_0^1 dx \int_{x+1}^{-x+3} (x - y) \, dy.$

6. $\displaystyle\int_1^3 dy \int_0^{\log y} 2y \, dx.$

7. $\displaystyle\int_{-2}^0 dx \int_2^5 dy \int_{-3}^2 (x - 2y + 3z) \, dz.$

8. $\displaystyle\int_{-2}^2 dx \int_{-\sqrt{4-x^2}}^{\sqrt{4-x^2}} dy \int_0^{2-y} 2x \, dz.$

9. $\displaystyle\int_0^2 dx \int_0^{2+3x} dy \int_0^{2-y+3x} x \, dz.$

In Exercises 10 through 14 evaluate the given iterated integral and sketch the region of integration. Then compute the corresponding double integral by means of an iterated integral with the opposite order of integration.

10. $\displaystyle\int_{0}^{2} dx \int_{x^2-4}^{4-x^2} x^2\, dy.$

11. $\displaystyle\int_{-3}^{3} dy \int_{0}^{9-y^2} (2x + y^2)\, dx.$

12. $\displaystyle\int_{-1}^{1} dx \int_{0}^{|x|} (2x^2 y + y^3)\, dy.$

13. $\displaystyle\int_{-2}^{3} dx \int_{0}^{e^x} xy^2\, dy.$

14. $\displaystyle\int_{0}^{1} dy \int_{y}^{\sin^{-1} y} dx.$

In Exercises 15 through 19 sketch S and evaluate $\displaystyle\int_{S} f(x, y)\, dx\, dy.$

15. $f(x, y) = x$; S is the triangle with vertices $(1, 2)$, $(4, -1)$, and $(4, 5)$.

16. $f(x, y) = 4x$; S is bounded by $y = 3 - x^2$ and $y = 2x$.

17. $f(x, y) = 2x$; S is bounded by $y = x^2$ and $x = y^2$.

18. $f(x, y) = xy$; S is bounded by $y = |x|$ and $y = 6 - x^2$.

19. $f(x, y) = x + y$; S is bounded by $x = 0$, $x = \pi/6$, $y = \sin x$, and $y = \cos x$.

In Exercises 20 through 23 sketch S and evaluate $\displaystyle\int_{S} f(x, y, z)\, dx\, dy\, dz.$

20. $f(x, y, z) = yz$, S is the solid in the first octant bounded by $z = \sqrt{x^2 + y^2}$, $x^2 + y^2 = 9$, $x = 0$, $y = 0$, and $z = 0$.

21. $f(x, y, z) = x + y + z$; S is bounded by $x + y + z = 1$ and the coordinate planes.

22. $f(x, y, z) = x^2$; S is bounded by $4x^2 + y^2 = 4$, $z + x = 2$, and $z = 0$.

23. $f(x, y, z) = 2xz$; S is in the first octant, bounded by $z = y$ and $y = 4 - x^2$.

In Exercises 24 through 30 find the area of the region bounded by the given curves.

24. $y = x^2 + 4$, $\quad y = x^2 - 4$, $\quad x = -1$, $\quad x = 1$.

25. $y = x + 1$, $\quad y = 3 - x$, $\quad x = 0$.

26. $y = 4 - x^2$, $\quad y = 2 - x$.

27. $y = x$, $\quad y = \dfrac{1}{x}$, $\quad y = 0$, $\quad x = 2$.

28. $y = \sqrt{x}$, $\quad y = x$, $\quad y = \frac{1}{2}x$.

29. $x = y^2 - 9$, $\quad x = 9 - y^2$.

$y = e^x$, $\quad y = -x$, $\quad x = 3$.

In Exercises 31 through 38 find the volume of S.

31. $S = \{(x, y, z) \mid 0 \le y \le x^2, \ 0 \le x \le 2, \ 0 \le z \le y^2\}$.

32. $S = \{(x, y, z) \mid x \ge 0, \ y \ge 0, \ x^2 + y^2 \le 16, \ 0 \le z \le x + y + 5\}$.

33. $S = \{(x, y, z) \mid x \ge 0, \ y \ge 0, \ 0 \le z \le 4 - 4x^2 - y^2\}$.

34. $S = \{(x, y, z) \mid x \ge 0, \ y \ge 0, \ z \ge 0, \ x^2 + y^2 \le 4, \ y^2 + z^2 \le 4\}$.

35. $S = \{(x, y, z) \mid 0 \le z \le x^2 + y^2,$ and $(x, y, 0)$ is in the triangle with vertices $(0, 1, 0)$, $(0, 0, 0)$, and $(1, 0, 0)\}$.

36. $S = \{(x, y, z) \mid y^2 \le z \le 16 - 4x^2 - y^2\}$.

37. $S = \{(x, y, z) \mid x^2 + y^2 \le z \le 8 - x^2 - y^2\}$.

38. Find the volume of the solid T defined in Example 1.3.

THEORETICAL EXERCISES

T-1. Show that

$$\int_0^1 dx_1 \int_0^{x_1} dx_2 \cdots \int_0^{x_{n-1}} dx_n = \frac{1}{n!}$$

T-2. Show that

$$\int_0^x dx_1 \int_0^{x_1} dx_2 \cdots \int_0^{x_{n-1}} f(x_n) \, dx_n = \frac{1}{(n-1)!} \int_0^x (x - t)^{n-1} f(t) \, dt.$$

T-3. Show that

$$\int_{a_1}^{b_1} dx_1 \int_{a_2}^{b_2} dx_2 \cdots \int_{a_n}^{b_n} f_1(x_1) f_2(x_2) \cdots f_n(x_n) \, dx_n$$

$$= \left(\int_{a_1}^{b_1} f_1(x_1) \, dx_1 \right) \left(\int_{a_2}^{b_2} f_2(x_2) \, dx_2 \right) \cdots \left(\int_{a_n}^{b_n} f_n(x_2) \, dx_2 \right).$$

(Compare this result with the result of Exercise T-4, Section 4.1.)

4.3 Change of Variables

In evaluating ordinary integrals it is often convenient to introduce a new variable of integration. The justification for this is provided by the following theorem.

Theorem 3.1 Let g be continuously differentiable in $[\alpha, \beta]$ and suppose that

$$(1) \qquad\qquad a = g(\alpha) \leq g(u) \leq g(\beta) = b$$

for $\alpha \leq u \leq \beta$. Let f be defined on $[a, b]$. Then f is integrable on $[a, b]$ if and only if $(f \circ g)g'$ is integrable on $[\alpha, \beta]$, and

$$(2) \qquad\qquad \int_a^b f(x)\, dx = \int_\alpha^\beta f(g(u))g'(u)\, du.$$

If (1) is replaced by

$$(3) \qquad\qquad a = g(\beta) \leq g(u) \leq g(\alpha) = b,$$

then (2) must be replaced by

$$(4) \qquad\qquad \int_a^b f(x)\, dx = - \int_\alpha^\beta f(g(u))g'(u)\, du.$$

In either case we say that the integral on the right is obtained from the one on the left by the *change of variable* $x = g(u)$.

Example 3.1 Let

$$I = \int_0^1 \sqrt{1 - x^2}\, dx$$

and

$$g_1(u) = \sin u;$$

then, if $0 \leq u \leq \pi/2$,

$$0 = g_1(0) \leq g_1(u) \leq g_1\left(\frac{\pi}{2}\right) = 1.$$

Thus (1) is satisfied and, since

$$g_1'(u) = \cos u,$$

the change of variable $x = \sin u$ yields

$$I = \int_0^{\pi/2} \sqrt{1 - \sin^2 u}\, \cos u\, du$$

$$= \int_0^{\pi/2} \cos^2 u\, du$$

$$= \frac{1}{2} \int_0^{\pi/2} (1 + \cos 2u)\, du$$

$$= \frac{1}{2} \left(u + \frac{\sin 2u}{2} \right) \Big|_0^{\pi/2} = \frac{\pi}{4}.$$

A second change of variable is given by

$$x = g_2(u) = \cos u,$$

for which

$$0 = g_1\left(\frac{\pi}{2}\right) \le g_1(u) \le g_1(0) = 1$$

if $0 \le u \le \pi/2$ and

$$g_2'(u) = -\sin u;$$

hence (3) is satisfied and, from (4),

$$I = -\int_0^{\pi/2} \sqrt{1 - \cos^2 u}\,(-\sin u)\, du$$

$$= \int_0^{\pi/2} \sin^2 u\, du$$

$$= \frac{1}{2} \int_0^{\pi/2} (1 - \cos 2u)\, du$$

$$= \frac{1}{2} \left(u - \frac{\sin 2u}{2} \right) \Big|_0^{\pi/2} = \frac{\pi}{4}.$$

If we require that $x = g(u)$ be one-to-one (Definition 6.1, Section 3.6), then it must be strictly increasing if it satisfies (1) or strictly decreasing if it satisfies (3); in either case g' cannot change sign in $[\alpha, \beta]$ and both (2) and (4) can be replaced by

$$(5) \qquad \int_a^b f(x) \ dx = \int_\alpha^\beta f(g(u)) \ |g'(u)| \ du.$$

We now generalize this form of the rule for change of variable to multiple integrals.

Theorem 3.2 Let T be a domain in R^n and $\mathbf{g}\colon R^n \to R^n$ be a continuously differentiable function which is one-to-one on T and maps T onto a domain S in R^n. Let

$$D_\mathbf{g}(\mathbf{U}) = \left. \frac{\partial(g_1, \dots, g_n)}{\partial(u_1, \dots, u_n)} \right|_\mathbf{U}$$

be the Jacobian of $\mathbf{g}$ (Definition 5.4, Section 3.5). Then $f\colon R^n \to R$ is integrable on S if and only if $(f \circ \mathbf{g}) \ |D_\mathbf{g}|$ is integrable on T, and

$$(6) \qquad \int_S f(\mathbf{X}) \ dx_1 \cdots dx_n = \int_T f(\mathbf{g}(\mathbf{U})) \ |D_\mathbf{g}(\mathbf{U})| \ du_1 \cdots du_n.$$

(Note that the *absolute value* $|D_\mathbf{g}(\mathbf{U})|$ appears in the integral on the right.) We say that the integral on the right is obtained from the one on the left by the *change of variable* $\mathbf{X} = \mathbf{g}(\mathbf{U})$.

The proof of this theorem is quite difficult. Hence, we shall simply illustrate it and make is plausible by means of examples.

Example 3.2 If $n = 1$ and we take $S = [a, b]$ and $T = [\alpha, \beta]$, then $D_g = g'$ and (6) reduces to (5).

Example 3.3 Let $f = 1$, T be a rectangle

$$T = \{(u, v) \mid u_0 \le u \le u_1, \quad v_0 \le v \le v_1\}$$

(Fig. 3.1), and $\mathbf{g}$ be the affine function

$$(7) \qquad \begin{bmatrix} x \\ y \end{bmatrix} = \begin{bmatrix} x_0 \\ y_0 \end{bmatrix} + \begin{bmatrix} a & b \\ c & d \end{bmatrix} \begin{bmatrix} u \\ v \end{bmatrix};$$

if $\mathbf{g}$ is one-to-one, then its Jacobian matrix,

$$\mathbf{J}_\mathbf{g} = \begin{bmatrix} a & b \\ c & d \end{bmatrix},$$

is nonsingular and

$$D_\mathbf{g} = ad - bc \ne 0.$$

To find S, we solve (7) for $\begin{bmatrix} u \\ v \end{bmatrix}$:

$$\begin{bmatrix} u \\ v \end{bmatrix} = \frac{1}{D_g} \begin{bmatrix} d & -b \\ -c & a \end{bmatrix} \begin{bmatrix} x - x_0 \\ y - y_0 \end{bmatrix}.$$

The lines $u = u_0$ and $u = u_1$ are transformed into the parallel lines

$$L_1: \quad d(x - x_0) - b(y - y_0) = D_g u_0$$

and

$$L_2: \quad d(x - x_0) - b(y - y_0) = D_g u_1;$$

also, $v = v_0$ and $v = v_1$ are transformed into the parallel lines

$$M_1: \quad -c(x - x_0) + a(y - y_0) = D_g v_0$$

and

$$M_2: \quad -c(x - x_0) + a(y - y_0) = D_g v_1.$$

Thus S is a parallelogram. The vertices P_1, P_2, P_3, and P_4 of T are mapped onto the vertices P_1', P_2', P_3' and P_4', respectively, of S (Fig. 3.1).

According to (6) the area of S is

$$(8) \quad A(S) = \int_S dx\, dy = \int_T |\, ad - bc\, |\, du\, dv$$

$$= |\, ad - bc\, | \int_S du\, dv$$

$$= |\, ad - bc\, |\, (u_1 - u_0)(v_1 - v_0),$$

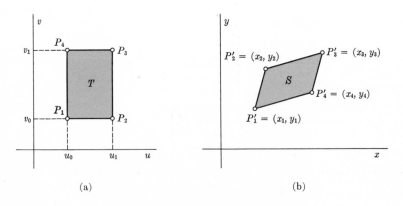

(a)

(b)

FIGURE 3.1

which can be verified independently as follows. Let $P_1' = (x_i, y_i)$ $(1 \leq i \leq 4)$; then

$$(9) \qquad A(S) = |(x_3 - x_2)(y_2 - y_1) - (x_2 - x_1)(y_3 - y_2)|$$

(See Section 1.3). From (7),

$$\begin{bmatrix} x_3 - x_2 \\ y_3 - y_2 \end{bmatrix} = \begin{bmatrix} a & b \\ c & d \end{bmatrix} \begin{bmatrix} u_1 - u_0 \\ v_1 - v_1 \end{bmatrix}$$

$$= (u_1 - u_0) \begin{bmatrix} a \\ c \end{bmatrix},$$

$$\begin{bmatrix} x_2 - x_1 \\ y_2 - y_1 \end{bmatrix} = \begin{bmatrix} a & b \\ c & d \end{bmatrix} \begin{bmatrix} u_0 - u_0 \\ v_1 - v_0 \end{bmatrix}$$

$$= (v_1 - v_0) \begin{bmatrix} b \\ d \end{bmatrix},$$

and substitution into (9) yields

$$A(S) = |ad - bc|(u_1 - u_0)(v_1 - v_0),$$

in agreement with (8).

Example 3.4 Again let $f = 1$ and let T be a rectangle,

$$T = \{(u, v) \mid \alpha \leq u \leq \beta, \quad \gamma \leq v \leq \delta\},$$

and suppose **g** (not necessarily an affine function) maps T one-to-one on a region S in R^2. According to (6),

$$(10) \qquad \int_S dx\, dy = \int_T \left| \frac{\partial(g_1, g_2)}{\partial(u, v)} \right| du\, dv.$$

Thus the integral on the right must equal $A(S)$, the area of S. To see that this is so, let Δ be a partition of T defined by

$$\alpha = u_0 < u_1 < \cdots < u_m = \beta$$

and

$$\gamma = v_0 < v_1 < \cdots < v_m = \delta,$$

and let T_{ij} be a typical subrectangle of T:

$$T_{ij} = \{(u, v) \mid u_{i-1} \leq u \leq u_i, \quad v_{j-1} \leq v \leq v_j\}.$$

Denote by S_{ij} the image of T_{ij} under $\mathbf{g}$ and let its area be $A(S_{ij})$ (Fig. 3.2a); then

$$A(S) = \sum_{i=1}^{m} \sum_{j=1}^{n} A(S_{ij}).$$

Now let us obtain an approximate value for $A(S_{ij})$. Let $\mathbf{U}_{ij}$ be a point in T_{ij}; since $\mathbf{g}$ is differentiable the transformation $\mathbf{X} = \mathbf{g}(\mathbf{U})$ can be approximated near $\mathbf{U}_{ij}$ (in particular, on T_{ij} if the latter is sufficiently small) by the affine function

$$(11) \qquad \mathbf{X} = \mathbf{h}(\mathbf{U}) = \mathbf{g}(\mathbf{U}_{ij}) + (d_{\mathbf{U}_{ij}}\mathbf{g})(\mathbf{U} - \mathbf{U}_{ij}),$$

which has Jacobian determinant

$$D_{\mathbf{g}}(\mathbf{U}_{ij}) = \left.\frac{\partial(g_1, g_2)}{\partial(u_1, u_2)}\right|_{\mathbf{U}_{ij}}.$$

According to Example 3.3 the transformation (11) maps T_{ij} onto a parallelogram S'_{ij} with area

$$(12) \qquad A(S'_{ij}) = |D_{\mathbf{g}}(\mathbf{U}_{ij})|\,(u_i - u_{i-1})(v_j - v_{j-1}).$$

If T_{ij} is sufficiently small, then S_{ij} and S'_{ij} are approximately equal (Fig. 3.2b) and therefore have approximately the same area; thus from (12)

$$A(S_{ij}) \cong |D_g(\mathbf{U}_{ij})|\,(u_i - u_{i-1})(v_j - v_{j-1}),$$

and

$$\sum_{i=1}^{m} \sum_{j=1}^{n} A(S_{ij}) \cong \sum_{i=1}^{m} \sum_{j=1}^{n} |D_{\mathbf{g}}(\mathbf{U}_{ij})|\,(u_i - u_{i-1})(v_j - v_{j-1}).$$

The sum on the left is the area of S, which is equal to

$$\int_{S} dx\,dy,$$

while the sum on the right is a Riemann sum associated with

$$\int_{T} |D_g(\mathbf{U})|\,du\,dv.$$

Hence (10) follows by letting $\|\Delta\|$ approach zero.

Example 3.5   Let $(u, v) = (r, \theta)$ and $\mathbf{g}$ be the transformation from polar to rectangular coordinates:

$$\begin{bmatrix} x \\ y \end{bmatrix} = \mathbf{g}(r, \theta) = \begin{bmatrix} r\cos\theta \\ r\sin\theta \end{bmatrix}.$$

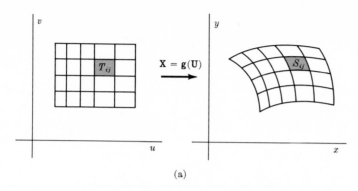

(a)

(b)

FIGURE 3.2

If

$$T = \{(r, \theta) \mid 0 < r_0 \leq r \leq r_1, \quad \theta_0 \leq \theta \leq \theta_1\},$$

(Fig. 3.3a), then **g** maps T one to one on the region S shown in Fig. 3.3b. Since

(13) $$\frac{\partial(x, y)}{\partial(r, \theta)} = \begin{vmatrix} \cos \theta & \sin \theta \\ -r \sin \theta & r \cos \theta \end{vmatrix} = r,$$

the area of S is given, according to (6), by

$$A(S) = \int_T r \, dr \, d\theta$$

$$= \int_{\theta_0}^{\theta_1} d\theta \int_{r_0}^{r} r \, dr$$

$$= (\theta_1 - \theta_0) \frac{r_1^2 - r_0^2}{2}$$

$$= (\theta_1 - \theta_0)(r_1 - r_0) \frac{r_1 + r_0}{2}.$$

Now let $\theta_1 - \theta_0 = \Delta\theta$ and $r_1 - r_0 = \Delta r$; if Δr is small compared to r_0,

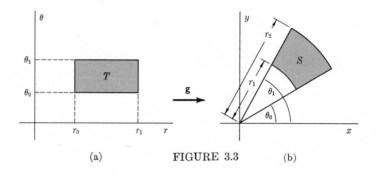

(a) FIGURE 3.3 (b)

we can write

$$\Delta(S) \cong r \, \Delta r \, \Delta\theta,$$

where r is any number satisfying $r_1 \leq r \leq r_2$. For this reason we call the symbol $r \, dr \, d\theta$ the *differential element of area, in polar coordinates, at the point with polar coordinates* (r, θ).

Example 3.6 Let Q be the part of the closed unit ball that lies in the first octant; thus

$$Q = \{(x, y, z) \mid 0 \leq x^2 + y^2 + z^2 \leq 1, \quad x \geq 0, \quad y \geq 0, \quad z \geq 0\}.$$

(Fig. 3.4). The volume of Q is

$$(14) \qquad V(Q) = \int_S \sqrt{1 - x^2 - y^2} \, dx \, dy,$$

where S is the quarter of the unit circle in the first quadrant of the (x, y) plane. The integral (14) can be evaluated by means of the iterated integral

$$V(S) = \int_0^1 dx \int_0^{\sqrt{1-x^2}} \sqrt{1 - x^2 - y^2} \, dy,$$

but this requires the evaluation of two rather formidable ordinary integrals. This complication can be avoided by thinking of S in terms of polar coordinates; thus S is the image of the rectangle

$$T = \left\{(r, \theta) \mid 0 \leq r \leq 1, \quad 0 \leq \theta \leq \frac{\pi}{2}\right\}$$

under the transformation

$$\begin{bmatrix} x \\ y \end{bmatrix} = \mathbf{g}(r, \theta) = \begin{bmatrix} r \cos \theta \\ r \sin \theta \end{bmatrix}.$$

According to Theorem 3.2,

$$V(Q) = \int_S \sqrt{1 - x^2 - y^2} \, dx \, dy = \int_T \sqrt{1 - r^2 \cos^2 \theta - r^2 \sin^2 \theta} \, r \, dr \, d\theta.$$

The integral in polar coordinates can be evaluated by means of a simple iterated integral:

$$V(Q) = \int_T \sqrt{1 - r^2} \, r \, dr \, d\theta$$

$$= \int_0^{\pi/2} d\theta \int_0^1 \sqrt{1 - r^2} \, r \, dr$$

$$= \frac{\pi}{2} \left[-\frac{1}{3} (1 - r^2)^{3/2} \right]\Big|_0^1 = \frac{\pi}{6}.$$

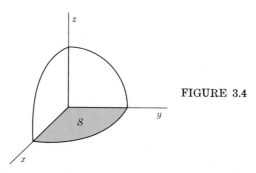

FIGURE 3.4

   In this example we applied Theorem 3.2 and obtained a correct result even though the hypotheses of the theorem are not satisfied: **g** is not one-to-one on T, since it maps every point of the form $(0, \theta)$ onto $(x, y) = (0, 0)$. The fact is that (6) holds even if the hypotheses of Theorem 3.2 are relaxed so that **g** fails to be one-to-one on a finite number of smooth surfaces (also curves or points) of dimension less than n in T.
   The rule for the change of variables from rectangular to polar coordinates is so important that we state it separately. The next theorem follows from Theorem 3.2 and (13).

Theorem 3.3 Let S be a region in R^2 and T be the set of polar coordinates of points in S:

$$T = \{ (r, \theta) \mid (x, y) = (r \cos \theta, r \sin \theta) \quad \text{is in } S \}.$$

Then

$$\int_S f(x, y) \, dx \, dy = \int_T f(r \cos \theta, r \sin \theta) r \, dr \, d\theta,$$

provided f is integrable over S. In particular, the area of S is given by

$$A(S) = \int_T r \, dr \, d\theta.$$

Example 3.7 Find the area of the region S in R^2 bounded by the curve whose polar coordinates satisfy

$$r = 1 - \cos \theta \qquad (0 \le \theta \le 2\pi).$$

(This heart-shaped curve, shown in Fig. 3.5a, is called a *cardioid*.) According to Theorem 3.3,

$$A(S) = \int_T r \, dr \, d\theta$$

where T is the region of the (θ, r) plane shown in Fig. 3.5b. The area can be calculated by the iterated integral

$$A(S) = \int_0^{2\pi} d\theta \int_0^{1-\cos\theta} r \, dr$$

$$= \int_0^{2\pi} \left(\frac{r^2}{2} \Big|_0^{1-\cos\theta} \right) d\theta$$

$$= \frac{1}{2} \int_0^{2\pi} (1 - \cos \theta)^2 \, d\theta$$

$$= \frac{1}{2} \int_0^{2\pi} (1 - 2 \cos \theta + \cos^2 \theta) \, d\theta$$

$$= \frac{1}{2} \int_0^{2\pi} \left(\frac{3}{2} - 2 \cos \theta + \frac{\cos 2\theta}{2} \right) d\theta = \frac{3}{4} (2\pi) = \frac{3}{2} \pi.$$

The next example illustrates Theorem 3.2 for another set of curvilinear coordinates in the plane.

Example 3.8 Let

$$I = \int_S (x^4 - y^4) e^{xy} \, dx \, dy$$

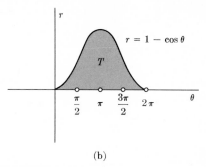

(a)   (b)

FIGURE 3.5

where S is the region in the first quadrant bounded by the hyperbolas $xy = 1$, $xy = 2$, $x^2 - y^2 = 2$, and $x^2 - y^2 = 3$ (Fig. 3.6a). Through each point in S there is exactly one hyperbola of the form

(15) $x^2 - y^2 = u$ $(2 \le u \le 3)$

and exactly one of the form

(16) $xy = v$ $(1 \le v \le 2)$;

hence u and v define a coordinate system in S. The change of variable

$$\begin{bmatrix} x \\ y \end{bmatrix} = \mathbf{g}(u, v)$$

maps the rectangle

$$T = \{(u, v) \mid 2 \le u \le 3, \ \ 1 \le v \le 2\}$$

in the (u, v) plane one-to-one onto S (Fig. 3.6b); hence

(17) $\displaystyle I = \int_T (x^4 - y^4) e^{xy} \left| \frac{\partial(x, y)}{\partial(u, v)} \right| du \, dv,$

where, of course, the integrand must be expressed in terms of u and v. From (15) and (16) we know $\mathbf{g}^{-1}$, not $\mathbf{g}$; however, this suits our purpose, since

$$\frac{\partial(x, y)}{\partial(u, v)} = \frac{1}{\dfrac{\partial(u, v)}{\partial(x, y)}} = \frac{1}{\begin{vmatrix} 2x & -2y \\ y & x \end{vmatrix}} = \frac{1}{2(x^2 + y^2)} \, .$$

Substituting this into (17) yields

$$I = \frac{1}{2} \int_T u e^v \, du \, dv = \frac{1}{2} \int_2^3 u \, du \int_1^2 e^v \, dv = \frac{1}{2} \left(\frac{u^2}{2} \right) \Big|_2^3 \left(e^v \Big|_1^2 \right) = \frac{5}{4} e(e - 1).$$

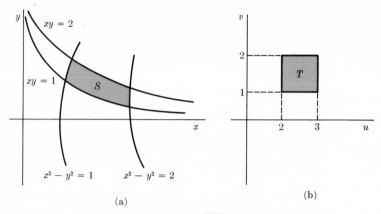

(a)

(b)

FIGURE 3.6

Example 3.9   Let

$$I = \int_S xe^{y^2-x^2}\, dx\, dy$$

where S is the rectangle bounded by the lines $x + y = 0$, $x + y = 2$, $x - y = -4$, and $x - y = -2$ (Fig. 3.7a). Let

$$\begin{bmatrix} u \\ v \end{bmatrix} = \mathbf{g}^{-1}(x, y) = \begin{bmatrix} x + y \\ x - y \end{bmatrix};$$

then S is the image of the rectangle

$$T = \{(u, v) \mid 0 \le u \le 2, \quad -4 \le v \le -2\}$$

(Fig. 3.7b) under the transformation

$$\begin{bmatrix} x \\ y \end{bmatrix} = \mathbf{g}(u, v) = \begin{bmatrix} \dfrac{u + v}{2} \\ \dfrac{u - v}{2} \end{bmatrix}.$$

Since

$$\frac{\partial(x, y)}{\partial(u, v)} = \begin{vmatrix} \tfrac{1}{2} & \tfrac{1}{2} \\ \tfrac{1}{2} & -\tfrac{1}{2} \end{vmatrix} = -\tfrac{1}{2},$$

we have

$$I = \int_T \frac{u + v}{2} e^{uv} \left(\frac{1}{2}\right) du\, dv$$

$$= \frac{1}{4} \int_{-4}^{-2} dv \int_0^2 (u + v) e^{uv}\, du.$$

We leave the evaluation of the iterated integral for the student.

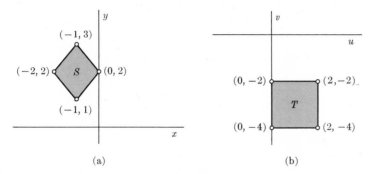

(a)   (b)

FIGURE 3.7

The two most common changes of variables for triple integrals are from rectangular to cylindrical coordinates and from rectangular to spherical coordinates. We first consider the former.

Theorem 3.4 Let S be a region in R^3 and T be the set of cylindrical coordinates of points in S:

$$T = \{(r, \theta, z) \mid (x, y, z) = (r \cos \theta, r \sin \theta, z) \quad \text{is in } S\}.$$

Then

$$\int_S f(x, y, z)\, dx\, dy\, dz = \int_T f(r \cos \theta, r \sin \theta, z) r\, dr\, d\theta\, dz$$

provided f is integrable over S. In particular, the volume of S is given by

$$V(R) = \int_T r\, dr\, d\theta\, dz.$$

Proof.   The transformation

(18)
$$\begin{bmatrix} x \\ y \\ z \end{bmatrix} = \begin{bmatrix} r \cos \theta \\ r \sin \theta \\ z \end{bmatrix}$$

maps T onto S. Since

$$\frac{\partial (x, y, z)}{\partial (r, \theta, z)} = \begin{vmatrix} \cos \theta & -r \sin \theta & 0 \\ \sin \theta & r \cos \theta & 0 \\ 0 & 0 & 1 \end{vmatrix} = r,$$

the result follows from Theorem 3.2. (The fact that (18) fails to be one-to-one when $r = 0$ is unimportant; see the remark following Example 3.6.)

Example 3.10   Let S be the region in R^3 whose cylindrical coordinates satisfy $r_0 \le r \le r_1$, $\theta_0 \le \theta \le \theta_1$, and $z_0 \le z \le z_1$ (Fig. 3.8a). Thus, S is the image of a rectangular parallelepiped T in R^3, shown in Fig. 3.8b, under the transformation (18); hence the volume of S is

$$V(S) = \int_S dx \, dy \, dz = \int_S r \, dr \, d\theta \, dz$$

$$= \int_{z_0}^{z_1} dz \int_{\theta_0}^{\theta_1} d\theta \int_{r_0}^{r_1} r \, dr$$

$$= \frac{r_0 + r_1}{2} (r_1 - r_0)(\theta_1 - \theta_0)(z_1 - z_0).$$

Now let $r_1 - r_0 = \Delta r$, $\theta_1 - \theta_0 = \Delta \theta$, and $z_1 - z_0 = \Delta z$; if Δr is small compared to r_0, we have

$$V(S) \approx r \, \Delta r \, \Delta \theta \, \Delta z,$$

provided $r_0 \le r \le r_1$. For this reason we call the symbol $r \, dr \, d\theta \, dz$ the *differential element of volume, in cylindrical coordinates, at the point with cylindrical coordinates* (r, θ, z).

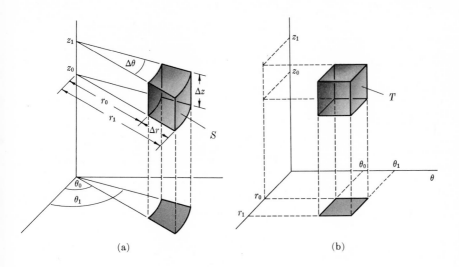

(a)   (b)

FIGURE 3.8

Example 3.11   Let

$$I = \int_S z \, dx \, dy \, dz,$$

where S is the hemisphere defined by

$$x^2 + y^2 + z^2 \leq a^2, \qquad z \geq 0$$

(Fig. 3.9a). The transformation (18) maps

$$T = \{ (r, \theta, z) \mid 0 \leq \theta \leq 2\pi, \;\; 0 \leq z \leq \sqrt{a^2 - r^2}, \;\; 0 \leq r \leq a \}$$

(Fig. 3.9b) onto S; hence

$$I = \int_T zr \, dr \, d\theta \, dz = \int_0^{2\pi} d\theta \int_0^a r \, dr \int_0^{\sqrt{a^2-r^2}} z \, dz$$

$$= 2\pi \int_0^a r \, dr \int_0^{\sqrt{a^2-r^2}} z \, dz = 2\pi \int_0^a r \left(\frac{z^2}{2} \Big|_0^{\sqrt{a^2-r^2}} \right) dr$$

$$= \pi \int_0^a (a^2 r - r^3) \, dr = \pi \left(\frac{a^2 r^2}{2} - \frac{r^4}{4} \right) \Big|_0^a = \frac{\pi a^4}{4}.$$

Now let us consider the rule for change of variable from rectangular to spherical coordinates.

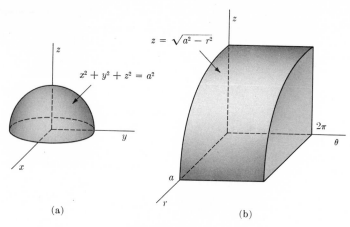

$z = \sqrt{a^2 - r^2}$

$x^2 + y^2 + z^2 = a^2$

2π

(a)

(b)

FIGURE 3.9

Theorem 3.5 Let S be a region in R^3 and T be the set of spherical co-ordinates of points in S:

$$T = \{(r, \theta, \phi) \mid (x, y, z) = (r \cos \theta \sin \phi, r \sin \theta \sin \phi, r \cos \phi) \quad \text{is in } S\}.$$

Then

$$\int_S f(x, y, z) \, dx \, dy \, dz = \int_T f(r \cos \theta \sin \phi, r \sin \theta \sin \phi, r \cos \phi) r^2 \sin \phi \, dr \, d\theta \, d\phi$$

provided f is integrable over S. In particular, the volume of S is given by

$$(19) \qquad V(S) = \int_T r^2 \sin \phi \, dr \, d\theta \, d\phi.$$

    Proof. The transformation

$$(20) \qquad \begin{bmatrix} x \\ y \\ z \end{bmatrix} = \begin{bmatrix} r \cos \theta \sin \phi \\ r \sin \theta \sin \phi \\ r \cos \phi \end{bmatrix}$$

maps T onto S. Since

$$\frac{\partial(x, y, z)}{\partial(r, \theta, \phi)} = \begin{vmatrix} \cos \theta \sin \phi & -r \sin \theta \sin \phi & r \cos \theta \cos \phi \\ \sin \theta \sin \phi & r \cos \theta \sin \phi & r \sin \theta \cos \phi \\ \cos \phi & 0 & -r \sin \phi \end{vmatrix}$$

$$= -r^2 \sin \phi,$$

the result follows from Theorem 3.2.

Because of (19) we call the symbol $r^2 \sin \phi \, dr \, d\theta \, d\phi$ *the differential element of volume in, spherical coordinates, at the point with spherical coordinates* (r, θ, ϕ).

By a computation similar to that of Example 3.10, it can be shown that if S is the region in R^3 defined by $r_0 \leq r \leq r_0 + \Delta r$, $\theta_0 \leq \theta \leq \theta_0 + \Delta \theta$, $\phi_0 \leq \phi \leq \phi_0 + \Delta \phi$, then

$$V(S) = r_0^2 \sin \phi_0 \, \Delta r \, \Delta \theta \, \Delta \phi$$

for small Δr, $\Delta \theta$, and $\Delta \phi$ (Fig. 3.10).

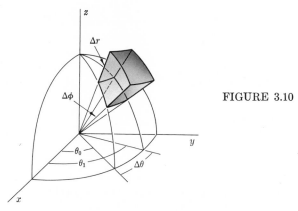

FIGURE 3.10

Example 3.12 Let S be the sphere defined by

$$x^2 + y^2 + z^2 \leq a^2$$

(Fig. 3.11a). The polar coordinates of points in S fill the rectangular parallelepiped

$$T = \{(r, \theta, \phi) \mid 0 \leq r \leq a, \quad 0 \leq \theta \leq 2\pi, \quad 0 \leq \phi \leq \pi\}$$

(Fig. 3.11b); hence

$$V(S) = \int_T r^2 \sin \phi \, dr \, d\theta \, d\phi$$

$$= \int_0^{2\pi} d\theta \int_0^{\pi} \sin \phi \, d\phi \int_0^{a} r^2 \, dr$$

$$= \left(\theta \Big|_0^{2\pi} \right) \left(-\cos \phi \Big|_0^{\pi} \right) \left(\frac{r^3}{3} \Big|_0^{a} \right)$$

$$= (2\pi)\,(2)\,\frac{a^3}{3} = \frac{4}{3}\pi a^3.$$

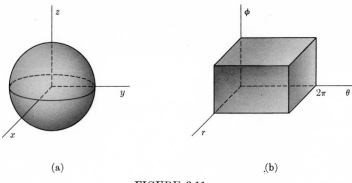

(a)   .(b)

FIGURE 3.11

Example 3.13   To evaluate the integral I of Example 3.11 by transforming it to spherical coordinates, observe that the transformation (20) maps

$$T = \left\{ (r, \theta, \phi) \mid 0 \le r \le a, \quad 0 \le \theta \le 2\pi, \quad 0 \le \phi \le \frac{\pi}{2} \right\}$$

onto S. Hence

$$\int_S z \, dx \, dy \, dz = \int_T (r \cos \phi) r^2 \sin \phi \, dr \, d\theta \, d\phi$$

$$= \int_0^{2\pi} d\theta \int_0^{\pi/2} \cos \phi \sin \phi \, d\phi \int_0^a r^3 \, dr$$

$$= \left(\theta \Big|_0^{2\pi} \right) \left(\frac{\sin^2 \phi}{2} \Big|_0^{\pi/2} \right) \left(\frac{r^4}{4} \Big|_0^a \right)$$

$$= (2\pi) \left(\frac{1}{2} \right) \left(\frac{a^4}{4} \right) = \frac{\pi a^4}{4}.$$

EXERCISES 4.3

1. Evaluate:

(a) $\displaystyle \int_0^1 \frac{dx}{4 + x^2}$;

(b) $\displaystyle \int_0^1 \sqrt{9 - x^2} \, dx$;

(c) $\displaystyle \int_2^4 \frac{\sqrt{x^2 - 4}}{x^2} \, dx$;

(d) $\displaystyle \int_0^1 \frac{e^{\sqrt{1 + 2x}}}{\sqrt{1 + 2x}} \, dx$.

2. Find S, the image of

$$T = \{(u, v) \mid 0 \leq u \leq 1, \quad 2 \leq v \leq 3\}$$

under the transformation

$$\begin{bmatrix} x \\ y \end{bmatrix} = \begin{bmatrix} 3 \\ -2 \end{bmatrix} + \begin{bmatrix} 1 & 2 \\ -1 & 3 \end{bmatrix} \begin{bmatrix} u \\ v \end{bmatrix}.$$

· Also, find the area of S.

3. Repeat Exercise 2 for

$$T = \{(u, v) \mid -2 \leq u \leq 2, \quad -3 \leq v \leq 0\}$$

and

$$\begin{bmatrix} x \\ y \end{bmatrix} = \begin{bmatrix} -1 \\ 3 \end{bmatrix} + \begin{bmatrix} 0 & 2 \\ -1 & 1 \end{bmatrix} \begin{bmatrix} u \\ v \end{bmatrix}.$$

In Exercises 4 through 6 find the image of T under the given change of variables, and, find the area of the image.

4. $T = \left\{ (r, \theta) \,\middle|\, 1 \leq r \leq 3, \ \dfrac{\pi}{6} \leq \theta \leq \dfrac{\pi}{4} \right\}$; $\begin{bmatrix} x \\ y \end{bmatrix} = \begin{bmatrix} r \cos \theta \\ r \sin \theta \end{bmatrix}.$

5. $T = \left\{ (u, v) \,\middle|\, 0 \leq u \leq 1, \ 0 \leq v \leq \dfrac{\pi}{2} \right\}$; $\begin{bmatrix} x \\ y \end{bmatrix} = \begin{bmatrix} e^u \cos v \\ e^u \sin v \end{bmatrix}.$

6. $T = \{ (u, v) \mid 0 \leq u \leq 1, \ 0 \leq v \leq 1 \}$; $\begin{bmatrix} x \\ y \end{bmatrix} = \begin{bmatrix} 2uv \\ u^2 - v^2 \end{bmatrix}.$

In Exercises 7 through 9 find the image of T under the given change of variables, and find the volume of the image.

7. $T = \left\{ (r, \theta, z) \,\middle|\, 1 \leq r \leq 3, \ \dfrac{\pi}{4} \leq \theta \leq \dfrac{3\pi}{2}, \ -2 \leq z \leq 3 \right\}$;

$$\begin{bmatrix} x \\ y \\ z \end{bmatrix} = \begin{bmatrix} r \cos \theta \\ r \sin \theta \\ z \end{bmatrix}.$$

8. $T = \left\{ (r, \theta, \phi) \,\middle|\, 2 \leq r \leq 5, \ \dfrac{\pi}{3} \leq \theta \leq \dfrac{3\pi}{4}, \ \dfrac{\pi}{6} \leq \phi \leq \dfrac{3\pi}{4} \right\};$

$$\begin{bmatrix} x \\ y \\ z \end{bmatrix} = \begin{bmatrix} r \sin \phi \cos \theta \\ r \sin \phi \sin \theta \\ r \cos \phi \end{bmatrix}.$$

9. $T = \{ (u, v, w) \mid 0 \leq u \leq 1, \ 0 \leq v \leq \pi, \ 0 \leq w \leq 2\pi \};$

$$\begin{bmatrix} x \\ y \\ z \end{bmatrix} = \begin{bmatrix} 2u \sin v \cos w \\ u \sin v \sin w \\ 3u \cos v \end{bmatrix}.$$

10. Evaluate by changing variables to polar coordinates in the equivalent double integrals:

(a) $\displaystyle \int_0^2 dx \int_{-\sqrt{4-x^2}}^{\sqrt{4-x^2}} x^2 y^2 \, dy;$ (b) $\displaystyle \int_0^2 dy \int_{-\sqrt{2y-y^2}}^{\sqrt{2y-y^2}} \sqrt{x^2 + y^2} \, dx.$

11. Evaluate by changing variables to polar coordinates in the equivalent double integrals:

(a) $\displaystyle \int_0^2 dx \int_0^{\sqrt{4-x^2}} e^{(x^2+y^2)} \, dy;$ (b) $\displaystyle \int_0^{\sqrt{2}} dy \int_y^{\sqrt{4-y^2}} \dfrac{dx}{1 + x^2 + y^2}.$

12. Evaluate

$$\int_S [(y - x^2)^4 + 6xy](y + 2x^2) \, dx \, dy,$$

where S is the region in the first quadrant bounded by $xy = 1$, $xy = 2$, $y = x^2$, and $y = x^2 + 1$. (*Hint*: See Example 3.8.)

13. Evaluate

$$\int_S x^2 y^6 \, dx \, dy,$$

where S is the region in the first quadrant bounded by $xy = 1$, $xy = 2$, $y = x$, and $y = 2x$. (*Hint*: See Example 3.8.)

14. Find the area of the region inside the circle $x^2 + y^2 - 2y = 0$ and outside the circle $x^2 + y^2 = 2$.

15. Find the area of the region in the first quadrant outside $r = 2(1 - \cos\theta)$ and inside $r = 2\sin\theta$.

16. Find the area of the region bounded by $r = 3\sin\theta$ and $r = 4\cos\theta$.

17. Find the area of the region bounded by $x^2 + y^2 = 2$ and $y = x^2$.

18. Find the volume of the solid bounded by $z^2 = x^2 + y^2$ and $x^2 + y^2 = 4$.

19. Find the volume of the solid bounded by $z = x^2 + y^2$ and $z^2 = 4x^2 + 4y^2$.

20. Find the volume of the part of the cylinder $x^2 + y^2 = 4$ which is contained in the sphere $x^2 + y^2 + z^2 = 16$.

21. Find the volume of the solid bounded above by $x^2 + y^2 + z^2 = 12$ and below by $z = x^2 + y^2$.

22. Evaluate $\displaystyle\int_S (x^2 + y^2)\, dx\, dy\, dz$, where S is the solid bounded by

$x^2 + y^2 = 4$, $z = 0$, and $z = 3$.

23. Evaluate $\displaystyle\int_S (x^2 + y^2 + z^2)\, dx\, dy\, dz$, where S is the solid bounded by

$$z = \frac{x^2 + y^2}{2}, \qquad z = 0, \qquad \text{and} \qquad x^2 + y^2 = 16.$$

24. Evaluate $\displaystyle\int_S \frac{e^{x^2+y^2+z^2}}{\sqrt{x^2 + y^2 + z^2}}\, dx\, dy\, dz$, where

$$S = \{(x, y, z) \mid 3 \le x^2 + y^2 + z^2 \le 5\}.$$

25. Evaluate $\displaystyle\int_S (x^2 + z^2)\, dx\, dy\, dz$, where

$$S = \{(x, y, z) \mid 3 \le x^2 + y^2 + z^2 \le 5\}.$$

26. Find the volume of the ellipsoid

$$\frac{x^2}{a^2} + \frac{y^2}{b^2} + \frac{z^2}{c^2} = 1.$$

27. Find the volume of the cylinder defined by

$$\frac{x^2}{a^2} + \frac{y^2}{b^2} = 1, \qquad |z| \le 1.$$

28. Find the volume of the parallelepiped bounded by the planes $x + y + z = 1$, $x + y + z = 2$, $x - y + z = 0$, $x - y + z = 1$, $x + y - z = 0$, and $x + y - z = 2$.

29. Evaluate $\displaystyle\int_S (3x + 2y + z)\ dx\ dy\ dz$ where S is the solid defined by

$$|x - y| \le 1, \qquad |y - z| \le 1, \qquad \text{and} \qquad |z + x| \le 1.$$

30. Evaluate $\displaystyle\int_S 4xyz(x^4 - y^4)\ dx\ dy\ dz$ where S is the region in the first octant defined by $0 \le x^2 - y^2 \le 1, 0 \le xy \le 1, 0 \le z^2 + y^2 \le 1$.

31. Evaluate $\displaystyle\int_S xy\ dx\ dy\ dz$, where S is defined by $0 \le z \le 4 - x^2 - 2y^2$.

32. Find the volume of the right circular cone of radius r and height h.

In Exercises 33 and 34, (r, θ, z) are cylindrical coordinates.

33. Find the volume of the solid defined by $0 \le z \le 4 - r^2$.

34. Find the volume of the solid bounded by $z = 0$, $z = r$, and $r = 2$.

THEORETICAL EXERCISES

T-1. Let $\mathbf{V}_1, \ldots, \mathbf{V}_n$ be a basis for R^n. Find the volume of

$$S = \{\mathbf{X} \mid a_i \le \mathbf{X} \cdot \mathbf{V}_i \le b_i \qquad (1 \le i \le n)\}.$$

T-2. Let $u_1, \ldots, u_n$ be constants. Describe how you would evaluate

$$\int_S (u_1 x_1 + u_2 x_2 + \cdots + u_n x_n)\ dV,$$

where S is defined in Exercise T-1.

T-3. Suppose V_n is the volume of the n-ball defined by

$$x_1^2 + x_2^2 + \cdots + x_n^2 \le 1.$$

Find the volume of the region defined by $a_1^2 x_1^2 + a_2^2 x_2^2 + \cdots + a_n^2 x_n^2 \le \rho^2$.

4.4 Physical Applications

Much of elementary mechanics is devoted to the study of systems of point masses; that is, physical systems in which the dimensions of the various parts are of no importance, so that they can be viewed as points to which nonzero masses are assigned. The multiple integral allows us to study mechanical systems comprised of parts with nonzero dimensions.

Mass Distributions

Definition 4.1 Let S be a region in R^3. A real valued function ρ is said to define a *mass distribution in* S if $\rho(\mathbf{X}) \geq 0$ for all $\mathbf{X}$ in S and

$$m(S) = \int_S \rho(\mathbf{X}) \, dx \, dy \, dz$$

exists. The integral $m(S)$ is the *mass of* S; more generally, if S_0 is a subset of S, the mass of S_0 is defined by

$$m(S_0) = \int_{S_0} \rho(\mathbf{X}) \, dx \, dy \, dz.$$

The function ρ is called the *mass density function of* S, and its value, $\rho(\mathbf{X})$, at a point $\mathbf{X}$ in S is *the density of* S *at* $\mathbf{X}$. If ρ is constant throughout S, then S is said to be *homogeneous*.

Throughout this section we shall assume that ρ is continuous.
Suppose $\mathbf{X}_0$ is an interior point of S and

$$S_\delta = \{ (x, y, z) \mid |x - x_0| \leq \delta, \quad |y - y_0| \leq \delta, \quad |z - z_0| \leq \delta \},$$

where δ is so small that the cube S_δ is contained in S (Fig. 4.1). The mass of S_δ is

$$m(S_\delta) = \int_{S_\delta} \rho(\mathbf{X}) \, dx \, dy \, dz.$$

For any $\epsilon > 0$ there is a $\delta > 0$ such that

$$| \rho(\mathbf{X}) - \rho(\mathbf{X}_0) | < \epsilon$$

if $\mathbf{X}$ is in S_δ; this follows from the continuity of ρ. Now,

$$m(S_\delta) = \rho(\mathbf{X}_0) \int_{S_\delta} dx \, dy \, dz + \int_{S_\delta} [\rho(\mathbf{X}) - \rho(\mathbf{X}_0)] \, dx \, dy \, dz$$

$$= \rho(\mathbf{X}_0) V(S_\delta) + \int_{S_\delta} [\rho(\mathbf{X}) - \rho(\mathbf{X}_0)] \, dx \, dy \, dz,$$

where $V(S_\delta)$ is the volume of S_δ. Hence

$$\left| \frac{m(S_\delta)}{V(S_\delta)} - \rho(X_0) \right| = \left| \frac{1}{V(S_\delta)} \int_{S_\delta} [\rho(\mathbf{X}) - \rho(\mathbf{X}_0)] \, dx \, dy \, dz \right|$$

$$\le \frac{1}{V(S_\delta)} \int_{S_\delta} |\rho(\mathbf{X}) - \rho(\mathbf{X}_0)| \, dx \, dy \, dz$$

$$< \frac{\epsilon}{V(S_\delta)} \int_{S_\delta} dx \, dy \, dz = \epsilon.$$

Thus, $\rho(\mathbf{X}_0)$ is approximately equal to the ratio of the mass of S_δ to its volume, and the error of approximation approaches zero as δ approaches zero; that is, the mass $m(S_\delta)$ is approximated by the product $\rho(\mathbf{X}_0) V(S_\delta)$ when δ is small.

FIGURE 4.1

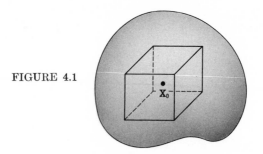

Example 4.1   Let S be the rectangular parallelepiped

$$S = \{(x, y, z) \mid a \le x \le b, \quad c \le x \le d, \quad e \le z \le f\}$$

and let Δ be a partition of S into subparallelepipeds $S_1, \ldots, S_n$. The Riemann sum

$$\sigma = \sum_{i=1}^{n} \rho(\mathbf{X}_i) V(S_i),$$

where $V(S_i)$ is the volume of S_i and $\mathbf{X}_i$ is a point in S_i, is on the one hand an approximation to $m(S)$ and, on the other, the mass of a system of particles of mass $\rho(\mathbf{X}_1) V(S_1), \ldots, \rho(\mathbf{X}_n) V(S_n)$, situated at $\mathbf{X}_1, \ldots, \mathbf{X}_n$, respectively. We shall use this observation to motivate the definitions of first moment, center of mass, and moment of inertia of a region with a continuous mass distribution. We first recall their definitions for systems of point masses.

Center of Mass of a System of Point Masses

Let Γ be the plane through $\mathbf{X}_0$ and normal to

$$N = \begin{bmatrix} A \\ B \\ C \end{bmatrix} \quad (A^2 + B^2 + C^2 \neq 0);$$

thus Γ has the equation

(1) $$A(x - x_0) + B(y - y_0) + C(z - z_0) = 0.$$

If $\mathbf{X}_1$ is an arbitrary point in R^3, then the directed distance from $\mathbf{X}_1$ to Γ is

$$d(\mathbf{X}_1) = \frac{A(x_1 - x_0) + B(y_1 - y_0) + C(z_1 - z_0)}{\sqrt{A^2 + B^2 + C^2}}$$

(Section 1.3), and the *moment about* Γ of a particle P_1 situated at $\mathbf{X}_1$ is

$$M_\Gamma = m_1 d(\mathbf{X}_1),$$

where m_1 is the mass of the particle. If P is a system of N particles of mass $m_1, \ldots, m_n$ located at $\mathbf{X}_1, \ldots, \mathbf{X}_n$, then the *moment* of P *about* Γ is the sum of the individual moments:

(2) $$M_\Gamma = \frac{A\left(\sum_{i=1}^{n} m_i x_i - m x_0\right) + B\left(\sum_{i=1}^{n} m_i y_i - m y_0\right) + C\left(\sum_{i=1}^{n} m_i z_i - m z_0\right)}{\sqrt{A^2 + B^2 + C^2}}$$

where

$$m = \sum_{i=1}^{n} m_i$$

is the total mass of the system.

The moment of P about the yz-plane is denoted by M_{yz}; setting $A = 1$ and $x_0 = B = C = 0$ in (2) yields

$$M_{yz} = \sum_{i=1}^{n} m_i x_i.$$

Similarly, the moments about the zx- and xy-planes are

$$M_{zx} = \sum_{i=1}^{n} m_i y_i$$

and

$$M_{xy} = \sum_{i=1}^{n} m_i z_i.$$

In terms of these moments, (2) can be rewritten as

(3) $$M_\Gamma = \frac{A(M_{yz} - mx_0) + B(M_{zx} - my_0) + C(M_{xy} - mz_0)}{\sqrt{A^2 + B^2 + C^2}}.$$

The *center of mass* of P is $\bar{\mathbf{X}} = (\bar{x}, \bar{y}, \bar{z})$, where

$$\bar{x} = \frac{M_{yz}}{m}, \qquad \bar{y} = \frac{M_{zx}}{m}, \qquad \bar{z} = \frac{M_{xy}}{m};$$

from (3), P has zero moment about every plane through $\mathbf{X}_0$ if and only if $\mathbf{X}_0 = \bar{\mathbf{X}}$.

Example 4.2   Let P consist of point masses of 2, 3, and 4 units at $\mathbf{X}_1 = (1, -1, 2)$, $\mathbf{X}_2 = (2, 1, 0)$, and $\mathbf{X}_3 = (-1, 0, 1)$, respectively (Fig. 4.2). Then

$$M_{yz} = 2(1) + 3(2) + 4(-1) = 4,$$

$$M_{zx} = 2(-1) + 3(1) + 4(0) = 1,$$

$$M_{xy} = 2(2) + 3(0) + 4(1) = 8,$$

and the total mass of P is

$$m = 2 + 3 + 4 = 9.$$

The moment of P about the plane Γ_0 defined by

$$2(x - 4) + (y + 1) - 3z = 0$$

is

$$M_{\Gamma_0} = \frac{2(M_{yz} - 4m) + (M_{zx} + m) - 3M_{xy}}{\sqrt{14}}$$

$$= \frac{2[4 - 4(9)] + [1 + 9] + 3[8]}{\sqrt{14}} = -\frac{78}{\sqrt{14}}.$$

The center of mass of P has coordinates

$$\bar{x} = \frac{M_{yz}}{m} = \frac{4}{9}, \qquad \bar{y} = \frac{M_{zx}}{m} = \frac{1}{9},$$

and

$$\bar{z} = \frac{M_{xy}}{m} = \frac{8}{9}.$$

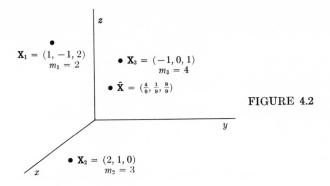

FIGURE 4.2

Center of Mass of a Solid

Now consider the rectangular parallelepiped of Example 4.1, again partitioned into subparallelepipeds $S_1, \ldots, S_n$, and let P_Δ be a system of particles situated at points $\mathbf{X}_1, \ldots, \mathbf{X}_n$ in $S_1, \ldots, S_n$, with masses $\rho(\mathbf{X}_1) V(S_1), \ldots, \rho(\mathbf{X}_n) V(S_n)$. The moment of P_Δ about the yz-plane is

$$M_{yz}(P_\Delta) = \sum_{i=1}^{n} x_i \rho(\mathbf{X}_i) V(S_i),$$

which is a Riemann sum for the integral

$$\int_S x\rho(\mathbf{X}) \, dx \, dy \, dz.$$

As $\| \Delta \|$ decreases the sum approaches the integral and P_Δ approaches S, in an intuitive sense. This and similar reasoning applied to the other moments motivates the following definition.

Definition 4.2 Let S be a region in R^3 with mass density function ρ and total mass

$$m = \int_S \rho(\mathbf{X}) \, dx \, dy \, dz.$$

The *moments of S about the yz-, zx- and xy-planes* are defined by

$$M_{yz} = \int_S x\rho(\mathbf{X}) \, dx \, dy \, dz,$$

$$M_{zx} = \int_S y\rho(\mathbf{X}) \, dx \, dy \, dz$$

and

$$M_{xy} = \int_S z\rho(\mathbf{X}) \, dx \, dy \, dz;$$

the *moment of S about the arbitrary plane* Γ defined by (1) is

$$M_\Gamma = \frac{A(M_{yz} - mx_0) + B(M_{zx} - my_0) + C(M_{xy} - mz_0)}{\sqrt{A^2 + B^2 + C^2}}.$$

The *center of mass of P* is the unique point $\bar{\mathbf{X}}$ with the property that $M_\Gamma = 0$ for every plane Γ which contains $\bar{\mathbf{X}}$; its coordinates are

$$\bar{x} = \frac{M_{yz}}{m}, \qquad \bar{y} = \frac{M_{zx}}{m}, \qquad \bar{z} = \frac{M_{xy}}{m}.$$

The center of mass is called the *centroid* of S if $\rho(\mathbf{X}) = 1$ for all $\mathbf{X}$ in S.

Example 4.3 Let S be the rectangular parallelepiped bounded by the coordinate axes and the three planes defined by $x = 1$, $y = 1$, and $z = 1$ (Fig. 4.3). Thus

$$S = \{(x, y, z) \mid 0 \le x \le 1, \ \ 0 \le y \le 1, \ \ 0 \le z \le 1\}.$$

Let the mass density function of S be $\rho(x, y, z) = z$. Then the mass of S is

$$m = \int_S z \, dx \, dy \, dz$$

$$= \int_0^1 z \, dz \int_0^1 dy \int_0^1 dx = \tfrac{1}{2}.$$

The moments of S about the coordinate planes are

$$M_{xy} = \int_S z^2 \, dx \, dy \, dz$$

$$= \int_0^1 z^2 \, dz \int_0^1 dy \int_0^1 dx = \tfrac{1}{3},$$

$$M_{yz} = \int_S xz \, dx \, dy \, dz$$

$$= \int_0^1 z \, dz \int_0^1 dy \int_0^1 x \, dx = \tfrac{1}{4},$$

and

$$M_{zx} = \int_S yz \, dx \, dy \, dz$$

$$= \int_0^1 z \, dz \int_0^1 y \, dy \int_0^1 dx = \tfrac{1}{4}.$$

The coordinates of the center of mass are

$$\bar{x} = \frac{M_{yz}}{m} = \frac{\frac{1}{4}}{\frac{1}{2}} = \frac{1}{2},$$

$$\bar{y} = \frac{M_{zx}}{m} = \frac{\frac{1}{4}}{\frac{1}{2}} = \frac{1}{2},$$

and

$$\bar{z} = \frac{M_{xy}}{m} = \frac{\frac{1}{3}}{\frac{1}{2}} = \frac{2}{3}.$$

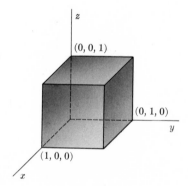

FIGURE 4.3

Example 4.4   To find the centroid of the region S considered in Example 4.3, take $\rho = 1$. Then

$$m = \int_S dx \, dy \, dz = 1,$$

$$M_{xy} = \int_S z \, dx \, dy \, dz = \tfrac{1}{2},$$

$$M_{yz} = \int_S x \, dx \, dy \, dz = \tfrac{1}{2},$$

and

$$M_{zx} = \int_S y \, dx \, dy \, dz = \tfrac{1}{2};$$

hence the centroid is

$$\bar{\mathbf{X}} = (\tfrac{1}{2}, \tfrac{1}{2}, \tfrac{1}{2}).$$

Example 4.5 Let it be required to find the centroid of a homogeneous right circular cone S with height h and base radius a. We first introduce a coordinate system so that the base of the cone is in the xy-plane and its longitudinal axis is along the positive z-axis (Fig. 4.4). The mass of S is

$$m = \rho \int_S dx \, dy \, dz$$

$$= \rho \int_T r \, dr \, d\theta \, dz,$$

where, in cylindrical coordinates,

$$(4) \quad T = \left\{ (r, \theta, z) \,\middle|\, 0 \le r \le a \left(1 - \frac{z}{h} \right), \quad 0 \le \theta \le 2\pi, \quad 0 \le z \le h \right\};$$

thus

$$m = \rho \int_0^{2\pi} d\theta \int_0^h dz \int_0^{a(1-z/h)} r \, dr = \frac{\pi \rho a^2 h}{3}.$$

The symmetry of S about the z-axis suggests that $\bar{x} = \bar{y} = 0$, which is easily verified by observing, for instance, that

$$M_{yz} = \rho \int_S x \, dx \, dy \, dz$$

can be evaluated by means of an interated integral where the first integration is with respect to x, between symmetric limits. To obtain $\bar{z}$, we calculate

$$M_{xy} = \rho \int_S z \, dx \, dy \, dz$$

$$= \rho \int_0^{2\pi} d\theta \int_0^h z \, dz \int_0^{a(1-z/h)} r \, dr = \frac{\pi \rho a^2 h^2}{12},$$

and

$$\bar{z} = \frac{M_{xy}}{m} = \frac{h}{4}.$$

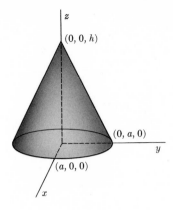

FIGURE 4.4

Example 4.6 Suppose S is the portion of a homogeneous solid sphere of radius 1 that lies in the first octant (Fig. 4.5). Without loss of generality, we can assume $\rho = 1$; then the mass of S is $m = \pi/6$ (Example 3.6). By symmetry, $\bar{x} = \bar{y} = \bar{z}$; hence, to find the centroid of S we need only calculate

$$M_{xy} = \int_S z \, dx \, dy \, dz.$$

This is best accomplished in spherical coordinates:

$$M_{xy} = \int_T r^3 \sin \phi \cos \phi \, dr \, d\theta \, d\phi,$$

where

$$T = \left\{ (r, \theta, \phi) \,\middle|\, 0 \leq r \leq 1, \quad 0 \leq \theta \leq \frac{\pi}{2}, \quad 0 \leq \phi \leq \frac{\pi}{2} \right\}.$$

Thus

$$M_{xy} = \int_0^1 r^3 \, dr \int_0^{\pi/2} d\theta \int_0^{\pi/2} \sin \phi \cos \phi \, d\phi$$

$$= \left(\frac{1}{4}\right)\left(\frac{\pi}{2}\right)\left(\frac{1}{2}\right) = \frac{\pi}{16},$$

and

$$\bar{x} = \bar{y} = \bar{z} = \frac{\dfrac{\pi}{16}}{\dfrac{\pi}{6}} = \frac{3}{8}.$$

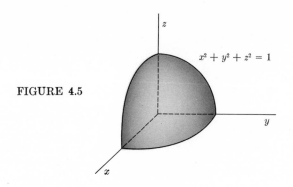

FIGURE 4.5

$x^2 + y^2 + z^2 = 1$

Moments of Inertia

Again let P be a system of N particles with masses $m_1, \ldots, m_n$, situated at $\mathbf{X}_1, \ldots, \mathbf{X}_n$, respectively. If the entire system rotates about a line L with instantaneous angular velocity ω then the velocity of the ith particle is

$$v_i = \omega d_i,$$

where d_i is the distance from $\mathbf{X}_i = (x_i, y_i, z_i)$ to L. The kinetic energy of the system is

$$(5) \qquad\qquad \text{K.E.} = \frac{1}{2} \sum_{i=1}^{n} m_i v_i^2$$

$$= \frac{\omega^2}{2} \sum_{i=1}^{n} m_i d_i^2.$$

Suppose we have chosen the coordinate system so that L is a line through the origin, defined by

$$(6) \qquad\qquad \frac{x}{A} = \frac{y}{B} = \frac{z}{C} \qquad (A^2 + B^2 + C^2 \neq 0);$$

then

$$d_i^2 = \frac{A^2(y_i^2+z_i^2) + B^2(x_i^2+z_i^2) + C^2(x_i^2+y_i^2) - 2ABx_iy_i - 2BCy_iz_i - 2CAz_ix_i}{A^2 + B^2 + C^2}$$

(Section 1.3). From this and (5)

$$(7) \qquad\qquad \text{K.E.} = \tfrac{1}{2} I_L \omega^2,$$

where

$$(8) \qquad I_L = \frac{A^2 I_x + B^2 I_y + C^2 I_z - 2AB I_{xy} - 2BC I_{yz} - 2CA I_{zx}}{A^2 + B^2 + C^2}$$

and

$$I_x = \sum_{i=1}^{n} m_i(y_i^2 + z_i^2),$$

$$I_y = \sum_{i=1}^{n} m_i(z_i^2 + x_i^2),$$

$$I_z = \sum_{i=1}^{n} m_i(x_i^2 + y_i^2),$$

$$I_{xy} = \sum_{i=1}^{n} m_i x_i y_i,$$

$$I_{yz} = \sum_{i=1}^{n} m_i y_i z_i,$$

$$I_{zx} = \sum_{i=1}^{n} m_i z_i x_i.$$

The quantity I_L is the *moment of inertia of P about L*; I_x, I_y, and I_z are the *moments of inertia of P about the x-, y- and z-axes*, respectively, and I_{xy}, I_{yz}, and I_{zx} are the *products of inertia of P*. From (8) it is clear that the moment of inertia about any line through the origin can be calculated from these six quantities.

Example 4.7 For the system P defined in Example 4.2,

$$I_x = 2[(-1)^2 + 2^2] + 3[1^2 + 0^2] + 4[0^2 + 1^2] = 17;$$

$$I_y = 2[2^2 + 1^2] + 3[0^2 + 2^2] + 4[1^2 + (-1)^2] = 30;$$

$$I_z = 2[1^2 + (-1)^2] + 3[2^2 + 1^2] + 4[(-1)^2 + 0^2] = 23;$$

$$I_{xy} = 2(1)(-1) + 3(2)(1) + 4(-1)(0) = 4;$$

$$I_{yz} = 2(-1)(2) + 3(1)(0) + 4(0)(1) = -4;$$

and

$$I_{zz} = 2(2)(1) + 3(0)(2) + 4(1)(-1) = 0.$$

The moment of inertia of P about the line L_0 defined by

$$\frac{x}{2} = \frac{y}{-1} = \frac{z}{0}$$

is

$$I_L = \frac{\begin{aligned}2^2(17) + (-1)^2(30) + 0^2(23) - 2(2)(-1)(4)\\- 2(-1)(0)(-4) - 2(0)(2)(0)\end{aligned}}{2^2 + (-1)^2 + 0^2}$$

$$= \frac{114}{5}.$$

Definition 4.3   Let S be a region in R^3 with mass density function ρ. The *moments of inertia of S about the x-, y- and z-axes* are defined by

$$I_x = \int_S \rho(\mathbf{X})\,(y^2 + z^2)\ dx\ dy\ dz,$$

$$I_y = \int_S \rho(\mathbf{X})\,(z^2 + x^2)\ dx\ dy\ dz,$$

and

$$I_z = \int_S \rho(\mathbf{X})\,(x^2 + y^2)\ dx\ dy\ dz.$$

The *products of inertia* are defined by

$$I_{xy} = \int_S \rho(\mathbf{X})xy\ dx\ dy\ dz,$$

$$I_{yz} = \int_S \rho(\mathbf{X})yz\ dx\ dy\ dz,$$

and

$$I_{zx} = \int_S \rho(\mathbf{X})zx\ dx\ dy\ dz.$$

In terms of these quantities the *moment of inertia I_L of S about the line L,* defined by (6), is given by (8).

If S rotates about L with instantaneous angular velocity ω, then its kinetic energy is given by (7).

Example 4.8   Let L be the line

$$\frac{x}{1} = \frac{y}{2} = \frac{z}{0}$$

and let S be the solid studied in Example 4.3. Then

$$I_L = \frac{I_x + 4I_y - 4I_{xy}}{5},$$

where

$$I_x = \int_S \rho(\mathbf{X})\,(y^2 + z^2)\, dx\, dy\, dz$$

$$= \int_S z\,(y^2 + z^2)\, dx\, dy\, dz = \tfrac{5}{12},$$

$$I_y = \int_S \rho(\mathbf{X})\,(x^2 + z^2)\, dx\, dy\, dz$$

$$= \int_S z\,(x^2 + z^2)\, dx\, dy\, dz = \tfrac{5}{12},$$

and

$$I_{xy} = \int_S \rho(\mathbf{X})\,xy\, dx\, dy\, dz$$

$$= \int_S xyz\, dx\, dy\, dz = \tfrac{1}{8}.$$

Thus

$$I_L = \frac{\tfrac{5}{12} + 4(\tfrac{5}{12}) - 4(\tfrac{1}{8})}{5} = \frac{19}{60}.$$

Example 4.9 Suppose we wish to find the moment of inertia I of a solid sphere S of radius a and uniform density ρ, about a line through its center. By symmetry I is independent of the particular line chosen. Consider a coordinate system with origin at the center of S. Then

$$I = I_z = \rho \int_S (x^2 + y^2)\, dx\, dy\, dz,$$

which is best evaluated in spherical coordinates:

$$I = \rho \int_T (r^2 \sin^2 \phi \cos^2 \theta + r^2 \sin^2 \phi \sin^2 \theta) r^2 \sin \phi\, dr\, d\theta\, d\phi$$

$$= \rho \int_T r^4 \sin^3 \phi\, dr\, d\theta\, d\phi,$$

where

$$T = \{(r, \theta, \phi) \mid 0 \leq r \leq a, \ \ 0 \leq \theta \leq 2\pi, \ \ 0 \leq \phi \leq \pi\}.$$

Thus

$$I = \rho \int_0^{2\pi} d\theta \int_0^{\pi} \sin^3 \phi \, d\phi \int_0^a r^4 \, dr$$

$$= \rho(2\pi) \left(\frac{4}{3}\right) \left(\frac{a^5}{5}\right) = \frac{8}{15} \pi \rho a^5.$$

Example 4.10.   The moment of inertia of the right circular cone of Example 4.5 about its longitudinal axis is given by

$$I = \rho \int_S (x^2 + y^2) \, dx \, dy \, dz,$$

which is best evaluated in cylindrical coordinates over T as defined by (4):

$$I = \rho \int_T (r^2 \cos^2 \theta + r^2 \sin^2 \theta) r \, dr \, d\theta \, dz$$

$$= \rho \int_T r^3 \, dr \, d\theta \, dz$$

$$= \rho \int_0^{2\pi} d\theta \int_0^h dz \int_0^{a(1-z/h)} r^3 \, dr = \frac{\pi a^4 h \rho}{10}.$$

Laminas

A *lamina* is a region S in R^2 on which a mass density function ρ, having the dimensions of mass per unit area, is defined. It can be thought of as the idealization of a thin metal sheet of uniform cross section. In analogy with the corresponding definitions given above, the *total mass of S* is

$$m = \int_S \rho(x, y) \, dx \, dy$$

and its *first moments about the x- and y-axes* are defined by

$$M_x = \int_S \rho(x, y) y \, dx \, dy$$

and

$$M_y = \int_S \rho(x, y)\, x \, dx \, dy.$$

The *moment of S about a line L* given by

$$A(x - x_0) + B(y - y_0) = 0$$

is defined to be

$$M_L = \frac{A(M_y - mx_0) + B(M_x - my_0)}{\sqrt{A^2 + B^2}} ;$$

thus S has zero moment about any line through its *center of mass* $\bar{\mathbf{X}}$, which has coordinates

$$\bar{x} = \frac{M_y}{m}, \qquad \bar{y} = \frac{M_x}{m}.$$

If ρ is constant throughout S, $\mathbf{X}$ is again called the *centroid* of S.

Example 4.11   Let S be the region bounded by the parabola $y = x^2$ and the line $y = 2x$ (Fig. 4.6), with density proportional to the distance from the x-axis; thus

$$\rho(x, y) = ky.$$

Then

$$m = \int_S ky \, dx \, dy$$

$$= k \int_0^2 dx \int_{x^2}^{2x} y \, dy = \frac{32k}{15}.$$

The moments of S with respect to the axes are

$$M_x = \int_S kxy \, dx \, dy$$

$$= k \int_0^2 x \, dx \int_{x^2}^{2x} y \, dy = \tfrac{8}{3} k$$

and

$$M_y = \int_S kx^2 \, dx \, dy$$

$$= k \int_0^2 x^2 \, dx \int_{x^2}^{2x} dy = \tfrac{8}{5} k.$$

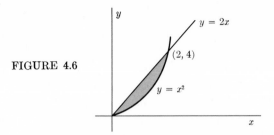

FIGURE 4.6

Thus the center of mass has coordinates

$$\bar{x} = \frac{\frac{8}{5}k}{\frac{32}{15}k} = \frac{3}{4}, \qquad \bar{y} = \frac{\frac{8}{3}k}{\frac{32}{15}k} = \frac{5}{4}.$$

Example 4.12   To find the centroid of the lamina of Example 4.11, take $\rho = 1$. Then

$$m = \int_0^2 dx \int_{x^2}^{2x} dy = \tfrac{4}{3},$$

$$M_x = \int_0^2 dx \int_{x^2}^{2x} y\, dy = \tfrac{32}{15},$$

and

$$M_y = \int_0^2 x\, dx \int_{x^2}^{2x} dy = \tfrac{4}{3};$$

hence

$$\bar{x} = \frac{\frac{4}{3}}{\frac{4}{3}} = 1$$

and

$$\bar{y} = \frac{\frac{32}{15}}{\frac{4}{3}} = \frac{8}{5}.$$

Example 4.13   Let S be the half disk defined by $x^2 + y^2 \le a^2$, $y \ge 0$ (Fig. 4.7), with density proportional to the distance from the origin; thus

$$\rho(x, y) = k\sqrt{x^2 + y^2}.$$

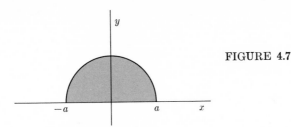

FIGURE 4.7

The mass of S is

$$m = \int_S k\sqrt{x^2 + y^2}\, dx\, dy$$

$$= \int_T kr^2\, dx\, d\theta,$$

where

$$T = \{(r, \theta) \mid 0 \le r \le a, \quad \theta \le 0 \le \pi\};$$

thus

$$m = k\int_0^\pi d\theta \int_0^a r^2\, dr = \frac{k\pi a^3}{3}.$$

The moments of S with respect to the axes are

$$M_x = \int_S ky\sqrt{x^2 + y^2}\, dx\, dy$$

$$= \int_T kr^3 \sin\theta\, dr\, d\theta$$

$$= k\int_0^\pi \sin\theta\, d\theta \int_0^a r^3\, dr = \frac{ka^4}{2}$$

and

$$M_y = \int_S kx\sqrt{x^2 + y^2}\, dx\, dy$$

$$= \int_T kr^3 \cos\theta\, dr\, d\theta$$

$$= k\int_0^\pi \cos\theta\, d\theta \int_0^a r^3\, dr = 0.$$

The coordinates of the center of mass are $\bar{x} = 0$ and

$$\bar{y} = \frac{\dfrac{ka^4}{2}}{\dfrac{k\pi a^3}{3}} = \frac{3a}{2\pi}.$$

Now suppose the entire xy-plane rotates with angular velocity ω about the line L defined by

$$\frac{x}{A} = \frac{y}{B},$$

carrying the lamina S with it (Fig. 4.8). Then it can be shown that the kinetic energy of S is

$$\text{K.E.} = \frac{I_L \omega^2}{2},$$

where

$$I_L = \frac{A^2 I_x + B^2 I_y - 2AB I_{xy}}{A^2 + B^2}$$

is *the moment of inertia of S about L,*

$$I_x = \int_S \rho(x, y) y^2 \, dx \, dy$$

and

$$I_y = \int_S \rho(x, y) x^2 \, dx \, dy$$

are the *moments of inertia of S about the x- and y-axes*, respectively, and

$$I_{xy} = \int_S \rho(x, y) xy \, dx \, dy$$

is its *product of inertia.*

If the xy-plane rotates about a line perpendicular to itself and through the origin, carrying S with it (Fig. 4.9), then the kinetic energy of S is

$$\text{K.E.} = \tfrac{1}{2} I_0 \omega^2,$$

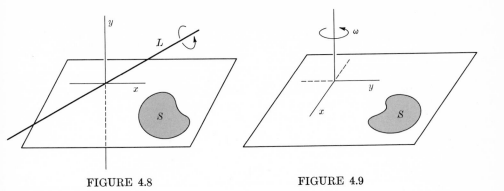

FIGURE 4.8   FIGURE 4.9

where

$$I_0 = \int_S \rho(x,\, y)\,(x^2 + y^2)\; dx\, dy$$

$$= I_y + I_x$$

is the *polar moment of inertia of S with respect to the origin.*

Example 4.14   For the lamina of Example 4.11,

$$I_x = \int_S kxy^2\; dx\, dy$$

$$= k \int_0^2 x\, dx \int_{x^2}^{2x} y^2\; dy = \frac{32}{5}\, k,$$

$$I_y = \int_S kx^3\; dx\, dy$$

$$= k \int_0^2 x^3\, dx \int_{x^2}^{2x} dy = \frac{32}{15}\, k,$$

and

$$I_{xy} = \int_S x^2 y\; dx\, dy$$

$$= k \int_0^2 x^2\, dx \int_{x^2}^{2x} y\; dy = \frac{128}{35}\, k;$$

hence the moment of inertia of S about the line $x = y/2$ is

$$I_L = \frac{1^2(\frac{32}{5}k) + 2^2(\frac{32}{15}k) - 2(1)(2)\frac{128}{35}k}{1^2 + 2^2} = \frac{32k}{525}.$$

Its polar moment of inertia is

$$I_0 = I_x + I_y$$

$$= \frac{32}{5}k + \frac{32}{15}k = \frac{128}{15}k.$$

EXERCISES 4.4

1. Let P consist of point masses of 4 and 5 units at $(-1, 0, 2)$ and $(0, 1, 2)$. Let Γ_0 be the plane $2x - 3y + z = 0$ and let L_0 be the line

$$\frac{x}{3} = \frac{y}{-1} = \frac{z}{0}.$$

Find: (a) M_{xy}, M_{yz}, and M_{zz}; (b) M_{Γ_0}; (c) $(\bar{x}, \bar{y}, \bar{z})$; (d) I_x, I_y, and I_z; (e) I_{xy}, I_{yz}, and I_{xz}; (f) I_{L_0}.

2. Repeat Exercise 1 for the system of point masses of 2, 3, and 5 units at $(-1, 0, 2)$, $(0, 1, 2)$, $(3, 2, -1)$. Let Γ_0 be the plane $3x + 2y + z = -2$ and L_0 be the line

$$\frac{x}{0} = \frac{y}{3} = \frac{z}{0}.$$

In Exercises 3 through 11 let ρ be a mass density function for S. Find: (a) the mass of S; (b) M_{xy}, M_{yz}, and M_{zz}; (c) M_{Γ_0}; (d) the centroid and center of mass of S; (e) I_x, I_y, and I_z; (f) I_{xy}, I_{yz}, and I_{zz}; (g) I_{L_0}.

3. $S = \{(x, y, z) \mid 0 \leq a \leq \sqrt{x^2 + y^2} \leq b, \ |z| < c\}$; $\rho(x, y, z) = |x|$; Γ_0 is defined by $x + y + z = 0$ and L_0 by $x/0 = y/1 = -z/2$.

4. S is bounded by $x + 3y = 3$, $z = 2$ and the coordinate planes; $\rho(x, y, z) = kz$; Γ_0 is defined by $x + y - z = 0$ and L_0 by $x = y = z/2$.

5. $S = \{(x, y, z) \mid 0 \leq z \leq 1 - x^2/4 - y^2/4\}$; $\rho(x, y, z) = kz$; Γ is defined by $x + y + z = 1$ and L_0 by $x = y = (z - 1)/0$.

6. S is a cube in the first octant with sides of length a, with a vertex at the origin and one face in the xy-plane; $\rho(x, y, z) = k(x^2 + y^2 + z^2)$; Γ_0 is the plane $z = a$ and L_0 is the diagonal of S that passes through the origin.

7. $S = \{(x, y, z) \mid x^2 + y^2 + z^2 \leq 4, x \geq 0, y \geq 0, z \geq 0\}$; $\rho(x, y, z) = kxyz$; Γ_0 is defined by $z = 2$ and L_0 by $y = x = z/0$.

8. S is bounded by $x^2 + y^2 + z^2 = 16$ and $x^2 + y^2 = 4$; $\rho(x, y, z) = k$; Γ_0 is defined by $x + y + z = 3$ and L_0 by $y/2 = x/1 = z/0$.

9. S is bounded above by $r = 4$ and below by $\phi = \pi/3$ (spherical coordinates); $\rho(r, \theta, \phi) = kr$; Γ_0 is defined by $x + y + 3z = 3$ and L_0 by $x = y/2 = z/2$.

10. S is bounded by $r = 2$, $r = 4$, $z = r/2$ (cylindrical coordinates) and the xy-plane; $\rho(r, \theta, \phi) = kr$; Γ_0 is defined by $x + y - z = 0$ and L_0 by $x = y/0 = z/0$.

11. $S = \{(x, y, z) \mid x^2/a^2 + y^2/b^2 + z^2/c^2 \leq 1\}$; $\rho(x, y, z) = k|z|$; Γ is defined by $x + y - z = 0$ and L_0 by $x/2 = y/3 = z/0$.

12. A homogeneous cylinder with a mass density ρ_0 has length h and radius a. Find its moment of inertia about its axis.

13. A right circular cone with mass density ρ_0 has height h and base radius a. Find its moment of inertia about a diameter of its base.

14. Find I_x, I_y, I_z, I_{xy}, I_{yz}, and I_{zz} for

$$S = \left\{(x, y, z) \;\middle|\; \frac{x^2}{a^2} + \frac{y^2}{b^2} + \frac{z^2}{c^2} \leq 1\right\};$$

take $\rho(x, y, z) = \rho_0$.

In Exercises 15 through 23, S is the given lamina with mass density function ρ. Find: (a) M_x and M_y; (b) M_{L_0}; (c) the centroid and center of mass of S; (d) I_x, I_y, I_{xy}, and I_0; (e) $\hat{I}_{L_0}$.

15. The rectangle with vertices $(0, 0)$, $(a, 0)$, $(0, b)$, and (a, b), where $a > 0$ and $b > 0$; $\rho(x, y) = kxy$; L_0 is the line $y = (b/a)x$.

16. $\{(x, y) \mid x^2 + y^2 \leq 4, \; y \geq 0\}$; $\rho(x, y) = ky$; L_0 is the line $y = -2x$.

17. $S = \{(x, y) \mid x^2 \leq y \leq x^2 + 4, \; 0 \leq x \leq 2\}$; $\rho(x, y) = kx$; L_0 is the line $2y = x$.

18. S is bounded by the coordinate axes and the line $y = x + 2$; $\rho(x, y) = 2ky$; L_0 is the line $3y = x$.

19. S is bounded by $y = x$, $y = 1/x$, $y = 0$, and $x = 2$; $\rho(x, y) = ky$; L_0 is the line $y = x$.

20. S is bounded by $x = 0$, $y = x + 1$, and $y = 3 - x$; $\rho(x, y) = kx$; L_0 is the line $y = 2x$.

21. S is bounded by $y = x^2$ and $y^2 = x$; $\rho(x, y) = kx$; L_0 is the line $y = x$.

22. S is bounded by $r = 2 \sin \theta$; $\rho(r, \theta) = kr$; L_0 is the line $\theta = 3\pi/4$.

23. S is bounded by $r = 1 + \cos \theta$; $\rho(x, y) = k$; L_0 is the line $\theta = \pi/4$.

THEORETICAL EXERCISES

T-1. Let S_1 and S_2 be two solids with no points in common. *Prove:* The centroid of the system comprised of S_1 and S_2 is on the line segment connecting the centroids of S_1 and S_2.

T-2. Let S be a solid with mass m. Let L_0 be a line through the center of gravity of S_1 and suppose L is a line parallel to L_0 and at a distance h from it. Show that

$$I_L = I_{L_0} + mh^2.$$

In Exercises T-3 and T-4 let $f(x) \geq 0$ in $[a, b]$ and suppose S is the solid swept out when the lamina

$$S_0 = \{ (x, y) \mid 0 \leq y \leq f(x), \quad a \leq x \leq b \}$$

is rotated through 2π radians about the x-axis. Let $\rho(x, y, z) = 1$.

T-3. Show that the mass (volume) of S is

$$m = \pi \int_a^b (f(x))^2 \, dx.$$

T-4. Show that $M_{xy} = 0$, $M_{zx} = 0$, and

$$M_{yz} = \pi \int_a^b x(f(x))^2 \, dx.$$

In Exercises T-5 and T-6 let S_0 be the lamina bounded by $y = f(x)$, the x-axis and the lines $x = a$ and $x = b$. Let $\rho(x, y) = 1$.

T-5.   Show that the mass (area) of S is

$$m = \int_a^b |f(x)| \, dx.$$

T-6.   Show that

$$M_x = \frac{1}{2}\int_a^b f(x) \, |f(x)| \, dx, \qquad M_y = \int_a^b x \, |f(x)| \, dx.$$

4.5   Line Integrals

In this section we consider the line integral, which has its physical origins in the notion of work done in moving a particle along a curve in a force field.

Work

To motivate our definition of the line integral, let us recall the notion of work from elementary physics. The work done by a constant force $\mathbf{F}$ in moving an object along a line segment from $\mathbf{X}_1$ to $\mathbf{X}_2$ in R^2 is

$$W = |\mathbf{F}| \, |\mathbf{X}_2 - \mathbf{X}_1| \cos\theta = \mathbf{F} \cdot (\mathbf{X}_2 - \mathbf{X}_1)$$

where θ is the angle between $\mathbf{F}$ and the line segment (Fig. 5.1).

FIGURE 5.1

Now suppose an object moves from $\mathbf{X}_0$ to $\mathbf{X}_f$, along a curve C in R^2, under the influence of a force $\mathbf{F} = \mathbf{F}(\mathbf{X})$ which varies continuously with $\mathbf{X}$ along C. We wish to define the work done on the object by $\mathbf{F}$. Let $\mathbf{X} = \mathbf{X}(t)$ $(a \le t \le b)$ be a smooth parametric representation of C and consider the partition Δ of $[a, b]$:

$$\Delta: \quad a = t_0 < t_1 < \cdots < t_n = b.$$

For the moment replace C by the path C' consisting of line segments $L_1, \ldots, L_n$ connecting $\mathbf{X}(t_{j-1})$ to $\mathbf{X}(t_j)$, $j = 1, \ldots, n$ (Fig. 5.2) and suppose

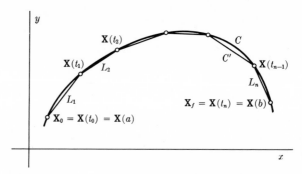

FIGURE 5.2

that the object is moved across C' by a "piecewise constant" force whose value is $\mathbf{F}(\mathbf{X}(\tau_j))$ on L_j, where τ_j is some point in $[t_{j-1}, t_j]$. Then the total work done on the object as it traverses C' is

$$W(\Delta) = \sum_{j=1}^{n} \mathbf{F}(\mathbf{X}(\tau_j)) \cdot (\mathbf{X}(t_j) - \mathbf{X}(t_{j-1})).$$

Applying the mean value theorem to the components of $\mathbf{X}(t)$, we can write

$$\mathbf{X}(t_j) - \mathbf{X}(t_{j-1}) = (t_j - t_{j-1}) \begin{bmatrix} x_1'(\tau_{1j}) \\ x_2'(\tau_{2j}) \\ \vdots \\ x_n'(\tau_{nj}) \end{bmatrix},$$

where $\tau_{1j}, \ldots, \tau_{nj}$ are in the interval (t_{j-1}, t_j). This can be rewritten as

$$\mathbf{X}(t_j) - \mathbf{X}(t_{j-1}) = (t_j - t_{j-1}) \begin{bmatrix} x_1'(\tau_j) \\ x_2'(\tau_j) \\ \vdots \\ x_n'(\tau_j) \end{bmatrix} + (t_j - t_{j-1}) \mathbf{E}_j$$

$$= \mathbf{X}'(\tau_j)(t_j - t_{j-1}) + (t_j - t_{j-1}) \mathbf{E}_j,$$

where

$$\mathbf{E}_j = \begin{bmatrix} x_1'(\tau_{1j}) \\ x_2'(\tau_{2j}) \\ \vdots \\ x_n'(\tau_{nj}) \end{bmatrix} - \begin{bmatrix} x_1'(\tau_j) \\ x_2'(\tau_j) \\ \vdots \\ x_n'(\tau_j) \end{bmatrix}.$$

Thus the work done in moving the object along C' can also be written as

$$W(\Delta) = \sum_{j=1}^{n} \mathbf{F}(\mathbf{X}(\tau_j)) \cdot \mathbf{X}'(\tau_j)(t_j - t_{j-1})$$

$$+ \sum_{j=1}^{n} \mathbf{F}(\mathbf{X}(\tau_j)) \cdot \mathbf{E}_j(t_j - t_{j-1}).$$

From the continuity of $\mathbf{X}'$ on $[a, b]$ it can be shown that the second sum on the right approaches zero as

$$\| \Delta \| = \max_{1 \leq y \leq n} (t_j - t_{j-1})$$

approaches zero. However, as this occurs, the first sum on right approaches

$$W = \int_a^b \mathbf{F}(\mathbf{X}(t)) \cdot \mathbf{X}'(t)\, dt;$$

consequently we define this integral to be the work done in moving the object from $\mathbf{X}_0 = \mathbf{X}(a)$ to $\mathbf{X}_f = \mathbf{X}(b)$ along C. It is an example of a line integral, defined next.

Henceforth all curves and parametric representations are assumed to be piecewise smooth (Definition 1.5, Section 1.5), and we shall also refer to curves as *paths*. When we say that a path (curve) has endpoints $\mathbf{X}_1$ and $\mathbf{X}_2$, we shall mean that the first named endpoint is the initial point.

Definition 5.1 Let $\mathbf{X} = \mathbf{X}(t)$ $(a \leq t \leq b)$ define a path C in the domain of a continuous vector field

$$\mathbf{F} = \begin{bmatrix} F_1 \\ \vdots \\ F_n \end{bmatrix}.$$

Then

(1)
$$I = \int_a^b \mathbf{F}(\mathbf{X}(t)) \cdot \mathbf{X}'(t)\, dt$$

$$= \int_a^b \left(\sum_{i=1}^n F_i(\mathbf{X}(t)) x_i'(t) \right) dt$$

is *the line integral of* $\mathbf{F}$ *along* C.

Example 5.1 Let C be given by

$$\mathbf{X} = \mathbf{X}(t) = \begin{bmatrix} t \\ 2t^2 \\ t^3 \end{bmatrix} \qquad (0 \leq t \leq 1)$$

and

$$\mathbf{F}(\mathbf{X}) = \begin{bmatrix} x \\ xz + 1 \\ yz + x \end{bmatrix};$$

then

$$\mathbf{X}'(t) = \begin{bmatrix} 1 \\ 4t \\ 3t^2 \end{bmatrix}, \qquad \mathbf{F}(\mathbf{X}(t)) = \begin{bmatrix} t \\ t^4 + 1 \\ 2t^5 + t \end{bmatrix},$$

and

$$I = \int_0^1 [(t)(1) + (t^4 + 1)(4t) + (2t^5 + t)(3t^2)] \, dt$$

$$= \int_0^1 (6t^7 + 4t^5 + 3t^3 + 5t) \, dt = \tfrac{14}{3}.$$

Properties of Line Integrals

Even though Definition 5.1 appears to depend upon a particular parametric representation of C, the line integral is in fact independent of the choice of parameter. Let $\mathbf{X} = \mathbf{X}(t)$ $(a \le t \le b)$ and $\mathbf{X} = \mathbf{Y}(\tau)$ $(c \le \tau \le d)$ be equivalent representations of C, related by a piecewise smooth change of parameter $t = \alpha(\tau)$; thus

$$(2) \qquad\qquad \mathbf{Y}(\tau) = \mathbf{X}(\alpha(\tau)) \qquad (c \le \tau \le d).$$

Applying the change of variable $t = \alpha(\tau)$ in (1) yields

$$(3) \qquad\qquad I = \int_c^d \mathbf{F}(\mathbf{X}(\alpha(\tau))) \cdot \mathbf{X}'(\alpha(\tau))\alpha'(\tau) \, d\tau;$$

this integral exists because α' is piecewise continuous. From the chain rule and (2),

$$\mathbf{X}'(\alpha(\tau))\alpha'(\tau) = \mathbf{Y}'(\tau)$$

at all except possibly a finite number of values of τ where one of the derivatives shown may fail to exist; hence (3) can be written as

$$I = \int_c^d \mathbf{F}(\mathbf{Y}(\tau)) \cdot \mathbf{Y}'(\tau) \, d\tau,$$

which is precisely the definition we would have given for I if we had originally represented C by $\mathbf{X} = \mathbf{Y}(\tau)$. This shows that the line integral depends only on $\mathbf{F}$ and the directed curve C, and not on the parametric representation chosen for C.

When we wish to denote a line integral without reference to a particular parametric function we shall write symbolically

$$I = \int_C \mathbf{F} \cdot d\mathbf{X}$$

$$= \int_C (F_1\, dx_1 + F_2\, dx_2 + \cdots + F_n\, dx_n).$$

Example 5.2   The functions

$$\mathbf{X} = \mathbf{X}(t) = \begin{bmatrix} \cos t \\ \sin t \end{bmatrix} \qquad (0 \le t \le \pi)$$

and

$$\mathbf{X} = \mathbf{Y}(\tau) = \begin{bmatrix} -\sin 2\tau \\ \cos 2\tau \end{bmatrix} \qquad \left(-\frac{\pi}{4} \le \tau \le \frac{\pi}{4}\right)$$

are equivalent representations of C, the upper half of the unit circle traversed from $(1, 0)$ to $(0, 1)$ (Fig. 5.3). If

$$\mathbf{F}(x, y) = \begin{bmatrix} \sqrt{x^2 + y^2} \\ \sqrt{1 - x^2} \end{bmatrix}$$

then

$$\mathbf{F}(\mathbf{X}(t)) = \begin{bmatrix} 1 \\ \sin t \end{bmatrix}$$

and

$$\mathbf{F}(\mathbf{Y}(\tau)) = \begin{bmatrix} 1 \\ \cos 2\tau \end{bmatrix}.$$

In terms of $\mathbf{X} = \mathbf{X}(t)$,

$$\int_C \mathbf{F} \cdot d\mathbf{X} = \int_0^\pi [(1)(-\sin t) + (\sin t)\cos t]\, dt$$

$$= (\cos t + \tfrac{1}{2}\sin^2 t)\Big|_0^\pi = -2,$$

while in terms of $\mathbf{X} = \mathbf{Y}(\tau)$

$$\int_C \mathbf{F} \cdot d\mathbf{X} = \int_{-\pi/4}^{\pi/4} [(1)(-2\cos 2\tau) + (\cos 2\tau)(-2\sin 2\tau)] \, d\tau$$

$$= -\left(\sin 2\tau + \frac{\sin^2 2\tau}{2}\right)\Bigg|_{-\pi/4}^{\pi/4} = -2.$$

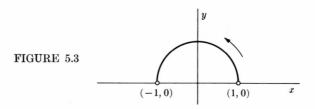

FIGURE 5.3

Theorem 5.1 Let $-C$ be the path obtained by reversing the order in which C is traversed. Then

$$\int_{-C} \mathbf{F} \cdot d\mathbf{X} = -\int_C \mathbf{F} \cdot d\mathbf{X}.$$

Proof. If C is given by $\mathbf{X} = \mathbf{X}(t)$ $(a \le t \le b)$ then $-C$ is given by $\mathbf{X} = \mathbf{Y}(\tau) = \mathbf{X}(-\tau)$ $(-b \le \tau \le -a)$; hence

$$\int_{-C} \mathbf{F} \cdot d\mathbf{X} = \int_{-b}^{-a} \mathbf{F}(\mathbf{Y}(\tau)) \cdot \mathbf{Y}'(\tau) \, d\tau$$

$$= \int_{-b}^{-a} \mathbf{F}(\mathbf{X}(-\tau)) \cdot \frac{d}{d\tau}(\mathbf{X}(-\tau)) \, d\tau.$$

Since

$$\frac{d}{d\tau}(\mathbf{X}(-\tau)) = -\mathbf{X}'(-\tau),$$

we have

$$\int_{-C} \mathbf{F} \cdot d\mathbf{X} = -\int_{-b}^{-a} \mathbf{F}(\mathbf{X}(-\tau)) \cdot \mathbf{X}'(-\tau) \, d\tau,$$

and the change of variable $t = -\tau$ yields

$$\int_{-C} \mathbf{F} \cdot d\mathbf{X} = \int_{b}^{a} \mathbf{F}(\mathbf{X}(t)) \cdot \mathbf{X}'(t) \ dt$$

$$= - \int_{a}^{b} \mathbf{F}(\mathbf{X}(t)) \cdot \mathbf{X}'(t) \ dt$$

$$= - \int_{C} \mathbf{F} \cdot d\mathbf{X},$$

which completes the proof.

Example 5.3   Let $\mathbf{F}$ and C be as in Example 5.1. Then $-C$ is given by

$$\mathbf{X} = \mathbf{X}(-\tau) = \begin{bmatrix} -\tau \\ 2\tau^2 \\ -\tau^3 \end{bmatrix} \quad (-1 \leq \tau \leq 0);$$

hence

$$\frac{d}{d\tau} \mathbf{X}(-\tau) = \begin{bmatrix} -1 \\ 4\tau \\ -3\tau^2 \end{bmatrix},$$

$$F(\mathbf{X}(-\tau)) = \begin{bmatrix} -\tau \\ \tau^4 + 1 \\ -2\tau^5 - \tau \end{bmatrix},$$

and

$$\int_{-C} \mathbf{F} \cdot d\mathbf{X} = \int_{-1}^{0} \left[(-\tau)(-1) + (\tau^4 + 1)(4\tau) + (-2\tau^5 - \tau)(-3\tau^2) \right] d\tau$$

$$= \int_{-1}^{0} (6\tau^7 + 4\tau^5 + 3\tau^3 + 5\tau) \ d\tau = -\tfrac{14}{3}$$

$$= - \int_{C} \mathbf{F} \cdot d\mathbf{X}.$$

The next theorem can be proved by parametrizing C and invoking properties of ordinary integrals (Exercise T-1).

Theorem 5.2 If α and β are real numbers and **F** and **G** are continuous vector fields whose domains contain C, then

$$\int_C (\alpha \mathbf{F} + \beta \mathbf{G}) \cdot d\mathbf{X} = \alpha \int_C \mathbf{F} \cdot d\mathbf{X} + \beta \int_C \mathbf{G} \cdot d\mathbf{X}.$$

It is sometimes convenient to subdivide a curve C into smaller segments; for example, a piecewise smooth curve can be divided into smooth parts (Fig. 5.4). Let C be given by $\mathbf{X} = \mathbf{X}(t)$ $(a \leq t \leq b)$ and partition $[a, b]$:

$$a = t_0 < t_1 < \cdots < t_n = b.$$

If **F** is a continuous vector field on C, then

$$(4) \qquad \int_C \mathbf{F} \cdot d\mathbf{X} = \int_a^b \mathbf{F}(\mathbf{X}(t)) \cdot \mathbf{X}'(t) \, dt$$

$$= \sum_{i=1}^n \int_{t_{i-1}}^{t_i} \mathbf{F}(\mathbf{X}(t)) \cdot \mathbf{X}'(t) \, dt$$

$$= \sum_{i=1}^n \int_{C_i} \mathbf{F} \cdot d\mathbf{X},$$

where, for $i = 1, \ldots, n$, the curve C_i is given by

$$(5) \qquad \mathbf{X} = \mathbf{X}(t) \qquad (t_{i-1} \leq t \leq t_i).$$

The terminal point of C_{i-1} coincides with the initial point of C_i. Symbolically we write

$$C = C_1 + C_2 + \cdots + C_n;$$

then (4) can be rewritten

$$\int_{C_1 + \ldots + C_n} \mathbf{F} \cdot d\mathbf{X} = \sum_{i=1}^n \int_{C_i} \mathbf{F} \cdot d\mathbf{X}.$$

Since the integrals on the right do not depend on the parametric representations chosen for $C_1, \ldots, C_n$, we need not use the representations (5) inherited from C to evaluate them; rather, we may choose any convenient representations for $C_1, \ldots, C_n$.

FIGURE 5.4

Example 5.4   Let C be the curve from $(0, 0)$ to $(1/\sqrt{2}, 1/\sqrt{2})$ along the parabola $y = \sqrt{2}x^2$, followed by the arc of the unit circle from $(1/\sqrt{2}, 1/\sqrt{2})$ to $(-1, 0)$ (Fig. 5.5). Then $C = C_1 + C_2$, where C_1 is given by

$$\mathbf{X} = \begin{bmatrix} t \\ \sqrt{2}t^2 \end{bmatrix} \qquad \left(0 \le t \le \frac{1}{\sqrt{2}} \right)$$

and C_2 by

$$\mathbf{X} = \begin{bmatrix} \cos t \\ \sin t \end{bmatrix} \qquad \left(\frac{\pi}{4} \le t \le \pi \right).$$

If

$$\mathbf{F} = \begin{bmatrix} P \\ Q \end{bmatrix}$$

then

$$\int_C \mathbf{F} \cdot d\mathbf{X} = \int_0^{1/\sqrt{2}} \left[P(t, \sqrt{2}t^2) + 2\sqrt{2}tQ(t, \sqrt{2}t^2) \right] dt$$

$$+ \int_{\pi/4}^{\pi} \left[-P(\cos t, \sin t) \sin t + Q(\cos t, \sin t) \cos t \right] dt.$$

FIGURE 5.5

Independence of Path

Let $\mathbf{F}$ be a continuous vector field in a region D and suppose that C_1 and C_2 are two paths in D with common initial and terminal points $\mathbf{X}_1$ and $\mathbf{X}_2$ (Fig. 5.6). Our next objective is to develop conditions which guarantee that

(6) $$\int_{C_1} \mathbf{F} \cdot d\mathbf{X} = \int_{C_2} \mathbf{F} \cdot d\mathbf{X}.$$

If this equation holds whenever C_1 and C_2 are two such curves in D, we say that *line integrals of $\mathbf{F}$ are independent of their paths* in D. This is not always the case, as is shown by the following example.

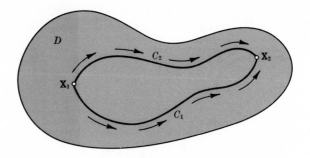

FIGURE 5.6

Example 5.5   Let C_1 be the line segment from $(0, 0)$ to $(1, 1)$ and C_2 be the section of the parabola $y = x^2$ connecting the same two points (Fig. 5.7). Thus C_1 is given by

$$\mathbf{X} = \begin{bmatrix} t \\ t \end{bmatrix} \qquad (0 \le t \le 1)$$

and C_2 by

$$\mathbf{X} = \begin{bmatrix} t \\ t^2 \end{bmatrix} \qquad (0 \le t \le 1).$$

If

$$\mathbf{F}(x, y) = \begin{bmatrix} x + y \\ y \end{bmatrix},$$

then

$$\int_{C_1} \mathbf{F} \cdot d\mathbf{X} = \int_0^1 \left[2t(1) + t(1) \right] dt$$

$$= \int_0^1 3t \, dt = \tfrac{3}{2}$$

and

$$\int_{C_2} \mathbf{F} \cdot d\mathbf{X} = \int_0^1 (t + t^2)(1) + t^2(2t) \,] \, dt$$

$$= \int_0^1 (t + t^2 + 2t^3) \, dt = \tfrac{4}{3};$$

hence (6) is not satisfied.

FIGURE 5.7

Example 5.6 Let C_1 and C_2 be as defined in Example 5.5 and

$$\mathbf{G}(x, y) = \begin{bmatrix} y \\ x \end{bmatrix}.$$

Then

$$\int_{C_1} \mathbf{G} \cdot d\mathbf{X} = \int_0^1 [t(1) + t(1)] \, dt$$

$$= \int_0^1 2t \, dt = 1$$

and

$$\int_{C_2} \mathbf{G} \cdot d\mathbf{X} = \int_0^1 [t^2(1) + t(2t)] \, dt$$

$$= \int_0^1 3t^2 \, dt = 1;$$

thus (6) is satisfied. This is not an accident, caused by a lucky choice of C_1 and C_2; if C is any smooth curve connecting $(0, 0)$ and $(1, 1)$ and parametrized by $\mathbf{X} = \mathbf{X}(t)$ $(a \le t \le b)$, then

$$\int_C \mathbf{G} \cdot d\mathbf{X} = \int_a^b [y(t)x'(t) + x(t)y'(t)] \, dt$$

$$= \int_a^b \frac{d}{dt} (y(t)x(t)) \, dt$$

$$= y(b)x(b) - y(a)x(a) = 1,$$

since $x(a) = y(a) = 0$ and $x(b) = y(b) = 1$.

The following definition is necessary for a careful discussion of conditions guaranteeing that line integrals of a vector field are independent of their paths.

Definition 5.2 A set D of points in R^n is said to be *connected* if any two points in D can be joined by a piecewise smooth curve which lies entirely in D. A *region* is an open connected set.

Prior to this we have used "region" in a much broader sense. Henceforth we shall use it in the sense of Definition 5.2.

Example 5.7 In R^2 the disk

$$D = \{(x, y) \mid x^2 + y^2 < 1\}$$

is a region. If we add to D its boundary, we obtain

$$\bar{D} = \{(x, y) \mid x^2 + y^2 \leq 1\},$$

which is still connected, but no longer open; hence $\bar{D}$ is not a region. The set consisting of the two open disks

$$D_1 = \{(x, y) \mid (x - 1)^2 + y^2 < \tfrac{1}{2}\}$$

and

$$D_2 = \{(x, y) \mid x^2 + y^2 < \tfrac{1}{2}\}$$

(Fig. 5.8) is open but not connected, since any curve joining points in D_1 and D_2 must contain at least one point which is in neither. If S_1 is obtained from S by adding the point $(\tfrac{1}{2}, 0)$, then S_1 is connected. However, S_1 is not open and therefore not a region.

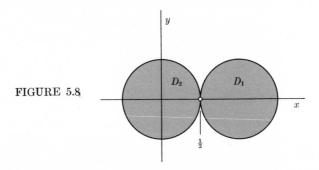

FIGURE 5.8

Theorem 5.3 Let $\mathbf{F}$ be a continuous vector field in a region D. Then the following statements are equivalent:

(a) For any two paths C_1 and C_2 in D with common initial and terminal points,

(7)
$$\int_{C_1} \mathbf{F} \cdot d\mathbf{X} = \int_{C_2} \mathbf{F} \cdot d\mathbf{X}.$$

(b) The vector field $\mathbf{F}$ is the gradient of a continuously differentiable real valued function f in D.

Proof. We first show that (b) implies (a). If $\mathbf{X} = \mathbf{X}(t)$ defines a smooth curve C in D and f is a real valued function such that

$$\mathbf{F} = \nabla f = \begin{bmatrix} \dfrac{\partial f}{\partial x_1} \\ \vdots \\ \dfrac{\partial f}{\partial x_n} \end{bmatrix},$$

then

$$\int_C \mathbf{F} \cdot d\mathbf{X} = \int_a^b \left[F_1(\mathbf{X}(t)) x_1'(t) + \cdots + F_n(\mathbf{X}(t)) x_n'(t) \right] dt$$

$$= \int_a^b \left[\frac{\partial f}{\partial x_1} (\mathbf{X}(t)) x_1'(t) + \cdots + \frac{\partial f}{\partial x_n} (\mathbf{X}(t)) x_n'(t) \right] dt$$

$$= \int_a^b \frac{d}{dt} (f(\mathbf{X}(t))) \, dt = f(\mathbf{X}(b)) - f(\mathbf{X}(a)).$$

If $\mathbf{U}$ and $\mathbf{V}$ are the initial and terminal points of C then $\mathbf{X}(a) = \mathbf{U}$ and $\mathbf{X}(b) = \mathbf{V}$; hence

$$(8) \qquad \int_C \mathbf{F} \cdot d\mathbf{X} = f(\mathbf{V}) - f(\mathbf{U}).$$

This result also holds if C is piecewise smooth but not smooth, for then there are points $a = t_0 < t_1 < \cdots < t_n = b$ such that $\mathbf{X}'$ is continuous in $[t_{i-1}, t_i]$ $(i = 1, \ldots, n)$ (continuous from the interior at the endpoints). Then $(d/dt)f(\mathbf{X}(t))$ is also continuous in these subintervals and

$$\int_{t_{i-1}}^{t_i} \frac{d}{dt} f(\mathbf{X}(t)) \, dt = f(\mathbf{X}(t_i)) - f(\mathbf{X}(t_{i-1}));$$

hence

$$\int_C \mathbf{F} \cdot d\mathbf{X} = \sum_{i=1}^n \int_{t_{i-1}}^{t_i} \frac{d}{dt} f(\mathbf{X}(t)) \, dt$$

$$= \sum_{i=1}^n (f(\mathbf{X}(t_i)) - f(\mathbf{X}(t_{i-1})))$$

$$= f(\mathbf{X}(t_n)) - f(\mathbf{X}(t_0)) = f(\mathbf{V}) - f(\mathbf{U}).$$

Thus (8) holds for all piecewise smooth curves in D. It follows that (7) holds if C_1 and C_2 have common endpoints.

Now suppose (a) holds. Let $\mathbf{X}_0$ be fixed and $\mathbf{X}$ arbitrary in D, and define

$$(9) \qquad\qquad f(\mathbf{X}) = \int_C \mathbf{F} \cdot d\mathbf{X},$$

where C is any piecewise smooth curve in D with initial point $\mathbf{X}_0$ and terminal point $\mathbf{X}$. The assumption is that all such integrals are the same; hence (9) defines a function on D which we denote by

$$f(\mathbf{X}) = \int_{\mathbf{X}_0}^{\mathbf{X}} \mathbf{F} \cdot d\mathbf{X}.$$

Now let $\mathbf{X} = (x_1, \ldots, x_n)$ and $\hat{\mathbf{X}} = (x_1 + \Delta x_1, x_2, \ldots, x_n)$ and suppose that Δx_1 is so small that the line segment L from $\mathbf{X}$ to $\hat{\mathbf{X}}$ is in D. If C is an arbitrary path from $\mathbf{X}_0$ to $\mathbf{X}$ then $C + L$ is a path from $\mathbf{X}_0$ to $\hat{\mathbf{X}}$ (Fig. 5.9 depicts the situation for $n = 2$) and therefore

$$(10) \qquad\qquad f(\hat{\mathbf{X}}) = \int_{\mathbf{X}_0}^{\mathbf{X}} \mathbf{F} \cdot d\mathbf{X} + \int_L \mathbf{F} \cdot d\mathbf{X}$$

$$= f(\mathbf{X}) + \int_L \mathbf{F} \cdot d\mathbf{X}.$$

Since L can be represented by

$$\mathbf{X}(t) = \begin{bmatrix} x_1 + t\,\Delta x_1 \\ x_2 \\ \vdots \\ x_n \end{bmatrix} \qquad (0 \le t \le 1),$$

with

$$\mathbf{X}'(t) = \begin{bmatrix} \Delta x_1 \\ 0 \\ \vdots \\ 0 \end{bmatrix},$$

(10) implies that

$$f(\hat{\mathbf{X}}) - f(\mathbf{X}) = \Delta x_1 \int_0^1 F_1(x_1 + t\,\Delta x_1, x_2, \ldots, x_n)\, dt.$$

From the continuity of F_1, this integral approaches $F_1(\mathbf{X})$ as Δx_1 approaches zero (Exercise T-3); hence

$$\lim_{\Delta x_i \to 0} \frac{f(\hat{\mathbf{X}}) - f(\mathbf{X})}{\Delta x_1} = F_1(\mathbf{X})$$

and therefore

$$\frac{\partial f}{\partial x_1}(\mathbf{X}) = F_1(\mathbf{X}).$$

The same argument applied to the other variables implies that

$$\frac{\partial f}{\partial x_i}(\mathbf{X}) = F_i(\mathbf{X}) \qquad (i = 1, \dots, n);$$

therefore (b) follows.

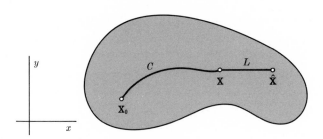

FIGURE 5.9

Conservative Fields; Potential Functions

Let us recall from Section 3.4 that a vector field $\mathbf{F}$ is *conservative* in a region D if $\mathbf{F} = \nabla f$ where f is a continuously differentiable real valued function on D; f is the *scalar field* and $-f$ the *potential function* associated with $\mathbf{F}$.

If $\mathbf{F}$ is conservative, then, formally,

$$\mathbf{F} \cdot d\mathbf{X} = F_1 \, dx_1 + \cdots + F_n \, dx_n$$

$$= \frac{\partial f}{\partial x_1} \, dx_1 + \cdots + \frac{\partial f}{\partial x_n} \, dx_n = df;$$

thus $\mathbf{F} \cdot d\mathbf{X}$ is formally the differential of f. For this reason we say that $\mathbf{F} \cdot d\mathbf{X}$ is an *exact differential* if $\mathbf{F}$ is conservative.

The next theorem follows from these definitions and the proof of Theorem 5.3.

Theorem 5.4 If $\mathbf{F}$ is a conservative vector field on a region D, f the associated scalar field, and C a path in D with endpoints $\mathbf{X}_1$ and $\mathbf{X}_2$, then

$$\int_C \mathbf{F} \cdot d\mathbf{X} = f(\mathbf{X}_2) - f(\mathbf{X}_1);$$

in particular, the integral of $\mathbf{F}$ around any closed path in D is zero.

Example 5.8   We saw in Section 3.4 that the inverse square law force field,

$$\mathbf{F}(\mathbf{X}) = \frac{-Gm_0}{|\mathbf{X}|^3}\mathbf{X},$$

is the gradient of

$$f(\mathbf{X}) = \frac{Gm_0}{|\mathbf{X}|}.$$

Thus $-f(\mathbf{X})$ is the *potential energy* of a unit mass at $\mathbf{X}$. Theorem 5.4 states that the work done by gravity on a unit mass which moves along a path C in this field is equal to the change in the potential energy of the particle.

The following example illustrates the significance of the adjective *conservative*.

Example 5.9   Let $\mathbf{X} = \mathbf{X}(t)$ be the position of a particle of mass m as a function of time, and let $\mathbf{X}'(t)$ be its velocity vector. Suppose that, as t varies from t_0 to t_1, the particle moves from $\mathbf{X}_0$ to $\mathbf{X}_1$ along a path C which lies in a domain D in R^3, in which a conservative force field $\mathbf{F}$ is defined. From Theorem 5.4 the work done on the particle by $\mathbf{F}$ is

$$(11) \qquad W = \int_C \mathbf{F} \cdot d\mathbf{X} = f(\mathbf{X}_1) - f(\mathbf{X}_0),$$

where $\mathbf{F} = \nabla f$. On the other hand, using Newton's second law of motion,

$$\mathbf{F}(\mathbf{X}(t)) = m\mathbf{X}''(t),$$

we can write

$$W = \int_{t_0}^{t_1} \mathbf{F}(\mathbf{X}(t)) \cdot \mathbf{X}'(t) \; dt$$

$$= m \int_{t_0}^{t_1} \mathbf{X}''(t) \cdot \mathbf{X}'(t) \; dt$$

$$= \frac{m}{2} \int_{t_0}^{t_1} \frac{d}{dt} (\mathbf{X}'(t) \cdot \mathbf{X}'(t)) \; dt$$

$$= \frac{m}{2} \int_{t_0}^{t_1} \frac{d}{dt} |\mathbf{X}'(t)|^2 \; dt$$

$$= \frac{m}{2} |\mathbf{X}'(t_1)|^2 - \frac{m}{2} |\mathbf{X}'(t_0)|^2.$$

Comparing this with (11), we see that

$$(12) \qquad \frac{m}{2} \mid \mathbf{X}'(t_1) \mid^2 - f(\mathbf{X}_1) = \frac{m}{2} \mid \mathbf{X}'(t_0) \mid^2 - f(\mathbf{X}_0).$$

The quantity $-f(\mathbf{X})$ is the potential energy of the particle at $\mathbf{X}$, while $(m/2) \mid \mathbf{X}'(t) \mid^2$ is its kinetic energy. From (12) it follows that their sum, the total energy, remains constant as the particle moves through a conservative force field; that is, energy is *conserved*.

Theorem 5.3 says that a vector field $\mathbf{F}$ is conservative if and only if it has a potential function; however, it does not specify how to find a potential function, nor does it provide a practical test for actually deciding whether one exists. The following theorem gives necessary conditions for a continuously differentiable vector field to be conservative.

Theorem 5.5 If

$$\mathbf{F} = \begin{bmatrix} F_1 \\ \vdots \\ F_n \end{bmatrix}$$

is a continuously differentiable, conservative vector field in a region D in R^n, then

$$(13) \qquad \frac{\partial F_i}{\partial x_j} = \frac{\partial F_j}{\partial x_i} \qquad (i, j = 1, \ldots, n).$$

In particular, if $n = 2$, and

$$\mathbf{F} = \begin{bmatrix} P \\ Q \end{bmatrix},$$

then

$$(14) \qquad \frac{\partial P}{\partial y} = \frac{\partial Q}{\partial x};$$

if $n = 3$ and

$$\mathbf{F} = \begin{bmatrix} P \\ Q \\ R \end{bmatrix},$$

then

$$(15) \qquad \frac{\partial P}{\partial y} = \frac{\partial Q}{\partial x}, \qquad \frac{\partial Q}{\partial z} = \frac{\partial R}{\partial y}, \qquad \frac{\partial R}{\partial x} = \frac{\partial P}{\partial z}.$$

Proof. If $n = 2$ and $\mathbf{F} = \nabla f$ then

$$P = \frac{\partial f}{\partial x}$$

and

$$Q = \frac{\partial f}{\partial y}.$$

Therefore

$$(16) \qquad \frac{\partial P}{\partial y} = \frac{\partial^2 f}{\partial y \, \partial x}$$

and

$$(17) \qquad \frac{\partial Q}{\partial x} = \frac{\partial^2 f}{\partial x \, \partial y}.$$

Since $\partial P/\partial y$ and $\partial Q/\partial x$ are continuous, the mixed second partial derivatives in (16) and (18) are equal (Section 2.2) and (14) follows.

The idea of the proof is the same for $n > 2$. If $\mathbf{F} = \nabla f$ and $1 \le i < j \le n$ then

$$(18) \qquad F_i = \frac{\partial f}{\partial x_i}$$

and

$$(19) \qquad F_j = \frac{\partial f}{\partial x_j};$$

differentiating (18) with respect to x_j and (19) with respect to x_i yields

$$\frac{\partial F_i}{\partial x_j} = \frac{\partial^2 f}{\partial x_j \, \partial x_i}$$

and

$$\frac{\partial F_j}{\partial x_i} = \frac{\partial^2 f}{\partial x_i \, \partial x_j},$$

and again the continuity of the $\partial F_i/\partial x_j$ and $\partial F_j/\partial x_i$ implies that the mixed second partial derivatives of f are equal. Thus (13) follows.

Example 5.10 Let

$$\mathbf{F}(x, y) = \begin{bmatrix} y \\ x \end{bmatrix};$$

then $P(x, y) = y$ and $Q(x, y) = x$, so that

$$\frac{\partial P}{\partial y} = \frac{\partial Q}{\partial x} = 1.$$

Thus $\mathbf{F}$ satisfies the necessary conditions for a conservative vector field and therefore *may* have a potential function. It does, since

$$\begin{bmatrix} y \\ x \end{bmatrix} = \nabla f,$$

where

$$f(x, y) = xy.$$

Example 5.11 The inverse square law field of Example 5.8 satisfies (15) if $\mathbf{X} \neq \mathbf{0}$ (Verify!); we exhibited its scalar field in Example 5.7.

We emphasize that Theorem 5.5 gives necessary, but not sufficient, conditions for a vector field to be conservative. The following example shows that a vector field may satisfy these conditions but fail to be conservative.

Example 5.12 The vector field

$$\mathbf{F}(x, y) = \begin{bmatrix} \dfrac{-y}{x^2 + y^2} \\[2ex] \dfrac{x}{x^2 + y^2} \end{bmatrix}$$

satisfies (14) in the domain D consisting of all $(x, y) \neq (0, 0)$; however, if C is the unit circle,

$$\mathbf{X} = \begin{bmatrix} \cos t \\ \sin t \end{bmatrix} \qquad (0 \leq t \leq 2\pi),$$

then

$$\int_C \mathbf{F} \cdot d\mathbf{X} = \int_0^{2\pi} [(-\sin t)(-\sin t) + (\cos t)(\cos t)] \, dt = 2\pi.$$

Therefore $\mathbf{F}$ is not conservative in D, by Theorem 5.4.

The conditions of Theorem 5.5 are also sufficient if the region D is simply connected, which we now define.

Definition 5.3 Consider a closed curve C in a region D to be constructed of a loop of elastic material. If it is possible to deform C to a single point in D without breaking the loop or leaving D, then we say that C *can be contracted to a point within* D. A region D is *simply connected* if any closed curve in D has this property.

This definition is intuitive, but adequate for our purposes; to formulate it in an entirely precise and rigorous fashion would take us beyond the scope of this book.

Example 5.13 In R^2 the interiors of a circle and a half plane are simply connected, as is R^2 itself. However the exterior of the unit disc is not. In fact no domain in R^2 with a "hole" in it is simply connected (Fig. 10); thus a region in R^2 obtained by removing a single point from a simply connected region is not simply connected.

Simply Connected   Not Simply Connected

FIGURE 5.10

Example 5.14 In R^n $(n > 3)$ the region between two concentric n-spheres, say

$$D = \{\mathbf{X} \mid 1 < |\mathbf{X}| < 2\}$$

is simply connected.

Example 5.15 In R^3 the region between two cylinders,

$$D = \{(x, y, z) \mid 1 < x^2 + y^2 < 4, \quad 0 < z < 1\},$$

is not simply connected, since a closed curve C which encloses the axis of D cannot be contracted to a point within D (Fig. 5.11).

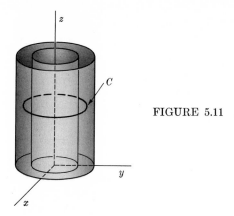

FIGURE 5.11

We state the following useful theorem without proof.

Theorem 5.6 If **F** satisfies the hypotheses of Theorem 5.5 in a simply connected domain D, then **F** is conservative in D.

Example 5.16 We saw in Example 5.12 that

$$\mathbf{F}(x, y) = \begin{bmatrix} \dfrac{-y}{x^2 + y^2} \\[2mm] \dfrac{x}{x^2 + y^2} \end{bmatrix} \qquad ((x, y) \neq (0, 0))$$

is not conservative in its domain of definition (which is not simple connected). However, Theorem 5.6 guarantees that **F** is conservative in any simply connected region which does not contain the origin.

Example 5.17 To see that **F** may be conservative even if D is not simply connected, consider

$$\mathbf{F}(x, y) = \begin{bmatrix} \dfrac{-x}{(x^2 + y^2)^{3/2}} \\[2mm] \dfrac{-y}{(x^2 + y^2)^{3/2}} \end{bmatrix} \qquad ((x, y) \neq (0, 0)),$$

whose domain D is again all of R^2 except the origin. It is conservative in D

since $\mathbf{G} = \boldsymbol{\nabla} g$, where

$$g(x, y) = \frac{1}{\sqrt{x^2 + y^2}}.$$

Theorem 5.6 has the following consequence in R^2.

Theorem 5.7 Let a vector field

$$\mathbf{F} = \begin{bmatrix} P \\ Q \end{bmatrix}$$

be continuously differentiable and satisfy

$$\frac{\partial P}{\partial y} = \frac{\partial Q}{\partial x}$$

in a domain D in R^2 which contains two simple closed curves C_1 and C_2, situated as shown in Fig. 5.12, and the region S bounded by them. Then

$$\int_{C_1} P\,dx + Q\,dy = \int_{C_2} P\,dx + Q\,dy,$$

where both curves are traversed counterclockwise.

Proof. Divide S into simply connected subregions S_1 and S_2 by inserting line segments L_1 and L_2 as shown in Fig. 5.13. By Theorem 5.6,

$$\int_{\Gamma_1} P\,dx + Q\,dy = \int_{\Gamma_2} P\,dx + Q\,dy = 0,$$

where Γ_1 and Γ_2 are the boundaries of S_1 and S_2, traversed counterclock-

FIGURE 5.12

FIGURE 5.13

wise. Hence

(20)
$$0 = \int_{\Gamma_1} P \, dx + Q \, dy + \int_{\Gamma_2} P \, dx + Q \, dy$$

$$= \int_{C_2} P \, dx + Q \, dy - \int_{C_1} P \, dx + Q \, dy$$

(Exercise T-4), which completes the proof.

Example 5.18   Let

$$\mathbf{F}(x, y) = \begin{bmatrix} P(x, y) \\ Q(x, y) \end{bmatrix} = \begin{bmatrix} y^2 \\ 2xy \end{bmatrix};$$

since $\partial P / \partial y = 2y = \partial Q / \partial x$ on all of R^2, which is simply connected, the value of the integral

$$I = \int_{(0,0)}^{(2,1)} y^2 \, dx + 2xy \, dy$$

does not depend upon the path of integration. The integral could be evaluated by means of Theorem 5.4, if we wished to find the potential function associated with **F**. However, this is not necessary; we can choose any convenient path. Let us integrate along the line segment L connecting $(0, 0)$ and $(2, 1)$ (Fig. 5.14). We parametrize L as

$$\mathbf{X}(t) = t \begin{bmatrix} 2 \\ 1 \end{bmatrix} \qquad (0 < t < 1);$$

that is, $x(t) = 2t$ and $y(t) = t$. Thus

$$I = \int_0^1 \left[(t^2)(2) + (2)(2t)(t)(1) \right] dt$$

$$= 6 \int_0^1 t^2 \, dt = 2.$$

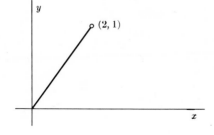

FIGURE 5.14

Finding the Scalar Field of a Conservative Vector Field

We now discuss a formal method for finding the scalar field associated with a conservative vector field.

If P is a real valued function defined on a region D in R^2, then we write

$$P_1(x, y) = \int P(x, y) \, dx$$

to indicate that P_1 is also defined on D and satisfies

$$(21) \qquad\qquad \frac{\partial P_1}{\partial x} = P;$$

thus P_1 is the "partial" indefinite integral of P with respect to x and is determined by holding y constant and integrating P as a function of x. It can be shown that if P_1 and P_2 are both continuously differentiable indefinite integrals (with respect to x) of P, then

$$(22) \qquad\qquad P_2(x, y) = P_1(x, y) + g(y),$$

where g is continuously differentiable with respect to y. Conversely, if P_1 satisfies (21), then the most general indefinite integral of P with respect to x is of the form (22); thus g is analogous to the constant of integration for ordinary indefinite integrals.

If P is a function of $\mathbf{X} = (x_1, \ldots, x_n)$, then P_1 is *an indefinite integral of P with respect to x_i* if

$$\frac{\partial P_1}{\partial x_i} = P;$$

the most general indefinite integral of P with respect to x_i is

$$P_2(\mathbf{X}) = P_1(\mathbf{X}) + g(x_1, \ldots, x_{i-1}, x_{i+1}, \ldots_i x_n).$$

Now suppose we wish to express

$$\mathbf{F} = \begin{bmatrix} P \\ Q \end{bmatrix}$$

in the form $\mathbf{F} = \nabla f$ in some region in R^2. Of course, we must first check to see that

$$(23) \qquad\qquad \frac{\partial P}{\partial y} = \frac{\partial Q}{\partial x}.$$

If this is so, let

$$P_1(x, y) = \int P(x, y)\, dx$$

be any indefinite integral of P with respect to x; since

$$\frac{\partial f}{\partial x} = P,$$

it follows that

$$f(x, y) = P_1(x, y) + g(y),$$

where g satisfies

$$\frac{\partial f}{\partial y}(x, y) = \frac{\partial P_1}{\partial y}(x, y) + g'(y) = Q(x, y),$$

or

(24) $$g'(y) = Q(x, y) - \frac{\partial P_1}{\partial y}(x, y).$$

Assumption (23) enters into the argument at this point, since it guarantees that the right side of (25) which seems to depend on x, is actually independent of it. To see this, consider

$$\frac{\partial}{\partial x}\left(Q - \frac{\partial P_1}{\partial y}\right) = \frac{\partial Q}{\partial x} - \frac{\partial}{\partial x}\left(\frac{\partial P_1}{\partial y}\right)$$

$$= \frac{\partial Q}{\partial x} - \frac{\partial^2 P_1}{\partial x\, \partial y}$$

$$= \frac{\partial Q}{\partial x} - \frac{\partial^2 P_1}{\partial y\, \partial x}$$

$$= \frac{\partial Q}{\partial x} - \frac{\partial}{\partial y}\left(\frac{\partial P_1}{\partial x}\right)$$

$$= \frac{\partial Q}{\partial x} - \frac{\partial P}{\partial y} = 0.$$

Example 5.19 Let

$$\mathbf{F}(x, y) = \begin{bmatrix} P(x, y) \\ Q(x_i\ y) \end{bmatrix} = \begin{bmatrix} xy^2 + x^2 \\ x^2 y + y^2 \end{bmatrix};$$

since

$$\frac{\partial P}{\partial y} = 2xy = \frac{\partial Q}{\partial x}$$

for all (x, y), there is a scalar field f such that

(25)
$$\frac{\partial f}{\partial x}(x, y) = xy^2 + x^2$$

and

(26)
$$\frac{\partial f}{\partial y}(x, y) = x^2 y + y^2$$

(Theorem 5.6). Integrating both sides of (25) with respect to x (treating y as a constant) yields

$$f(x, y) = \frac{x^2 y^2}{2} + \frac{x^3}{3} + g(y).$$

From this,

$$\frac{\partial f}{\partial y}(x, y) = x^2 y + g'(y),$$

which, when compared with (26), yields

$$g'(y) = y^2.$$

Therefore

$$g(y) = \frac{y^3}{3} + C,$$

where C is a constant, and

$$f(x, y) = \frac{x^2 y^2}{2} + \frac{x^3}{3} + \frac{y^3}{3} + C.$$

It is easy to verify that $\mathbf{F} = \nabla f$.

We emphasize that this is a formal method. It can happen that the indefinite integrals P_1 and g are not defined in a particular region. Indeed, we must expect the method to break down in some cases, since it assumes only the hypotheses of Theorem 5.5 (for $n = 2$), which are not sufficient to guarantee the existence of a potential function in an arbitrary region in R^2.

An analogous procedure is available for $n > 2$. We give an example for $n = 3$.

Example 5.20 Let $n = 3$ and

$$\mathbf{F}(\mathbf{X}) = \begin{bmatrix} P(x, y, z) \\ Q(x, y, z) \\ R(x, y, z) \end{bmatrix} = \begin{bmatrix} x + z \\ y + z \\ x + y \end{bmatrix};$$

then

$$\frac{\partial P}{\partial y} = \frac{\partial Q}{\partial x} = 0, \qquad \frac{\partial Q}{\partial z} = \frac{\partial R}{\partial y} = 1,$$

and

$$\frac{\partial R}{\partial x} = \frac{\partial P}{\partial z} = 1$$

in R^3. Theorem 5.6 implies that there is a function $f \colon R^3 \to R$ such that

(27)
$$\frac{\partial f}{\partial x}(\mathbf{X}) = x + z,$$

(28)
$$\frac{\partial f}{\partial y}(\mathbf{X}) = y + z,$$

and

(29)
$$\frac{\partial f}{\partial z}(\mathbf{X}) = x + y.$$

Integrating (27) with respect to x yields

(30)
$$f(\mathbf{X}) = \frac{x^2}{2} + xz + g(y, z),$$

where g is a "constant of integration"; that is, a function which does not depend upon x. Differentiating (30) with respect to y yields

$$\frac{\partial f}{\partial y}(\mathbf{X}) = \frac{\partial g}{\partial y}(y, z);$$

thus, from (28)

$$\frac{\partial g}{\partial y}(y, z) = y + z$$

and

$$g(y, z) = \frac{y^2}{2} + zy + h(z).$$

Substituting this into (30) yields

$$f(\mathbf{X}) = \frac{x^2}{2} + xz + \frac{y^2}{2} + yz + h(z),$$

and therefore

$$\frac{\partial f(\mathbf{X})}{\partial z} = x + y + h'(z).$$

From (29), $h'(z) = 0$ and h is constant; thus

$$f(\mathbf{X}) = \frac{x^2}{2} + xz + \frac{y^2}{2} + yz + C,$$

where C is an arbitrary constant.

Line Integrals with Respect to Arc Length

These is another type of line integral that we have not yet mentioned.

Definition 5.4 Let C be a smooth curve in R^n, given parametrically by $\mathbf{X} = \mathbf{X}(t)$ $(a \le t \le b)$, and suppose f is a continuous scalar field defined on C. Then the integral

$$\int_a^b f(\mathbf{X}(t)) \mid \mathbf{X}'(t) \mid dt$$

is called the *line integral of f, with respect to arc length, along C.* It is denoted by

(31) $$\int_C f \, ds.$$

If C is a piecewise smooth curve with smooth segment $C_1, \ldots, C_n$, we define

$$\int_C f \, ds = \int_{C_1} f \, ds + \cdots + \int_{C_n} f \, ds.$$

In view of the notation (31), the student should not be surprised to learn that the value of this new line integral is independent of the parametrization chosen for C. We leave it to the student to verify this (Exercise T-7), and also that

$$\int_C f \, ds = \int_{-C} f \, ds$$

(Exercise T-8).

If a thin wire is represented as a curve in R^3 with mass density (mass per unit length) function $\rho = \rho(x, y, z)$, then its mass is

$$m = \int_C \rho \, ds.$$

The moments, center of gravity, and moments and products of inertia of the wire can be defined and computed as in Section 4.4.

Example 5.21   Let a wire be in the shape of a helical curve C defined by

$$\mathbf{X}(t) = 2 \cos t \, \mathbf{i} + 2 \sin t \, \mathbf{j} + \mathbf{k} \qquad \left(0 \le t \le \frac{\pi}{2}\right),$$

with mass density function $\rho(x, y, z) = ax$. Then

$$|\, \mathbf{X}'(t) \,| = |\, -2 \sin t \, \mathbf{i} + 2 \cos t \, \mathbf{j} \,| = 2,$$

and the mass of the wire is

$$m = \int_C \rho \, ds = a \int_C x \, ds$$

$$= a \int_0^{\pi/2} (2 \cos t) 2 \, dt = 4a.$$

The first moments with respect to the coordinate planes are

$$M_{yz} = \int_C x\rho \, ds = a \int_C x^2 \, ds$$

$$= a \int_0^{\pi/2} (2 \cos t)^2 (2) \, dt = 2a\pi,$$

$$M_{zx} = \int_C y\rho \, ds = a \int_C xy \, ds$$

$$= a \int_0^{\pi/2} (2 \cos t) (2 \sin t) (2) \, dt = 4a,$$

and

$$M_{xy} = \int_C z\rho \, ds = a \int_C zx \, ds$$

$$= a \int_0^{\pi/2} (1) (2 \cos t) (2) \, dt = 4a.$$

The coordinates of the center of mass are

$$\bar{x} = \frac{M_{yz}}{m} = \frac{\pi}{2}, \qquad \bar{y} = \frac{M_{zx}}{m} = 1, \qquad \bar{z} = M_{xy} = 1.$$

We leave the computation of the moments and products of inertia to the student.

The line integral of a vector function $\mathbf{F}$ over a curve C can also be viewed as an integral, with respect to arc length, of a certain scalar valued function. If C has the smooth representation $\mathbf{X} = \mathbf{X}(t)$ $(a \le t \le b)$, we can assign to every $\mathbf{X}$ on C a unit tangent vector $\mathbf{T}(\mathbf{X})$, defined by

$$\mathbf{T}(\mathbf{X}(t)) = \frac{\mathbf{X}'(t)}{|\mathbf{X}(t)|}.$$

This vector function is independent of the parametric representation chosen for C, except for sign. If $\mathbf{F}$ is a continuous vector valued function on C, then

$$\int_C \mathbf{F} \cdot d\mathbf{X} = \int_a^b \mathbf{F}(\mathbf{X}(t)) \cdot \mathbf{X}'(t) \, dt$$

$$= \int_a^b \mathbf{F}(\mathbf{X}(t)) \cdot \frac{\mathbf{X}'(t)}{|\mathbf{X}'(t)|} \, |\mathbf{X}'(t)| \, dt,$$

which can be rewritten as

$$\int_C \mathbf{F} \cdot d\mathbf{X} = \int_C (\mathbf{F} \cdot \mathbf{T}) \, ds.$$

EXERCISES 4.5

In Exercises 1 through 9 compute the line integral of $\mathbf{F}$ along the given path.

1. $\mathbf{F}(\mathbf{X}) = \begin{bmatrix} x^2 + y \\ y^2 - x \end{bmatrix}$; $\quad \mathbf{X} = \begin{bmatrix} t^2 \\ t^3 \end{bmatrix} \quad (0 \le t \le 1).$

2. $\mathbf{F}(\mathbf{X}) = \begin{bmatrix} xy + z \\ yz - x \\ xyz + x^2 \end{bmatrix}$; $\quad \mathbf{X} = \begin{bmatrix} t \\ t \\ t^2 \end{bmatrix} \quad (0 \le t \le 2).$

3. $\mathbf{F}(\mathbf{X}) = \begin{bmatrix} xy \\ y^2 - x^2 \end{bmatrix}$; $\qquad y = x^2 \qquad (-1 \le x \le 1).$

4. $\mathbf{F}(\mathbf{X}) = \begin{bmatrix} x + 1 \\ y^2 - x \end{bmatrix}$ from $(0, 2)$ to $(2, 0)$ along the arc of the circle $x^2 + y^2 = 4$ in the first quadrant.

5. $\mathbf{F}(\mathbf{X}) = \begin{bmatrix} x + y \\ xy \\ z^2 \end{bmatrix}$ along the line segment joining $(0, 0, 0)$ and $(1, 2, 1)$.

6. $\mathbf{F}(\mathbf{X}) = \begin{bmatrix} x^2 y + x \\ x^2 - y \end{bmatrix}$ along the path consisting of $y = x^2$ from $(0, 0)$ to $(1, 1)$ and $x = y^2$ from $(1, 1)$ to $(0, 0)$.

7. $\mathbf{F}(\mathbf{X}) = \begin{bmatrix} x^2 y \\ y + z^2 \\ xz \end{bmatrix}$ along the path consisting of the line segment from $(0, 0, 0)$ to $(1, 2, 1)$, followed by the line segment from $(1, 2, 1)$ to $(2, 3, 4)$, followed by the line segment from $(2, 3, 4)$ to $(4, 2, 1)$.

8. $\mathbf{F}(\mathbf{X}) = \begin{bmatrix} x^2 - 3y \\ 2xy^2 \end{bmatrix}$ along the parabola $y = x^2 + 1$ from $(0, 1)$ to $(2, 5)$.

9. $\mathbf{F}(\mathbf{X}) = \begin{bmatrix} x^2 + y \\ x^2 + y^2 \end{bmatrix}$ counterclockwise around the ellipse $\dfrac{x^2}{a^2} + \dfrac{y^2}{b^2} = 1.$

In Exercises 10 through 15 compute the work done by the force $\mathbf{F}$ in moving a particle along the given path.

10. $\mathbf{F}(\mathbf{X}) = \begin{bmatrix} x + 2y \\ x - y \end{bmatrix}$; $\qquad \mathbf{X} = \begin{bmatrix} 2 \cos t \\ 3 \sin t \end{bmatrix} \quad \left(0 \le t \le \dfrac{\pi}{4}\right).$

11. $\mathbf{F}(\mathbf{X}) = \begin{bmatrix} x - y \\ x + z \\ y - z \end{bmatrix}$; $\qquad \mathbf{X} = \begin{bmatrix} a \cos t \\ a \sin t \\ bt \end{bmatrix} \quad \left(0 \le t \le \dfrac{\pi}{3}\right).$

12. $\mathbf{F}(\mathbf{X}) = \begin{bmatrix} xy + x^2 \\ x^2y - y \end{bmatrix}$; $\quad y = \dfrac{1}{x} \quad (1 \leq x \leq 3)$.

13. $\mathbf{F}(\mathbf{X}) = \begin{bmatrix} xz + y^2 \\ yz - x^2 \\ xy - z^2 \end{bmatrix}$

along the path consisting of the line segment from $(1, 2, 3)$ to $(4, 3, 5)$, followed by the line segment from $(4, 3, 5)$ to $(1, 1, 1)$.

14. $\mathbf{F}(\mathbf{X}) = \begin{bmatrix} x^2 - y^2 + y \\ y^2 + x^2 - x \end{bmatrix}$

along the circular arc $y = \sqrt{4 - x^2}$ from $(2, 0)$ to $(1, \sqrt{3})$, followed by the line segment from $(1, \sqrt{3})$ to $(0, 0)$.

15. $\mathbf{F}(\mathbf{X}) = \begin{bmatrix} x^2y^2 - 2x \\ y + x \end{bmatrix}$

along the parabola $x = 4 - y^2$ from $(0, 2)$ to $(4, 0)$.

16. Show that

$$\boldsymbol{\Phi}(t) = \begin{bmatrix} t \\ t^2 \\ t^3 \end{bmatrix} \quad (1 \leq t \leq 2)$$

and

$$\boldsymbol{\Psi}(s) = \begin{bmatrix} s^2 \\ s^4 \\ s^6 \end{bmatrix} \quad (1 \leq s \leq \sqrt{2})$$

are equivalent parametric functions, and let C be the curve that they

represent. Evaluate $\displaystyle\int_C \mathbf{F} \cdot d\mathbf{X}$, using both parametrizations, for

$$\mathbf{F}(\mathbf{X}) = \begin{bmatrix} xy + z^2 \\ x^2y - z^2 \\ yz \end{bmatrix}.$$

17. Repeat Exercise 16 with

$$\Phi(t) = \begin{bmatrix} e^t \\ e^{-t} \\ 2\cosh t \end{bmatrix} \quad (0 \le t \le 1),$$

$$\Psi(s) = \begin{bmatrix} s \\ \dfrac{1}{s} \\ s + \dfrac{1}{s} \end{bmatrix} \quad (1 \le s \le e),$$

and

$$\mathbf{F}(\mathbf{X}) = \begin{bmatrix} x^2 + xy \\ y^2 + xz \\ z^2 + yz \end{bmatrix}.$$

In Exercises 18 through 25 determine whether **F** is conservative. If it is conservative find a potential function.

18. $\mathbf{F}(\mathbf{X}) = \begin{bmatrix} e^{xy} + y \\ e^{xy} + x \end{bmatrix}.$

19. $\mathbf{F}(\mathbf{X}) = \begin{bmatrix} x^2 + \dfrac{y^2}{2} \\ xy + \dfrac{z^2}{2} \\ yz \end{bmatrix}.$

20. $\mathbf{F}(\mathbf{X}) = \begin{bmatrix} xyz \\ \dfrac{x^2 z}{2} + y \\ xz \end{bmatrix}.$

21. $\mathbf{F}(\mathbf{X}) = \begin{bmatrix} x\log y + y \\ y\log x + x \end{bmatrix}.$

22. $\mathbf{F}(\mathbf{X}) = \begin{bmatrix} 2xe^y - e^xy^2 \\ x^2e^y - 2ye^x \end{bmatrix}$.

23. $\mathbf{F}(\mathbf{X}) = \begin{bmatrix} 2xye^z + z^2e^y + y^2ze^x \\ x^2e^z + xz^2e^y + 2yze^x \\ x^2ye^z + 2xze^y + y^2e^x \end{bmatrix}$.

24. $\mathbf{F}(\mathbf{X}) = \begin{bmatrix} xy - z \\ \dfrac{x^2}{2} + \dfrac{z^3}{3} \\ z^2y + x \end{bmatrix}$.

25. $\mathbf{F}(\mathbf{X}) = \begin{bmatrix} \cos y + y \cos x \\ -x \sin y + \sin x \end{bmatrix}$.

In Exercises 26 through 29 find $\int \mathbf{F} \cdot d\mathbf{X}$ for any path C from $\mathbf{X}_0$ to $\mathbf{X}_1$.

26. $\mathbf{F}(\mathbf{X}) = \begin{bmatrix} e^x \cos y + e^y \cos x \\ e^y \sin x - e^x \sin y \end{bmatrix}$; $\quad \mathbf{X}_0 = \left(0, \dfrac{\pi}{2}\right), \quad \mathbf{X}_1 = \left(\dfrac{\pi}{4}, \pi\right)$.

27. $\mathbf{F}(\mathbf{X}) = \begin{bmatrix} 2xy + yz \\ x^2 + z^2 + xz \\ 2yz + xy \end{bmatrix}$; $\quad \mathbf{X}_0 = (1, 2, 1), \quad \mathbf{X}_1 = (-2, 3, 0)$.

28. $\mathbf{F}(\mathbf{X}) = \begin{bmatrix} x^2y^2 + \dfrac{y^3}{3} + 1 \\ \frac{2}{3}x^3y + xy^2 + 1 \end{bmatrix}$; $\quad \mathbf{X}_0 = (1, 1), \quad \mathbf{X}_1 = (-2, 3)$.

29. $\mathbf{F}(\mathbf{X}) = \begin{bmatrix} x^2y^3 + xy^2 \\ x^3y^2 + x^2y + yz^2 \\ y^2z \end{bmatrix}$; $\quad \mathbf{X}_0 = (0, 1, 2), \quad \mathbf{X}_1 = (-1, 3, -4)$.

In Exercises 30 through 34 compute the line integral of f with respect to arc length along the given curve.

30. $f(\mathbf{X}) = x^2 + y^2 + z^2$; $\quad \mathbf{X} = a \cos t\,\mathbf{i} + a \sin t\,\mathbf{j} + bt\mathbf{k} \quad (0 \leq t \leq 2\pi)$.

31. $f(\mathbf{X}) = 3x^2 - 2xy + y^2$; $\quad$ along the semicircle $x^2 + y^2 = 4$, $\quad y \geq 0$.

32. $f(x, y, z) = z^2$; $\quad \mathbf{X} = \cos t\,\mathbf{i} + \sin t\,\mathbf{j} + e^t\mathbf{k} \quad \left(0 \leq t \leq \dfrac{\pi}{2}\right)$.

33. $f(x, y, z) = xy;$ $\mathbf{X} = e^t \cos t\,\mathbf{i} + e^t \sin t\,\mathbf{j}$ $\left(0 \le t \le \dfrac{\pi}{4}\right).$

34. $f(x, y, z) = xy;$ $\mathbf{X} = \dfrac{t^2}{2}\mathbf{i} + \dfrac{\sqrt{2}}{3} t^3\,\mathbf{j} + \dfrac{t^0}{4}\mathbf{k}$ $(0 \le t \le 1).$

35.   Find the moment of inertia of a circular wire of radius R about a diameter. Assume that the mass density is ρ_0, a constant.

36.   Find the first moment of the wire of Exercise 35 with respect to a plane perpendicular to the wire and tangent to it.

37.   Let a wire be in the form of an equilateral triangle with sides of length L. Suppose the mass density of the base is ρ_0, a constant, and that the density of the other two sides varies linearly with altitude, from ρ_0 at the base to $\rho_0/2$ at the opposite vertex. Find the mass and center of gravity of the triangle, and its moments of inertia with respect to its three sides.

THEORETICAL EXERCISES

T-1.   Prove Theorem 5.2.

T-2.   Prove Theorem 5.4.

T-3.   Let f be a real valued function, continuous in a neighborhood of $\mathbf{X}_0 = (c_1, \ldots, c_n)$. Show that

$$\lim_{y \to 0} \int_0^1 f(c_1 + ty, c_2, \ldots, c_n)\, dt = f(\mathbf{X}_0).$$

T-4.   Verify Eq. (20).

T-5.   If C is given in R^2 by $y = g(x)$ $(a \le x \le b)$, show that

$$\int_C f\, ds = \int_a^b f(x, g(x))\, \sqrt{1 + (g'(x))^2}\, dx.$$

T-6.   Let C be any piecewise smooth simple closed curve in R^2 and let $\mathbf{X}_0 = (x_0, y_0)$ be any point in the interior of C. Show that

(a) $\displaystyle \int_C \frac{(y - y_0)\, dx - (x - x_0)\, dy}{(x - x_0)^2 + (y - y_0)^2} = -2\pi$

and

(b)
$$\int_C \frac{(x - x_0)\,dx + (y - y_0)\,dy}{(x - x_0)^2 + (y - y_0)^2} = 0.$$

(*Hint*: Evaluate the integrals for a small circle centered at $\mathbf{X}_0$, and apply Theorem 5.6.)

T-7. *Prove*: If $\mathbf{X} = \boldsymbol{\Phi}(t)$ $(a \leq t \leq b)$ and $\mathbf{X} = \boldsymbol{\Psi}(s)$ $(c \leq s \leq d)$ are equivalent parametrizations of a curve C, then

$$\int_a^b f(\boldsymbol{\Phi}(t)) \mid \boldsymbol{\Phi}'(t) \mid dt = \int_c^d f(\boldsymbol{\Psi}(s)) \mid \boldsymbol{\Psi}'(s) \mid ds.$$

T-8. *Prove*:

$$\int_C f\,ds = \int_{-C} f\,ds.$$

4.6   Surface Integrals of Scalar Fields

In this section we discuss surface integrals. We consider only integrals over *smooth* surfaces in R^3. Roughly speaking, a smooth surface is one which has a tangent plane at each point.

Definition 6.1   A *smooth surface* S in R^3 is the range of a parametric function

$$\boldsymbol{\Phi}(u, v) = \phi_1(u, v)\mathbf{i} + \phi_2(u, v)\mathbf{j} + \phi_1(u, v)\mathbf{k}$$

with the following properties:

(a) $\boldsymbol{\Phi}$ is continuous on a domain D in R^2 consisting of a connected open set D^o along with the boundary $B(D)$, which is made up of finitely many piecewise smooth curves.

(b)   If (u, v) is in D^o and (u_1, v_1) is a point in D distinct from (u, v), then $\boldsymbol{\Phi}(u, v) \neq \boldsymbol{\Phi}(u_1, v_1)$; thus $\boldsymbol{\Phi}$ is one-to-one on D^o.

(c)   The partial derivatives $\partial\boldsymbol{\Phi}/\partial u$ and $\partial\boldsymbol{\Phi}/\partial v$ are continuous and

(1)
$$\mathbf{N}(u, v) = \frac{\partial\boldsymbol{\Phi}}{\partial u}(u, v) \times \frac{\partial\boldsymbol{\Phi}}{\partial v}(u, v)$$

is nonzero in D^o. (We shall see that $\mathbf{N}(u, v)$ is perpendicular to the tangent plane to S at the point $\mathbf{X} = \boldsymbol{\Phi}(u, v)$.)

(d)   If (u_0, v_0) is a point on $B(D)$ then

$$\lim_{(u, v) \to (u_0, v_0)} \frac{\mathbf{N}(u, v)}{|\mathbf{N}(u, v)|}$$

exists.

We say that S is defined by $\mathbf{X} = \boldsymbol{\Phi}(u, v)$, and that $\boldsymbol{\Phi}$ is a *smooth parametric function* of two variables.

Example 6.1   The surface S defined by

$$\mathbf{X} = \boldsymbol{\Phi}(\theta, z) = a \cos \theta \, \mathbf{i} + a \sin \theta \, \mathbf{j} + z \mathbf{k} \qquad (a > 0)$$

for (θ, z) in

$$D = \{(\theta, z) \mid 0 \le \theta \le 2\pi, \; |z| \le 1\}$$

is the part of the cylinder $x^2 + y^2 = a^2$ bounded by $z = 1$ and $z = -1$ (Fig. 6.1a). Clearly $\boldsymbol{\Phi}$ satisfies (a) and (b). Its partial derivatives are continuous and

$$(2) \qquad \mathbf{N}(\theta, z) = \begin{vmatrix} \mathbf{i} & \mathbf{j} & \mathbf{k} \\ -a \sin \theta & a \cos \theta & 0 \\ 0 & 0 & 1 \end{vmatrix}$$

$$= a \cos \theta \, \mathbf{i} + a \sin \theta \, \mathbf{j}$$

for (θ, z) in D°. The boundary $B(D)$ of D consists of points of the forms $(0, z)$ and $(2\pi, z)$ with $|z| < 1$ (Fig. 6.1b). From (2),

$$\frac{\mathbf{N}(\theta, z)}{|\mathbf{N}(\theta, z)|} = \frac{a \cos \theta \, \mathbf{i} + a \sin \theta \, \mathbf{j}}{\sqrt{a^2 \cos^2 \theta + a^2 \sin^2 \theta}} = \cos \theta \, \mathbf{i} + \sin \theta \, \mathbf{j};$$

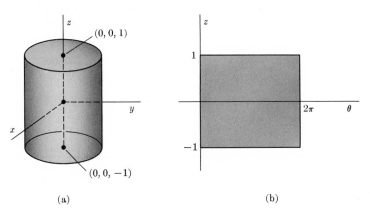

(a)   (b)

FIGURE 6.1

hence

$$\lim_{(\theta, z) \to (0, z_0)} \frac{\mathbf{N}(\theta, z)}{|\mathbf{N}(\theta, z)|} = \lim_{(\theta, z) \to (2\pi, z_0)} \frac{\mathbf{N}(\theta, z)}{|\mathbf{N}(\theta, z)|} = \mathbf{i},$$

and therefore (d) is satisfied. Hence S is smooth.

Example 6.2 The hemisphere S defined by $x^2 + y^2 + z^2 = a^2$, $z > 0$ can be represented by

$$\mathbf{X} = \mathbf{\Phi}(\phi, \theta) = a \sin \phi \cos \theta \, \mathbf{i} + a \sin \phi \sin \theta \, \mathbf{j} + a \cos \phi \mathbf{k}$$

for (ϕ, θ) in

$$D = \left\{ (\phi, \theta) \mid 0 \le \phi \le \frac{\pi}{2}, \quad 0 \le \theta \le 2\pi \right\};$$

here ϕ and θ are simply the spherical coordinate angles (Fig. 6.2). Now

$$(3) \quad \mathbf{N}(\phi, \theta) = \begin{vmatrix} \mathbf{i} & \mathbf{j} & \mathbf{k} \\ a \cos \phi \cos \theta & a \cos \phi \sin \theta & -a \sin \phi \\ -a \sin \phi \sin \theta & a \sin \phi \cos \theta & 0 \end{vmatrix}$$

$$= a^2 \sin^2 \phi \cos \theta \, \mathbf{i} + a^2 \sin^2 \phi \sin \theta \, \mathbf{j} + a^2 \sin \phi \cos \phi \, \mathbf{k}$$

$$= a \sin \phi \, \mathbf{\Phi}(\phi, \theta),$$

from which it is easily verified that S is smooth.

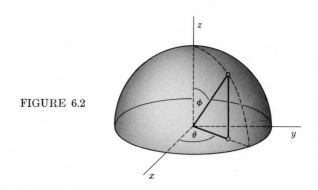

FIGURE 6.2

Example 6.3 If in Example 6.1 we replace the domain of Φ by

$$D_1 = \{(\theta, z) \mid 0 \le \theta \le 3\pi, \mid z \mid \le 1\},$$

then Φ no longer satisfies (b). For example,

$$\Phi\left(\frac{\pi}{2}, \frac{1}{2}\right) = \Phi\left(\frac{5\pi}{2}, \frac{1}{2}\right) = a\mathbf{j} + \tfrac{1}{2}\mathbf{k}.$$

Hence (b) is not satisfied. In fact any point on S for which $y > 0$ is the image of distinct points in D_1^o. Some authors would still say that S is smooth, but not *simple*.

For the cylindrical surface of Example 6.1,

$$\mathbf{N}(\theta, z) = a \cos \theta \, \mathbf{i} + a \sin \theta \, \mathbf{j} = x\mathbf{i} + y\mathbf{j};$$

and for the hemisphere of Example 6.2,

$$\mathbf{N}(\phi, \theta) = a \sin \phi \, \Phi(\phi, \theta) = a \sin \phi \, (x\mathbf{i} + y\mathbf{j} + z\mathbf{k}).$$

In both cases, $\mathbf{N}(u, v)$ is normal to the surface in question (Verify!). This is no accident, as we shall now see. If (u_0, v_0) is in D^o then $\mathbf{X} = \Phi(u, v_0)$ defines a curve C_u on S which passes through $\mathbf{X}_0 = \Phi(u_0, v_0)$ (Fig. 6.3). The vector $(\partial\Phi/\partial u)(u_0, v_0)$ is tangent to C_u at $\mathbf{X}_0$ and therefore lies in the plane Γ which is tangent to S at $\mathbf{X}_0$. Similarly, $\mathbf{X} = \Phi(u_0, v)$ defines a curve C_v in S, and $(\partial\Phi/\partial v)(u_0, v_0)$ is in Γ. Thus

$$\mathbf{N}(u_0, v_0) = \frac{\partial\Phi}{\partial u}(u_0, v_0) \times \frac{\partial\Phi}{\partial v}(u_0, v_0)$$

is perpendicular to Γ; that is, $\mathbf{N}(u_0, v_0)$ is normal to S at $\mathbf{X}_0$.

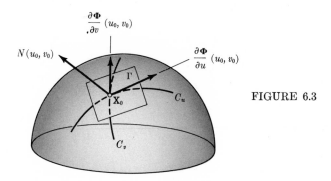

FIGURE 6.3

The following definition of surface area is motivated by the formula for the area of a parallelepiped, derived in Section 1.3.

Definition 6.2 Let the surface S in R^3 be given parametrically by $\mathbf{X} = \mathbf{X}(u, v)$ for (u, v) in a domain D in R^3. Then the *area* of S is defined by

$$(4) \qquad A(S) = \int_D |\, \mathbf{N}(u, v)\, |\; du\; dv,$$

provided the integral exists. We shall refer to the symbol

$$dS = |\, \mathbf{N}(u, v)\, |\; du\; dv$$

as the *differential element of surface area with respect to the parameters u and v.*

Example 6.4 Suppose S is a parallelogram with vertices $\mathbf{X}_1$, $\mathbf{X}_2$, $\mathbf{X}_3$, and $\mathbf{X}_4$ (Fig. 6.4). In Section 1.3 we showed that the area of S is

$$(5) \qquad A(S) = |\, (\mathbf{X}_2 - \mathbf{X}_1) \times (\mathbf{X}_3 - \mathbf{X}_1)\, |.$$

Now we wish to show that Definition 6.2 assigns the same area to S.

The parallelogram can be represented parametrically by

$$\mathbf{X} = \mathbf{X}_1 + u(\mathbf{X}_2 - \mathbf{X}_1) + v(\mathbf{X}_3 - \mathbf{X}_4)$$

for (u, v) in

$$D = \{(u, v) \mid 0 \le u \le 1, \quad 0 \le v \le 1\}.$$

Therefore

$$\frac{\partial \mathbf{X}}{\partial u} = (X_2 - X_1), \qquad \frac{\partial \mathbf{X}}{\partial v} = (X_3 - X_1),$$

and

$$|\, \mathbf{N}(u, v)\, | = |\, (\mathbf{X}_2 - \mathbf{X}_1) \times (\mathbf{X}_3 - \mathbf{X}_1)\, | = \text{constant}.$$

From Definition 6.2,

$$A(S) = \int_D |\, \mathbf{N}(u, v)\, |\; du\; dv$$

$$= |\, (\mathbf{X}_2 - \mathbf{X}_1) \times (\mathbf{X}_3 - \mathbf{X}_1)\, | \int_D du\; dv$$

$$= |\, (\mathbf{X}_2 - \mathbf{X}_1) \times (\mathbf{X}_3 - \mathbf{X}_1)\, | \int_0^1 dv \int_0^1 du$$

$$= |\, (\mathbf{X}_2 - \mathbf{X}_1) \times (\mathbf{X}_3 - \mathbf{X}_1)\, |,$$

which agrees with (5).

FIGURE 6.4

To motivate Definition 6.2 for more general surfaces, let S be defined by $\mathbf{X} = \mathbf{X}(u, v)$. For simplicity we take the domain of the parametric function to be a rectangle:

$$D = \{(u, v) \mid a \le u \le b, \ c \le z \le d\}.$$

Let $a = a_0 < a_1 < \cdots < a_m = b$ and $c = c_0 < c_1 < \cdots < c_n = d$, and let Δ be the partition of D into subrectangles

$$D_{ij} = \{(u, v) \mid a_{i-1} \le u \le a_i, \ b_{j-1} \le v \le b_j \ \ (1 \le i \le m, \ 1 \le j \le n)\}$$

(Fig. 6.5). Denote by S_{ij} the portion of S given by $\mathbf{X} = \mathbf{X}(u, v)$ $((u, v)$ in $D_{ij})$, suppose $\mathbf{U}_{ij} = (u_{ij}, v_{ij})$ is a point in D_{ij}, and let $\mathbf{X}_{ij} = \mathbf{X}(u_{ij}, v_{ij})$. Finally, let T_{ij} be that portion of the tangent plane to S at $\mathbf{X}_{ij}$ which is given parametrically by

$$\mathbf{X} = \mathbf{X}_{ij} + \frac{\partial \mathbf{X}}{\partial u}\,(\mathbf{U}_{ij})\,(u - u_{ij}) + \frac{\partial \mathbf{X}}{\partial v}\,(\mathbf{U}_{ij})\,(v - v_{ij}) \qquad ((u, v) \text{ in } D_{ij});$$

then T_{ij} is a parallelogram in R^3 (Fig. 6.6). By an argument similar to that of Example 6.4, it can be shown (Exercise T-5) that the area of T_{ij} is

$$A(T_{ij}) = |\mathbf{N}(u_{ij}, v_{ij})|\,(a_i - a_{i-1})\,(b_j - b_{j-1}).$$

FIGURE 6.5

FIGURE 6.6

The total area of all the parallelograms is

$$A_\Delta = \sum_{i=1}^{m} \sum_{j=1}^{n} A(T_{ij}).$$

which approaches the integral on the right side of (4) as $\| \Delta \|$ approaches zero.

Example 6.5 Let S be the cylindrical surface of Example 6.1. From (2),

$$dS = a \, d\theta \, dz;$$

hence

$$A(S) = \int_D a \, d\theta \, dz$$

$$= a \int_0^{2\pi} d\theta \int_{-1}^{1} dz = a(2\pi)(2) = 4\pi a.$$

This result is easily checked: if we cut S along a line parallel to the z-axis and flatten it onto the page, we obtain a rectangle with dimensions 2 and $2\pi a$. (Fig. 6.7). The result is then evident.

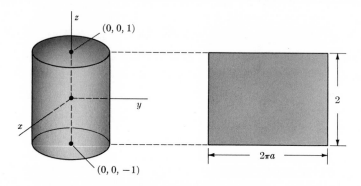

FIGURE 6.7

Example 6.6 For the hemisphere of Example 6.2

$$| \mathbf{N}(\phi, \theta) | = a^2 \sin \phi$$

(see (3)); hence

$$dS = a^2 \sin \phi \, d\theta \, d\phi$$

and

$$A(S) = \int_D a^2 \sin \phi \, d\theta \, d\phi$$

$$= a^2 \int_0^{2\pi} d\theta \int_0^{\pi/2} \sin \phi \, d\phi = 2\pi a^2.$$

By symmetry the surface area of the sphere $x^2 + y^2 + z^2 = a^2$ is $4\pi a^2$.

If S is defined explicitly by

$$z = f(x, y) \qquad ((x, y) \text{ in } D),$$

it can be given the parametric representation

$$\mathbf{X} = \mathbf{X}(x, y) = x\mathbf{i} + y\mathbf{j} + f(x, y)\mathbf{k}.$$

Then

$$\mathbf{N}(x, y) = \begin{vmatrix} \mathbf{i} & \mathbf{j} & \mathbf{k} \\ 1 & 0 & f_x \\ 0 & 1 & f_y \end{vmatrix}$$

$$= -f_x\mathbf{i} - f_y\mathbf{j} + \mathbf{k},$$

and

$$dS = \sqrt{1 + f_x^2 + f_y^2} \, dx \, dy.$$

Example 6.7 Let S be the part of the paraboloid

$$z = f(x, y) = x^2 + y^2$$

which is bounded above by $z = 1$ (Fig. 6.8). It can be represented parametrically by

$$\mathbf{X} = \mathbf{X}(x, y) = x\mathbf{i} + y\mathbf{j} + (x^2 + y^2)\mathbf{k}$$

where (x, y) is in

$$D = \{(x, y) \mid x^2 + y^2 \leq 1\}.$$

Since $f_x(x, y) = 2x$ and $f_y(x, y) = 2y$,

$$dS = \sqrt{1 + 4x^2 + 4y^2}$$

and

$$A(S) = \int_D \sqrt{1 + 4x^2 + 4y^2} \, dx \, dy.$$

To evaluate this integral it is natural to change to polar coordinates, which yields

$$A(S) = \int_0^{2\pi} d\theta \int_0^1 \sqrt{1 + 4r^2}\, r\, dr$$

$$= 2\pi \left[\frac{1}{12} (1 + 4r^2)^{3/2} \right]\Bigg|_0^1 = \frac{\pi}{6} (5^{3/2} - 1).$$

FIGURE 6.8

Equivalent Parametrizations

It can be shown that any two parametric representations of an orientable surface S which meet the requirements of Definition 6.1 are equivalent in the following sense.

Definition 6.3 Two parametric functions $\Phi_1 = \Phi_1(u, v)$ $((u, v)$ in $D_1)$ and $\Phi_2 = \Phi_2(\xi, \eta)$ $((\xi, \eta)$ in $D_2)$ are said to be *equivalent* if they are related by

(6) $\Phi_2(g_1(u, v), g_2(u, v)) = \Phi_1(u, v),$

where

$$\begin{bmatrix} \xi \\ \eta \end{bmatrix} = \begin{bmatrix} g_1(u, v) \\ g_2(u, v) \end{bmatrix}$$

is a continuously differentiable one to one mapping of D_1 onto D_2 for which

$$\frac{\partial(g_1, g_2)}{\partial(u, v)} \neq 0$$

if (u, v) is in D_1^0.

Example 6.8   Let

$$\Phi_1(\phi, \theta) = a \sin \phi \cos \theta \, \mathbf{i} + a \sin \phi \sin \theta \, \mathbf{j} + a \cos \phi \, \mathbf{k},$$

$$D_1 = \left\{ (\phi, \theta) \mid 0 \le \phi \le \frac{\pi}{2}, \quad 0 \le \theta \le 2\pi \right\},$$

$$\Phi_2(x, y) = x \mathbf{i} + y \mathbf{j} + \sqrt{a^2 - x^2 - y^2} \, \mathbf{k}$$

and

$$D_2 = \{ (x, y) \mid 0 \le x^2 + y^2 \le a^2 \}.$$

The transformation

$$\begin{bmatrix} x \\ y \end{bmatrix} = \begin{bmatrix} a \sin \phi \cos \theta \\ a \sin \phi \sin \theta \end{bmatrix}$$

is one-to-one on D_1° and

$$\frac{\partial(x, y)}{\partial(\phi, \theta)} = a^2 \sin \phi \cos \phi$$

is positive on D_1°. Furthermore

$$\Phi_2(a \sin \phi \cos \theta, a \sin \phi \sin \theta) = \Phi_1(\phi, \theta) ;$$

hence $\mathbf{X} = \Phi_1(\phi, \theta)$ and $\mathbf{X} = \Phi_2(x, y)$ are equivalent parametric representations of the sphere.

The following theorem states an expected result; namely, that the area of a surface S does not depend upon the particular parametric representation chosen for S.

Theorem 6.1   Let $\mathbf{X} = \Phi_2(u, v)$ and $\mathbf{X} = \Phi_2(\xi, \eta)$ be equivalent smooth parametric representations of a surface S, in accordance with Definition 6.3. Then

(7)
$$\int_{D_1} \left| \frac{\partial \Phi_1}{\partial u} \times \frac{\partial \Phi_1}{\partial v} \right| du \, dv = \int_{D_2} \left| \frac{\partial \Phi_2}{\partial \xi} \times \frac{\partial \Phi_2}{\partial \eta} \right| d\xi \, d\eta.$$

Proof.   First we observe that the Jacobian determinant does not change sign since it is continuous and nonvanishing. From (6) and the chain rule,

$$\frac{\partial \Phi_1}{\partial u} = \frac{\partial g_1}{\partial u} \frac{\partial \Phi_2}{\partial \xi} + \frac{\partial g_2}{\partial u} \frac{\partial \Phi_2}{\partial \eta}$$

and

$$\frac{\partial \mathbf{\Phi}_1}{\partial v} = \frac{\partial g_1}{\partial v}\frac{\partial \mathbf{\Phi}_2}{\partial \xi} + \frac{\partial g_2}{\partial v}\frac{\partial \mathbf{\Phi}_2}{\partial \eta},$$

where the partial derivatives of $\mathbf{\Phi}_2$ are evaluated at $(g_1(u, v), g_2(u, v))$, and those of g_1, g_2, and $\mathbf{\Phi}_1$ are evaluated at (u, v). Therefore

$$(8) \quad \frac{\partial \mathbf{\Phi}_1}{\partial u} \times \frac{\partial \mathbf{\Phi}_1}{\partial v} = \frac{\partial g_1}{\partial u}\frac{\partial g_1}{\partial v}\left(\frac{\partial \mathbf{\Phi}_2}{\partial \xi} \times \frac{\partial \mathbf{\Phi}_2}{\partial \xi}\right) + \frac{\partial g_1}{\partial u}\frac{\partial g_2}{\partial v}\left(\frac{\partial \mathbf{\Phi}_2}{\partial \xi} \times \frac{\partial \mathbf{\Phi}_2}{\partial \eta}\right)$$

$$+ \frac{\partial g_2}{\partial u}\frac{\partial g_1}{\partial v}\left(\frac{\partial \mathbf{\Phi}_2}{\partial \eta} \times \frac{\partial \mathbf{\Phi}_2}{\partial \xi}\right) + \frac{\partial g_2}{\partial u}\frac{\partial g_2}{\partial v}\left(\frac{\partial \mathbf{\Phi}_2}{\partial \eta} \times \frac{\partial \mathbf{\Phi}_2}{\partial \eta}\right).$$

Since

$$\frac{\partial \mathbf{\Phi}_2}{\partial \xi} \times \frac{\partial \mathbf{\Phi}_2}{\partial \xi} = \frac{\partial \mathbf{\Phi}_2}{\partial \eta} \times \frac{\partial \mathbf{\Phi}_2}{\partial \eta} = 0$$

and

$$\frac{\partial \mathbf{\Phi}_2}{\partial \eta} \times \frac{\partial \mathbf{\Phi}_2}{\partial \xi} = -\left(\frac{\partial \mathbf{\Phi}_2}{\partial \xi} \times \frac{\partial \mathbf{\Phi}_2}{\partial \eta}\right),$$

it follows from (8) that

$$\frac{\partial \mathbf{\Phi}_1}{\partial u} \times \frac{\partial \mathbf{\Phi}_1}{\partial v} = \left(\frac{\partial g_1}{\partial u}\frac{\partial g_2}{\partial v} - \frac{\partial g_1}{\partial v}\frac{\partial g_2}{\partial u}\right)\left(\frac{\partial \mathbf{\Phi}_2}{\partial \xi} \times \frac{\partial \mathbf{\Phi}_2}{\partial \eta}\right)$$

$$= \frac{\partial(g_1, g_2)}{\partial(u, v)}\left(\frac{\partial \mathbf{\Phi}_2}{\partial \xi} \times \frac{\partial \mathbf{\Phi}_2}{\partial \eta}\right).$$

Therefore

$$\int_{D_1} \left|\frac{\partial \mathbf{\Phi}_1}{\partial u} \times \frac{\partial \mathbf{\Phi}_1}{\partial v}\right| du\, dv = \int_{D_1} \left|\frac{\partial \mathbf{\Phi}_2}{\partial \xi} \times \frac{\partial \mathbf{\Phi}_2}{\partial \eta}\right| \left|\frac{\partial(g_1, g_2)}{\partial(u, v)}\right| du\, dv,$$

and (7) follows from the rule for change of variables in multiple integrals (Theorem 3.2, Section 4.3).

Surface Integrals of Scalar Functions

Let S be a smooth surface defined by $\mathbf{X} = \mathbf{\Phi}(u, v)$ for (u, v) in a domain D, and suppose f is a real valued function defined on S. Then the *integral of f over S* is defined by

(9)
$$\int_D f(\Phi(u, v)) \left| \frac{\partial \Phi}{\partial u}(u, v) \times \frac{\partial \Phi}{\partial v}(u, v) \right| du\, dv$$

if the latter exists. It can be shown (Exercise T-1) that if $X = \Phi_1(\xi, \eta)$ $((\xi, \eta)$ in $D_1)$ is an equivalent representation of S, then (9) equals

$$\int_{D_1} f(\Phi_1(\xi, \eta)) \left| \frac{\partial \Phi_1}{\partial \xi}(\xi, \eta) \times \frac{\partial \Phi_2}{\partial \eta}(\xi, \eta) \right| d\xi\, d\eta;$$

hence the integral of f over S depends only on f and S, and not on the particular parametrization chosen for S. We indicate this by writing the integral as

(10)
$$\int_S f\, dS.$$

If $f = 1$, then (10) is simply the area of S.

If ρ is defined and nonnegative for all X on S, we can regard it as defining a mass distribution. This is analogous to the corresponding definition of mass distribution for a solid (Definition 4.1, Section 4.4), except that here $\rho(X)$ has the dimensions of mass per unit area. The mass of S is given by

(11)
$$m = \int_S \rho\, dS.$$

The moments, center of gravity, and moments and products of inertia of S, as defined in Section 4.4, can be computed by means of surface integrals.

Example 6.9   Let S be the hemisphere

$$x^2 + y^2 + z^2 = a^2 \qquad (z > 0),$$

with mass density function

$$\rho(x, y, z) = (x^2 + y^2)z.$$

Again we represent S by

$$X = a \sin \phi \cos \theta\, i + a \sin \phi \sin \theta\, j + a \cos \phi\, k,$$

where (ϕ, θ) is in

$$D = \left\{ (\phi, \theta) \mid 0 \leq \phi \leq \frac{\pi}{2}, \quad 0 \leq \theta \leq 2\pi \right\}.$$

From (11),

$$m(S) = \int_S z(x^2 + y^2) \, dS$$

$$= \int_D (a \cos \phi)(a^2 \sin^2 \phi)(a^2 \sin \phi) \, d\theta \, d\phi$$

$$= a^5 \int_0^{2\pi} d\theta \int_0^{\pi/2} \cos \phi \sin^3 \phi \, d\phi$$

$$= a^5(2\pi) \left[\frac{\sin^4 \phi}{4} \Big|_0^{\pi/2} \right] = \frac{\pi a^5}{2}.$$

By symmetry $M_{yz} = M_{xz} = 0$ (verify this by setting up the appropriate surface integrals), and

$$M_{xy} = \int_S z^2(x^2 + y^2) \, dS$$

$$= \int_D (a \cos \phi)^2 (a^2 \sin^2 \phi)(a^2 \sin \phi) \, d\theta \, d\phi$$

$$= a^6 \int_0^{2\pi} d\theta \int_0^{\pi/2} \cos^2 \phi \sin^3 \phi \, d\phi$$

$$= 2\pi a^6 \int_0^{\pi/2} \cos^2 \phi (1 - \cos^2 \phi) \sin \phi \, d\varphi$$

$$= 2\pi a^6 \left[\left(-\frac{\cos^3 \phi}{3} + \frac{\cos^5 \phi}{5} \right) \Big|_0^{\pi/2} \right] = \frac{4\pi a^6}{15}.$$

The center of mass of S has coordinates $\bar{x} = \bar{y} = 0$ and

$$\bar{z} = \frac{\dfrac{4\pi a^6}{15}}{\dfrac{\pi a^5}{2}} = \frac{8a}{15}.$$

The moment of inertia of S about the z-axis is

$$I_z = \int_S z(x^2 + y^2)^2 \, dS$$

$$= \int_D (a \cos \phi)(a^4 \sin^4 \phi) a^2 \sin \phi \, d\theta \, d\phi$$

$$= a^7 \int_0^{2\pi} d\theta \int_0^{\pi/2} \cos \phi \sin^5 \phi \, d\phi$$

$$= 2\pi a^7 \left[\frac{\sin^6 \phi}{6} \Big|_0^{\pi/2} \right] = \frac{\pi a^7}{3} \,.$$

Many commonly occurring surfaces do not meet the requirements of Definition 6.1. For example, the surface of the parallelepiped in Fig. 6.9 is not smooth since no continuous normal vector can be defined so as to be continuous on its edges. The same can be said for the surface comprised of the hemisphere $x^2 + y^2 + z^2 = 1$ $(z \leq 0)$ and the section of the paraboloid $z = 1 - x^2 - y^2$ for which $z \leq 0$, joined together by the unit circle on the xy-plane (Fig. 6.10).

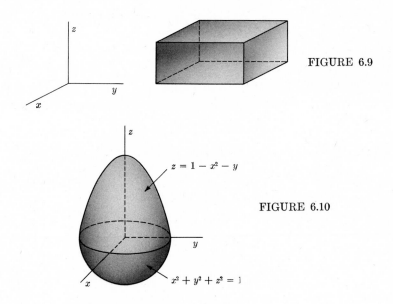

FIGURE 6.9

$z = 1 - x^2 - y$

FIGURE 6.10

$x^2 + y^2 + z^2 = 1$

Let us consider the surface of the parallelepiped further. Clearly it can be viewed as consisting of six smooth surfaces $S_1, \ldots, S_6$ (its faces) joined together. It is surely not surprising that we define

$$\int_S f \, dS = \sum_{i=1}^{6} \int_{S_i} f \, dS.$$

Definition 6.4 If f is a real valued function defined on a surface S made up of smooth sections $S_1, \ldots, S_k$, then

$$\int_S f \, dS = \sum_{i=1}^{k} \int_{S_i} f \, dS,$$

provided the integrals on the right exist.

Example 6.10 Let S be the surface shown in Fig. 6.10 and $f(x, y, z) = z$. We represent S_1, the upper half of S, by

$$\mathbf{X} = \mathbf{\Phi}_1(r, \theta) = r \cos \theta \, \mathbf{i} + r \sin \theta \, \mathbf{j} + (1 - r^2) \mathbf{k}$$

for (r, θ) in

$$D_1 = \{ (r, \theta) \mid 0 \leq r \leq 1, \quad 0 \leq \theta \leq 2\pi \}.$$

Then

$$\mathbf{N}_1(r, \theta) = 2r^2 \cos \theta \, \mathbf{i} + 2r^2 \sin \theta \, \mathbf{j} + r \mathbf{k}$$

and

$$dS = | \mathbf{N}_1(r, \theta) | \, dr \, d\theta = r \sqrt{1 + 4r^2} \, dr \, d\theta.$$

On S_1, $f(x, y, z) = z = 1 - r^2$; hence

$$\int_{S_1} f \, dS = \int_{D_1} (1 - r^2) \, r \sqrt{1 + 4r^2} \, dr \, d\theta$$

$$= \int_0^{2\pi} d\theta \int_0^1 (1 - r^2) \, r \sqrt{1 + 4r^2} \, dr.$$

We leave the evaluation of the integral to the student. The result is

(12) $$\int_{S_1} f \, dS = \frac{\pi}{60} (25\sqrt{5} - 11).$$

The lower half S_2 of S can be parametrized by

$$\mathbf{X} = \mathbf{\Phi}_2(\phi, \theta) = \sin \phi \cos \theta \, \mathbf{i} + \sin \phi \sin \theta \, \mathbf{j} + \cos \phi \mathbf{k}$$

for (ϕ, θ) in

$$D_2 = \left\{ (\phi, \theta) \mid -\frac{\pi}{2} \leq \phi \leq 0, \quad 0 \leq \theta \leq 2\pi \right\}.$$

Then

$$\mathbf{N}_2(\phi, \theta) = \sin \phi \, (\sin \phi \cos \theta \, \mathbf{i} + \sin \phi \sin \theta \, \mathbf{j} + \cos \phi \, \mathbf{k})$$

and

$$dS = \sin \phi \, d\phi \, d\theta.$$

On S_2, $f(x, y, z) = \cos \phi$; hence

$$\int_{S_2} f \, dS = \int_{D_2} \cos \phi \sin \phi \, d\phi \, d\theta$$

$$= \int_0^{2\pi} d\theta \int_{-\pi/2}^0 \cos \phi \sin \phi \, d\phi$$

$$= 2\pi \left(\frac{\sin^2 \phi}{2} \Big|_{-\pi/2}^0 \right) = -\pi.$$

From this and (12) it follows that

$$\int_S f \, dS = \frac{\pi}{60} (25\sqrt{5} - 71).$$

EXERCISES 4.6

In Exercises 1 through 11 compute the area of the given surface.

1. The part of the surface $z^2 = x^2 + y^2$ between the planes $z = 1$ and $z = 3$.

2. The spiral surface given by

    $$\mathbf{X} = \mathbf{X}(u, v) = u \cos v \, \mathbf{i} + u \sin v \, \mathbf{j} + v\mathbf{k}$$

    $$(0 \le u \le 1, \quad 0 \le v \le 2\pi).$$

3. The surface defined by

    $$\mathbf{X} = \mathbf{X}(u, v) = (u + v)\mathbf{i} + (u - v)\mathbf{j} + u\mathbf{k}$$

    $$(0 \le u \le 2, \quad 1 \le v \le 3).$$

4. The part of the sphere $x^2 + y^2 + z^2 = 1$ between the planes $z = 1/\sqrt{2}$ and $z = -1/\sqrt{2}$.

5. The part of the plane $x + y + z = 2$ inside the cylinder $x^2 + y^2 = 4$.

6. The part of the cylinder $x^2 + y^2 = 2x$ bounded above by the cone $x^2 + y^2 = z^2$ and below by the xy-plane.

7. The part of the sphere $x^2 + y^2 + z^2 = 1$ inside the upper nappe of the cone $x^2 + y^2 = z^2$.

8. The part of the cylinder $x^2 + y^2 = 4$ between the planes $z = x$ and $z = 2x + 3$.

9. The triangle determined by the plane

$$\frac{x}{a} + \frac{y}{b} + \frac{z}{c} = 1$$

and the coordinate planes.

10. The part of the cylinder $x^2 + z^2 = a^2$ inside the cylinder $y^2 + z^2 = a^2$.

11. The part of the cylinder $x^2 + y^2 = ax$ inside the sphere $x^2 + y^2 + z^2 = a^2$.

In Exercises 12 through 17 compute $\int_S f \, dS$ over the given surface.

12. $f(x, y, z) = x^2$; the cylinder defined by $x^2 + y^2 = a^2, |z| \leq 1$, including the top and bottom.

13. $f(x, y, z) = y$; the part of the plane $2x + y + z = 2$ inside the cylinder $x^2 + y^2 = 4$.

14. $f(x, y, z) = x^2$; the part of the paraboloid $z = 1 - x^2 - y^2$ for which $z \geq 0$.

15. $f(x, y, z) = x^2$; the part of the cone $x^2 + y^2 = z^2$ between the planes $z = 0$ and $z = 2$.

16. $f(x, y, z) = x$; the part of the sphere $x^2 + y^2 + z^2 = a^2$ inside the upper nappe of the cone $x^2 + y^2 = z^2$, together with the upper part of the cone contained within the sphere.

17. $f(x, y, z) = x$; the part of the cylinder $x^2 + y^2 = 1$ between the planes $z = 0$ and $z = x + 2$, together with the top and bottom.

In Exercises 18 through 22 compute the center of mass and I_z for the given surface S with density ρ.

18. $\rho(x, y, z) = \rho_0$; the surface of Exercise 17.

19. $\rho(x, y, z) = x^2$; the surface of Exercise 16.

20. $\rho(x, y, z) = x^2 + y^2$; the surface of Exercise 4.

21. $\rho(x, y, z) = x^2$; the surface of Exercise 5.

22. $\rho(x, y, z) = \sqrt{x^2 + y^2}$; the surface of Exercise 6.

THEORETICAL EXERCISES

T-1. Let $\mathbf{X} = \mathbf{\Phi}(u, v)$ $((u, v)$ in $D)$ and $\mathbf{X} = \mathbf{\Phi}_1(\xi, y)$ $((\xi, y)$ in $D_1)$ be equivalent parametrizations of an orientable surface S. Show that

$$\int_D f(\mathbf{\Phi}(u, v)) \left| \frac{\partial \mathbf{\Phi}}{\partial u} (u, v) \times \frac{\partial \mathbf{\Phi}}{\partial v} (u, v) \right| du\, dv$$

$$= \int_{D_1} f(\mathbf{\Phi}_1(\xi, y)) \left| \frac{\partial \mathbf{\Phi}_1}{\partial \xi} (\xi, y) \times \frac{\partial \mathbf{\Phi}_1}{\partial y} (\xi, y) \right| d\xi\, dy$$

if f is a scalar valued function for which either integral exists. *Hint*: See the proof of Theorem 6.1.

T-2. Let the surface S be given explicitly by $z = f(x, y)$ $((x, y)$ in $D)$.
(a) Show that the area of S is given by

$$A(S) = \int\int_D \sqrt{1 + \left(\frac{\partial f}{\partial x}\right)^2 + \left(\frac{\partial f}{\partial y}\right)^2}\, dx\, dy.$$

(b) If γ is the acute angle between the z-axis and the normal to S, show that

$$A(S) = \int\int_D \sec \gamma\, dx\, dy.$$

T-3. Let the surface S be given by $\mathbf{X} = \mathbf{X}(u, v)$ $((u, v)$ in $D)$ and define

$$E = \left|\frac{\partial \mathbf{X}}{\partial u}\right|^2, \qquad F = \frac{\partial \mathbf{X}}{\partial u} \cdot \frac{\partial \mathbf{X}}{\partial v}, \qquad G = \left|\frac{\partial \mathbf{X}}{\partial v}\right|^2.$$

Show that

$$A(S) = \int\int_D \sqrt{EG - F^2}\, du\, dv.$$

T-4. Prove the following theorem of Pappus:
If a path of length L is revolved about the y-axis to obtain a surface S, then the area of S is $A(S) = 2\pi \bar{x} L$, where $(\bar{x}, \bar{y})$ is the centroid of the path.

T-5. Show that $A(T_{ij}) = |\mathbf{N}(u_{ij}, v_{ij})|\, (a_i - a_{i-1})(b_j - b_{j-1})$.

4.7   Surface Integrals of Vector Fields

We now consider surface integrals of vector functions over *orientable* surfaces in R^3.

Definition 7.1   An *orientable* surface S in R^3 is a surface which has a smooth parametric representation

$$\mathbf{X} = \mathbf{\Phi}(u, v) \qquad ((u, v) \text{ in } D)$$

that satisfies

$$(1) \qquad \lim_{(u, v) \to (u_0, v_0)} \frac{\mathbf{N}(u, v)}{|\mathbf{N}(u, v)|} = \lim_{(u, v) \to (u_1, v_1)} \frac{\mathbf{N}(u, v)}{|\mathbf{N}(u, v)|}$$

whenever (u_0, v_0) and (u_1, v_1) are points on the boundary $B(D)$ of D which correspond to the same point on S; that is,

$$(2) \qquad \mathbf{X}_0 = \mathbf{\Phi}(u_0, v_0) = \mathbf{\Phi}(u_1, v_1).$$

A point $\mathbf{X}_0$ on S which satisfies (2) with $(u_0, v_0) \neq (u_1, v_1)$ is a *multiple point* of S. From the definition of a smooth surface (specifically, part (b) of Definition 6.1, Section 4.6, (2) can be satisfied with $(u_0, v_0) \neq (u_1, v_1)$ only if both (u_0, v_0) and (u_1, v_1) are in $B(D)$.

Example 7.1   Consider the cylindrical surface S defined by $x^2 + y^2 = a^2$, $|z| \leq 1$, and represented parametrically by

$$(3) \qquad \mathbf{X} = \mathbf{\Phi}(\theta, z)$$

$$= a \cos \theta \, \mathbf{i} + a \sin \theta \, \mathbf{j} + z\mathbf{k}$$

for (θ, z) in

$$D = \{(\theta, z) \mid 0 \leq \theta \leq 2\pi, \quad |z| \leq 1\}.$$

We have seen (Example 6.1, Section 4.6) that

$$\mathbf{N}(\theta, z) = a \cos \theta \, \mathbf{i} + a \sin \theta \, \mathbf{j};$$

hence (assuming that $a > 0$),

$$\frac{\mathbf{N}(\theta, z)}{|\mathbf{N}(\theta, z)|} = \cos \theta \, \mathbf{i} + \sin \theta \, \mathbf{j}.$$

To find the multiple points of S, we need only examine the image of $B(D)$ under (3). They are of the form $(a, 0, z_0)$ $(|z_0| \leq 1)$ since

$$\mathbf{\Phi}(0, z_0) = \mathbf{\Phi}(2\pi, z_0) = a\mathbf{i} + z_0\mathbf{k}.$$

However,

$$\lim_{(\theta,\,z)\to(0,\,z_0)} \frac{\mathbf{N}(\theta,\,z)}{|\,\mathbf{N}(\theta,\,z)\,|} = \lim_{(\theta,\,z)\to(2\pi,\,z_0)} \frac{\mathbf{N}(\theta,\,z)}{|\,\mathbf{N}(\theta,\,z)\,|} = \mathbf{i};$$

thus (1) is satisfied, and S is orientable.

Example 7.2 Consider the hemisphere S defined by $x^2 + y^2 + z^2 = a^2$, $z > 0$. We have seen (Example 6.2, Section 4.6) that S can be represented by

(4) $\mathbf{X} = \mathbf{\Phi}(\phi, \theta) = a \sin \phi \cos \theta \, \mathbf{i} + a \sin \phi \sin \theta \, \mathbf{j} + a \cos \phi \, \mathbf{k}$

for (ϕ, θ) in

$$D = \left\{ (\phi, \theta) \mid 0 \le \phi \le \frac{\pi}{2}, \quad 0 \le \theta \le 2\pi \right\},$$

and that

$$\mathbf{N}(\phi, \theta) = a^2 \sin \phi \, (\sin \phi \cos \theta \, \mathbf{i} + \sin \phi \sin \theta \, \mathbf{j} + \cos \phi \, \mathbf{k}).$$

Therefore (assuming $a > 0$),

$$\frac{\mathbf{N}(\phi, \theta)}{|\,\mathbf{N}(\phi, \theta)\,|} = \sin \phi \cos \theta \, \mathbf{i} + \sin \phi \sin \theta \, \mathbf{j} + \cos \phi \, \mathbf{k}.$$

Again we examine the image of $B(D)$ under (4) to find the multiple points of S. All points of the form $(0, \theta_0)$ map onto the "north pole" since

$$\mathbf{\Phi}(0, \theta_0) = a\mathbf{k};$$

however, for any such point,

$$\lim_{(\phi,\theta)\to(0,\theta_0)} \frac{\mathbf{N}(\phi, \theta)}{|\,\mathbf{N}(\phi, \theta)\,|} = \mathbf{k}.$$

Also, $\mathbf{\Phi}(\phi_0, 0) = \mathbf{\Phi}(\phi_0, 2\pi)$ and

$$\lim_{(\sigma,\theta)\to(\sigma_0,0)} \frac{\mathbf{N}(\phi, \theta)}{|\,\mathbf{N}(\phi, \theta)\,|} = \lim_{(\sigma,\theta)\to(\sigma_0,2\pi)} \frac{\mathbf{N}(\phi, \theta)}{|\,\mathbf{N}(\phi, \theta)\,|} = \sin \phi_0 \, \mathbf{i} + \cos \phi_0 \, \mathbf{k}.$$

Therefore S is orientable.

Example 7.3 Any smooth surface without multiple points is orientable since (2) then implies that $(u_0, v_0) = (u_1, v_1)$ and (1) follows automatically.

Example 7.4 The surface S defined by

$$\mathbf{X} = \mathbf{\Phi}(\theta, v)$$

$$= \left(1 + v \sin \frac{\theta}{2}\right) \cos \theta \, \mathbf{i} + \left(1 + v \sin \frac{\theta}{2}\right) \sin \theta \, \mathbf{j} + v \cos \frac{\theta}{2} \, \mathbf{k}$$

for (θ, v) in

$$D = \{(\theta, v) \mid 0 \le \theta \le 2\pi, \ |v| \le 1\}$$

is called a *Möbius strip*. A model of S can be constructed by subjecting one end of a narrow strip of paper to a half twist and pasting it to the other end (Fig. 7.1). By straightforward calculation (Exercise T-1) it can be shown that

(5) $$\lim_{(\theta,v)\to(0,0)} \frac{\mathbf{N}(\theta, v)}{|\mathbf{N}(\theta, v)|} = - \lim_{(\theta,v)\to(2\pi,0)} \frac{\mathbf{N}(\theta, v)}{|\mathbf{N}(\theta, v)|} = \mathbf{i}.$$

Since $\mathbf{\Phi}(0, 0) \ne \mathbf{\Phi}(2\pi, 0)$, S is not orientable. (Actually, we have not proved that the Möbius strip *cannot* be represented by a parametric function which has the properties of Definition 7.1. However, this can be shown.)

FIGURE 7.1

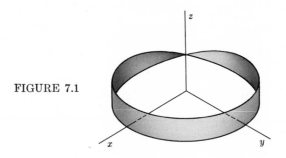

To each point $\mathbf{X}$ on an orientable surface S we can assign a unit vector $\mathbf{n}(\mathbf{X})$ as follows: if $\mathbf{X} = \mathbf{\Phi}(u, v)$ for (u, v) in D°, then

$$\mathbf{n}(\mathbf{X}) = \frac{\mathbf{N}(u, v)}{|\mathbf{N}(u, v)|},$$

while if $\mathbf{X} = \mathbf{\Phi}(u_0, v_0)$ for some (u_0, v_0) in $B(D)$, then

(6) $$\mathbf{n}(\mathbf{X}) = \lim_{(u,v)\to(u_0,v_0)} \frac{\mathbf{N}(u, v)}{|\mathbf{N}(u, v)|}.$$

Since $\mathbf{N}(u, v)$ is normal to S at $\mathbf{X} = \mathbf{\Phi}(u, v)$, it follows that $\mathbf{n}(\mathbf{X})$ is a unit normal to S at $\mathbf{X} = \mathbf{\Phi}(u, v)$. It follows from (1) that (6) is well defined even if $\mathbf{X}$ is a multiple point of S.

Example 7.5   For the cylindrical section of Example 7.1,

$$\mathbf{N}(\theta, z) = a \cos \theta \, \mathbf{i} + a \sin \theta \, \mathbf{j}$$

$$= x\mathbf{i} + y\mathbf{j};$$

hence

$$\mathbf{n}(\mathbf{X}) = \frac{x\mathbf{i} + y\mathbf{j}}{\sqrt{x^2 + y^2}}.$$

For the hemisphere of Example 7.2,

$$\mathbf{N}(\phi, \theta) = a \sin \phi \, \mathbf{X}(\phi, \theta),$$

and therefore

$$\mathbf{n}(\mathbf{X}) = \frac{\mathbf{X}}{|\mathbf{X}|} = \frac{x\mathbf{i} + y\mathbf{j} + z\mathbf{k}}{\sqrt{x^2 + y^2 + z^2}}.$$

We shall regard $\mathbf{n}$ as a vector field defined on S. It can be represented visually by drawing an arrow in the direction of $\mathbf{n}(\mathbf{X})$ at each point $\mathbf{X}$ of S (Fig. 7.2).

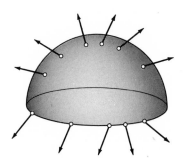

FIGURE 7.2

We shall call the vector field $\mathbf{n}$ the *orientation of S induced by the parametric representation* $\mathbf{X} = \boldsymbol{\Phi}(u, v)$. Along with $\mathbf{n}$, the vector $-\mathbf{n}$ is also an orientation of S. It is induced by $\mathbf{X} = \boldsymbol{\Psi}(v, u)$, where $\boldsymbol{\Psi}(v, u) = \boldsymbol{\Phi}(u, v)$ (Exercise T-2).

The student knows that some surfaces—such as hemispheres—have boundaries, while others—such as spheres—do not. We shall not attempt to give a precise definition of the boundary of a surface. Rather, we shall be guided by geometric intuition.

An orientable surface S which has no boundary is said to be *closed*. A closed surface determines a bounded set, *the inside of S*, and an unbounded set, the *outside* of S, such that any continuous curve with initial point inside and terminal point outside of S must pass through S. It is customary to designate the positive orientation of a closed surface to be the one for which the arrow representing $\mathbf{n}(\mathbf{X})$ is directed toward the outside of S at every point of S.

Theorem 7.1 Let $\mathbf{X} = \mathbf{\Phi}_1(u, v)$ and $\mathbf{X} = \mathbf{\Phi}_2(u, v)$ be equivalent smooth parametric representations of a surface S (Definition 6.3, Section 4.6), related by

$$\mathbf{\Phi}_2(g_1(u, v), g_2(u, v)) = \mathbf{\Phi}_1(u, v),$$

and let $\mathbf{n}_1$ and $\mathbf{n}_2$ be the orientations induced on S by $\mathbf{\Phi}_1$ and $\mathbf{\Phi}_2$, respectively. Then $\mathbf{n}_1 = \mathbf{n}_2$ if the Jacobian determinant $\partial(g_1, g_2)/\partial(u, v)$ is positive, and $\mathbf{n}_1 = -\mathbf{n}_2$ if the Jacobian determinant is negative.

We leave the proof of this theorem to the student (Exercise T-3).

Surface Integrals of Vector Fields

Let the surface S be oriented by the unit vector function $\mathbf{n}$ and suppose

$$\mathbf{F} = F_1\mathbf{i} + F_2\mathbf{j} + F_3\mathbf{k}$$

is a vector field defined on S. Then $\mathbf{F} \cdot \mathbf{n}$ is a scalar function on S, and the integral

(7) $$\int_S \mathbf{F} \cdot \mathbf{n} \, dS,$$

if it exists, is called the *surface integral* of $\mathbf{F}$ over S. If $\mathbf{X} = \mathbf{X}(u, v)$ $((u, v)$ in $D)$ is a parametric representation of S which induces $\mathbf{n}$, then

$$\int_S \mathbf{F} \cdot \mathbf{n} \, dS = \int_D \left(\mathbf{F}(\mathbf{X}(u, v)) \cdot \frac{\mathbf{N}(u, v)}{|\mathbf{N}(u, v)|} \right) |\mathbf{N}(u, v)| \, du \, dv$$

$$= \int_D \mathbf{F}(\mathbf{X}(u, v)) \cdot \mathbf{N}(u, v) \, du \, dv$$

$$= \int_D \mathbf{F}(\mathbf{X}(u, v)) \cdot \left(\frac{\partial \mathbf{X}}{\partial u}(u, v) \times \frac{\partial \mathbf{X}}{\partial v}(u, v) \right) du \, dv.$$

Notice that (7) depends upon which of the two possible orientations of S occurs in the integrand. If S is closed it is to be understood, in the absence of a statement to the contrary, that $\mathbf{n}(\mathbf{X})$ is directed toward the outside of S. For a surface which is not closed the orientation must be stated explicitly.

Example 7.6 Let S be the sphere $x^2 + y^2 + z^2 = a^2$ $(a > 0)$ and

$\mathbf{F}(x, y, z) = x\mathbf{i} + y\mathbf{j}$, and suppose we wish to calculate $\int_S \mathbf{F} \cdot \mathbf{n} \, dS$, where $\mathbf{n}$

is the outward normal. We parametrize S by

(8)
$$X = \mathbf{X}(\phi, \theta)$$
$$= a \sin \phi \cos \theta \, \mathbf{i} + a \sin \phi \sin \theta \, \mathbf{j} + a \cos \phi \mathbf{k}$$

for (ϕ, θ) in

$$D = \{ (\phi, \theta) \mid 0 \le \phi \le \pi, \ \ 0 \le \theta \le 2\pi \}.$$

From Example 7.5,

(9)
$$\frac{\mathbf{N}(\phi, \theta)}{\mid \mathbf{N}(\phi, \theta) \mid} = \frac{\mathbf{X}(\phi, \theta)}{\mid \mathbf{X}(\phi, \theta) \mid},$$

which is a positive multiple of the position vector of the point $\mathbf{X} = \mathbf{X}(\phi, \theta)$ with respect to the origin. Hence the unit vector (9) points outward from the sphere S at every point (Fig. 7.3), and therefore (8) induces the positive orientation on S. Hence,

$$\int_S \mathbf{F} \cdot \mathbf{n} \, dS = \int_D (a \sin \phi \cos \theta \, \mathbf{i} + a \sin \phi \sin \theta \, \mathbf{j}) \cdot (a \sin \phi \, \mathbf{X}(\phi, \theta) \, d\phi \, d\theta$$

$$= \int_D [(a \sin \phi \cos \theta)^2 + (a \sin \phi \sin \theta)^2] a \sin \phi \, d\phi \, d\theta$$

$$= \int_D a^3 \sin^3 \phi \, d\phi \, d\theta = a^3 \int_0^{2\pi} d\theta \int_0^\pi \sin^3 \phi \, d\phi$$

$$= 2\pi a^3 \int_0^\pi (1 - \cos^2 \phi) \sin \phi \, d\phi$$

$$= 2\pi a^3 \left(-\cos \phi + \frac{\cos^3 \phi}{3} \right) \Big|_0^\pi = 2\pi a^3 \left(\frac{4}{3} \right) = \frac{8\pi a^3}{3}.$$

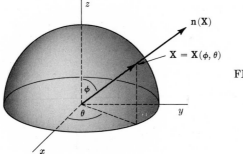

n(X)

X = $\mathbf{X}(\phi, \theta)$

FIGURE 7.3

Example 7.7   Let S be the part of the paraboloid

$$z = f(x, y) = x^2 + y^2$$

which is bounded above by $z = 1$ (Fig. 6.8). It can be represented parametrically by

(10) $$\mathbf{X} = \mathbf{X}(x, y) = x\mathbf{i} + y\mathbf{j} + (x^2 + y^2)\mathbf{k},$$

where (x, y) is in

$$D = \{(x, y) \mid x^2 + y^2 \le 1\}.$$

Suppose we wish to integrate

$$\mathbf{F}(x, y, z) = y\mathbf{i} - x\mathbf{j} + z^2\mathbf{k}$$

over S, where the orientation is chosen so that $\mathbf{n}(\mathbf{X})$ has a negative z component. It is easy to verify that the representation (10) induces $-\mathbf{n}$; hence

$$\int_S \mathbf{F} \cdot \mathbf{n} \, dS = -\int_D \mathbf{F}(\mathbf{X}(x, y)) \cdot \mathbf{N}(x, y) \, dx \, dy$$

$$= -\int_D [y\mathbf{i} - x\mathbf{j} + (x^2 + y^2)^2\mathbf{k}] \cdot [-2x\mathbf{i} - 2y\mathbf{j} + \mathbf{k}] \, dx \, dy$$

$$= -\int_D (x^2 + y^2)^2 \, dx \, dy$$

Changing to polar coordinates yields

$$\int_S \mathbf{F} \cdot \mathbf{n} \, dS = -\int_0^{2\pi} d\theta \int_0^1 r^5 \, dr = (2\pi) \left(\frac{1}{6}\right) = -\frac{\pi}{3}.$$

Piecewise Orientable Surfaces

Many commonly occurring surfaces do not meet the requirements of Definition 7.1. For example, the surface of the parallelepiped in Fig. 7.4 is not orientable since it is not smooth, as noted in Section 4.6. The same can be said for the surface comprised of the hemisphere $x^2 + y^2 + z^2 = 1$ ($z \le 0$) and the section of the paraboloid $z = 1 - x^2 - y^2$ for which $z \le 0$, joined together by the unit circle on the xy-plane (Fig. 7.5).

Let us consider the surface of the parallelepiped further. Clearly it can be viewed as consisting of six orientable surfaces $S_1, \ldots, S_6$ (its faces)

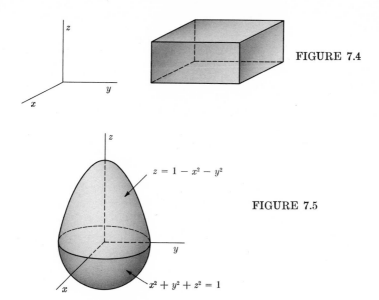

FIGURE 7.4

FIGURE 7.5

joined together. For surface integrals of scalar functions we defined

$$\int_S f \, dS = \sum_{i=1}^{6} \int_{S_i} f \, dS$$

for any real valued function f such that the integrals on the right exist. However, the situation is not so simple if we wish to define the integral of a vector field over S. To write simply

(11) $$\int_S \mathbf{F} \cdot \mathbf{n} \, dS = \sum_{i=1}^{6} \int_{S_i} \mathbf{F} \cdot \mathbf{n} \, dS$$

would be nonsense, since there is as yet no orientation $\mathbf{n}$ defined over all six faces. We could remedy this by choosing orientations $\mathbf{n}_1, \ldots, \mathbf{n}_6$ for $S_1, \ldots, S_6$ and replacing the right side of (11) by

$$\sum_{i=1}^{6} \int_{S_i} \mathbf{F} \cdot \mathbf{n}_i \, dS.$$

However, this expression has in general $2^6 = 64$ distinct values, since each $\mathbf{n}_i$ can be chosen in two ways.

Clearly we must adopt a convention regarding the way in which the orientable pieces of surfaces such as those in Figs. 7.4 and 7.5 are to be joined together. To do this we first introduce the notion of a *directed boundary* of a surface.

Suppose an orientable surface S has a boundary which we denote by

$B(S)$. It can be shown that if S is represented by $\mathbf{X} = \mathbf{\Phi}(u, v)$ ((u, v) in D) then every point on $B(S)$ is the image of a point on $B(D)$. However, there may be points on $B(D)$ whose images are not on $B(S)$. These statements are easily verified for the surfaces of Examples 7.1 and 7.2 (Exercise T-4).

If S has a boundary then $B(S)$ can be shown to consist of a finite number of piecewise smooth curves. For example, the boundary of the cylindrical surface of Example 7.1 consists of the two circles

$$C_1 = \{ (x, y, z) \mid x^2 + y^2 = a^2, \quad z = 1 \}$$

and

$$C_2 = \{x, y, z) \mid x^2 + y^2 = a^2, \quad z = -1 \}.$$

Let C be a smooth curve in $B(S)$, parametrized by $\mathbf{X} = \mathbf{\Psi}(t)$. If, whenever $\dot{\mathbf{\Psi}}(t)$ exists, the vector $\mathbf{n}(\mathbf{\Psi}(t)) \times \dot{\mathbf{\Psi}}(t)$ points toward S, we say that C is *positively directed with respect to the orientation* $\mathbf{n}$, or simply *positively directed*. If every piecewise smooth curve in $B(S)$ is positively directed we say that $B(S)$ *is positively directed* (with respect to $\mathbf{n}$). In this case a person walking in the positive direction around $B(S)$ with his head in the direction of $\mathbf{n}(\mathbf{X})$ would always see S to his left (Fig. 7.6). (To see this, recall that $\dot{\mathbf{\Psi}}(t)$ is tangent to $B(S)$ at the point $\mathbf{X} = \mathbf{\Psi}(t)$, and points in the direction of motion.)

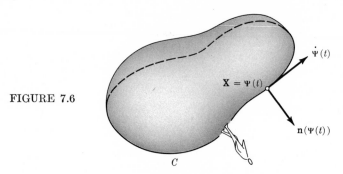

FIGURE 7.6

Example 7.8 Consider the hemisphere S of Example 7.2 with the orientation

$$\mathbf{n}(\mathbf{X}) = \frac{\mathbf{X}}{|\mathbf{X}|}$$

induced by (4). The boundary of S is the circle

$$C = \{ (x, y, z) \mid x^2 + y^2 = a^2, \quad z = 0 \}.$$

We define a direction on C by means of the parametrization

$$\mathbf{\Psi}(t) = a \cos t\, \mathbf{i} + a \sin t\, \mathbf{j} \qquad (0 \leq t \leq 2\pi).$$

Then the tangent vector

$$\dot{\mathbf{\Psi}}(t) = -a \sin t\, \mathbf{i} + a \cos t\, \mathbf{j}$$

has the direction shown in Fig. 7.7a. Since

$$\mathbf{n}(\mathbf{\Psi}(t)) = \frac{\dot{\mathbf{\Psi}}(t)}{|\dot{\mathbf{\Psi}}(t)|} = \cos t\, \mathbf{i} + \sin t\, \mathbf{j},$$

it follows that

$$\mathbf{n}(\mathbf{\Psi}(t)) \times \dot{\mathbf{\Psi}}(t) = \begin{vmatrix} \mathbf{i} & \mathbf{j} & \mathbf{k} \\ \cos t & \sin t & 0 \\ -a \sin t & a \cos t & 0 \end{vmatrix} = a\mathbf{k},$$

which points in the direction of S (Fig. 7.7a); therefore C is positively directed with respect to $\mathbf{n}$.

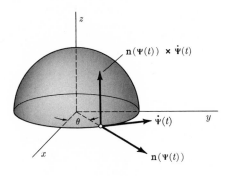

(a)

FIGURE 7.7

(b)

If we parametrize C by

(12) $X = \Psi_1(t) = a \sin t\, \mathbf{i} + a \cos t\, \mathbf{j}$ $(0 \le t \le 2\pi)$

then

$$\dot{\Psi}_1(t) = a \cos t\, \mathbf{i} - a \sin t\, \mathbf{j},$$

$$\mathbf{n}(\Psi_1(t)) \times \dot{\Psi}_1(t) = \begin{vmatrix} \mathbf{i} & \mathbf{j} & \mathbf{k} \\ \sin t & \cos t & 0 \\ a \cos t & -a \sin t & 0 \end{vmatrix} = -a\mathbf{k},$$

which points away from S (Fig. 7.7b). Thus (12) does not direct C positively with respect to $\mathbf{n}$.

Example 7.9   The boundaries of the cylinders in Figs. 7.8a and 7.8b, traversed in the directions of the arrows, are positively directed with respect to the indicated orientations. In Fig. 7.8c, C_1 is positively directed, but C_2 is not.

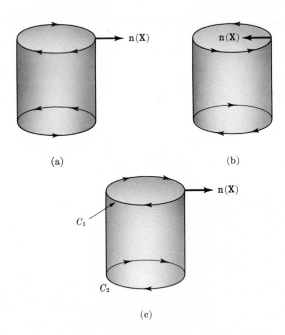

(a)   (b)

(c)

FIGURE 7.8

Definition 7.2 A surface S is *piecewise orientable* if it is orientable or can be constructed by joining together along sections of their boundaries a finite number of orientable surfaces $S_1, \ldots, S_k$ which have the following property: It is possible to choose orientations $\mathbf{n}_1, \ldots, \mathbf{n}_k$ on $S_1, \ldots, S_k$ so that if γ_{ij} is any curve common to $B(S_i)$ and $B(S_j)$ ($i \neq j$) then the positive direction of γ_{ij} with respect to S_i is the opposite of the positive direction of γ_{ij} with respect to S_j. If this is done we say that S is *oriented* and that $S_1, \ldots, S_k$ are *sections* of S.

Example 7.10 The surface S in Fig. 7.5 is orientable. In Figs. 7.9a and 7.9b, it has the outward (positive) and inward (negative) orientations, respectively; in Fig. 7.9c it is not oriented.

The orientation chosen for a section of an orientable surface determines the orientations for all of its sections. Thus, an orientable surface has exactly two orientations.

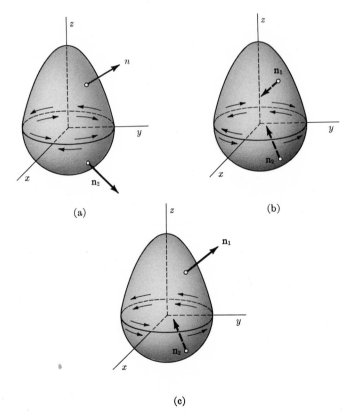

FIGURE 7.9

We can now define the integral of a vector field over a piecewise orientable surface.

Definition 7.3 Let S be a piecewise orientable surface with sections $S_1, \ldots, S_k$, and suppose that the sections are oriented in the manner described in Definition 7.2. Let $\mathbf{F}$ be a vector field defined on S. Then

$$\int_S \mathbf{F} \cdot \mathbf{n} \, dS = \sum_{i=1}^{k} \int_{S_i} (\mathbf{F} \cdot \mathbf{n}_i) \, dS$$

if the integrals on the right exist.

Example 7.11 Let S be the surface of Fig. 7.5. As in Example 6.10, Section 4.6, we represent S_1, the upper half of S, by

$$(13) \qquad \mathbf{X} = \mathbf{\Phi}_1(r, \theta) = r \cos \theta \, \mathbf{i} + r \sin \theta \, \mathbf{j} + (1 - r^2) \mathbf{k}$$

for (r, θ) in

$$D_1 = \{ (r, \theta) \mid 0 \leq r \leq 1, \ \ 0 \leq \theta \leq 2\pi \},$$

and S_2, the lower half of S, by

$$\mathbf{X} = \mathbf{\Phi}_2(\phi, \theta) = \sin \phi \cos \theta \, \mathbf{i} + \sin \phi \sin \theta \, \mathbf{j} + \cos \phi \, \mathbf{k}$$

for (ϕ, θ) in

$$(14) \qquad D_2 = \left\{ (\phi, \theta) \mid -\frac{\pi}{2} \leq \phi \leq 0, \ \ 0 \leq \theta \leq 2\pi \right\}.$$

Suppose we wish to integrate

$$F(\mathbf{X}) = \frac{y\mathbf{i} - y\mathbf{j} + \mathbf{k}}{\sqrt{x^2 + y^2}}$$

over S. The parametrizations (13) and (14) induce the outward (positive) orientations on S_1 and S_2; hence

$$(15) \qquad \int_S \mathbf{F} \cdot \mathbf{n} \, dS = \int_{D_1} \mathbf{F}(\mathbf{\Phi}_1(r, \theta)) \cdot \mathbf{N}_1(r, \theta) \, dr \, d\theta$$

$$+ \int_{D_2} \mathbf{F}(\mathbf{\Phi}_2(\phi, \theta)) \cdot \mathbf{N}_2(\phi, \theta) \, d\phi \, d\theta,$$

where $\mathbf{N}_1$ and $\mathbf{N}_2$ are defined by

$$\mathbf{N}_1(r, \theta) = 2r^2 \cos \theta \, \mathbf{i} + 2r^2 \sin \theta \, \mathbf{j} + r\mathbf{k}$$

and

$$\mathbf{N}_2(\phi, \theta) = \sin \phi \, (\sin \phi \cos \theta \, \mathbf{i} + \sin \phi \sin \theta \, \mathbf{j} + \cos \phi \, \mathbf{k}).$$

Easy calculations show that

$$\mathbf{F}(\boldsymbol{\Phi}_1(r, \theta)) \cdot \mathbf{N}_1(r, \theta) = 1$$

and

$$\mathbf{F}(\boldsymbol{\Phi}_2(\phi, \theta)) \cdot \mathbf{N}_2(\phi, \theta) = \cos \phi;$$

hence, from (15)

$$\int_S \mathbf{F} \cdot \mathbf{n} \, dS = \int_0^{2\pi} d\theta \int_0^1 dr + \int_0^{2\pi} d\theta \int_{-\pi/2}^0 \cos \varphi \, d\varphi$$

$$= 2\pi + 2\pi = 4\pi.$$

Example 7.12   Let V be the parallelepiped bounded by the planes $x = x_1, x = x_2, y = y_1, y = y_2, z = z_1$, and $z = z_2$ (Fig. 7.10). Let S be the surface of V with the positive (outward) orientation, and suppose $\mathbf{F} = F_1\mathbf{i} + F_2\mathbf{j} + F_3\mathbf{k}$ is a continuous vector function on S. Then

$$\int_S \mathbf{F} \cdot \mathbf{n} \, dS = \sum_{i=1}^6 \int_{S_i} \mathbf{F} \cdot \mathbf{n} \, dS,$$

where $S_1, \ldots, S_6$ are the faces of V.

Let S_1 be the face on which $x = x_2$; it can be represented by

$$\mathbf{X} = x_2\mathbf{i} + y\mathbf{j} + z\mathbf{k}$$

for (y, z) in

$$D = \{(y, z) \mid y_1 \leq y \leq y_2, \quad z_1 \leq z \leq z_2\}.$$

Clearly $\mathbf{n}_1 = \mathbf{i}$ and therefore

$$(16) \qquad \int_{S_1} \mathbf{F} \cdot \mathbf{n} \, dS = \int_D F_1(x_2, y, z) \, dy \, dz.$$

Similarly, the face S_2 on which $x = x_1$ can be represented by

$$\mathbf{X} = x_1\mathbf{i} + y\mathbf{j} + z\mathbf{k}$$

for (y, z) in D. Here $\mathbf{n}_2 = -\mathbf{i}$ and

$$(17) \qquad \int_{S_2} \mathbf{F} \cdot \mathbf{n} \, dS = -\int_D F_1(x_1, y, z) \, dy \, dz.$$

Now suppose $\mathbf{F}$ is defined and has continuous first partial derivatives on some open set in R^3 which contains V; then (16) and (17) imply that

$$(18) \quad \int_{S_1} \mathbf{F} \cdot \mathbf{n}\, dS + \int_{S_2} \mathbf{F} \cdot \mathbf{n}\, dS = \int_{D} (F_1(x_2, y, z) - F_1(x_1, y, z))\, dy\, dz$$

$$= \int_{D} \left(\int_{x_1}^{x_2} \frac{\partial F_1}{\partial x} (x, y, z)\, dx \right) dy\, dz$$

$$= \int_{V} \frac{\partial F_1}{\partial x}\, dx\, dy\, dz.$$

If S_3 and S_4 are the faces for which $y = y_2$ and $y = y_1$, respectively, a similar argument yields

$$(19) \qquad \int_{S_3} \mathbf{F} \cdot \mathbf{n}\, dS + \int_{S_4} \mathbf{F} \cdot \mathbf{n}\, dS = \int_{V} \frac{\partial F_2}{\partial y}\, dx\, dy\, dz;$$

also,

$$(20) \qquad \int_{S_5} \mathbf{F} \cdot \mathbf{n}\, dS + \int_{S_6} \mathbf{F} \cdot \mathbf{n}\, dS = \int_{V} \frac{\partial F_3}{\partial z}\, dx\, dy\, dz,$$

where S_5 and S_6 are the faces for which $z = z_2$ and $z = z_1$, respectively. Adding (18), (19), and (20) yields

$$\int_{S} \mathbf{F} \cdot \mathbf{n}\, dS = \int_{V} \left(\frac{\partial F_1}{\partial x} + \frac{\partial F_2}{\partial y} + \frac{\partial F_3}{\partial z} \right) dV = \int_{V} \operatorname{div} \mathbf{F}\, dV.$$

This is a remarkable relationship. We shall explore its consequences in the next section.

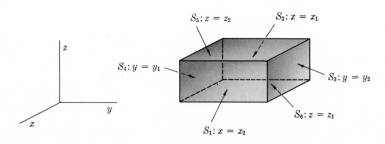

$S_5: z = z_2$ $S_2: x = x_1$

$S_4: y = y_1$

$S_3: y = y_2$

$S_6: z = z_1$

$S_1: x = x_2$

FIGURE 7.10

EXERCISES 4.7

In Exercises 1 through 14, compute $\int_S \mathbf{F} \cdot \mathbf{n}\, dS$ for the given $\mathbf{F}$ and S.

1. $\mathbf{F}(\mathbf{X}) = y\mathbf{i} - x\mathbf{j} + xy^2\mathbf{k}$; the surface of Exercise 1, Section 4.6; $\mathbf{n}(\mathbf{X})$ has a negative $\mathbf{k}$ component.

2. $\mathbf{F}(\mathbf{X}) = y\mathbf{i} - x\mathbf{j} + \mathbf{k}$; the surface of Exercise 2, Section 4.6, with the orientation induced by the given parametrization.

3. $\mathbf{F}(\mathbf{X}) = x^2\mathbf{i} + y^2\mathbf{j} + z^2\mathbf{k}$; the surface of Exercise 3, Section 4.6, with the orientation induced by the given parametrization.

4. $\mathbf{F}(\mathbf{X}) = \mathbf{X}$; the surface of Exercise 4, Section 4.6; $\mathbf{n}(\mathbf{X})$ pointing away from the origin.

5. $\mathbf{F}(\mathbf{X}) = (x^2 - y^2)\mathbf{i} + (y^2 - z^2)\mathbf{j} + (z^2 - x^2)\mathbf{k}$; the surface of Exercise 5, Section 4.6; $\mathbf{n}(\mathbf{X})$ has a positive $\mathbf{i}$ component.

6. $\mathbf{F}(\mathbf{X}) = (x - 1)\mathbf{i} + y\mathbf{j} + z\mathbf{k}$; the surface of Exercise 6, Section 4.6; $\mathbf{n}(\mathbf{X})$ points away from the axis of the cylinder.

7. $\mathbf{F}(\mathbf{X}) = x\mathbf{i}$; the surface of Exercise 7, Section 4.6; $\mathbf{n}(\mathbf{X})$ points away from the origin.

8. $\mathbf{F}(\mathbf{X}) = yz\mathbf{k}$; the surface of Exercise 8, Section 4.6; $\mathbf{n}(\mathbf{X})$ points toward the axis of the cylinder.

9. $\mathbf{F}(\mathbf{X}) = x\mathbf{i} + y^2\mathbf{j} + z\mathbf{k}$; the surface of Exercise 9, Section 4.6; $\mathbf{n}(\mathbf{X})$ points away from the origin. (Assume a, b, and $c > 0$.)

10. $\mathbf{F}(\mathbf{X}) = x\mathbf{i} + y\mathbf{j} + z^2\mathbf{k}$; the surface of Exercise 12, Section 4.6, positively oriented.

11. $\mathbf{F}(\mathbf{X}) = y\mathbf{i} - x\mathbf{j} + xyz\mathbf{k}$; the surface of Exercise 17, Section 4.6, positively oriented.

12. $\mathbf{F}(\mathbf{X}) = |\mathbf{X}|^2\mathbf{X}$; the surface consisting of

$$S_1 = \{\mathbf{X} \mid x^2 + y^2 + z^2 = 1, \quad x \geq 0, \quad y \geq 0, \quad z \geq 0\}$$

and

$$S_2 = \{\mathbf{X} \mid x^2 + y^2 \leq 1, \quad x \geq 0, \quad y \geq 0, \quad z = 0\},$$

oriented so that $n(\mathbf{X}) = \mathbf{k}$ on S_2.

13. $F(\mathbf{X}) = 2\mathbf{i} + 3\mathbf{j} + 4\mathbf{k}$; the surface consisting of

$$S_1 = \{\mathbf{X} \mid 0 \le z = 1 - x^2 - y^2\},$$

$$S_2 = \{\mathbf{X} \mid x^2 + y^2 = 1, \quad -1 \le z \le 0\}$$

and

$$S_3 = \{\mathbf{X} \mid x^2 + y^2 \le 1, \quad z = -1\},$$

oriented so that $\mathbf{n}(\mathbf{X}) = -\mathbf{k}$ on S_3.

14. $\mathbf{F}(\mathbf{X}) = x^2\mathbf{i} + y^2\mathbf{j} + z^2\mathbf{k}$; the surface of the parallelepiped

$$V = \{\mathbf{X} \mid 0 \le x \le 1, \quad 0 \le y \le 1, \quad 0 \le z \le 1\},$$

positively oriented.

THEORETICAL EXERCISES

T-1. Verify Eq. (4) for the Möbius strip.

T-2. Prove that if $\mathbf{n}$ is induced on S by $\mathbf{X} = \boldsymbol{\Phi}(u, v)$ then $-\mathbf{n}$ is induced by $\mathbf{X} = \boldsymbol{\Psi}(v, u)$, where $\boldsymbol{\Psi}(v, u) = \boldsymbol{\Phi}(u, v)$.

T-3. Prove Theorem 7.1. *Hint*: See the proof of Theorem 6.1, Section 4.6.

T-4. For the parametric representations of the cylindrical and hemispherical surfaces of Examples 7.1 and 7.2, verify that all points of $B(S)$ are images of $B(D)$. Show also that the images of some points of $B(D)$ are not in $B(S)$.

T-5. For the surfaces of Exercises 7.1 through 7.9, specify the positive directions of the boundaries with respect to the given orientations.

4.8 The Divergence Theorem; Green's and Stokes's Theorems

In Example 7.12 of Section 4.7 we saw that

(1)
$$\int_S \mathbf{F} \cdot \mathbf{n} \, dS = \int_V \operatorname{div} \mathbf{F} \, dV$$

if S is the surface of a parallelepiped V with faces parallel to the coordinate planes, and $\mathbf{F} = F_1\mathbf{i} + F_2\mathbf{j} + F_3\mathbf{k}$ is a continuously differentiable vector field on an open set containing V. In this section we shall extend this result and consider some of its consequences.

The next theorem, which is known as *Gauss's theorem*, or the *divergence theorem*, extends (1) to more general regions.

Theorem 8.1   Let V be a region in R^3 enclosed by an orientable surface S, and let $\mathbf{n}$ be the outward normal on S. Suppose also that $\mathbf{F}$ is continuously differentiable in an open set containing V and S. Then

$$(2) \qquad \int_S \mathbf{F} \cdot \mathbf{n} \, dS = \int_V \operatorname{div} \mathbf{F} \, dV.$$

Proof.   We shall prove this theorem under the additional assumption that V is convex (Exercise T-12, Section 2.1), and simply indicate how it can be extended to more complicated regions. The convexity of V implies that any line which intersects S meets S in exactly two points, or is tangent to S. Let R_{xy} be the projection of V into the xy-plane (Fig. 8.1), and for each (x, y) in R_{xy} let $z_1(x, y)$ and $z_2(x, y)$ be the least and greatest values of z such that $(x, y, z_1(x, y))$ and $(x, y, z_2(x, y))$ are on S. Now V can be described as

$$V = \{ (x, y, z) \mid (x, y) \text{ in } R_{xy}, \quad z_1(x, y) \leq z \leq z_2(x, y) \},$$

and S can be divided into three parts:
  (a)   an upper surface S_2, given by

$$(3) \qquad \mathbf{X} = \mathbf{X}_2(x, y) = x\mathbf{i} + y\mathbf{j} + z_2(x, y)\mathbf{k} \qquad ((x, y) \text{ in } R_{xy}) ;$$

  (b)   a lower surface S_1, given by

$$(4) \qquad \mathbf{X} = \mathbf{X}_1(x, y) = x\mathbf{i} + y\mathbf{j} + z_1(x, y)\mathbf{k} \qquad ((x, y) \text{ in } R_{xy}) ;$$

  (c)   vertical sections on which the normal is parallel to the xy-plane.
  It may happen that part (c) contains only boundary curves common to S_1 and S_2, as in Fig. 8.1; however, it may also contain sections of S with nonzero area, such as the shaded section in Fig. 8.2.

We shall first show that

$$\int_S (F_3\mathbf{k}) \cdot \mathbf{n} \, dS = \int_V \frac{\partial F_3}{\partial z} \, dV.$$

First we observe that

$$(5) \qquad \int_S (F_3\mathbf{k}) \cdot \mathbf{n} \, dS = \int_{S_2} (F_3\mathbf{k}) \cdot \mathbf{n} \, dS + \int_{S_1} (F_3\mathbf{k}) \cdot \mathbf{n} \, dS$$

even if S has sections of type (c) with positive area, since on such a section $\mathbf{n}$ would be perpendicular to $\mathbf{k}$. Let us assume that z_1 and z_2 are continuously differentiable; from (3) and (4),

(6)
$$\mathbf{N}_2 = -\frac{\partial z_2}{\partial x}\mathbf{i} - \frac{\partial z_2}{\partial y}\mathbf{j} + \mathbf{k}$$

and

(7)
$$\mathbf{N}_1 = -\frac{\partial z_1}{\partial x}\mathbf{i} - \frac{\partial z_1}{\partial y}\mathbf{j} + \mathbf{k}.$$

Since (6) induces the positive orientation on S_2 and (7) induces the negative orientation on S_1, it follows from (5) that

$$\int_S (F_3\mathbf{k}) \cdot \mathbf{n}\, dS = \int_{R_{xy}} F_3(x, y, z_2(x, y))\mathbf{k} \cdot \mathbf{N}_2(x, y)\, dx\, dy$$

$$- \int_{R_{xy}} F_3(x, y, z_1(x, y))\mathbf{k} \cdot \mathbf{N}_1(x, y)\, dx\, dy$$

$$= \int_{R_{xy}} F_3(x, y, z_2(x, y))\, dx\, dy - \int_{R_{xy}} F_3(x, y, z_1(x, y))\, dx\, dy.$$

$$= \int_{R_{xy}} \left(\int_{z_1(x,y)}^{z_2(x,y)} \frac{\partial F_3}{\partial z}(x, y, z)\, dz \right) dx\, dy.$$

Therefore

(8)
$$\int_S (F_3\mathbf{k}) \cdot \mathbf{n}\, dS = \int_V \frac{\partial F}{\partial z}\, dV.$$

A similar argument with x, y, z, and F_3 replaced by y, z, x, and F_1 yields

(9)
$$\int_S (F_1\mathbf{i}) \cdot \mathbf{n}\, dS = \int_V \frac{\partial F_1}{\partial x}\, dV.$$

Another permutation of the variables yields

(10)
$$\int_S (F_2\mathbf{j}) \cdot \mathbf{n}\, dS = \int_V \frac{\partial F_2}{\partial y}\, dV$$

(Exercise T-1), and (2) can be obtained by adding (8), (9), and (10).

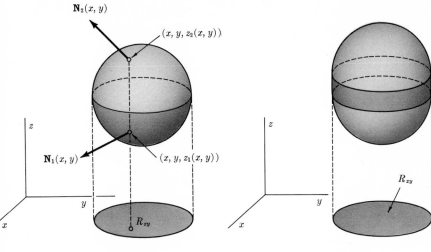

$\mathbf{N}_2(x, y)$

$(x, y, z_2(x, y))$

z

$\mathbf{N}_1(x, y)$

$(x, y, z_1(x, y))$

y

x

R_{xy}

FIGURE 8.1

z

y

x

R_{xy}

FIGURE 8.2

Now (2) can be extended to any region which can be broken down into convex subregions, and also to regions bounded by more than one surface, such as the punctured ball

(11) $$V = \{\mathbf{X} \mid 0 < \rho_1 < |\mathbf{X}| < \rho_2\}.$$

To see how this is accomplished for the punctured ball, we write

$$\int_V \operatorname{div} \mathbf{F} \, dV = \sum_{r=1}^{8} \int_{V_r} \operatorname{div} \mathbf{F} \, dV$$

where V_r is the part of V in the rth octant. Since V_r is convex,

(12) $$\int_{V_r} \operatorname{div} \mathbf{F} \, dV = \int_{\sigma_r} \mathbf{F} \cdot \mathbf{n} \, dS,$$

where σ_r comprises one spherical and three planar sections (Fig. 8.3). Every planar section appears twice among $\sigma_1, \ldots, \sigma_8$, with opposite signs; hence adding (12) for $r = 1, \ldots, 8$ yields

(13) $$\int_V \operatorname{div} \mathbf{F} \, dV = \int_{S_2} \mathbf{F} \cdot \mathbf{n} \, dS - \int_{S_1} \mathbf{F} \cdot \mathbf{n} \, dS,$$

where S_1 and S_2 are the positively oriented spheres

$$S_i = \{\mathbf{X} \mid |\mathbf{X}| = \rho_i\} \qquad (i = 1, 2).$$

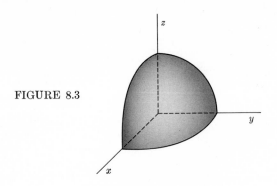

FIGURE 8.3

Example 8.1 Let V be the tetrahedron bounded by the coordinate planes and the plane $x + y + z = 1$ (Fig. 8.4), and let

$$\mathbf{F}(\mathbf{X}) = 3x^2\mathbf{i} + xy\mathbf{j} + z\mathbf{k}.$$

Then

$$(\operatorname{div}\mathbf{F})(\mathbf{X}) = 6x + x + 1 = 7x + 1$$

and

$$\int_V \operatorname{div}\mathbf{F}\, dV = \int_0^1 (7x + 1)\, dx \int_0^{1-x} dy \int_0^{1-x-y} dz$$

$$= \int_0^1 (7x + 1)\, dx \int_0^{1-x} (1 - x - y)\, dy$$

$$= \frac{1}{2} \int_0^1 (7x + 1)(x - 1)^2\, dx = \frac{1}{8}.$$

Let S_1, S_2, and S_3 be the faces of V in the xy-, yz- and zx-planes, respectively. Then

$$\int_{S_1} \mathbf{F} \cdot \mathbf{n}\, dS = \int_{S_1} (3x^2\mathbf{i} + xy\mathbf{j} + 0\mathbf{k}) \cdot (-\mathbf{k})\, dS = 0,$$

$$\int_{S_2} \mathbf{F} \cdot \mathbf{n}\, dS = \int_{S_2} (0\mathbf{i} + 0\mathbf{j} + z\mathbf{k}) \cdot (-\mathbf{i})\, dS = 0,$$

and

$$\int_{S_3} \mathbf{F} \cdot \mathbf{n}\, dS = \int_{S_3} (3x^2\mathbf{i} + 0\mathbf{j} + z\mathbf{k}) \cdot (-\mathbf{j})\, dS = 0.$$

The remaining face S_4 can be represented by

$$\mathbf{X} = \mathbf{X}(x, y) = x\mathbf{i} + y\mathbf{j} + (1 - x - y)\mathbf{k} \qquad (0 \le y \le 1 - x, \quad 0 \le x \le 1);$$

hence

$$\mathbf{F}(\mathbf{X}(x, y)) = 3x^2\mathbf{i} + xy\mathbf{j} + (1 - x - y)\mathbf{k}$$

and

$$\mathbf{N}(x, y) = \mathbf{i} + \mathbf{j} + \mathbf{k}.$$

Therefore

$$\int_{S_4} \mathbf{F} \cdot \mathbf{n} \, dS = \int_0^1 dx \int_0^{1-x} (3x^2 + xy + 1 - x - y) \, dy$$

$$= \int_0^1 \left[\frac{(x-1)^3}{2} - 3x^2(x-1) \right] dx = \frac{1}{8}$$

and

$$\int_V \operatorname{div} \mathbf{F} \, dV = \int_S \mathbf{F} \cdot \mathbf{n} \, dS = \frac{1}{8},$$

as predicted by the divergence theorem.

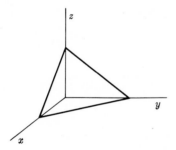

FIGURE 8.4

Example 8.2   Let $\mathbf{F}$ be the inverse square law force of gravity:

$$\mathbf{F}(\mathbf{X}) = -Gm_0 \frac{\mathbf{X}}{|\mathbf{X}|^3};$$

then div $\mathbf{F} = 0$ (Exercise T-9, Section 3.4) and therefore

$$\int_S \mathbf{F} \cdot \mathbf{n} \, dS = 0$$

if S is any closed surface. In particular, if V is the punctured ball (11), it follows from (13) that the integrals of $\mathbf{F}$ over any two concentric spheres about the origin are equal. To verify this, let S_ρ be the sphere defined by $|\mathbf{X}| = \rho$. The outward normal to S is

$$\mathbf{n}(\mathbf{X}) = \frac{\mathbf{X}}{|\mathbf{X}|},$$

and therefore

$$\int_{S_\rho} \mathbf{F} \cdot \mathbf{n} \, dS = -Gm_0 \int_{S_\rho} \left(\frac{\mathbf{X}}{|\mathbf{X}|^3} \right) \cdot \left(\frac{\mathbf{X}}{|\mathbf{X}|} \right) dS$$

$$= -Gm_0 \int_{S_\rho} \frac{|\mathbf{X}|^2}{|\mathbf{X}|^4} \, dS.$$

$$= -\frac{Gm_0}{\rho^2} \int_{S_\rho} dS = -\frac{Gm_0}{\rho^2} (4\pi\rho^2) = -4\pi Gm_0.$$

Green's Theorem in R^3

Let u and v be continuously differentiable scalar valued function in a region T in R^3 and let v_{xx}, v_{yy}, and v_{zz} exist and be continuous there. From Theorem 8.1 with

$$\mathbf{F} = u \, \boldsymbol{\nabla} v = u \frac{\partial v}{\partial x} \mathbf{i} + u \frac{\partial v}{\partial y} \mathbf{j} + u \frac{\partial v}{\partial z} \mathbf{k},$$

it follows that

(14)
$$\int_S (u \, \boldsymbol{\nabla} v) \cdot \mathbf{n} \, dS = \int_V \operatorname{div}(u \, \boldsymbol{\nabla} v) \, dV,$$

for any volume V in T with orientable surface S. Now

(15)
$$\boldsymbol{\nabla} v \cdot \mathbf{n} = \frac{\partial v}{\partial \mathbf{n}},$$

where $\partial v / \partial \mathbf{n}$ stands for the directional derivative in the direction of the outward normal to S (Section 2.2). Direct calculation yields

$$\operatorname{div}(u \, \boldsymbol{\nabla} v) = \boldsymbol{\nabla} u \cdot \boldsymbol{\nabla} v + u \, \boldsymbol{\nabla}^2 v$$

(Exercise T-12, Section 3.4), where $\boldsymbol{\nabla}^2 v$ is the Laplacian of v, defined in Section 3.4 as

(16)
$$\boldsymbol{\nabla}^2 v = v_{xx} + v_{yy} + v_{zz}.$$

Using (15) and (16) we can rewrite (14) as

(17)
$$\int_S u \frac{\partial v}{\partial \mathbf{n}} \, dS = \int_V (\boldsymbol{\nabla} u \cdot \boldsymbol{\nabla} v + u \, \boldsymbol{\nabla}^2 v) \, dV.$$

This relation is known as *Green's first theorem*.

If we also assume that u_{xx}, u_{yy}, and u_{zz} are continuous in T then we can interchange u and v in (17) to obtain

$$\int_S v \frac{\partial u}{\partial \mathbf{n}} \, dS = \int_V (\boldsymbol{\nabla} v \cdot \boldsymbol{\nabla} u + v \nabla^2 u) \, dV;$$

subtracting this from (17) yields.

$$\int_S \left(u \frac{\partial v}{\partial \mathbf{n}} - v \frac{\partial u}{\partial \mathbf{n}} \right) dS = \int_V (u \, \nabla^2 v - v \, \nabla^2 u) \, dV,$$

which is known as *Green's second theorem.*

Green's theorems have important applications in the theory of *harmonic functions;* that is, functions which satisfy Laplace's equation,

$$\nabla^2 u = 0.$$

(See Section 3.4).

Stokes's Theorem

Suppose $\mathbf{F}$ has continuous second partial derivatives in a region T. Since $\operatorname{div}(\mathbf{curl\ F}) = 0$ (Example 11, Section 3.4), it follows from the divergence theorem that

(18)
$$\int_S (\mathbf{curl\ F}) \cdot \mathbf{n} \, dS = 0,$$

if S is a closed orientable surface which encloses a volume contained in T.

Now suppose S is made up of two nonintersecting surfaces S_1 and S_2 with common boundary C. (For example, let S be the sphere $x^2 + y^2 + z^2 = 1$, C the circle $x^2 + y^2 = 1$, and S_1 and S_2 the upper and lower halves of S, as in Fig. 8.5.) Suppose S_1 and S_2 are oriented in opposite senses, so that C is positively directed with respect to both (Fig. 8.5). Then from (18),

$$\int_{S_2} (\mathbf{curl\ F}) \cdot \mathbf{n}_1 \, dS - \int_{S_1} (\mathbf{curl\ F}) \cdot \mathbf{n}_2 \, dS = 0;$$

that is, the integrals of **curl F** over all orientable surfaces in T with a given boundary C are equal. It is therefore reasonable that the common value of these integrals can be computed from a knowledge of $\mathbf{F}$ on C alone. The following theorem, which we shall not prove, tells how this can be done.

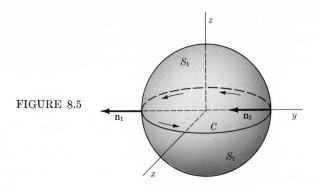

FIGURE 8.5

Theorem 8.2 (Stokes's Theorem)   Suppose S is an orientable surface and $\mathbf{F}$ is a continuously differentiable vector field on an open set containing S, and let $B(S)$ be the boundary of S. Then

$$(19) \qquad \int_S (\mathbf{curl\ F}) \cdot \mathbf{n}\, dS = \int_{B(S)} \mathbf{F} \cdot d\mathbf{X},$$

provided $B(S)$ is positively directed with respect to $\mathbf{n}$. If $B(S)$ consists of piecewise smooth curves $C_1, \ldots, C_n$, the right side of (19) is to be interpreted as

$$\int_{B(S)} \mathbf{F} \cdot d\mathbf{X} = \int_{C_1} \mathbf{F} \cdot d\mathbf{X} + \cdots + \int_{C_n} \mathbf{F} \cdot d\mathbf{X},$$

where $C_1, \ldots, C_n$ are positively directed with respect to S.

Example 8.3   Let S be the part of the sphere $x^2 + y^2 + z^2 = 1$ bounded by $z = 0$ and $z = \frac{1}{2}$, with the orientation shown in Figure 8.6, and let

$$F = z\mathbf{i} + x\mathbf{j} + y\mathbf{k}.$$

According to Stokes's theorem

$$(20) \qquad \int_S (\mathbf{curl\ F}) \cdot \mathbf{n}\, dS = \int_{C_1} \mathbf{F} \cdot d\mathbf{X} - \int_{C_2} \mathbf{F} \cdot d\mathbf{X},$$

where C_1 and C_2 are the directed circles shown in Fig. 8.6. Let us verify this.
We can represent S by

$$\mathbf{X} = \sin \phi \cos \theta\, \mathbf{i} + \sin \phi \sin \theta\, \mathbf{j} + \cos \phi \mathbf{k}$$

for (ϕ, θ) in

$$D = \left\{ (\phi, \theta) \ \middle|\ \frac{\pi}{3} \le \phi \le \frac{\pi}{2}, \quad 0 \le \theta \le 2\pi \right\}.$$

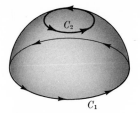

FIGURE 8.6

Then

(21) $\displaystyle\int_S (\text{curl }\mathbf{F})\cdot\mathbf{n}\,dS$

$\displaystyle = \int_D (\mathbf{i}+\mathbf{j}+\mathbf{k})\cdot(\sin\phi\,\mathbf{X}(\phi,\theta))\,d\phi\,d\theta$

$\displaystyle = \int_{\pi/3}^{\pi/2} d\phi \int_0^{2\pi} (\sin^2\phi\cos\theta + \sin^2\phi\sin\theta + \sin\phi\cos\phi)\,d\theta$

$\displaystyle = 2\pi \int_{\pi/3}^{\pi/2} \sin\phi\cos\phi\,d\phi = \frac{\pi}{4}\,.$

The circle C_1 can be represented by

$$\mathbf{X} = \cos t\,\mathbf{i} + \sin t\,\mathbf{j} \qquad (0 \le t \le 2\pi),$$

so that on C_1

$$\mathbf{F}(\mathbf{X}(t))\cdot\mathbf{X}'(t) = (0\mathbf{i} + \cos t\,\mathbf{j} + \sin t\,\mathbf{k})\cdot(-\sin t\,\mathbf{i} + \cos t\,\mathbf{j}) = \cos^2 t;$$

thus

(22) $$\int_{C_1} \mathbf{F}\cdot d\mathbf{X} = \int_0^{2\pi} \cos^2 t\,dt = \pi.$$

On C_2,

$$\mathbf{X} = \frac{\sqrt{3}}{2}\cos t\,\mathbf{i} + \frac{\sqrt{3}}{2}\sin t\,\mathbf{j} + \frac{1}{2}\mathbf{k}$$

and

$$\mathbf{F}(\mathbf{X}(t))\cdot\mathbf{X}'(t) = \left(\frac{1}{2}\mathbf{i} + \frac{\sqrt{3}}{2}\cos t\,\mathbf{j} + \frac{\sqrt{3}}{2}\sin t\,\mathbf{k}\right)\cdot\left(-\frac{\sqrt{3}}{2}\sin t\,\mathbf{i} + \frac{\sqrt{3}}{2}\cos t\,\mathbf{j}\right)$$

$$= -\frac{\sqrt{3}}{4}\sin t + \frac{\sqrt{3}}{4}\cos^2 t;$$

hence

$$(23) \qquad \int_{C_2} \mathbf{F} \cdot d\mathbf{X} = \int_0^{2\pi} \left(-\frac{\sqrt{3}}{4} \sin t + \frac{3}{4} \cos^2 t \right) dt = \tfrac{3}{4}\pi.$$

Now (20) follows from (21), (22), and (23).

Green's Theorem in the Plane

We now obtain the following theorem, known as *Green's theorem in the plane*, as a special case of Stokes's theorem. (If one wished to prove Stokes's theorem, then it would be necessary to establish Green's theorem in the plane by other means, since the latter is an essential part of the proof of the former.)

Theorem 8.3   Let S be a region in the xy-plane bounded by piecewise smooth simple closed curves $C_1, \ldots, C_n$, no two of which intersect. Suppose further that P and Q are continuously differentiable functions in a region containing S and $C_1, \ldots, C_n$. Then

$$\int_S \left(\frac{\partial Q}{\partial x} - \frac{\partial P}{\partial y} \right) dx\, dy = \sum_{i=1}^n \int_{C_i} P\, dx + Q\, dy,$$

where each C_i is positively directed with respect to the orientation $\mathbf{n} = \mathbf{k}$ on S (Fig. 8.7); that is, the interior of S is to the left of C_i.

To obtain this result we take

$$\mathbf{F}(x, y, z) = P(x, y)\mathbf{i} + Q(x, y)\mathbf{j}$$

and parametrize S by

$$\mathbf{X} = x\mathbf{i} + y\mathbf{j} \qquad ((x, y) \text{ in } S);$$

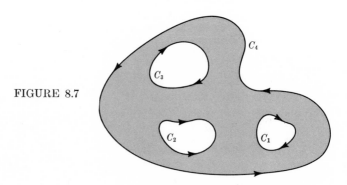

FIGURE 8.7

then $\mathbf{n}(\mathbf{X}) = \mathbf{k}$,

$$\operatorname{curl} \mathbf{F} = \left(\frac{\partial Q}{\partial x} - \frac{\partial P}{\partial y}\right) \mathbf{k},$$

and the result follows from Stokes's theorem.

Example 8.4 Let us verify Green's theorem in the plane for $P(x, y) = x^2 + y$ and $Q(x, y) = 2x + 2y$, where S is the region outside the circle $x^2 + y^2 = 1$ and inside the square with vertices at $(1, 1), (-1, 1), (-1, -1)$, and $(1, -1)$ (Fig. 8.8). In this case the double integral is simply the area of S:

(24) $$\int_S \left(\frac{\partial Q}{\partial x} - \frac{\partial P}{\partial y}\right) dx \, dy = \int_S (2 - 1) \, dx \, dy = 4 - \pi.$$

The boundary of S consists of five smooth segments, identified as $C_1, \ldots, C_5$ in Fig. 8.8. The integrals over these segments are as follows:

$$\int_{C_1} P \, dx + Q \, dy = -\int_0^{2\pi} [(\cos^2 t + \sin t)(-\sin t)$$

$$+ (2 \cos t + 2 \sin t)(\cos t)] \, dt = -\pi;$$

$$\int_{C_2} P \, dx + Q \, dy = -\int_{-1}^1 (x^2 + 1) \, dx = -\tfrac{8}{3};$$

$$\int_{C_3} P \, dx + Q \, dy = -\int_{-1}^1 (-2 + 2y) \, dy = 4;$$

$$\int_{C_4} P \, dx + Q \, dy = -\int_{-1}^1 (x^2 - 1) \, dx = -\tfrac{4}{3},$$

and

$$\int_{C_5} P \, dx + Q \, dy = \int_{-1}^1 (2 + 2y) \, dy = 4.$$

Thus

$$\int_C P \, dx + Q \, dy = \sum_{i=1}^5 \int_{C_i} P \, dx + Q \, dy = 4 - \pi,$$

which agrees with (24).

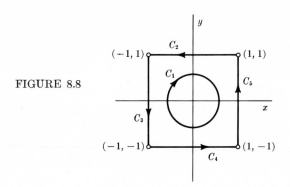

FIGURE 8.8

EXERCISES 4.8

In Exercises 1 through 11 verify the divergence theorem for the given vector function $\mathbf{F}$ and volume V.

1. $\mathbf{F}(\mathbf{X}) = x\mathbf{i} + y\mathbf{j} + z\mathbf{k};$ $V = \{\mathbf{X} \mid 0 \le z \le 4 - x^2 - y^2\}.$

2. $\mathbf{F}(\mathbf{X}) = x^2\mathbf{i} + y^2\mathbf{j} + z^2\mathbf{k};$ $V = \{\mathbf{X} \mid x^2 + y^2 \le 4, \ \ 0 \le z \le x - 2\}.$

3. $\mathbf{F}(\mathbf{X}) = z\mathbf{i} + z\mathbf{j};$ $V = \{\mathbf{X} \mid \sqrt{x^2 + y^2} \le z \le 2\}.$

4. $\mathbf{F}(\mathbf{X}) = (x + y)\mathbf{i} + (y - z)\mathbf{j} + (x + y + z)\mathbf{k};$ the tetrahedron bounded by the plane $2x + y + z = 2$ and the coordinate planes.

5. $\mathbf{F}(\mathbf{X}) = x\mathbf{i} + y\mathbf{j} + z\mathbf{k};$ $V = \{\mathbf{X} \mid 0 \le |\mathbf{X}| \le a\}.$

6. $\mathbf{F}(\mathbf{X}) = x\mathbf{i} + y^2\mathbf{j} + z\mathbf{k};$ $V = \{\mathbf{X} \mid 2 + x^2 + y^2 \le z \le 6\}.$

7. $\mathbf{F}(\mathbf{X}) = x^2\mathbf{i} + y^2\mathbf{j} + z^2\mathbf{k};$ $V = \{\mathbf{X} \mid x_1 \le x \le x_2, \ \ y_1 \le y \le y_2,$

   $z_1 \le z \le z_2\}.$

8. $\mathbf{F}(\mathbf{X}) = \dfrac{x\mathbf{i} + y\mathbf{j} + z\mathbf{k}}{\sqrt{x^2 + y^2 + z^2}};$ $V = \{\mathbf{X} \mid 3 \le |\mathbf{X}| \le 4\}.$

9. $\mathbf{F}(\mathbf{X}) = x^2\mathbf{i} + y\mathbf{j} + z^2\mathbf{k};$ the solid bounded by the cylinder $x^2 + y^2 = 4$ and the planes $z = 1$ and $z = 3$.

10. $\mathbf{F}(\mathbf{X}) = 2x\mathbf{i} + 3y\mathbf{j} + z\mathbf{k};$ the solid bounded by the cylinder $y^2 + z^2 = 1$ and the planes $x = 0$ and $2x + y + z = 2$.

11. $F(\mathbf{X}) = x\mathbf{i} + y\mathbf{j} + z\mathbf{k};$ the solid bounded by the paraboloid $z = x^2 + y^2$, the cylinder $x^2 + y^2 = 9$, and the plane $z = 0$.

In Exercises 12 through 17 verify Stokes's theorem for the given function $\mathbf{F}$ and surface S.

12. $\mathbf{F}(\mathbf{X}) = z\mathbf{i} + x\mathbf{j} + y\mathbf{k}$; $S = \{\mathbf{X} \mid 0 \leq z = 4 - x^2 - y^2\}$.

13. $\mathbf{F}(\mathbf{X}) = (x^2 + y)\mathbf{i} + yz\mathbf{j} + (x - z^2)\mathbf{k}$;

   $S = \{\mathbf{X} \mid 2x + y + 2z = 2, \quad x \geq 0, \quad y \geq 0, \quad z \geq 0\}$.

14. $\mathbf{F}(\mathbf{X}) = x\mathbf{i} + z\mathbf{j} - y\mathbf{k}$; $S = \{\mathbf{X} \mid |\mathbf{X}| = 2, \quad y \geq 0\}$.

15. $\mathbf{F}(\mathbf{X}) = x\mathbf{i} + y\mathbf{j}$;   the part of the paraboloid $z = x^2 + y^2$ inside the cylinder $x^2 + y^2 = 4$.

16. $\mathbf{F}(\mathbf{X}) = y\mathbf{i} + x\mathbf{j} + z\mathbf{k}$;   the part of the cylinder $x^2 + y^2 = 4$ between the planes $z = 1$ and $z = x + 4$.

17. $\mathbf{F}(\mathbf{X}) = (y + x)\mathbf{i} + (x + z)\mathbf{j} + z^2\mathbf{k}$;   the part of the cone $z^2 = x^2 + y^2$ between the planes $z = 0$ and $z = 1$.

In Exercises 18 through 21 compute $\displaystyle\int_{\partial S} \operatorname{curl} \mathbf{F} \cdot \mathbf{n} \, dS$ by using Stokes's theorem. Orient S so that $\mathbf{n}$ has a nonnegative $\mathbf{k}$ component.

18. $\mathbf{F}(\mathbf{X}) = y\mathbf{i} + z\mathbf{j} + x\mathbf{k}$; S is the triangle with vertices $(1, 0, 0)$, $(0, 1, 0)$, $(0, 0, 1)$.

19. $\mathbf{F}(\mathbf{X}) = (x + y)\mathbf{i} + (y - z)\mathbf{j} + (x + y + z)\mathbf{k}$; S is the hemisphere $x^2 + y^2 + z^2 = a^2, z \geq 0$.

20. $\mathbf{F}(\mathbf{X}) = x\mathbf{i} + z\mathbf{j} + y\mathbf{k}$; S is the paraboloid $z = 4 - x^2 - y^2, \quad z \geq 0$.

21. $\mathbf{F}(\mathbf{X}) = x\mathbf{i} + y\mathbf{j}$; S is the surface $z = 1 - \sqrt{x^2 + y^2}, z \geq 0$.

In Exercises 22 through 29 verify Green's theorem in the plane for the given functions P and Q and the given region S.

22. $P(x, y) = x^2$, $Q(x, y) = y^2$; $S = \{(x, y) \mid 0 \leq x \leq 1, \quad 1 \leq y \leq 3\}$.

23. $P(x, y) = x$, $Q(x, y) = y$; $S = \{(x, y) \mid 1 \leq x^2 + y^2 \leq 4\}$.

24. $P(x, y) = 2y$, $Q(x, y) = -3x$;   the triangle bounded by the line $2x + y = 2$ and the coordinate axes.

25. $P(x, y) = y^2$, $Q(x, y) = 2x$;   the square with vertices $(1, 0)$, $(0, 1)$, $(-1, 0)$, and $(0, -1)$.

26. $P(x, y) = 2x + y, Q(x, y) = 3y + x$;   the region bounded by $y = x^2$ and $y = x$.

27. $P(x, y) = y$, $Q(x, y) = 2x$; the region between the circles $x^2 + y^2 = 4$ and $x^2 + y^2 = 9$.

28. $P(x, y) = -y$, $Q(x, y) = 2x$; $S = \{(x, y) \mid 0 \leq y \leq \sin x,\ 0 \leq x \leq \pi\}$.

29. $P(x, y) = 2y$, $Q(x, y) = 3x$; $S = \{(x, y) \mid |x| \leq y \leq 2\}$.

In Exercises 30 through 34 compute the line integral of **F** along C using Green's theorem. Orient C in the positive sense.

30. $\mathbf{F(X)} = (x^2 y + y^3 - \log x)\mathbf{i} + (xy^2 + x^2 y + x^3 + \log y^2)\mathbf{j}$; C is the square with sides $x = 1$, $x = 2$, $y = 1$, $y = 2$.

31. $\mathbf{F(X)} = (\sin y - x^2 y)\mathbf{i} + (x \cos y + xy^2)\mathbf{j}$ over the circle $x^2 + y^2 = 1$.

32. $\mathbf{F(X)} = (e^x \sin y - xy^2)\mathbf{i} + (e^x \cos y + x^2 y)\mathbf{j}$ over the ellipse $x^2/4 + y^2/9 = 1$.

33. $\mathbf{F(X)} = (x^2 y + y^3 x)\mathbf{i} + (y^4 x^2 - x^3)\mathbf{j}$ over the boundary of the rectangle $1 \leq x \leq 3$, $0 \leq y \leq 1$.

34. $\mathbf{F(X)} = (3x^2 y^2 + \frac{1}{4}y^4 - 3xy)\mathbf{i} + (xy^3 + \frac{1}{3}x^3)\mathbf{j}$ over the boundary of the triangle with vertices $(0, 1)$, $(1, 0)$, and $(-1, 0)$.

THEORETICAL EXERCISES

T-1. Verify Eq. (10).

T-2. Let a region V be bounded by a closed orientable surface S, and suppose u is harmonic in an open set containing V and S. Show that

$$\int_S \frac{\partial u}{\partial \mathbf{n}}\, dS = 0$$

and

$$\int \frac{\partial (u^2)}{\partial \mathbf{n}}\, dS > 0,$$

where the inequality is strict if u is not constant.

T-3. Let S and V be as in Exercise T-2, and suppose that u and v are both harmonic in an open set containing S and V. Show that

$$\int_S u \frac{\partial v}{\partial \mathbf{n}}\, dS = \int_S v \frac{\partial u}{\partial \mathbf{n}}\, dS.$$

Answers to Selected Problems

Chapter 1

Section 1.1, page 19

1. (a) $z = 3, w = 4$ (b) $z = 5, w = 8$ (c) $z = 4, w = 5$

3. (a) $\begin{bmatrix} 2 \\ -2 \end{bmatrix}, \begin{bmatrix} -4 \\ 6 \end{bmatrix}, \begin{bmatrix} 3 \\ -6 \end{bmatrix}, \begin{bmatrix} -11 \\ 16 \end{bmatrix}$ (b) $\begin{bmatrix} 2 \\ 4 \\ 2 \end{bmatrix}, \begin{bmatrix} 2 \\ 2 \\ -2 \end{bmatrix},$

$\begin{bmatrix} -6 \\ -9 \\ 0 \end{bmatrix}, \begin{bmatrix} 4 \\ 3 \\ -6 \end{bmatrix}$ (c) $\begin{bmatrix} -5 \\ -3 \\ -2 \\ 5 \end{bmatrix}, \begin{bmatrix} 3 \\ 3 \\ -2 \\ 3 \end{bmatrix}, \begin{bmatrix} 3 \\ 0 \\ 6 \\ -12 \end{bmatrix}, \begin{bmatrix} 10 \\ 9 \\ -4 \\ 5 \end{bmatrix}$

7. (b) and (d)

9. (b) and (c)

13. (a) and (c)

15. (a)

17. (a) $\frac{9}{5}X_1 + \frac{1}{5}X_2$ (b) $X_1 + 2X_2$ (c) $5X_1 - 7X_2$

    (d) $2X_1 + 3X_2$

19. (a) 2   (b) 1   (c) 2

21. (a) $\sqrt{14}$ (b) 0   (c) $\sqrt{2}$ (d) $\sqrt{14}$

23. (a) -20 (b) 0   (c) -8 (d) 4

25. (a) $\dfrac{-20}{\sqrt{14}\,\sqrt{29}}$ (b) 0   (c) $-\dfrac{8}{\sqrt{17}\,\sqrt{29}}$

    (d) $\dfrac{4}{\sqrt{3}\,\sqrt{14}}$

Section 1.2, page 38

1. (a), (c), (d)

3. (a) $\dfrac{x_1 - 2}{0} = \dfrac{x_2 + 1}{-1} = \dfrac{x_3 - 3}{4} = \dfrac{x_4 - 3}{-3}$

    (b) $\dfrac{x - 2}{1} = \dfrac{y - 1}{-1} = \dfrac{z}{2}$

5. $X = \begin{bmatrix} 1 \\ 2 \\ -1 \end{bmatrix} + t \begin{bmatrix} 4 \\ -2 \\ 5 \end{bmatrix}$

7. $X = \begin{bmatrix} 2 \\ 1 \\ 0 \\ 3 \end{bmatrix} + t \begin{bmatrix} 3 \\ 2 \\ -2 \\ 3 \end{bmatrix}$

9. $X = \begin{bmatrix} 4 \\ -2 \\ 5 \\ 2 \end{bmatrix} + t \begin{bmatrix} 0 \\ 1 \\ -2 \\ 4 \end{bmatrix}$

11. yes

13. $(-12, 16, -26)$

17. no

21. yes

23. $X = \begin{bmatrix} 3 \\ 5 \\ 3 \end{bmatrix} + t \begin{bmatrix} 19 \\ 73 \\ 8 \end{bmatrix}$

25. $\dfrac{-1}{\sqrt{14}\ \sqrt{39}}$

27. $\begin{bmatrix} 6 \\ -4 \\ 3 \end{bmatrix}$

29. $2x_1 - x_2 + 4x_3 - 5x_4 = 16.$

31. $3y - 6z = -15$

33. $X = \begin{bmatrix} 1 \\ 3 \\ -2 \\ 4 \end{bmatrix} + t \begin{bmatrix} 2 \\ 3 \\ 3 \\ -2 \end{bmatrix}$

Section 1.3, page 63

1. (a), (c)

3. (a) $\begin{bmatrix} 2 \\ -3 \\ 0 \end{bmatrix}$ (b) $\begin{bmatrix} 1 \\ \dfrac{-2}{3} \\ \dfrac{4}{7} \end{bmatrix}$ (c) $\begin{bmatrix} 0 \\ 2 \\ 0 \end{bmatrix}$ (d) $\begin{bmatrix} 3 \\ -1 \\ 1 \end{bmatrix}$

5. (a) $4\mathbf{i} + 8\mathbf{j} + 4\mathbf{k}$ (b) $20\mathbf{j} + 8\mathbf{k}$ (c) $-8\mathbf{i} + 11\mathbf{j} + \mathbf{k}$
   (d) $-5\mathbf{i} - 10\mathbf{j} - 6\mathbf{k}$

9. $3/\sqrt{2}$

11. $\frac{1}{2} \sqrt{362}$

13. $x - 5y - 2z = 15$

15. $X = \begin{bmatrix} 1 \\ -1 \\ 3 \end{bmatrix} + t \begin{bmatrix} 1 \\ -23 \\ -8 \end{bmatrix}$

19. $X = \begin{bmatrix} \dfrac{20}{3} \\ \dfrac{-5}{3} \\ 0 \end{bmatrix} + t \begin{bmatrix} -2 \\ 5 \\ 3 \end{bmatrix}$

21.   distance $= 22/\sqrt{13}$; $\bar{X} = \left(\dfrac{25}{13}, -5, \dfrac{47}{13}\right)$

23. $3/\sqrt{5}$

25. $4/\sqrt{14}$

27. $4/\sqrt{6}$

29. $29x - 13y + 2z = 152$

31. $2y - z = 7$

33. $2x - 21y + 3z = -49$

35. $a[2(x - 2) + 3(y + 2)] + b[z - 5] = 0$; a and b arbitrary but not both zero

37. $x + 4y + 3z = 2$

39.   26

41.   (a) $-25\mathbf{i} + 5\mathbf{j} - 20\mathbf{k}$ (b) $-2\mathbf{i} - 4\mathbf{j} - 2\mathbf{k}$

Section 1.4, page 83

1. $\mathbf{V}(t) = 3t^2\mathbf{i} + 2t\mathbf{j},$ $\mathbf{A}(t) = 6t\mathbf{i} + 2\mathbf{j},$

$$\cos \theta = \frac{9t^2 + 2}{\sqrt{9t^2 + 4}\,\sqrt{9t^2 + 1}}, \qquad |\,\mathbf{V}(t)\,| = t\sqrt{9t^2 + 4}\,,$$

$$\mathbf{T}(t) = \frac{3t\mathbf{i} + 2\mathbf{j}}{\sqrt{9t^2 + 4}}$$

3. $\mathbf{V}(t) = e^t\mathbf{i} - \sin t\mathbf{j} + 3t^2\mathbf{k},$ $\mathbf{A}(t) = e^t\mathbf{i} - \cos t\mathbf{j} + 6t\mathbf{k}$

$$\cos \theta = \frac{e^{2t} + \sin t \cos t + 18t^3}{\sqrt{e^{2t} + \sin^2 t + 9t^4}\,\sqrt{e^{2t} + \cos^2 t + 36t^2}}$$

$$|\,\mathbf{V}(t)\,| = \sqrt{e^{2t} + \sin^2 t + 9t^4},$$

$$\mathbf{T}(t) = \frac{e^t\mathbf{i} - \sin t\mathbf{j} + 3t^2\mathbf{k}}{\sqrt{e^{2t} + \sin^2 t + 9t^4}}$$

7. $\dfrac{2t + 2t^3}{\sqrt{t^4 + 2t^2 + 2}}$; $\dfrac{A\mathbf{i} + B\mathbf{j} + C\mathbf{k}}{(t^4 + 2t^2 + 2)^{3/2}}$,

where

$$A = -t^4 - 2t^3 - 2t + 2, \qquad B = -t^4 - 2t^3 + 2t + 2,$$

$$C = 2t^3 + 4t^2$$

9. $e^t;$ $\cos t\,\mathbf{i} - \sin t\,\mathbf{k}$

11. $1;$ $\dfrac{1}{(t+1)^2}\left[\dfrac{1-t}{\sqrt{2t}}\,\mathbf{i} - \mathbf{j} - \mathbf{k}\right]$

13. $T\left(\dfrac{\pi}{3}\right) = \dfrac{1}{2}\mathbf{i} - \dfrac{\sqrt{3}}{2}\,\mathbf{j};$ $N\left(\dfrac{\pi}{3}\right) = \dfrac{-\sqrt{3}}{2}\,\mathbf{i} - \dfrac{1}{2}\,\mathbf{j};$

  $B\left(\dfrac{\pi}{3}\right) = -\mathbf{k},$ $z = -2$

15. $\mathbf{T}(0) = \dfrac{\mathbf{j} + \mathbf{k}}{\sqrt{2}},$ $\mathbf{N}(0) = \mathbf{i},$ $\mathbf{B}(0) = \dfrac{1}{\sqrt{2}}\,(\mathbf{j} - \mathbf{k});$ $y - z = 1$

17. $\frac{5}{2}$

19. $\kappa(t_0) = \frac{1}{2},$ $\rho(t_0) = 2,$ $\tau(t_0) = \dfrac{1}{\sqrt{2}},$ $\mathbf{A}_T(t_0) = \mathbf{0},$

  $\mathbf{A}_N(t_0) = -\frac{1}{2}\mathbf{i} - \dfrac{\sqrt{3}}{2}\,\mathbf{j}$

21. $\kappa(t_0) = 2\sqrt{2},$ $\rho(t_0) = \dfrac{1}{2\sqrt{2}},$ $\tau(t_0) = 0,$ $\mathbf{A}_T(t_0) = \mathbf{0},$

  $\mathbf{A}_N(t_0) = 2(\mathbf{i} + \mathbf{j})$

23. $\mathbf{A}_T(t) = \mathbf{0},$ $\mathbf{A}_N(t) = \dfrac{V_0^2}{r^2}\,\mathbf{X}(t)$

Section 1.5, page 94

3. both

5. both

7. neither

9. $\Phi'(t) = \begin{bmatrix} 2 \\ 4 \\ 4 \end{bmatrix}$

15. possible answers: $\Psi_1(s) = \begin{bmatrix} -\sin 2s \\ \cos 2s \\ \sin 2s \end{bmatrix}$ $\left(0 \leq s \leq \frac{\pi}{4}\right);$

$\Psi_2(s) = \begin{bmatrix} -\sin s^2 \\ \cos s^2 \\ \sin s^2 \end{bmatrix}$ $\left(0 \leq s \leq \sqrt{\frac{\pi}{2}}\right)$

17. $s = t + 3$

19. $\mathbf{X} = \begin{bmatrix} 2\cos t \\ -3\sin t \end{bmatrix}$ $(\pi \leq t \leq 3\pi)$

21. $2(e^\pi - 1)$

Section 1.6, page 107

1. (a) $\left(10, \frac{11\pi}{6}, 2\right)$ (b) $\left(2\sqrt{2}, \frac{5\pi}{4}, 5\right)$

(c) $\left(1, \frac{\pi}{2}, 0\right)$ (d) $\left(6, \frac{\pi}{3}, -3\right)$

3. (a) $\left(2\sqrt{2}, \frac{\pi}{4}, \frac{\pi}{4}\right)$ (b) $\left(6, \frac{2\pi}{3}, \frac{\pi}{2}\right)$

(c) $\left(2, \frac{5\pi}{3}, \frac{5\pi}{6}\right)$ (d) $\left(2, \frac{5\pi}{6}, \frac{5\pi}{6}\right)$

5. (a) $\left(\sqrt{13}, \frac{\pi}{3}, \cos^{-1}\frac{2}{\sqrt{13}}\right)$ (b) $\left(5, \frac{2\pi}{3}, \cos^{-1}-\frac{3}{5}\right)$

(c) $\left(2\sqrt{5}, \frac{5\pi}{6}, \cos^{-1}\frac{2}{\sqrt{5}}\right)$ (d) $\left(2, \frac{3\pi}{4}, 3\right)$

7. (a) $\left(0, \frac{3}{\sqrt{2}}, -\frac{1}{\sqrt{2}}\right)$ (b) $(-3, -\sqrt{2}, 0)$

(c) $(-1, 2\sqrt{2}, 0)$ (d) $\left(1, \frac{3}{\sqrt{2}}, \frac{-7}{\sqrt{2}}\right)$

Section 1.7, page 118

1. hyperbolic paraboloid

3. hyperboloid of one sheet

5. hyperbolic cylinder

7. y axis

9. parabolic cylinder

11. sphere

13. hyperboloid of one sheet

15. ellipsoid

17. hyperbolic paraboloid

19. hyperbolic cylinder

21. $x' = x + 1$, $y' = y - 1$, $z' = z - 2$; ellipsoid

23. $x' = x + 2$, $y' = y - 3$, $z' = z + 2$; hyperboloid of two sheets

25. $x' = x - 2$, $z' = z - 1$; pair of lines

27. $x' = x + 2$, $y' = y + 2$, $z' = z + 1$; elliptic paraboloid

29. $x' = x - 1$, $y' = y - 2$, $z' = z + 3$; sphere

Chapter 2

Section 2.1, page 134

1. All (x, y) such that   (a) $x^2 + y^2 \geq 1$ (b) $xy \geq 0$
   (c) $xy \neq 0$ (d) $x^2 - y^2 \neq \pi n$ (n = integer)

   (e) $(x^2 - y^2) \neq (2n + 1)\dfrac{\pi}{2}$ (n = integer)   (f) all (x, y)

3. The common domain of $f - g$, $f + g$, and fg is the set of all (x, y) such that $x^2 + y^2 \leq 1$ and $x > y$; the domain of f/g is the set of all (x, y) such that $x^2 + y^2 \leq 1$, $x > y$ and $x - y \neq 1$.

5. $-\dfrac{1}{\cos(x+y)}$; the set of all (x, y) such that $x + y \neq (2n + 1)\dfrac{\pi}{2}$

(n = integer)   (b) -1

7. $\rho - |\mathbf{X}_1 - \mathbf{X}_0|$

9. (b)   has no limit points. The limit points of the other sets are the points (x, y) such that   (a) $x^2 + y^2 \leq 1$ (c) $x^2 + y^2 \leq 1$ (d) $0 \leq x \leq y$ (e) x and y are arbitrary real numbers   (f) $x^2 + y^2 = 1$

11. (a)   9(b) and 9(f) have no interior points. The interior points of the other sets are those (x, y) such that: 9(a) $x^2 + y^2 < 1$;   9(c) $x^2 + y^2 < 1$, $x \neq 0$, $y \neq 0$;   9(d) $0 < x < y$;   9(e) $(x, y) \neq (0, 0)$

(b)   9(b) has no limit points. The limit points of the other sets which are not interior points are (x, y) such that: 9(a) $x^2 + y^2 = 1$;   9(c) $x^2 + y^2 = 1$, $(x,y) = (x,0)$ ($|y| < 1$), or $(x,y) = (0,y)$ ($|y| < 1$);   9(d) $x = 0$, $y \geq 0$ or $x = y \geq 0$;   9(e) $(x, y) = (0, 0)$;   9(f) $x^2 + y^2 = 1$

(c)   The isolated points of 9(b) are $(0, 0)$ and $(1, 1)$; $(1, 2)$ is the only isolated point of 9(f). The other sets have no isolated points.

13. (a)   8   (b)   no limit   (c)   0   (d)   2

15. (a)   0   (b)   2   (c)   no limit

17. 1

19. Each is continuous at every point of its domain.

21. All (x, y) such that $x^2 \geq y^2$

23. (a) (x, y, z) such that $3xyz \neq 1$ (b)   all (x, y) (c)   all $(x, y) \neq (0, 0)$

25. (a)   0   (b)   1   (c)   0

Section 2.2, page 148

1. (a) $3\sqrt{2}$ (b) $\dfrac{7}{\sqrt{2}}$ (c) $\dfrac{1}{\sqrt{2}}$ (d) $\dfrac{1}{9\sqrt{2}}$

3. (a) $f_x(\mathbf{X}) = 4x$, $f_y(\mathbf{X}) = 2y$, $f_z(\mathbf{X}) = 1$, $f_x(\mathbf{X}_0) = 4$, $f_y(\mathbf{X}_0) = 2$, $f_z(\mathbf{X}_0) = 1$;

(b) $f_x(\mathbf{X}) = e^x \cos(x^2 + y^2 + z^2) - 2xe^x \sin(x^2 + y^2 + z^2)$, $f_y(\mathbf{X}) =$

$-2y \sin(x^2 + y^2 + z^2), \quad f_z(\mathbf{X}) = -2z \sin(x^2 + y^2 + z^2),$
$f_x(\mathbf{X_0}) = -\sqrt{\pi}\, e^{\sqrt{\pi/4}} = f_y(\mathbf{X_0}), \quad f_z(\mathbf{X_0}) = 0$

(c) $f_x(\mathbf{X}) = \dfrac{2x}{1 + x^2 + y + z}, \quad f_y(\mathbf{X}) = f_z(\mathbf{X}) = \dfrac{1}{1 + x^2 + y + z},$

$f_X(\mathbf{X_0}) = 0, \quad f_y(\mathbf{X_0}) = f_z(\mathbf{X_0}) = 1$

(d) $f_x(\mathbf{X}) = f_y(\mathbf{X}) = 1$

5.   (a) $f_{xy}(-1,\ 2) = f_{yx}(-1,\ 2) = -14$ (b) $f_{xy}(-1,\ 2) = f_{yx}(-1,2) = 2$

7.   (a) $f_x(x, y, z) = ye^z - 3z \cos y,$
$f_y(x, y, z) = xe^z + 4yz + 3xz \sin y,$
$f_z(x, y, z) = xye^z + 2y^2 - 3x \cos y$

(b) $f_x(x, y) = \dfrac{4x^2 - 20x^3y + 4xy^2 - 15x^2y^3 - 6y^2}{(2x + y^2)^2},$

$f_y(x, y) = \dfrac{12xy + 5x^3y^2 - 4x^2y - 10x^4}{(2x + y^2)^2}$

(c) $f'(x) = 3$

11.   (a) (x, y) such that $x = -y$;   (b) (x, y) such that $x^2 + y^2 = 1$;
(c) (x, y) such that $y \sin x = -1$

Section 2.3, page 162

1.   (a) $a = 1, b = 0$ (b) $a = 3, b = 4$

3.   (a) $(d_{X_0}f)(\mathbf{X} - \mathbf{X_0}) = 10(x + 1) + 2y, \quad (d_{X_0}f)(\mathbf{X_1} - \mathbf{X_0}) = 44$

(b) $(d_{X_0}f)(\mathbf{X} - \mathbf{X_0}) = \dfrac{2}{\sqrt{3}}\left(x - \dfrac{1}{\sqrt{3}}\right) - \dfrac{2}{\sqrt{3}}\left(y + \dfrac{1}{\sqrt{3}}\right)$

$+ \dfrac{2}{\sqrt{3}}\left(z - \dfrac{1}{\sqrt{3}}\right),$

$(d_{X_0}f)(\mathbf{X_1} - \mathbf{X_0}) = -2.$

(c) $(d_{X_0}f)(\mathbf{X} - \mathbf{X_0}) = -\dfrac{\pi}{2}(x - 1) - \dfrac{\pi}{2}(y - 1) - \left(z - \dfrac{\pi}{2}\right),$

$$(d_{X_0}f)(\mathbf{X}_1 - \mathbf{X}_0) = -2\pi.$$

(d) $(d_{X_0}f)(\mathbf{X} - \mathbf{X}_0) = 10(z - 3);$ $(d_{X_0}f)(\mathbf{X}_1 - \mathbf{X}_0) = 140$

5. $\dfrac{1}{4} + \dfrac{5\pi\sqrt{3}}{72}$

7. (a) 0   (b) 1   (c) 2   (d) 2, 5, 6, 7, 9

9. (a) $d_{X_0}f = 10\,dx + 2\,dy$ (b) $d_{X_0}f = \dfrac{2}{\sqrt{3}}\,dx - \dfrac{2}{\sqrt{3}}\,dy + \dfrac{2}{\sqrt{3}}\,dz$

(c) $d_{X_0}f = -\dfrac{\pi}{2}\,dx - \dfrac{\pi}{2}\,dy - dz$ (e) $d_{X_0}f = 0.$

11. (a) $df = (9x^2 + 2xy + 1)\,dx + (x^2 + 1)\,dy$

(b) $df = \dfrac{2x\,dx + 2y\,dy + 2z\,dz}{x^2 + y^2 + z^2}$

(c) $df = -\sin xyz\; e^{\cos xyz}(yz\,dx + xz\,dy + xy\,dz)$

(d) $df = \dfrac{x^2 - y^2}{xy}\left(\dfrac{dx}{x} + \dfrac{dy}{y}\right)$

13. $d_{X_0}(2f) = \sqrt{2}(dx + dy + dz),$ $d_{X_0}(f + g) = 2\sqrt{2}(dx + dy + dz),$
 $d_{X_0}(fg) = dx + dy + dz,$ $d_{X_0}(f/g) = 0$

15. (a) $\dfrac{2}{7\sqrt{3}}$ (b) $\dfrac{1}{\sqrt{3}}$ (c) $\dfrac{e^6}{\sqrt{3}}$ (d) $-\dfrac{1}{49\sqrt{3}}$

Section 2.4, page 169

1. (a) 0   (b) $-5\left(\dfrac{\pi}{2}\right)^{4/5}$ (c) 0   (d) 2   (e) $-\pi^2$

3. $\dfrac{1}{9\sqrt{2}}$

5. (a) 3   (b) 0

7. $e^y \log y\,(\tan y + \sec^2 y) + \dfrac{e^y \tan y}{y}$

Section 2.5, page 176

9. $18x + 16y + \sqrt{11}\, z = 61$

11. $2x + 3y - z = 6$

13. $2x - 8y - z = 9$

Section 2.6, page 185

1. (a) domain $= \{(1, 5, 1), (1, 1, 2), (1, 1, 1), (1, 1, 3)\}$; range $=$ $\{1, 2, 3\}$

    (c) domain $= \{(2, 7, 5), (1, 7, 5), (3, 6, 2), (7, 5, 3)\}$; range $=$ $\{1, 3\}$

    (d) domain $= \{(1, 2, 3), (2, 3, 4), (3, 4, 1), (4, 1, 2)\}$; range $=$ $\{1, 2, 3, 4\}$

3. (c) domain $= \{x \,|\, |x| \le 1\}$; range $= \{y \,|\, 0 \le y \le \pi\}$
    (d) domain $= \{x \,|\, |x| \le 1\}$; range $= \{y \,|\, \pi \le y \le 2\pi\}$

5. (a) domain $= \{(5, 1, 1), (1, 2, 2), (1, 1, 3), (1, 3, 2)\}$; range $= \{1\}$
    (b) domain $= \{(1, 1, 1), (1, 2, 2), (1, 1, 2), (1, 3, 2)\}$; range $= \{1\}$
    (d) domain $= \{(2, 3, 4), (3, 4, 1), (4, 1, 2), (1, 2, 3)\}$; range $=$ $\{1, 2, 3, 4\}$

7. (c) domain $= \{(x, z) \,|\, x^2 + z^2 \ge 1\}$; range $= \{y \,|\, y \ge 0\}$

11. $g(0) = 1,\;\; g'(0) = -\tfrac{2}{3};\;\; h(0) = 2,\;\; h'(0) = -\tfrac{10}{3}$

13. $g_x(1, 1, 1) = \tfrac{5}{9},\;\; g_z(1, 1, 1) = \tfrac{4}{9},\;\; g_w(1, 1, 1) = -\tfrac{8}{9}$

Section 2.7, page 195

1. (a) $\begin{bmatrix} a_1 \\ a_2 \\ \vdots \\ a_n \end{bmatrix}$ (b) $\begin{bmatrix} 0 \\ -2 \\ 4 \end{bmatrix}$ (c) $\begin{bmatrix} \tfrac{1}{6} \\ \tfrac{1}{6} \\ \tfrac{1}{6} \end{bmatrix}$ (d) $\begin{bmatrix} 2e \\ 0 \\ 0 \end{bmatrix}$

3. (a) $\dfrac{1}{\sqrt{a_1^2 + \cdots + a_n}}\begin{bmatrix} a_1 \\ a_2 \\ \vdots \\ a_n \end{bmatrix}$; $\sqrt{a_1^2 + \cdots + a_n^2}$

(b) $\dfrac{1}{\sqrt{5}}\begin{bmatrix} 0 \\ -1 \\ 2 \end{bmatrix}$; $2\sqrt{5}$ (c) $\begin{bmatrix} \dfrac{1}{\sqrt{3}} \\ \dfrac{1}{\sqrt{3}} \\ \dfrac{1}{\sqrt{3}} \end{bmatrix}$; $\dfrac{1}{2\sqrt{3}}$

(d) $\begin{bmatrix} 1 \\ 0 \\ 0 \end{bmatrix}$; $2e$

5. (a) 0   (b) $2e/\sqrt{3}$

9. $2x + 4y - 3z = -2$ (plane).

11. $\{(x, y, z)\,|\,x^2 + y^2 + z^2 = \pi\}$

13. $7x + 7y + z = 20$

15. $x = \sqrt{\pi}$

17. $x = 0$

19. $\begin{bmatrix} \dfrac{1}{\sqrt{2}} \\ -\dfrac{1}{\sqrt{2}} \end{bmatrix}$

21. $\begin{bmatrix} \dfrac{1}{\sqrt{2}} \\ \dfrac{1}{\sqrt{2}} \\ 0 \end{bmatrix}$

Section 2.8, page 208

1. (a) $e(1 + (t - 1) + \tfrac{1}{2}(t - 1)^2)$ (b) $1 + \tfrac{1}{2}(t - 1) - \tfrac{1}{8}(t - 1)^2$
   (c) $1 + \tfrac{1}{2}t^2$

3. $8 + 10(t - 1) + 5(t - 1)^2 + (t - 1)^3$

5.   (a) 6   (b) 4   (c) 6   (d) 5

7.   1.6487

9.   (a) $(d_{X_0}^0 f)\,(\mathbf{X} - \mathbf{X}_0) = 0,\qquad (d_{X_0} f)\,(\mathbf{X} - \mathbf{X}_0) = 2x + (y - 1),$
      $(d_{X_0}^2 f)\,(\mathbf{X} - \mathbf{X}_0) = -4x^2 - 4x(y - 1) - (y - 1)^2$

     (b) $(d_{X_0}^0 f)\,(\mathbf{X} - \mathbf{X}_0) = 0,\qquad (d_{X_0} f)\,(\mathbf{X} - \mathbf{X}_0) = 0,$
      $(d_{X_0}^2 f)\,(\mathbf{X} - \mathbf{X}_0) = 2(x^2 + y^2).$

11. (a) $(d_{X_0}^0 f)\,(\mathbf{X} - \mathbf{X}_0) = 1,\qquad (d_{X_0} f)\,(\mathbf{X} - \mathbf{X}_0) = 0,$
      $(d_{X_0}^2 f)\,(\mathbf{X} - \mathbf{X}_0) =$

$$-\left[\left(x - \frac{\pi}{6}\right)^2 + \left(y - \frac{\pi}{6}\right)^2 + \left(z - \frac{\pi}{6}\right)^2 + 2\left(x - \frac{\pi}{6}\right)\left(y - \frac{\pi}{6}\right)\right.$$

$$\left. + 2\left(x - \frac{\pi}{6}\right)\left(z - \frac{\pi}{6}\right) + 2\left(y - \frac{\pi}{6}\right)\left(z - \frac{\pi}{6}\right)\right]$$

     (b) $(d_{X_0}^0 f)(\mathbf{X} - \mathbf{X}_0) = 1,\qquad (d_{X_0} f)\,(\mathbf{X} - \mathbf{X}_0) = x - 1,$
      $d_{X_0}^2 f = y^2 + z^2$

13. (a) $2x + (y - 1) - 2x^2 - 2x(y - 1) - \frac{1}{2}(y - 1)^2$ (b) $x^2 + y^2$

15. $1 - \dfrac{1}{2}\left(x - \dfrac{\pi}{6}\right)^2 - \dfrac{1}{2}\left(y - \dfrac{\pi}{6}\right)^2 - \dfrac{1}{2}\left(z - \dfrac{\pi}{6}\right)^2 - \left(x - \dfrac{\pi}{6}\right)\left(y - \dfrac{\pi}{6}\right)$

$$-\left(x - \frac{\pi}{6}\right)\left(z - \frac{\pi}{6}\right) - \left(y - \frac{\pi}{6}\right)\left(z - \frac{\pi}{6}\right)$$

17. $6 + 2(x - 2) + 2(y - 3) + (x - 2)^2 - (x - 2)(y - 3)$

Section 2.9, page 221

1.   (a) $f(-\tfrac{3}{2}) = -39,\ \ f(1) = 11,\ \ f(\tfrac{3}{2}) = 21/2$

     (b) $f(0) = 0,\ \ f(1) = 1$

     (c) $f(1) = f(-1) = 0,\ \ f(0) = 1$

     (d) $f(1) = f(-1) = 0,\ \ f(0) = 1$

     (e) $f(x, -x) = 1$

     (f) $f(x, y) = 0$ if $xy = k\pi$ (k = integer); $f(x, y) = 1$ if $xy =$

   $(2k + 1)\dfrac{\pi}{2}$ (k = integer)

3. (a) $(0, 0)$ (b) $(0, 0), (0, 1), (0, -1), (1, 0), (-1, 0), (1, 1),$

$(-1, 1), (1, -1), (-1, -1)$ (c) $\left(\dfrac{1}{e}, \dfrac{1}{e}\right)$ (d) $(0, 0)$

5. $\frac{7}{3}$

7. $l = w = (2v)^{1/3}, \quad h = (v/4)^{1/3}$

9. $\sqrt{10/3}$

11. $a/3, b/3, c/3$

13. (a) neither   (b) $(0, 0)$ is a relative minimum; $(1, 1), (-1, 1),$ $(1, -1)$ and $(-1, -1)$ are relative maxima; $(0, 1), (0, -1), (1, 0)$ and $(-1, 0)$ are not extreme points   (c) relative minimum (d) neither

15. $8abc/3\sqrt{3}$

17. (a) $(0, 0)$ and (x, y) such that $(x^2 + y^2) = (2k + \frac{3}{2})\pi$ $(k = 0, 1, 2, \ldots)$ are relative minima; (x, y) such that $x^2 + y^2 = (2k + \frac{1}{2})\pi$ $(k = 0, 1, 2, \ldots)$ are relative maxima.   (b) $(0, 0)$ is a critical point, but not an extreme point.   (c) $(0, 0)$ is a relative minimum.   (d) $(0, 0)$ is a critical point, but not an extreme point. (e) $(0, 0)$ is a relative minimum.

Section 2.10, page 235

1. $l = \sqrt{A} = w$

3. $\left(\dfrac{2}{\sqrt{5}}, \dfrac{4}{\sqrt{5}}\right), \; \left(-\dfrac{2}{\sqrt{5}}, \dfrac{-4}{\sqrt{5}}\right)$

5. $(0, 2, 2), (0, -2, 2)$

7. $a/3, b/3, c/3$

9. $(0, 2)$ and $(0, -2)$ are closest; $(0, 3)$ and $(0, -3)$ are farthest

11. $8abc/3\sqrt{3}$

13. $f\left(\dfrac{10}{29}, \dfrac{15}{29}, \dfrac{-20}{29}\right) = \dfrac{25}{29}$

15. $\sqrt{2} - 1$

17. $Q(0, 0, 0, 1) = 1$

Chapter 3

Section 3.1, page 247

1. (a) $f_1(x, y) = \cos xy$, $f_2(x, y) = \sqrt{x^2 + y^2 - 4}$, $f_3(x, y) = 1/xy$;
$D = \{(x, y)\,|\,x^2 + y^2 \geq 4, \quad x \neq 0, \quad y \neq 0\}$

   (b) $f_1(x, y) = e^{x^2+y^2}$, $\quad f_2(x, y) = \dfrac{x}{y^2 - 1}$;

   $D = \{(x, y)\,|\,y \neq \pm 1\}$

   (c) $f_1(x, y) = x^2 + y^2 + z^2$, $f_2(x, y) = x^2$, $f_3(x, y, z) = -y^2$;
$D = R^3$.

   (d) $f_1(x, y) = \sqrt{4 - x^2}$, $f_2(x, y) = \sqrt{4 - y^2}$, $f_3(x, y) = x + y$;
$D = \{(x, y)\,|\,|x| \leq 2, |y| \leq 2\}$.

3. (a) $n = 2, m = 3$ (b) $n = 2, m = 2$
   (c) $n = 3, m = 3$ (d) $n = 2, m = 3$.

5. (a) $D = \{(x, y)\,|\,x^2 + y^2 \leq 2, \quad x \neq 0, \quad y \neq 0\}$

   (b) $\begin{bmatrix} \frac{9}{2} \\ 4 + \dfrac{\sqrt{3}}{2} \\ 1 \end{bmatrix}$, $\begin{bmatrix} -\frac{7}{2} \\ 4 - \dfrac{\sqrt{3}}{2} \\ 0 \end{bmatrix}$ (c) $\begin{bmatrix} 8 \\ 3 \\ 5 \end{bmatrix}$

7. (a) $\begin{bmatrix} -12 \\ 0 \\ 1 \end{bmatrix}$ (b) $\begin{bmatrix} -3x + 6y \\ 3x + 6y \\ 3x + 5y \end{bmatrix}$ (d) no

9. (a) $\begin{bmatrix} \frac{1}{2} \\ 2 \end{bmatrix}$ (b) $\begin{bmatrix} x - y \\ z \\ \sqrt{z} \end{bmatrix}$ (c) $D = \{(x, y, z)\,|\,z > 0\}$

11. (a) $\begin{bmatrix} \frac{1}{2} \\ 1 \end{bmatrix}$ (b) $\begin{bmatrix} \dfrac{\cos^2 y}{x^2} \\ x \end{bmatrix}$

13. (a) $\begin{bmatrix} \frac{1}{5} \\ -6 \\ \frac{12}{5} \end{bmatrix}$ (b)   no limit   (c) $\begin{bmatrix} 0 \\ 0 \\ 0 \end{bmatrix}$

15. (a) $\begin{bmatrix} 2 \\ -3 \end{bmatrix}$ (b) $\sqrt{2(x+z)^2 + 2y^2}$

17. (a) $\begin{bmatrix} x+y+z \\ 2x+z \\ x-y \end{bmatrix}$, $\begin{bmatrix} 7 \\ 9 \\ 2 \end{bmatrix}$

19. (a) $\{(x, y, z)\,|\,y \neq \pm 3,\, z \neq 0\}$
     (b) $\{(x, y)\,|\,x > y\}$ (c) $\{(x, y, z)\,|\,x \neq y,\ \ x \neq z,\ \ y \neq \pm z\}$

     (d) $\{(x, y)\,|\,x^2 - y^2 \neq (2k+1)\,\dfrac{\pi}{2}\ \ (k = \text{integer}),\ \ |x| \geq |y|\}$

21. $\{(x, y)\,|\,x^2 + y^2 \leq 2,\ \ x \neq 0,\ \ y \neq 0\}$

23. R^2
25. $\{(x, y)\,|\,x \neq 0\}$,

     $\{(u, v)\,|\,u^2 + v^2 \neq 0,\ \ v \neq (2k+1)\,\dfrac{\pi}{2}\ \ (k = \text{integer})\}$

Section 3.2, page 261

1. (a) $\begin{bmatrix} \frac{1}{3}\,dx + \frac{2}{3}\,dy - \frac{2}{3}\,dz \\ 3\,dx + 2\,dy + 4\,dz \end{bmatrix}$, $\begin{bmatrix} -8 \\ 3 \\ 10 \end{bmatrix}$

    (b) $\begin{bmatrix} dx - dy \\ 12\,dx + 2\,dy \\ 8\,dx + dy \end{bmatrix}$, $\begin{bmatrix} 0 \\ 14 \\ 9 \end{bmatrix}$

    (c) $\begin{bmatrix} e^2\,dx + e^2\,dy + e^2\,dz \\[2mm] \dfrac{1}{2\sqrt{2}}\,dx + \dfrac{1}{2\sqrt{2}}\,dy \\[2mm] \frac{3}{2}\,dx + \frac{3}{2}\,dy + \frac{3}{2}\,dz \end{bmatrix}$, $\begin{bmatrix} e^2 \\ 0 \\ \frac{3}{2} \end{bmatrix}$

3. (a) $\begin{bmatrix} dx + dy + dz \\ 2\,dx + 2\,dy + 4\,dz \end{bmatrix}, \quad \begin{bmatrix} 2 \\ 6 \end{bmatrix}$

    (b) $\begin{bmatrix} \dfrac{1}{\sqrt{2}}\,(dx + dy) \\[2mm] 2\,(dx + dy) \\[2mm] -\dfrac{1}{\sqrt{2}}\,(dx + dy) \end{bmatrix}, \quad -\dfrac{\pi}{4\sqrt{2}} \begin{bmatrix} 1 \\ 2\sqrt{2} \\ -1 \end{bmatrix}$

    (c) $\begin{bmatrix} dx \\ 0 \\ dx \end{bmatrix}, \quad \begin{bmatrix} 1 \\ 0 \\ 1 \end{bmatrix}$

5. (a) $\begin{bmatrix} -\sin(x + y)\,(dx + dy) \\ \cos(x + y)\,(dx + dy) \end{bmatrix}$ (b) $\begin{bmatrix} dx \\ 2x\,dx \\ 3x^2\,dx \end{bmatrix}$

    (c) $\begin{bmatrix} e^x \sin yz\,dx + ze^x \cos yz\,dy + ye^x \cos yz\,dz \\ ze^y \cos xz\,dx + e^y \sin xz\,dy + xe^y \cos xz\,dz \\ ye^z \cos xy\,dx + xe^z \cos xy\,dy + e^z \sin xy\,dz \end{bmatrix}$

7. (a) $\begin{bmatrix} e^x(y + z + w)\,dx + e^x(dy + dz + dw) \\ \cos(x + y + z + w)\,(dx + dy + dz + dw) \\ yz\,dx + xz\,dy + xy\,dz \end{bmatrix}$

    (b) $\begin{bmatrix} -\cos(1 - x - y)\,(dx + dy) \\ (2 + y)\,dx + x\,dy \\ 3x^2\,dx + 2y\,dy \end{bmatrix}$

    (c) $e^x \begin{bmatrix} (\sin x + \cos x)\,dx \\ (\cos x - \sin x)\,dx \end{bmatrix}$

9. (a) $\begin{bmatrix} 6\,dx + 24\,dy \\ 6\,dx + 3\,dy \end{bmatrix}$ (b) $\begin{bmatrix} 4\,dx + 7\,dy \\ 5\,dx + 5\,dy \end{bmatrix}$

    (c) $\begin{bmatrix} -2\,dx + 19\,dy \\ -5\,dx - 10\,dy \end{bmatrix}$

11. (a) $\mathbf{g}(\mathbf{X}) = \begin{bmatrix} \frac{1}{3}x + \frac{2}{3}y - \frac{2}{3}z \\ -3x + 2y + 4z + 1 \end{bmatrix}$

    (b) $\mathbf{g}(\mathbf{X}) = \begin{bmatrix} x - y - 1 \\ 12x + 2y - 17 \\ 8x + y - 8 \end{bmatrix}$

(c) $\quad \mathbf{g(X)} = \begin{bmatrix} e^2 (x + y + z - 2) \\[2ex] \dfrac{1}{2\sqrt{2}} (x + y + 2) \\[2ex] \frac{3}{2}x + \frac{3}{2}y + \frac{3}{2}z - 2 \end{bmatrix}$

Section 3.3, page 274

1. (a) $\begin{bmatrix} 4\,dx - 3\,dy + dz \\ dy + dz \end{bmatrix}$ (b) $\begin{bmatrix} 2\,dx \\ 2\,dx \end{bmatrix}$

3. (a) $\begin{bmatrix} 2 & -1 & 1 \\ 2 & -2 & 0 \end{bmatrix}, \begin{bmatrix} 1 & 1 \\ 1 & -1 \end{bmatrix}, \begin{bmatrix} 4 & -3 & 1 \\ 0 & 1 & 1 \end{bmatrix}, \begin{bmatrix} 4 & -3 & 1 \\ 0 & 1 & 1 \end{bmatrix}$

   (b) $\begin{bmatrix} 1 & 0 \\ 0 & 1 \end{bmatrix}, \begin{bmatrix} 2 & 0 \\ 2 & 0 \end{bmatrix}, \begin{bmatrix} 2 & 0 \\ 2 & 0 \end{bmatrix}, \begin{bmatrix} 2 & 0 \\ 2 & 0 \end{bmatrix}$

5. (a) $\mathbf{JU} = \begin{bmatrix} \dfrac{x}{\sqrt{x^2+y^2+z^2}} & \dfrac{y}{\sqrt{x^2+y^2+z^2}} & \dfrac{z}{\sqrt{x^2+y^2+z^2}} \\[3ex] \dfrac{-y}{x^2+y^2} & \dfrac{x}{x^2+y^2} & 0 \\[3ex] \dfrac{-xz}{\sqrt{x^2+y^2}(x^2+y^2+z^2)} & \dfrac{-yz}{\sqrt{x^2+y^2}(x^2+y^2+z^2)} & \dfrac{\sqrt{x^2+y^2}}{(x^2+y^2+z^2)} \end{bmatrix},$

$\mathbf{Jg} = \begin{bmatrix} v & u & 0 \\ 0 & w & v \\ w & 0 & u \end{bmatrix},$

   (b) $\mathbf{J_{X_0} U} = \begin{bmatrix} \dfrac{1}{\sqrt{2}} & \dfrac{-1}{\sqrt{2}} & 0 \\[2ex] \dfrac{1}{\sqrt{2}} & \dfrac{1}{\sqrt{2}} & 0 \\[2ex] 0 & 0 & 1 \end{bmatrix}, \quad \mathbf{J_{U_0} g} = \begin{bmatrix} -\dfrac{\pi}{4} & 1 & 0 \\[2ex] 0 & 0 & -\dfrac{\pi}{4} \\[2ex] 0 & 0 & 1 \end{bmatrix},$

$$\mathbf{J_{x_0}h} = \begin{bmatrix} \frac{1}{\sqrt{2}}\left(1 - \frac{\pi}{4}\right) & \frac{1}{\sqrt{2}}\left(1 + \frac{\pi}{4}\right) & 0 \\ 0 & 0 & -\frac{\pi}{4} \\ 0 & 0 & 1 \end{bmatrix}$$

(c) $\mathbf{d_{x_0}u} = \begin{bmatrix} \frac{1}{\sqrt{2}}(dx - dy) \\ \frac{1}{\sqrt{2}}(dx + dy) \\ dz \end{bmatrix}$, $\mathbf{J_{u_0}g} = \begin{bmatrix} -\frac{\pi}{4}dx + dy \\ -\frac{\pi}{4}dz \\ dz \end{bmatrix}$,

$$\mathbf{J_{x_0}h} = \begin{bmatrix} \frac{1}{\sqrt{2}}\left(1 - \frac{\pi}{4}\right)dx + \frac{1}{\sqrt{2}}\left(1 + \frac{\pi}{4}\right)dy \\ -\frac{\pi}{4}dz \\ dz \end{bmatrix}$$

7. (a) $\sqrt{3}, \sqrt{3}\, dt$ (b) $1, dt$ (c) $\dfrac{3 - 2\pi}{4}, \dfrac{3 - 2\pi}{4}\, dt$

9. (a) $\left(2 + \dfrac{1}{\sqrt{2}}\right) dr + (\sqrt{2} - 4)\, d\theta$

(b) $\begin{bmatrix} \left(\dfrac{1}{\sqrt{2}} - 2\right) dr - (\sqrt{2} + 4)\, d\theta \\ \dfrac{3}{\sqrt{2}} dr - \sqrt{2}\, d\theta \end{bmatrix}$ (c) $\begin{bmatrix} 8\, dr \\ 4\, dr \end{bmatrix}$

11. (a) $d\theta + dz$ (b) $\begin{bmatrix} dr + 2\, dz \\ 2\, dr - 2\, dz \end{bmatrix}$ (c) $\begin{bmatrix} d\theta \\ d\theta \end{bmatrix}$

13.   (a) $\begin{bmatrix} -2\,d\theta \\ 2\,dr \end{bmatrix}$ (b) $\begin{bmatrix} \frac{1}{2}(dr - d\theta - d\phi) \\ \frac{1}{2}(dr + d\theta - d\phi) \\ dr + d\phi \end{bmatrix}$

(c) $-\frac{1}{2}\,d\theta$

15. $dx + \frac{1}{5}\,dy$

17. $\frac{7}{5}, \frac{3}{5}, \frac{3}{25}$

19. $h_{uv} = g_{xx}x_u x_v + g_{xy}(x_u y_v + x_v y_u) + g_{yy}y_u y_v + g_x x_{uv} + g_y y_{uv};$
      $h_{vv} = g_{xx}x_v^2 + 2g_{xy}x_v y_v + g_{yy}y_v^2 + g_x x_{vv} + g_y y_{vv}$

Section 3.4, page 289

1.   (a) $ye^{xy}\sin z\,\mathbf{i} + xe^{xy}\sin z\,\mathbf{j} + e^{xy}\cos z\,\mathbf{k}$

(b) $\dfrac{\mathbf{X}}{|\mathbf{X}|^2}$ (c) $z(x+1)\mathbf{i} + x(x+\frac{1}{2})\mathbf{k}$

5.   (a) $\dfrac{y}{x} + \dfrac{x}{y} + \dfrac{xy}{z}$ (b) x (c) $3x^2 + 2y + 3z^2$

(d) $6(x + y + z)$ (e) $-[(xy)^2 + (yz)^2 + (xz)^2]\cos xyz$

7.   (a) $z(x^2 + y^2)e^{xy}$ (b) $e^{yz}(y^2 + z^2 - 1)\cos x$

(c) $\dfrac{2(3 + x^2 + y^2 + z^2)}{(1 + x^2 + y^2 + z^2)^2}$ (d)   4

(e) $-[(xy)^2 + (yz)^2 + (xz)^2]\sin xyz$

11.   (a) $-\frac{1}{2}x^2\mathbf{i} + (xy - z)\mathbf{j} + 2xz\mathbf{k}$
(b) $2\mathbf{i} + \mathbf{j}$
(c) $2z(ye^x - xe^y)\mathbf{i} + y(x^2e^z - yze^x)\mathbf{j} + (z^2e^y - x^2e^z)\mathbf{k}$
(d) $-(e^y\sin x + e^x\cos y)\mathbf{k}$

Section 3.5, page 300

1. $u = x + \frac{3}{4}y, \quad v = -\frac{5}{4}y;$ $\begin{bmatrix} 1 & \frac{3}{4} \\ 0 & -\frac{5}{4} \end{bmatrix}$

3. $\begin{bmatrix} 1 & 2 \\ -2 & 1 \end{bmatrix}, \quad \begin{bmatrix} -1 & 1 \\ 2 & 2 \end{bmatrix}, \quad \begin{bmatrix} 1 & \frac{3}{4} \\ 0 & -\frac{5}{4} \end{bmatrix}$

5. $\mathbf{f}(\mathbf{X}_0) = \begin{bmatrix} 1 \\ 1 \end{bmatrix}, \quad \mathbf{J}_{\mathbf{X}_0}\mathbf{f} = \begin{bmatrix} 3 & -1 \\ 1 & 0 \end{bmatrix}, \quad d_{\mathbf{X}_0}\mathbf{f} = \begin{bmatrix} 3\,dx - dy \\ dx \end{bmatrix};$

$\mathbf{g}(\mathbf{X}_0) = -\begin{bmatrix} 1 \\ 1 \end{bmatrix}, \quad \mathbf{J}_{\mathbf{X}_0}\mathbf{g} = \begin{bmatrix} -3 & -1 \\ -1 & 0 \end{bmatrix}, \quad d_{\mathbf{X}_0}\mathbf{g} = \begin{bmatrix} -3\,dx - dy \\ -dx \end{bmatrix}$

7. 1

9. $\begin{bmatrix} 2 & -2 & 1 \\ -2 & 0 & -1 \\ -2 & 1 & -2 \end{bmatrix}$

11. $(x, y), \quad (x, u), \quad (x, v), \quad (y, u), \quad (u, v)$

Section 3.6, page 313

1. $\mathbf{f}^{-1}(u, v) = \begin{bmatrix} 3u - v \\ -2u + v \end{bmatrix}, \qquad \mathbf{Jf}^{-1}(u, v) = \begin{bmatrix} 3 & -1 \\ -2 & 1 \end{bmatrix}$

3. (b) $\begin{bmatrix} x \\ y \end{bmatrix} = \begin{bmatrix} \sqrt{\dfrac{2v - u}{3}} \\[4mm] \sqrt{\dfrac{2u - v}{3}} \end{bmatrix} ; \quad D = \{(u, v) \,|\, 0 \leq u \leq 2v, \ \ 0 \leq v \leq 2u\}$

   (c) $\begin{bmatrix} x \\ y \end{bmatrix} = \begin{bmatrix} -\sqrt{\dfrac{2v - u}{3}} \\[4mm] \sqrt{\dfrac{2u - v}{3}} \end{bmatrix} ; \quad D = \{(u, v) \,|\, 0 \leq u \leq 2v, \ \ 0 \leq v \leq 2u\}$

   (d)   no

5. $\dfrac{1}{\sqrt{3}} \begin{bmatrix} -\frac{1}{2} & 1 \\[2mm] \dfrac{1}{\sqrt{13}} & \dfrac{-1}{2\sqrt{13}} \end{bmatrix}$

7. (b) $\begin{bmatrix} x \\ y \end{bmatrix} = \begin{bmatrix} 1 + \frac{2}{3}u - \frac{1}{3}v \\ 1 - \frac{1}{2}u + \frac{1}{2}v \end{bmatrix}$

9. $\dfrac{\partial(r,\theta,\phi)}{\partial(x,y,z)} = \begin{bmatrix} \cos\theta\sin\phi & \sin\theta\sin\phi & \cos\phi \\[2ex] \dfrac{-\sin\theta}{r\sin\phi} & \dfrac{\cos\theta}{r\sin\phi} & 0 \\[2ex] \dfrac{\cos\theta\cos\phi}{r} & \dfrac{\sin\theta\cos\phi}{r} & \dfrac{-\sin\phi}{r} \end{bmatrix}$

$\dfrac{\partial(x,y,z)}{\partial(r,\theta,\phi)} = \begin{bmatrix} \cos\theta\sin\phi & -r\sin\theta\sin\phi & r\cos\theta\cos\phi \\ \sin\theta\sin\phi & r\cos\theta\sin\phi & r\sin\theta\cos\phi \\ \cos\phi & 0 & -r\sin\phi \end{bmatrix}$

11. (a) $\begin{bmatrix} 1 & 0 & 0 \\ 0 & 1 & 0 \\ 0 & 0 & 1 \end{bmatrix}$ (b) $\begin{bmatrix} 1 & 0 & 0 \\ 0 & 0 & 1 \\ 0 & 1 & 0 \end{bmatrix}$

(c) $\begin{bmatrix} 0 & 0 & 1 \\ 0 & 1 & 0 \\ 1 & 0 & 0 \end{bmatrix}$

Chapter 4

Section 4.1, page 330

1. 12   3. 12   5. 10

7. 16   13. 48

15. 48   17. 1

Section 4.2, page 344

1. $\dfrac{135}{2}$ 3. $-\dfrac{32}{3}$ 5. $-\dfrac{5}{3}$ 7. -285 9. $\dfrac{114}{3}$

11. 324   13. $\dfrac{1}{27}(8e^9 + 7e^{-6})$ 15. 27   17. $\dfrac{3}{10}$

19. $\dfrac{5\sqrt{3}}{8} - \dfrac{3}{2} + \dfrac{\pi}{12}(1 + \sqrt{3})$ 21. $\dfrac{1}{8}$ 23. $\dfrac{32}{3}$ 25. 1

27. $\frac{1}{2} + \log 2$ 29. 72   31. $\frac{128}{21}$ 33. π 35. $\frac{1}{6}$

37. 16π

Section 4.3, page 364

1. (a) $\frac{1}{2}\tan^{-1}\frac{1}{2}$,   (b) $\sqrt{2}+\frac{9}{2}\sin^{-1}\frac{1}{3}$
    (c) $-\frac{1}{2}\sqrt{3}+\log(2+\sqrt{3})$ (d) $e^{\sqrt{3}} - e$

3. 24   5. $\frac{\pi}{4}(e^2 - 1)$ 7. 25π 9. 8π

11. (a) $\frac{\pi}{4}(e^4 - 1)$ (b) $\frac{\pi}{8}\log 5$

13. $\frac{93}{20}$ 15. $4 - \pi$ 17. $\frac{1}{3} + \frac{\pi}{2}$ 19. $\frac{8\pi}{3}$

21. $\frac{\pi}{2}(32\sqrt{3} - 45)$ 23. $\frac{4096\pi}{3}$ 25. $\frac{23{,}056\pi}{15}$

27. $2\pi ab$ 29. 0   31. 0   33. 8π

Section 4.4, page 388

1. (a) $18, -4, 5$ (b) $-5/\sqrt{14}$ (c) $(-\frac{4}{9}, \frac{5}{9}, 2)$
    (d) $41, 40, 9$ (e) $0, 10, -8$ (f) $\frac{409}{10}$

3. (a) $\frac{8}{3}c(b^3 - a^3)$ (b) $0, 0, 0$ (c) 0   (d) $(0, 0, 0), (0, 0, 0)$
    (e) $\frac{8}{45}c(3b^5 + 5c^2b^3 - 3a^5 - 5c^2a^3)$, $\frac{8}{45}c(6b^5 + 5c^2b^3 - 6a^5 - 5c^2a^3)$,
         $\frac{8}{5}c(b^5 - a^5)$
    (f) $0, 0, 0$ (g) $\frac{8}{45}c(2b^5 + c^2b^3 - 2a^5 - c^2a^3)$

5. (a) $\frac{2}{3}k\pi$ (b) $\frac{1}{3}k\pi, 0, 0$ (c) $-\frac{k\pi}{3\sqrt{3}}$

    (d) $(0, 0, \frac{1}{3}), (0, 0, \frac{1}{2})$ (e) $\frac{8}{15}k\pi, \frac{8}{15}k\pi, \frac{2}{3}k\pi$
    (f) $0, 0, 0$ (g) $\frac{8}{15}k\pi$

7. (a) $\frac{4}{3}k$ (b) $\frac{128}{105}k, \frac{128}{105}k, \frac{128}{105}k$ (c) $-\frac{152}{105}k$

(d) $\left(\dfrac{3}{4}, \dfrac{3}{4}, \dfrac{3}{4}\right)$, $\left(\dfrac{32}{35}, \dfrac{32}{35}, \dfrac{32}{35}\right)$ (e) $\dfrac{8}{3}k, \dfrac{8}{3}k, \dfrac{8}{3}k$

(f) $\frac{1}{3}k\pi, \frac{1}{3}k\pi, \frac{1}{3}k\pi$ (g) $\frac{1}{3}k(8 - \pi)$

9. (a) $64k\pi$ (b) $\dfrac{768}{5}k\pi, 0, 0$ (c) $\dfrac{1344}{5\sqrt{11}}k\pi$

(d) $\left(0, 0, \dfrac{9}{4}\right)$, $\left(0, 0, \dfrac{12}{5}\right)$ (e) $\dfrac{4864}{9}k\pi, \dfrac{4864}{9}k\pi, \dfrac{2560}{9}k\pi$

(f) $0, 0, 0$ (g) $\dfrac{1280}{3}k\pi$

11. (a) $\frac{1}{2}abc^2k\pi$ (b) $0, 0, 0$ (c) 0 (d) $(0, 0, 0), (0, 0, 0)$
    (e) $\frac{1}{12}abc^2k\pi(b^2 + 2c^2), \frac{1}{12}abc^2k\pi(a^2 + 2c^2), \frac{1}{12}abc^2k\pi(a^2 + b^2)$

(f) $0, 0, 0$ (g) $\dfrac{1}{156}abc^2k\pi(9a^2 + 4b^2 + 26c^2)$

13. $\frac{1}{12}ah\rho_0\pi(a^2 + 2h^2)$

15. (a) $\frac{1}{6}ka^2b^3, \frac{1}{6}ka^3b^2$ (b) 0 (c) $(a/2, b/2), (\frac{2}{3}a, \frac{2}{3}b)$

(d) $\frac{1}{8}ka^2b^4, \frac{1}{8}ka^4b^2, \frac{1}{9}ka^3b^3, \frac{1}{8}ka^2b^2(a^2 + b^2)$ (e) $\dfrac{ka^4b^4}{36(a^2 + b^2)}$

17. (a) $32k, \dfrac{32}{3}k$ (b) $-\dfrac{160}{3\sqrt{5}}k$ (c) $\left(1, \dfrac{10}{3}\right), \left(\dfrac{4}{3}, 4\right)$

(d) $\dfrac{448}{3}k, 16k, \dfrac{704}{15}k, \dfrac{496}{3}k$ (e) $\dfrac{2128}{25}k$

19. (a) $\dfrac{5}{24}k, \dfrac{1}{2}k\left(\dfrac{1}{4} + \log 2\right)$ (b) $\dfrac{1}{24}k(12\log 2 - 17)$

(c) $\left(\dfrac{8}{3 + 6\log 2}, \dfrac{5}{6 + 12\log 2}\right), \left(\dfrac{3}{10} + \dfrac{6}{5}\log 2, \dfrac{1}{2}\right)$

(d) $\dfrac{59}{480}k, \dfrac{3}{5}k, \dfrac{7}{30}k, \dfrac{347}{480}k$ (e) $\dfrac{41}{320}k$

21. (a) $\dfrac{1}{12}k, \dfrac{3}{35}k$ (b) $\dfrac{k}{420\sqrt{2}}$ (c) $\left(\dfrac{9}{20}, \dfrac{9}{20}\right), \left(\dfrac{4}{7}, \dfrac{5}{9}\right)$

(d) $\dfrac{3}{56}\,k,\ \dfrac{1}{18}\,k,\ \dfrac{3}{56}\,k,\ \dfrac{55}{504}\,k$ (e) $\dfrac{k}{1008}$

23. (a) $0,\dfrac{5}{4}\,k\pi$ (b) $\dfrac{5}{4\sqrt{2}}\,k\pi$ (c) $\left(\dfrac{5}{6},0\right),\left(\dfrac{5}{6},0\right)$

(d) $\dfrac{21}{32}\,k\pi,\ \dfrac{49}{32}\,k\pi,\ 0,\ \dfrac{35}{16}\,k\pi$ (e) $\dfrac{35}{32}\,k\pi$

Section 4.5, page 420

1. $\dfrac{7}{15}$ 3. 0   5. $\dfrac{19}{6}$

7. $\dfrac{181}{4}$ 9. $-\pi ab$

11. $a^2\left(\dfrac{\pi}{3}-\dfrac{3}{8}\right)+\dfrac{\sqrt{3}}{6}\,\pi ab-\dfrac{\pi^2}{18}\,b^2$

13. $\dfrac{31}{2}$ 15. -2

17. $\dfrac{2}{3}\,e^3+2e+2e^{-1}+e^{-3}-\dfrac{17}{3}$

19. $-\left(\dfrac{1}{3}\,x^3+\dfrac{1}{2}\,xy^2+\dfrac{1}{2}\,yz^2+c\right)$

21. not conservative

23. $-(x^2ye^z+xz^2e^y+y^2ze^x+c)$

25. $-(x\cos y+y\sin x+c)$

27. 6   29. $\dfrac{131}{2}$ 31. 16π

33. $\dfrac{\sqrt{2}}{26}\,(3e^{3\pi/4}+2)$ 35. $\pi\rho R^3$

37. $\dfrac{5}{2}\,\rho_0 L$; center of gravity at a distance $\dfrac{2\sqrt{3}}{15}\,L$ from base, $\dfrac{5}{16}\,\rho_0 L^3$

Section 4.6, page 441

1. $8\pi\sqrt{2}$

3. $4\sqrt{6}$

5. $4\pi\sqrt{3}$

7. $\pi(2 - \sqrt{2})$

9. $\frac{1}{2}\sqrt{a^2b^2 + a^2c^2 + b^2c^2}$

11. $4a^2$

13. 0 15. $4\pi\sqrt{2}$

17. π 19. $\left(0, 0, \dfrac{27a}{160 - 85\sqrt{2}}\right), \dfrac{\pi a^6}{240}(128 - 81\sqrt{2})$

21. $(0, 0, 2), \dfrac{32}{3}\pi\sqrt{3}$

Section 4.7, page 459

1. 0

3. $\dfrac{104}{3}$

5. 0

7. $\dfrac{\pi}{12}(8 - 5\sqrt{2})$

9. $\dfrac{1}{12}abc(4 + b)$

11. 0

13. 0

Section 4.8, page 472

1. 24π 3. 0 5. $4\pi a^3$

7. $(x_2 - x_1)(y_2 - y_1)(z_2 - z_1)(x_1 + x_2 + y_1 + y_2 + z_1 + z_2)$

9. 40π

11. $\dfrac{243}{2}\pi$

13. $-\dfrac{13}{6}$

15. 0

17. 0

19. $-\pi a^2$

21. 0

23. 0

25. 4

27. 5π

29. 4 31. $\dfrac{\pi}{2}$ 33. $-\dfrac{556}{15}$

Index